MILITARY INJURY BIOMECHANICS
The Cause and Prevention of Impact Injuries

MILITARY INJURY BIOMECHANICS

The Cause and Prevention of Impact Injuries

Edited by
Melanie Franklyn
Peter Vee Sin Lee

The University of Melbourne
Victoria, Australia

CRC **CRC Press**
Taylor & Francis Group
Boca Raton London New York

CRC Press is an imprint of the
Taylor & Francis Group, an **informa** business

CRC Press
Taylor & Francis Group
6000 Broken Sound Parkway NW, Suite 300
Boca Raton, FL 33487-2742

First issued in paperback 2020

ISBN-13: 978-1-4987-4282-5 (hbk)
ISBN-13: 978-0-367-65794-9 (pbk)

Library of Congress Cataloging-in-Publication Data

Names: Franklyn, Melanie, editor. | Lee, Peter Vee Sin, editor.
Title: Military injury biomechanics : the cause and prevention of impact injuries / [edited by] Melanie Franklyn and Peter Vee Sin Lee.
Description: Boca Raton, FL : CRC Press/Taylor & Francis Group, 2017. | Includes bibliographical references.
Identifiers: LCCN 2016051977 | ISBN 9781498742825 (hardback : alk. paper)
Subjects: | MESH: War-Related Injuries--prevention & control | Biomechanical Phenomena | Military Personnel | Military Medicine--methods
Classification: LCC RC971 | NLM WO 700 | DDC 616.9/8023--dc23
LC record available at https://lccn.loc.gov/2016051977

Visit the Taylor & Francis Web site at
http://www.taylorandfrancis.com

and the CRC Press Web site at
http://www.crcpress.com

Contents

Preface

Military injury biomechanics has more recently emerged as a topical research area, with the wars in Afghanistan, Iraq and other regions resulting in a large number of fatalities and injured combatants. The nature of warfare is constantly evolving, with different types of munitions and battlefield scenarios resulting in new types of injury patterns. The altering landscape of recent conflicts has led to new types of vehicles, while body armour design has developed with the advanced materials now available to deal with current and emerging threats. As a result, there has been a need for contemporary tools and techniques to predict and assess injuries sustained in contemporary conflict scenarios.

In this book, we have covered the background and current status of military injury biomechanics research for different body regions. The focus is on acceleration and blunt impact injury, including backface deformation resulting from ballistic impacts and blunt injury sustained by soldiers in military vehicles subjected to landmines. We chose not to include injuries from blast waves, such as blast-induced traumatic brain injury, for several reasons. Firstly, there are currently other books available on the market covering this detailed topic. Secondly, we felt that this book could be more focused and provide more detailed information on each of the topics if the material covered the different facets of acceleration and blunt impact injuries in terms of their causation and treatment, assessment using mechanical surrogates and prediction using computational models. The book's first few chapters address the physics of ballistic projectiles and vehicle dynamics when subjected to a landmine, while subsequent chapters include topics ranging from the medical management of combat injuries to anthropomorphic test devices, and cover issues specific to each of the different body regions such as injuries sustained, test methods and emerging research. The authors are all specialists and leading researchers in their fields, and we are sure their knowledge and insights into the various topics will be invaluable to readers.

The aim of this book is to provide a contemporary standard reference manual for academic and military researchers, physicians and trauma researchers, and graduate students working or studying in the fields of military injuries and impact injury. It can also be a useful guide for engineers involved in testing defence products, such as combat helmets and military vehicle seats, in addition to individuals employed in defence-related industries, such as defence materiel manufacturers. Improved mitigation of combat-related injuries will help reduce the number of casualties in current warfare and lead to better prevention strategies, treatment options and rehabilitation measures for long-term impairments. We hope that this book will be a useful reference guide for readers in the discipline of military injury biomechanics.

Melanie Franklyn
The University of Melbourne

Peter Vee Sin Lee
The University of Melbourne

Editors

Dr. Melanie Franklyn has a PhD in Biomedical Engineering and is an internationally recognised expert in impact injury biomechanics, trauma, orthopaedics and sports injuries. She is currently employed as a research scientist in impact injury biomechanics at the Defence Science & Technology Group (DST Group) Australia (since 2010) and is also an honorary senior research fellow in the Department of Mechanical Engineering at The University of Melbourne (since 2010). Dr. Franklyn established and developed the capability in military injury biomechanics at DST Group, where her research includes military helmet evaluation, Behind Armour Blunt Trauma (BABT), brain injury prediction using finite element models, the development of injury criteria for the spine and pelvis, and injury prediction methodology for underbelly vehicle blast tests. She also conducts injury biomechanics work at The University of Melbourne, including research on tibial stress injuries, trauma and orthopaedics.

Dr. Franklyn was previously employed as a research fellow at the Monash University Accident Research Centre (2001–2009 inclusive) where her research focused on occupant injury in automotive crashes, crashworthiness, computer modelling, injury coding, trauma and sports injuries. She is a certified abbreviated injury scale specialist through the Association for the Advancement of Automotive Medicine (AAAM) and is a qualified injury coding instructor. Dr. Franklyn is on the editorial board of the *World Journal of Orthopaedics* and is a reviewer for various international journals and granting bodies. She has also supervised numerous graduate student projects and her research has been featured in various forums such as the ABC *Catalyst* TV programme.

Dr. Peter Lee is currently a professor in the Melbourne School of Engineering at The University of Melbourne. He obtained his BEng in Mechanical Engineering (1991) and PhD (1996) in bioengineering from the University of Strathclyde, UK, and continued his post-doc in the same university from 1996 to 1998. He was a research fellow with the Biomaterials Group at the Institute of Materials Research and Engineering, Singapore, from 1998 to 2001. In 2001, he joined the Defence Medical and Environmental Research Institute, DSO National Laboratories, Singapore, as the head of the Bioengineering Laboratory. He joined The University of Melbourne as a senior lecturer in 2008.

Professor Lee is recognised internationally in tissue biomechanics research, injury biomechanics and rehabilitation engineering. He has published more than 140 articles in journals, conference proceedings and books. He currently leads a research team focusing on understanding impact-type injuries to the body, leading to effective prevention strategies. His research spans all three levels – the human, organs and cells. These investigations also apply computational models to further understand diseases and injuries to the various joints in the body.

Dr. Melanie Franklyn's proceeds from this book have been donated to the RSL Defence Care charity, an organisation which helps current and former serving members of the Australian Defence Force suffering illness, injury or crisis.

Contributors

Cameron R. 'Dale' Bass
Department of Biomedical Engineering
Duke University
Durham, North Carolina

Lynne Bilston
Neuroscience Research Australia
Randwick, Australia
and
Prince of Wales Clinical School
UNSW Medicine
Sydney, Australia

Cynthia Bir
Keck School of Medicine
University of Southern California
Los Angeles, California

Johno Breeze
Academic Department of Military Surgery
    and Trauma
Royal Centre for Defence Medicine
Birmingham Research Park
Birmingham, United Kingdom

Anthony M.J. Bull
Department of Bioengineering
Imperial College London
London, United Kingdom

Debra J. Carr
Impact and Armour Group
Centre for Defence Engineering
Cranfield University
Defence Academy of the United Kingdom
Shrivenham, United Kingdom

Stephen J. Cimpoeru
Defence Science and Technology
    Group
Melbourne, Australia

Melanie Franklyn
Department of Mechanical Engineering
The University of Melbourne
Melbourne, Australia

Tom Gibson
Human Impact Engineering
and
University of Technology Sydney
Sydney, Australia

Grigoris Grigoriadis
Department of Bioengineering
Imperial College London
London, United Kingdom

Damian Keene
Department of Military Anesthesia and
    Critical Care
Royal Centre for Defence Medicine
Birmingham Research Park
Birmingham, United Kingdom

Michael Kleinberger
US Army Research Laboratory
Aberdeen Proving Ground, Maryland

Heow Pueh Lee
Department of Mechanical Engineering
National University of Singapore
Singapore

Peter Vee Sin Lee
Department of Biomedical Engineering
The University of Melbourne
Melbourne, Australia

Eluned A. Lewis
Defence Equipment and Support
UK Ministry of Defence
Bristol, United Kingdom

Jianfei Liu
Department of Mechanical Engineering
National University of Singapore
Singapore

Peter Mahoney
Department of Military Anesthesia and
    Critical Care
Royal Centre for Defence Medicine
Birmingham Research Park
Birmingham, United Kingdom

Spyros Masouros
Department of Bioengineering
Imperial College London
London, United Kingdom

Brian McKay
Department of Biomedical Engineering
Wayne State University
Detroit, Michigan

Paul Phillips
Defence Science and Technology Group
Melbourne, Australia

Frank A. Pintar
Department of Neurosurgery
Medical College of Wisconsin
and
Clement J. Zablocki Veterans Affairs
Medical Center Milwaukee, Wisconsin

Barbara R. Presley
Department of Biomedical Engineering
Wayne State University
Detroit, Michigan

Arul Ramasamy
Department of Bioengineering
Imperial College London
London, United Kingdom

Christine Read-Allsopp
Royal North Shore Hospital Trauma Service
Sydney, Australia

David V. Ritzel
Dyn-FX Consulting Ltd.
Amherstburg, Ontario, Canada

Tal Saif
Department of Mechanical Engineering
Wayne State University
Detroit, Michigan

Nicholas Shewchenko
Biokinetics and Associates Limited
Ottawa, Ontario, Canada

Victor P.W. Shim
Department of Mechanical Engineering
National University of Singapore
Singapore

Brian D. Stemper
Department of Neurosurgery
Medical College of Wisconsin
and
Clement J. Zablocki Veterans Affairs
Medical Center Milwaukee, Wisconsin

Long Bin Tan
Department of Mechanical Engineering
National University of Singapore
Singapore

Ee Chong Teo
Department of Mechanical Engineering
Nanyang Technological University
Singapore

Kwong Ming Tse
Department of Mechanical Engineering
The University of Melbourne
Melbourne, Australia
and
Department of Mechanical Engineering
National University of Singapore
Singapore

Tom Whyte
Human Impact Engineering
and
University of Technology Sydney
Sydney, Australia

King H. Yang
Department of Biomedical Engineering
Wayne State University
Detroit, Michigan

Narayan Yoganandan
Department of Neurosurgery
Medical College of Wisconsin
Milwaukee, Wisconsin

Feng Zhu
Department of Mechanical Engineering
Embry-Riddle Aeronautical University
Daytona Beach, Florida

1 Introduction

Melanie Franklyn and Peter Vee Sin Lee

Impact injury biomechanics involves the study of human tolerance to impact loading. Using the principles of mechanics, the discipline focuses on the mechanisms of injury in traumatic events such as military scenarios and automotive impacts. When a traumatic event occurs, the impact loading causes deformation of the tissues beyond their failure limits, resulting in anatomical lesions or altered physiological function. Injury prediction and prevention involves using tools and techniques to determine the specific causes of the trauma and the effect of variables such as the injury mechanism, impact surface, rate of impact and the material properties of the tissue in order to reduce the likelihood of injury and design better countermeasures to minimise trauma to the individual.

Injuries resulting from high-speed impact became an emerging problem with the introduction of motorised vehicles in the late 1880s, and as automobiles became more widespread, predominately in the early 1900s, the number of crashes started rising. This led to a dramatic increase in the prevalence of injuries and fatalities sustained in these events, particularly as these 'horseless carriages' had no safety features in addition to a multitude of other issues such as hard interiors and a non-compliant steering wheel. Ironically, although the safety belt was used in horse-drawn carriages as early as the 1850s, they would not be standard issue in motor vehicles until the early 1960s. In the 1920s, one of the early pioneers in crash investigation was Hugh De Haven. Having survived an aeroplane crash when he was serving as a combat pilot cadet in the Canadian Royal Flying Corps during WWI, De Haven commenced his investigations into injuries sustained in traumatic events such as aeroplane crashes and falls; in the latter case, he was particularly interested in individuals who had fallen from a great height and survived (Hurley et al. 2002). De Haven later established the Crash Injury Research project at Cornell in the 1940s for investigating motor vehicle crashes.

With the advent of WWII, there were significant developments in the field, predominately in the specialities of brain injuries and aviation medicine. Australian-born neurosurgeon Hugh W.B. Cairns, who was both an advisor on head injuries for the UK Ministry of Health and a consultant neurosurgeon to the British Army, became an influential advocate for the use of helmets by motorbike riders for preventing brain injuries, particularly after his experience as one of the attending physicians for Colonel T.E. Lawrence after the latter's fatal motorbike crash in 1935 (Lanska 2016; Marriott and Argent 1996). About the same time, two British neurologists, Majors D. Denny-Brown and W.R. Russell, both of the Royal Army Medical Corp and the University of Oxford, developed an animal model for brain injury, demonstrating that suddenly applied acceleration could result in concussion (Lanska 2016).

The results of this earlier work led A.H.S. Holbourn, a research physicist at the Department of Surgery, the University of Oxford, to postulate on the cause of brain injuries in 1943, stating that angular acceleration without impact was an important head injury mechanism. From 1939, Lissner and Gurdjian were also studying head trauma but at Wayne State University. Using cadaveric heads, they developed their theory of linear acceleration as a cause of brain injury, which later led to the Wayne State Tolerance Curve (Lanska 2016).

Whilst these important developments in brain injury research were transpiring, advances in aviation medicine were emerging when it became critical to address the problem of trauma in jet aircraft and aeroplane crashes. Both the UK and Germany recognised there was a need to understand the tolerance of the spine to injury and conducted some of the early experiments in the field. Much of this data was not published until after the war, in particular, a large body of German aviation medicine research.

The Americans realised the importance of this work and facilitated the publication of this data from German to English, thus making the spinal injury work and injury tolerance curves developed by researchers such as Ruff and Geertz accessible (German Aviation Medicine in World War II 1950).

In the late 1940s, the US military surgeon Colonel John Paul Stapp studied the effects of deceleration on the human body using a rocket sled, using himself as a test subject. Stapp was interested in applying the results to aircraft crashes in addition to studying the effect of seats and harnesses in a crash. This research continued through several organisations in the 1950s and 1960s and led to the development of a number of injury curves and injury models which are still in use today. This work includes that of Eiband of NASA, US (Eiband 1959), which led to the well-known Eiband curve for rapidly applied acceleration, and Ruff of Germany (Ruff 1950), who established an injury curve for whole-body tolerance under acceleration. In the latter part of the 1950s, Latham of the Royal Air Force Institute of Aviation Medicine, UK, introduced the mechanical spring-mass model for vertical loading to examine spinal injuries for pilots in ejection seat scenarios; the concept of which was later developed by Stech and Payne (1969) and Brinkley and Shaffer (1971) to become the Dynamic Response Index (DRI) model.

In order to further assess the effect of pilot ejection seats, pilot restraint harnesses and restraint systems for military vehicles, Sierra Sam, a 95th percentile male anthropomorphic test device (ATD), was created by Alderson Research Laboratories and Sierra Engineering for the military in 1949. This was later followed by other ATDs such as the VIP (Very Important Person) series in 1966, which was developed for Ford and General Motors and specifically designed for automotive safety. However, the automotive industry was becoming more concerned about reducing motor vehicle trauma. In the late 1960s, Stapp changed the focus of his experiments from military to automotive safety research through the initiative of the US National Highway Traffic Safety Administration (NHTSA), and in the early 1970s, General Motors commenced their own testing and research programme, developing the hybrid series of ATDs which are still in use today.

While there has been extensive research into automotive crashes and vehicle occupant safety over the last 50 years, this has not been the case for impact injuries sustained in military events. This has been due to, in part, the changing threats combatants have been exposed to. Since the 1970s, military injury research has predominately focused on issues like blast-related trauma and overuse injuries such as stress fractures, with only more recent work including the analysis of injury mechanisms of blunt trauma sustained by soldiers in military vehicles subjected to improvised explosive device (IED) events. Consequently, many of the standards currently used to assess military impact injuries were developed in the 1950s–1970s, while studies published at the time often presented limited data or omitted critical information, such as loading rates. Thus, injury criteria developed for ATDs used to assess impact injury potential in military vehicle tests are heavily based on pilot ejection studies or automotive data–situations they were not designed for. For example, the DRI originated from pilot ejection studies, while other measures such as the Head Injury Criterion (HIC) and the Viscous Criterion (VC) were developed for automotive impacts.

In addition to the limitations of injury criteria for assessing military impact injuries, the ATDs used to measure them were not designed for underbelly blast testing. Both the automotive Hybrid III and the Federal Aviation Administration (FAA) Hybrid III are currently used for live-fire testing and evaluation of military vehicles; however, the automotive Hybrid III is a regulatory frontal-impact crash test ATD, while the FAA Hybrid III was designed for assessing spinal injury potential in light aeroplane crashes. Nevertheless, to date, there have been no alternative forms of assessment, and these mannequins remain the only alternative for underbelly live-fire testing injury evaluation.

Due to the significant limitations in the current ATDs and injury criteria for assessing the type of injuries sustained in underbelly blast events, there have been, more recently, significant research efforts towards developing new ATD components and a new military ATD. For example, the Mil-Lx (Military Lower Extremity) leg, which is a straight leg optimised to measure vertical loading, was developed through a collaboration between Humanetics (Denton ATD) and Wayne State University and is currently used in live-fire testing and evaluation. More recently, the WIAMan (Warrior Injury

Assessment Manikin) initiative to develop an ATD specifically for underbelly blast testing has resulted in ongoing research and development to investigate the causes and mechanisms of the different types of injuries sustained in these events. In addition, there are a number of large research programmes to determine the effect of different variables such as loading rates and peak acceleration levels under vertical loading. Recent studies on the sizes of current military vehicle occupants are also being performed, as the anthropometry of the currently used ATDs are not representative of today's military combatant.

Other emerging research on military impact injuries includes computational modelling efforts such as finite element analysis, where parametric studies on different loading conditions are leading to a better understanding of the injury mechanisms involved in underbelly blast events. In the ballistics and behind armour blunt trauma (BABT) field, the rapid development of new materials has led to significant improvements in the design of helmets and body armour; however, the techniques used to assess them, which were developed in the 1960s, have very little correlation with brain or thorax injuries. Consequently, recent research in this area has focused on gaining a better understanding of the relationship between backface deformation and injury outcome and potentially establishing new standards for assessment.

Clearly, a substantial amount of historical military and automotive impact data is still in use today; however to date, this information has not been presented and discussed in any single publication. Thus, it is hoped that this book will be an invaluable learning and reference guide, both for researchers already working in the field of military injury biomechanics and individuals new to the discipline. While the chapters cover the historical background and development in the different areas of impact injury, current research is also discussed. Although the discipline is rapidly evolving, this book should provide researchers with an invaluable contemporary source of information which can be used as a study or reference guide.

REFERENCES

Brinkley, J.W. and Schaffer, J.T. 1971. *Dynamic Simulation Techniques for the Design of Escape Systems: Current Applications and Future Air Force Requirements.* AMRL-TR-71-291971. Aerospace Medical Research Laboratory, Wright-Patterson Air Force Base, OH.

Eiband, A.M. 1959. *Human Tolerance to Rapidly Applied Accelerations: A Summary of the Literature.* NASA-MEMO-5-19-59E. National Aeronautics and Space Administration, Cleveland OH.

German Aviation Medicine in World War II. 1950. Volumes I and II. Department of the Air Force, US Government Printing Office, Washington, DC.

Hurley, T.R., Vandenburg, J.M. and Labun, L.C. 2002. Introduction to Crashworthiness. In: Hurley, T.R. and Vandenburg, J.M. Eds. *Small Airplane Crashworthiness Design Guide.* Simula Technologies, TR-98099. Phoenix, AZ, pp. 1–9.

Lanska, D.J. 2016. Traumatic Brain Injury Studies in Britain during World War II. In: Tatu, L. and Bogousslavsley, J. *War Neurology. Frontiers of Neurology and Neuroscience.* Karger, Basel, Switzerland, pp. 68–76.

Marriott, P. and Argent, Y. 1996. The Long Wait, 14–19 May 1935. In: *The Last Days of T.E. Lawrence,* Chapter 13, The Alpha Press, Brighton, Great Britain.

Ruff, S. 1950. Brief Acceleration: Less than One Second. In: *German Aviation Medicine, World War II,* Volume I. Department of the Air Force, pp. 584–598.

Stech, E.L. and Payne. P.R. 1969. *Dynamic Models of the Human Body.* AMRL-TR-66-157. Aerospace Medical Research Laboratory, Wright-Patterson Air Force Base, OH.

2 Ballistic Threats and Body Armour Design

Johno Breeze, Eluned A. Lewis and Debra J. Carr

CONTENTS

2.1 TYPES OF BALLISTIC THREATS ON THE MODERN BATTLEFIELD

2.1.1 INTRODUCTION

The modern battlefield contains a wide variety of ballistic threats that military personnel may encounter, and these threats may change depending on the conflict and as a conflict matures. Intelligence as to the nature of the threat is essential for optimising protection. Ballistic threats not only come from enemy combatants but also include those utilised by the civilian populace as well as allies or own forces. The key to optimising protection lies with an ability to rapidly scale any personal armour worn both for its ballistic protective ability as well as the anatomical coverage provided based upon the threat. Since World War I (from a UK perspective), fragmentation wounds have generally outnumbered those caused by bullets, with the exception of particular smaller scale conflicts such as those involving jungle warfare (e.g. Borneo) and internal security operations (e.g. Northern Ireland) (Figure 2.1).

2.1.2 ENERGISED FRAGMENTS

Fragments vary greatly in nature among differing types of weapons (Table 2.1). Large calibre artillery rounds tend to produce larger high-velocity fragments and are highly lethal (Holmes 1952; Zecevic 2011). Grenades and small calibre mortars produce smaller fragments that tend to lose their momentum relatively quickly, with the specific aim to wound and not to kill, thereby affecting the logistics chain by managing multiple casualties (Zecevic 2011).

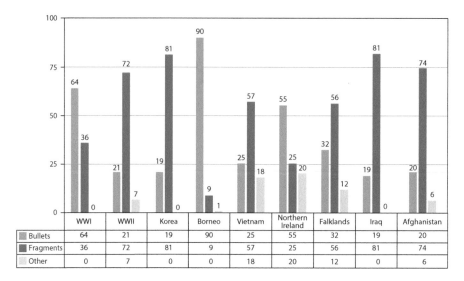

FIGURE 2.1 Incidence of battle injury (%) by wounding type (other injury causes include interpersonal assault and blunt trauma). (From Carey, M.E., *Journal of Trauma*, 40, S165–S169, 1991; Rees, D., *Korea: The Limited War*, MacMillan, London, 1964; Moffat, W.C., *Journal of the Royal Army Medical Corps*, 112, 3–8, 1976; Crew, F.A.E., *History of the Second World War*, The Army Medical Services, HMSO, London, 1953; Adamson, P.B., *Journal of the Royal Army Medical Corps*, 123, 93–103, 1977; Garfield, R.M., and A. Neugut, *JAMA*, 266, 688–692, 1991; Jackson, D.S., et al., *Annals of the Royal College of Surgeons of England*, 65, 281–286, 1983; Hardaway, R.M., *Journal of Trauma*, 18, 635–543, 1978; Reister, F.A., *Battle Causalities and Medical Statistics: US Experience in the Korean War*, The Surgeon General, Department of the Army, Washington, DC, 1973; Beebe, G.W., and M.E. DeBakey, *Battle Casualties*, Charles C. Thomas, Springfield, IL, 165–205, 1952.)

TABLE 2.1

Groups of Weaponry Producing Energised Fragmentation

Group	Examples of Specific Types
Improvised explosive device (IED)	Time-initiated, command-initiated, victim-operated
Antipersonnel mine	Blast, fragmentation, directional fragmentation, bounding fragmentation
Grenade	Fragmenting case, preformed fragments inside
Indirect fire	Artillery shell, mortar

Fragments which strike a human target will have different effects depending on fragment mass, shape, velocity and body part struck. Due to fragment air drag, most casualties occur at ranges less than 10 m (Galbraith 2001). Most individuals, if within a few metres of exploding ordnance, are more likely to be killed by the blast overpressure than from the fragments. Fragmenting munitions generally either utilise preformed fragments or the detonation breaks up the (usually) metallic casing (Figures 2.2 and 2.3). Fragmentation grenades can be hand thrown, under-slung from a rifle, or rocket propelled. Fragments from grenades are most commonly produced by notched wire breaking up the plastic or steel outer casing, by ball bearings included inside the grenade, or by depressions within the actual casing which create fragments by the expanding explosive force.

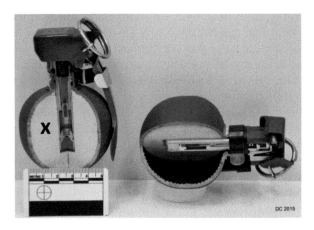

FIGURE 2.2 Cross-section of a high explosive fragmentation grenade in which the core (marked with 'x') contains an explosive which is ignited by the fuse and propels fragments, each formed by dimples in the inner surface of the steel casing (50 mm scale).

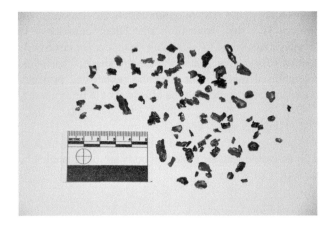

FIGURE 2.3 Range of sizes of energised fragments produced by a typical mortar (50 mm scale).

Improvised explosive devices (IEDs) represent the most common threat to soldiers worldwide involved in counter-insurgency operations and are the leading cause of injury and death of soldiers in modern conflicts (McFate 2005). IEDs can be manufactured using conventional weapons or may be completely homemade from a diverse range of commonly available objects and materials including metal cans, glass or polymer bottles and pipes, and thus fragments are not necessarily metallic (Thurman 2011; Steier et al. 2014). Any metal fragments produced are typically in the size range of <1 to 5 mm and have a mass of 0.1 to >20 g with non-uniform shapes (Ryan et al. 1991). However, spherical fragments (ball bearings) are commonly found in IEDs. In suicide bombings, human body parts can also act as fragments and therefore be incorporated into wounds (Patel et al. 2012).

2.1.3 AMMUNITION

Ammunition comprises four components: (1) the cartridge case, (2) a primer, (3) a propellant and (4) the projectile (bullet) (Heard 2008). The prevalent threat in recent conflicts is high-velocity (rifle) ammunition, with 7.62 mm calibre ammunition (e.g. 7.62 × 39) and 5.45 mm calibre ammunition (5.45 × 39) typically used by enemy combatants and 5.56 mm calibre ammunition used by NATO forces (e.g. 5.56 mm NATO; 5.56 × 45) (Figure 2.4). Low velocity (pistol) bullets (e.g. 9 mm Luger) generally occur in urban conflicts and internal security operations (Figure 2.4). A common misconception is that such pistol bullets are less lethal due to their relative velocity; as will be demonstrated later, the energy transfer and resultant tissue cavitation determines wounding and lethality. However, the effective range of pistols is significantly less than rifles.

Bullets which conform to the Hague Convention of 1899 have a full metal jacket (FMJ). Such bullets typically consist of a soft core (lead or mild steel), excluding armour-piercing bullets which contain a hardened core. The core is encased in a jacket of gilding metal (copper and zinc), gilding metal-clad steel, cupro-nickel (copper and nickel), or aluminium that typically covers the nose and sides with the rear part of the core left exposed (Di Maio 1998). The jacket engages with the rifling of the barrel of the weapon, resulting in higher muzzle velocities than if the projectile had no jacket and reduces expansion of the lead core on impact with the target, thereby causing a lower immediate energy deposition and potentially a decrease in lethality. Some bullets are designed (or tend) to fragment during the impact event; others are known to yaw before impact and/or during penetration.

2.1.4 SHOTGUNS AND LESS LETHAL KINETIC ENERGY DEVICES

The combat shotgun has seen resurgence in use in current conflicts. Shotguns are also common weapons used during civil incidents in the UK. Shotgun barrels are usually smooth bore; two types of ammunition are common: (1) a shell containing shot which consists of small spherical pellets (lead, steel, non-toxic equivalents) or (2) a solid slug (lead, copper-covered lead, steel). Less lethal kinetic energy ammunition may vary in material, size and shape. Typical examples include baton rounds, beanbags, fin stabilised rubber projectiles and rubber ball projectiles. Such ammunition types are designed to incapacitate people, and the effects should be temporary and reversible.

2.1.5 SUMMARY

- There is a need for a flexible scalable system of body armour capable of being worn worldwide to defeat a wide variety of threats.

FIGURE 2.4   Ammunition types (from left to right): shotgun cartridge, 7.62 mm NATO (7.62 × 51), 7.62 × 39, 5.45 × 39, 5.56 mm NATO (5.56 × 45) and 9 mm Luger (9 × 19) (50 mm scale).

- The ideal armour must be rapidly scalable in terms of its ballistic protective ability as well as its coverage.
- Fragments are the major cause of injury in modern warfare.
- Bullets are more likely to cause fatalities than fragments because of greater energy deposition into the body.
- In internal security operations, vulnerable areas of the body tend to be targeted; in conventional warfare, wounds are randomly dispersed over the body area.

2.2 PROJECTILE–TISSUE INTERACTION

2.2.1 INTRODUCTION

Projectiles can cause injury to living tissues through three potential mechanisms. The first mechanism is the crushing and cutting effect of the presented surface of the projectile, which is responsible for the production of a permanent wound cavity (PWC) (Figure 2.5) (Kneubuehl et al. 2011). In the second mechanism, the further passage of the projectile leads to radial acceleration of the tissue, which will expand, contract and oscillate after the projectile has passed through (Figure 2.6) (Kneubuehl et al. 2011). The maximum extent of this pulsating temporary cavity produced in tissues occurs several milliseconds after the projectile has passed. Very limited evidence exists that the third mechanism (termed either the pressure wave or shock wave) causes actual tissue damage in humans (Kneubuehl et al. 2011; Breeze, Sedman, and James 2013).

Energy loss along a wound track is not uniform, with variations occurring due to the behaviour of the projectile or changes in the structure of the tissues as the projectile traverses it.

Projectile factors which affect energy loss include:

- Shape and size: Projectiles with a greater presenting surface area (assuming mass and velocity are similar) result in higher energy deposition in the body.
- Yaw: Besides contributing towards 'size and shape' effects, yaw increases a projectile's cross-sectional area and therefore kinetic energy deposition by increased drag.
- Fragmentation: Projectile fragmentation in the body results in greater energy deposition.
- Deformation: The presence of an FMJ prevents bullet expansion with an associated reduction in kinetic energy deposited in the tissues.

FIGURE 2.5 Typical example of a PWC cause by a 9 mm Luger bullet passing through a swine thorax (PWC highlighted; 50 mm scale).

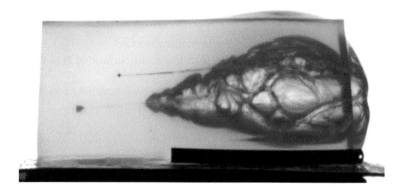

FIGURE 2.6   Typical example of a still taken from a high-speed video of a temporary cavity due to a high-velocity rifle projectile penetrating a 250 mm × 250 mm × 500 mm 10% (by mass) gelatine block. (From Mabbott, A.J., The overmatching of armour, PhD thesis, Centre for Defence Engineering, Cranfield University, Defence Academy of the United Kingdom, Shrivenham, UK, 2015.)

2.2.2   PERMANENT WOUND TRACT

The permanent wound tract is the clinical result of the crushing and cutting effect of a projectile in all tissues, in conjunction with the rapid radial displacement of the temporary cavity. It comprises a central PWC (Berlin, Gelin, and Janzon 1976; Orlowski, Piecuch, and Domaniecki 1982). Macroscopic damage can be reversible or irreversible, with the area immediately adjacent to the PWC generally having irreversible changes (contusion zone) and the outer layer (concussion zone) having reversible changes (Wang, Tang, and Chen 1988). Clinically, however, such discrete zones are not found, with damage being patchy and not necessarily correlated to distance from the path of the projectile. This suggests that the clinical effect is dependent on the highly complex interacting variables of both tissue type and architecture. Macroscopic damage will heal completely in some tissue types, but this effect is likely to be rare, especially in complicated military wounds and in the presence of infection and contamination. Although irreversible macroscopic tissue damage in muscle may lead to scarring, in many cases there will be little residual clinical effect should the area of scarring be small or other muscles may compensate.

2.2.3   DAMAGE PRODUCED SPECIFICALLY BY THE TEMPORARY CAVITY

The temporary cavity results in a transient, rapid strain of tissues that may, depending on the mechanical characteristics of the tissue, produce injury. The temporary cavity affects dense homogenous tissues, particularly if enclosed by a connective tissue capsule or casing such as the liver or brain, the greatest. Conversely, elastic tissues with a high strain to failure, such as lungs and large arteries, resist its effects. This extension is responsible for a small portion of the PWC as well as some macroscopic damage lateral to that it. Skin is damaged by both the direct crushing effect of the projectile and the rapid radial tissue displacement produced by the temporary cavity demonstrated by the stellate exit wounds in tumbling projectiles that can be greater than the largest dimensions of the projectile. Indirect bone fractures can occur at a distance remote from the projectile path due to temporary cavity formation (Kieser et al. 2013), although clinically their occurrence is rare (Figure 2.7). Indirect vertebral fractures may damage the adjacent spinal cord directly or through the production of secondary fragments (Amato, Syracuse, and Seaver 1989).

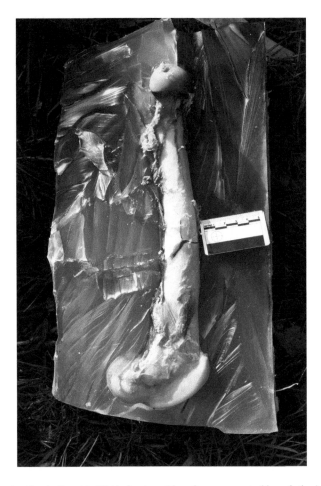

FIGURE 2.7 Example of an indirect ballistic fracture (deer femur mounted in gelatine). (From Kieser, D.C., Ballistic skeletal injuries: An experimental study of the orthopaedic, biomechanic and forensic characteristics, PhD thesis, Orthopaedic Surgery, Dunedin School of Medicine, The University of Otago, Dunedin, New Zealand, 2014.)

2.3 PERSONAL PROTECTIVE EQUIPMENT FOR MILITARY PERSONNEL

2.3.1 INTRODUCTION

Ballistic and sharp-weapon protective clothing is worn by military, police, other first responders (e.g. fire, ambulance personnel) and civilian personnel such as security guards. Such clothing is often referred to as personal armour. In the modern military context, the term personal armour includes body armour (waistcoat or vest-like garments covering the torso), helmets (covering the cranium), face and eye protection (visors, glasses, goggles), explosive ordnance disposal (EOD) suits and ballistic shields (UK Ministry of Defence 2005). More recently, the use of next-to-skin garments that incorporate ballistic protection have been discussed and developed, e.g. pelvic and neck protection (Sakaguchi et al. 2012; Lewis et al. 2013; Helliker et al. 2014; Breeze, Allanson-Bailey et al. 2015). Military body armour provides protection from fragmentation and high-velocity rifle bullets. In comparison, armour for the police protects from sharp-weapon attack and low-velocity handgun bullets (Horsfall 2012).

Body armour typically comprises soft and hard elements (Tobin and Iremonger 2006; Carr and Lewis 2014). The soft element of body armour is the familiar waistcoat or vest-like garment worn by the military and police, and is manufactured from high-performance man-made synthetic polymer fibres. These fibres are high-tenacity, high-stiffness and high-cost products. The fibres used in the manufacture of body armour include para-aramids (e.g. Kevlar®, Twaron®) and ultra high molecular weight polyethylene (e.g. Dyneema®, Spectra®). Para-aramids are usually used to manufacture a plain woven fabric, and UHMWPEs are usually used in a cross-ply arrangement. Many layers of fabric are used in body armour. Protection from high-velocity rifle bullets can be provided by the use of hard plates which are usually ceramic faced composite backed (e.g. alumina/para-aramid) or 100% composite (e.g. UHMWPE) (Tobin and Iremonger 2006). The resulting body armour can be heavy (typically in excess of 10 kg), restrictive and increases the thermophysiological burden on the wearer (Carr and Lewis 2014). Thus, efforts to identify new materials and/or systems with improved performance and a lower mass penalty are the primary focus of much research, e.g. the UK MOD initiative *Reducing the Burden on the Dismounted Soldier* (RBDS). The scope of RBDS was to reduce the load of the solider, whilst also enhancing effectiveness and survivability. Specifically, the Lightweight Personal Protection task aimed to provide a mass saving of 15%–30% (Bruton 2012).

The response of body armour and helmets to ballistic impact has been discussed in the peer-reviewed literature for over 60 years from both theoretical and experimental perspectives. High-set, plain woven, balanced fabrics reportedly provide the best protection from fragmentation threats (e.g. Cunniff 1992; Shim et al. 1995; Cheeseman and Bogetti 2003). When a woven fabric (such as that used in a typical body armour) is impacted, longitudinal waves pass along the impacted warp and weft yarns and can result in yarn pull-out; thus a pyramidal or cone-shaped through-thickness deformation occurs (e.g. Roylance et al., 1973; Cunniff 1992; Sakaguchi et al., 2012). Fibre and yarn rupture occurs when the strain at the impact point exceeds the failure strain (Horsfall 2012). At higher impact energies, fibres and yarns fail in a shear or 'plug' model (Carr 1999). Hard armour protects from high-velocity (rifle) bullets because the ceramic is harder than the core of the bullet. On impact, the bullet may mushroom and/or fracture; the ceramic fractures into a characteristic conoid pattern. The composite backing will deform with some elastic recovery, but a permanent deformation will remain. The transfer of the applied pressure through the plate (and the acceleration of the plate) into the body may result in behind armour blunt trauma (BABT) injury. Such injuries are typically minor to moderate (contusions, fractured ribs); there is no evidence of fatality due to BABT (Carr, Horsfall, and Malbon 2013).

A woven fabric pack (such as in body armour) has a higher ballistic protective level than a composite (such as a helmet) containing the same type and number of woven fabric layers due to the pinning of the yarn crossover point by the matrix materials in the composite and enhanced reflection of waves. The addition of the matrix material to form the composite minimises the transient deformation (and hence reduces the contribution of that mechanism to failure) that occurs during the impact event—clearly this is of interest when protecting vulnerable structures such as the brain. What is important is that a ballistic protective composite should have a high volume fraction of fibre with relatively weak inter-ply adhesion, i.e. not necessarily a structurally good composite; the matrix content may be as low as 10% by mass (e.g. Shephard 1987; Brown and Egglestone 1989; Segal 1991; Hsieh et al. 1990; Lin et al. 1990). This is because delamination is the major mechanism by which energy is dissipated away from the impact point in ballistic protective composites (e.g. Shephard 1987; Morye et al. 1999; Sharma et al. 1999; Taylor and Carr 1999; Cheeseman and Bogetti 2003).

If armour is defeated (i.e. the projectile perforates through the armour and penetrates into the body), secondary projectiles and debris may be found in the wound as well as the primary (original) projectile (Mabbott et al. 2012, 2014; Nisbet et al. 2013; Mabbott 2015). Such secondary projectiles/debris may include fragmented primary projectile, bone, parts of the armour and fabric debris from clothing (Figure 2.8).

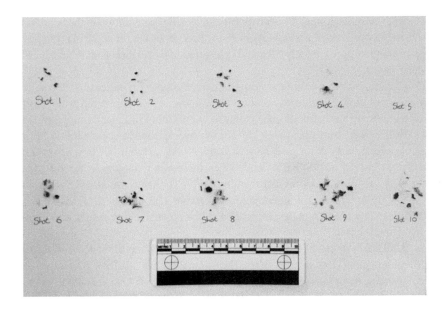

FIGURE 2.8 Example of debris (secondary projectile and body armour; n = 10) due to a high-velocity rifle bullets overmatching police body armour (100 mm scale). (From Mabbott, A.J., The overmatching of armour, PhD thesis, Centre for Defence Engineering, Cranfield University, Defence Academy of the United Kingdom, Shrivenham, UK, 2015.)

2.3.2 MILITARY BODY ARMOUR AND HELMETS

A typical military body armour is the UK OSPREY system, which is a tabard style armour, i.e. a sleeveless garment with front and back pieces and a hole for the head. OSPREY comprises layers of water repellent treated (WRT) para-aramid fabric encased in a water- and light-resistant cover to prevent ingress of water and ultraviolet radiation (UVR) and placed in a Cordura® outer carrier (Figure 2.9a). This soft armour provides protection from fragmentation.

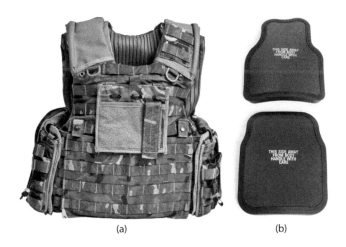

(a)　　　(b)

FIGURE 2.9 (a) OSPREY Body Armour Mk4 (without collar or brassards). (b) OSPREY plates (upper image front plate; lower image back plate.) (Courtesy of MOD Crown Copyright 2015.)

OSPREY plates provide protection from high-velocity rifle ammunition to the heart, mediastinum, liver, and spleen (Lewis 2006; Carr and Lewis 2014; Breeze, Lewis, et al. 2015) (Figure 2.9b). Each OSPREY plate weighs approximately 3 kg and is inserted in plate pockets in the front and rear of the OSPREY outer cover.

OSPREY is a modular body armour system that can be adapted to increase or decrease the level of protection dependent on the threat and the individual's role; for example, the armour system can be worn with or without the hard armour plates, thus only providing fragmentation protection. Wearing high or low collars and brassards increases the area of coverage and hence the level of protection, but also increases the total mass. Smaller enhanced combat body armour (ECBA; approximately 1 kg) plates can be used instead of OSPREY plates when there is a reduction in the threat level, giving the wearer slightly less protection but increased mobility. The complete OSPREY system, as issued in 2005, comprised the OSPREY fragmentation protective vest with integral load carriage system, two OSPREY plates, two ECBA plates, high collars, low collars and brassards, and weighs approximately 15 kg in total (Lewis 2006).

Since the OSPREY system was introduced in 2005, it has been modified and improved in response to the user's requirements. For example, where the original issued OSPREY protective vest had plate pockets situated on the outer face of the fragmentation protective vest; the most recent iterations of the OSPREY has internal plate pockets in order to reduce snagging (e.g. in vehicles), to improve integration with other equipment, and to streamline the front of the vest. Smaller side plates can now be inserted into integral side pockets of the OSPREY fragmentation protective vest, and a new collar has been designed specifically for dismounted operations.

UK military helmets are designed to protect the brain from fragments and non-ballistic impacts ('bump' protection). Modern UK military combat helmets typically have a composite shell, a foam liner, a suspension system and a retention system (Marsden 1994; Carr 1996a, 1996b; Carr et al. 2014). The Mk7 Combat Helmet was introduced in 2010 and has a para-aramid composite shell and a high-density, closed-cell polyethylene foam liner (Figure 2.10).

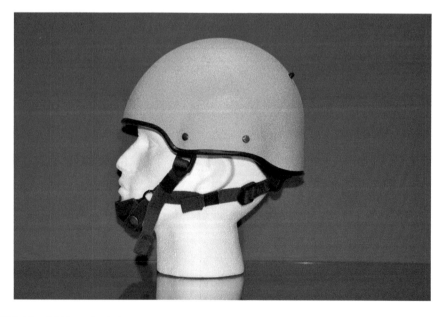

FIGURE 2.10 Mk7 combat helmet.

2.3.3 Essential and Desirable Medical Coverage

The coverage requirements for future UK military body armour systems has been refined recently to identify those anatomical structures requiring essential and desirable medical coverage with the ability to represent the geometry of internal anatomical structures within modern numerical models (Breeze, Lewis, et al. 2015). Essential and desirable medical coverage have been defined as:

- Essential medical coverage: those anatomical structures that if damaged would likely lead to death prior to definitive surgical intervention being available, e.g. bleeding from the thorax that cannot be compressed and requires surgical access (thoracotomy) to arrest it. Structures were chosen based upon a period of up to 60 minutes from injury to surgery. These structures have been identified as the heart, great vessels, liver and spleen (Breeze, Lewis, et al. 2015).
- Desirable medical coverage: those anatomical structures potentially responsible for mortality not fitting the requirement for essential coverage as well as those causing morbidity necessitating lifelong medical treatment or that result in significant disability. This includes both physical disability as well as psychological disability, e.g. damage to the lower parts of the spinal cord (lumbar or sacral parts) may result in significant loss of function in one or both the limbs (Breeze, Lewis, et al. 2015).

Work has also been conducted on considering the coverage of the UK military helmet from a military medical perspective (Breeze et al. 2013). This work confirmed that helmets should protect the brain and the brainstem due to their vulnerability and suggested that the nasion, external auditory meatus and superior nuchal line be used as anatomical landmarks to define the area of coverage.

REFERENCES

Adamson, P.B. 1977. A comparison of ancient and modern weapons in the effectiveness of producing battle casualties. *Journal of the Royal Army Medical Corps* 123:93–103.

Amato, J.J., D. Syracuse, and J.J. Seaver. 1989. Bone as a secondary missile: An experimental study in the fragmenting of bone by high-velocity missiles. *Journal of Trauma* 29:609–612.

Beebe, G.W., and M.E. DeBakey. 1952. *Battle Casualties*, 165–205. Springfield, IL: Charles C. Thomas.

Berlin, R.H., L.E. Gelin, and B. Janzon. 1976. Local effects of assault rifle bullets in live tissues. *Acta Chirurgica Scandinavica (Suppl)* 459:1–76.

Breeze, J., L.C. Allanson-Bailey, N.C. Hunt, R. Delaney, A.E. Hepper, and E.A. Lewis. 2015. Using computerised surface wound mapping to compare the potential medical effectiveness of Enhanced Protection Under Body Armour Combat Shirt collar designs. *Journal of the Royal Army Medical Corps* 161:22–26.

Breeze, J., D. Baxter, D.J. Carrr, and M.J. Midwinter. 2015. Defining combat helmet coverage for protection against explosively propelled fragments. *Journal of the Royal Army Medical Corps*. 161:9–13.

Breeze, J., E.A. Lewis, R. Fryer, A.E. Hepper, P. Mahoney, and J. Clasper. 2015. Defining the optimum anatomical coverage provided by military body armour against high velocity projectiles. *Journal of the Royal Army Medical Corps*. doi:10.1136/jramc-2015-000431.

Breeze, J., A.J. Sedman, and G.R. James. 2013. Determining the wounding effects of ballistic projectiles to inform future injury models: A systematic review. *Journal of the Royal Army Medical Corps*. doi: 10.1136/jramc-2013-000099.

Brown, J.R., and G.T. Egglestone. 1989. *Ballistic Properties of Composite Materials for Personnel Protection*. Melbourne, Australia: Materials Research Laboratory, DSTO.

Bruton, M. 2012. *Weight and Power*. Available online at http://www.soldiermod.com/volume-8/des.html [accessed 19 January 2016].

Carey, M.E. 1991. Analysis of wounds incurred by US Army Seventh Corps personnel treated in Corps hospitals during Operation Desert Storm. *Journal of Trauma* 40:S165–S169.

Carr, D.J. 1996a. The cause and severity of non-ballistic impact head injuries suffered by UK military personnel. Paper read at Personal Armour Systems Symposium (PASS 1996), 3–6 September, DCTA, Colchester, UK.

Carr, D.J. 1996b. Assessment of UHMWPE composites as candidate materials for UK military helmets. Paper read at European Composite Materials Conference-7, 14–16 May, London, UK.

Carr, D.J. 1999. Failure mechanisms of yarns subjected to ballistic impact. *Journal of Materials Science Letters* 18(7):585–588.

Carr, D.J., and E.A. Lewis. 2014. Ballistic protective clothing and body armour. In *Protective Clothing: Managing Thermal Stress*, edited by F. Wang and C. Gao, 146. Manchester, UK: Woodhead Publishing/The Textile Institute.

Carr, D.J., I. Horsfall, and C. Malbon. 2016. Is behind armour blunt trauma a real threat to users of body armour? A systematic review. *Journal of the Royal Army Medical Corps* 162:8–11.

Carr, D., G. Starling, T. de Wilton, and I. Horsfall. 2014. Tensile properties of military chin-strap webbing. *Textile Research Journal* 84(6):655–661.

Cheeseman, B.A., and T.A. Bogetti. 2003. Ballistic impact into fabric and compliant composite laminates. *Composite Structures* 61(1–2):161–173.

Crew, F.A.E. 1953. *History of the Second World War*. London, UK: The Army Medical Services, HMSO.

Cunniff, P.M. 1992. An analysis of the system effects in woven fabrics under ballistic impact. *Textile Research Journal* 56:45–60.

Di Maio, V.J.M. 1998. *Gunshot Wounds: Practical Aspects of Firearms, Ballistics, and Forensic Techniques*. 2nd ed. Boca Raton, FL: CRC Press.

Galbraith, K.A. 2001. Combat casualties in the first decade of the 21st century—New and emerging weapon systems. *Journal of the Royal Army Medical Corps* 147(1):7–14.

Garfield, R.M., and A. Neugut. 1991. Epidemiological analysis of warfare. *JAMA* 266:688–692.

Hardaway, R.M. 1978. Vietnam wound analysis. *Journal of Trauma* 18:635–543.

Heard, B.J. 2008. *Handbook of Firearms and Ballistics: Examining and Interpreting Forensic Evidence*. 2nd ed. Chichester: Wiley-Blackwell.

Helliker, M., D.J. Carr, C. Lankester, L. Fenton, E. Girvan, and I. Horsfall. 2014. Effect of domestic laundering on the fragment protective performance of fabrics used in personal protection. *Textile Research Journal* 84(12):1298–1306.

Holmes, R.H. 1952. Wound ballistics and body armour. *Journal of American Medical Association* 150:73–78.

Horsfall, I. 2012. Key issues in body armour: Threats, materials and design. In *Advances in Military Textiles and Equipment*, edited by E. Sparks, 3–20. Manchester, UK: Woodhead Publishing and The Textile Institute.

Hsieh, C.Y., A. Mount, B.Z. Jang, and R.H. Zee. 1990. Response of polymer composites to high and low velocity impact. In 22nd International SAMPE Technical Conference, SAMPE, Boston, MA.

Jackson, D.S., C.G. Batty, J.M. Ryan, and W.S. McGregor. 1983. The Falklands War: Army field surgical experience. *Annals of the Royal College of Surgeons of England* 65:281–286.

Kieser, D.C. 2014. Ballistic skeletal injuries: An experimental study of the orthopaedic, biomechanic and forensic characteristics. PhD thesis, Orthopaedic Surgery, Dunedin School of Medicine, The University of Otago, Dunedin, New Zealand.

Kieser, D.C., D.J. Carr, S.C.J. Leclair, I. Horsfall, J.C. Theis, M.V. Swain, and J.A. Kieser. 2013. Gunshot induced indirect femoral fracture: Mechanism of injury and fracture morphology. *Journal of the Royal Army Medical Corps* 159:294–299.

Kneubuehl, B.P., R.M. Coupland, M.A. Rothschild, and M.J. Thali. 2011. *Wound Ballistics Basics and Applications*. Berlin: Springer.

Lewis, E.A. 2006. Between Iraq and a hard plate: Recent developments in UK military personal armour. Paper read at Personal Armour Systems Symposium 2006, 18–22 September, The Royal Armouries, Leeds, UK.

Lewis, E.A., M.A. Pigott, A. Randall, and E.A. Hepper. 2013. The development and introduction of ballistic protection of the external genitalia and perineum. *Journal of the Royal Army Medical Corps* 159(Supp 1):i15–i17.

Lin, L.C., A. Bhatnager, and H.W. Chang. 1990. Ballistic energy absorption of composites. In 22nd International SAMPE Technical Conference, 6–8 November, SAMPE, Boston, MA.

Mabbott, A., D.J. Carr, S. Champion, and C. Malbon. 2014. Boney debris ingress into lungs due to gunshot. Paper read at International Symposium on Ballistics, 22–26 September, Atlanta, GA.

Mabbott, A., D.J. Carr, C. Malbon, C. Tickler, and S. Champion. 2012. Behind soft armour wounding by penetrating ammunition. Paper read at Personal Armour Systems Symposium, 17–21 September, Nuremburg, Germany.

Mabbott, A.J. 2015. The overmatching of armour. PhD thesis, Centre for Defence Engineering, Cranfield University at the Defence Academy of the United Kingdom, Shrivenham, UK.

Marsden, A. 1994. UK body armour and helmets. Paper read at Personal Armour Systems Symposium (PASS94), 21–25 June, Colchester, UK.

McFate, M. 2005. The social context of IEDs. *Military Review*. January–February: 37–40.

Moffat, W.C. 1976. British Forces causalities in Northern Ireland. *Journal of the Royal Army Medical Corps* 112:3–8.

Morye, S.S., P.J. Hine, R.A. Duckett, D.J. Carr, and I.M. Ward. 1999. A comparison of the properties of hot compacted gel-spun polyethylene fibre composites with conventional gel-spun polyethylene fibre composites. *Composites: Part A* 30:649–660.

Nisbet, W.J., D.J. Carr, S.M. Champion, T. Stuart, and M. Helliker. 2013. Effect of impact velocity on the wounding potential behind ceramic/composite plates. Paper read at International Symposium on Ballistics, 22–26 April, Freiberg, Germany.

Orlowski, T., T. Piecuch, and J. Domaniecki. 1982. Mechanisms of shot wounds caused by missiles of different initial velocity. *Acta Chirurgica Scandinavica Suppl* 508:123–127.

Patel, H.D.L., S. Dryden, A. Gupta, and N. Stewart. 2012. Human body projectiles implantation in victims of suicide bombings and implications for health and emergency care providers: The 7/7 experience. *Annals of the Royal College of Surgeons of England* 94(5):313–317.

Rees, D. 1964. *Korea: The Limited War*. London, UK: MacMillan.

Reister, F.A. 1973. *Battle Causalities and Medical Statistics: US Experience in the Korean War*. Washington, DC: The Surgeon General, Department of the Army.

Roylance, D., A. Wilde, and G. Tocci. 1973. Ballistic impact of textile structures. *Textile Research Journal* 43: 34–41.

Ryan, J.M., G.J. Cooper, I.R. Haywood, and S.M. Milner. 1991. Field surgery on a future conventional battlefield: Strategy and wound management. *Annals of the Royal College of Surgeons of England* 73(1):13–20.

Sakaguchi, S.M., D.J. Carr, I. Horsfall, and E. Girvan. 2012. Protecting the extremities of military personnel: Fragment protective performance of one- and two-layer ensembles. *Textile Research Journal* 82(12):1295–1303.

Segal, C.L. 1991. High performance organic fibres, fabrics and composites for soft and hard armour applications. In 23rd International SAMPE Technical Conference, 21–24 October, Kiamesha Lake, NY.

Sharma, N., D.J. Carr, P.M. Kelly, and C. Viney. 1999. Modelling and experimental investigation into the ballistic behaviour of an ultra-high molecular weight polyethylene/thermoplastic rubber matrix composite. In 12th International Conference on Composite Materials (ICCM-12), Paris, France.

Shephard, R.G. 1987. The use of polymers in personal ballistic protection. In *Polymers in Defence*. London, UK: The Plastics and Rubber Institute.

Shim, V.P.W., V.B.C. Tan, and T.E. Tay. 1995. Modelling deformation and damage characteristics of woven fabric under small projectile impact. *International Journal of Impact Engineering* 16(4):585–605.

Steier, V., D.J. Carr, C. Crawford, M. Teagle, D. Miller, S. Holden, and M. Helliker. 2014. Effect of FSP material on the penetration of a typical body armor fabric. In the proceedings of the International Symposium on Ballistics, 22–26 September, pp. 880–885. Atlanta, GA.

Taylor, S.A., and D.J. Carr. 1999. Post failure analysis of 0°/90° ultra-high molecular weight polyethylene composite after ballistic testing. *Journal of Microscopy* 196(2):249–256.

Thurman, J.T., ed. 2011. *Practical Bomb Scene Investigation*. 2nd ed. Boca Raton, FL: Taylor and Francis.

Tobin, L., and M. Iremonger. 2006. *Modern Body Armour and Helmets: An Introduction*. Canberra, Australia: Argros Press.

UK Ministry of Defence. 2005. Proof of ordnance, munitions, armour and explosives: Part 2–guidance, Defence Standard 05–10. 1 Part 2 Issue 1, Defence Procurement Agency, Glasgow, UK.

Wang, Z.G., C.G. Tang, and X.Y. Chen. 1988. Early pathomorphologic characteristics of the wound track caused by fragments. *Journal of Trauma* 28:S89–S95.

Zecevic, B. 2011. Characterization of distribution parameters of fragment mass and number for conventional projectiles. In proceedings of the 14th Seminar on New Trends in Research of Energetic Materials, 13–14 April, pp. 1026–1039. Pardubice, Czech Republic.

3 Blast Physics and Vehicle Response

Stephen J. Cimpoeru, Paul Phillips and David V. Ritzel

CONTENTS

3.1 INTRODUCTION

Explosive munitions, particularly landmines and roadside bombs, are a challenging threat for armoured vehicles. The rise in the number of buried improvised explosive device (IED) attacks over the last decade and the resultant number of casualties (icasualties 2014) has led Defence forces to increase vehicle protection and field new vehicles to counter blast threats, particularly underbelly (UB) blast threats.

Vehicle blast protection is complex and often has features specific to each particular vehicle design, and many existing vehicles, while blast resistant, have not been optimally designed for blast survivability. The blast-resistant design of vehicles has been compromised by an ill-defined, dynamically changing threat condition combined with an insufficient understanding by designers of factors such as the underlying physics of the blast output, how such loading is imparted and transmitted, how damage is sustained, and how occupant injuries are inflicted. In this chapter, the physics of the interaction between blast-induced loading and vehicle and occupant response is examined for both underbelly and side-blast attack of vehicles.

3.2   THE BLAST EVENT

The destructive power of explosives has been known since at least 220 BC, the first documented explosive accident injured early alchemists in China and led to the development of gunpowder. However, the fundamental scientific understanding of the underlying blast physics* only really developed in earnest in the 1940s due to the need to understand the blast generated by the first nuclear weapons. Understanding the complex nature of these blast processes is the key to devising the most effective protective technologies for vehicle occupants. This section first describes the basic phenomena of the blast flow-field loading conditions for air blast (relevant to side-blast attack on vehicles) before discussing UB blast.

3.2.1   Air Blast

The simplest case of an idealised blast from a bare spherical high-explosive charge in the air is first discussed. In reality, actual blast events are mostly non-ideal and are strongly affected by explosive type and many other factors such as charge shape, casing, burial and so forth, and particularly interactions with the ground. Figure 3.1 depicts the very early development of the blast-wave flow for an idealised centrally-detonated bare spherical charge of high explosive. Following initiation, a detonation wave sweeps through the unreacted explosive material at speeds typically ~6–8 km/s, effectively converting the solid explosive to hot and extremely high-pressure gases at about 3,000 K and 400,000 atm or 40 GPa. Due to their extreme pressure, the gaseous detonation products expand rapidly to about 4,000-fold the original charge volume and are visible as a radiant fireball; it is the hydrodynamics of this expansion process which generates the blast-wave flow.

The rapid expansion of this fireball of detonation products drives a shock wave into the surrounding air ahead of it, much like the action of a spherical piston. The edge of the expanding fireball is effectively a material front designated as a *contact surface* across which there is theoretically no mass or heat transfer. In reality, there is always some degree of turbulent mixing at this interface and consequent momentum, material and heat transfer. The most distinctive feature of the propagated air-blast wave is the shock front through which there is a nearly instantaneous step change in all gas-dynamic conditions of the air including static pressure[†], density, flow velocity and temperature.

A potentially significant and generally unrecognised aspect of the early blast flow development is that the air immediately surrounding the charge is 'shock heated' by the passage of the intense shock front to extreme temperatures of the order of 8,000 K. This process is not related to heat transfer from the fireball as might be expected; ironically, heat would only be transferred from the shock-heated air to the fireball. The presence of superheated air just beyond the periphery of the fireball after the passage of the shock is the cause for secondary combustion of afterburning material from the fireball which might have been mixed in by turbulence or projected into this zone of shock-heated air. Although the fireball is typically highly radiant and luminous, at its full expansion, it is much colder (~500 K) than the thick annular shell of air immediately beyond its perimeter, which will persist at temperatures up to ~3,000 K until dissipated some seconds after the passage of the shock wave.

The combined violent expansion of detonation-product gases and the resultant propagated air-shock wave constitutes the 'blast flow-field' loading condition, and it is important to recognise the dual nature of these blast flow conditions. That is, close to the charge within the region of the fireball expansion, not only is the amplitude of the blast forces more extreme as would be expected,

---

* Blast physics concerns the processes by which the energy of an explosion source propagates into its surrounding environment then interacts, loads and damages materials, structures and systems.

† Static pressure is the pressure experienced by a point which does not obstruct the air-blast wave, i.e. as would be measured on a surface parallel to the flow.

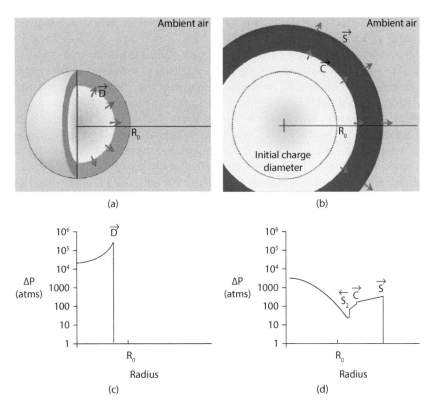

FIGURE 3.1   Schematic depiction of the very early development of the blast-wave flow for a centrally-initiated bare spherical charge of high explosive, (a) and (b), with spatial profiles of the change in static pressure, ΔP, with respect to charge radius, R, in the lower figures corresponding to the physical depiction above, (c) and (d). Figure 3.1 (a) shows the detonation front $\overrightarrow{D}$ prior to reaching the charge surface, with Figure 3.1 (c) showing the change in static pressure within the detonated explosive. Figure 3.1 (b) shows the generation of the air-shock wave $\overrightarrow{S}$ being driven by the expansion of the fireball contact surface $\overrightarrow{C}$. Figure 3.1 (d) shows that the flow within the expanding detonation products is further partitioned by an 'embedded' rearward-facing shock, $\overleftarrow{S_2}$. The various materials are shown with green representing unreacted explosive, yellow/orange representing gaseous detonation products (orange being the centre of detonation) and blue designating air; R_0 designates the original charge radius. [Images (a) and (b) courtesy of DR&DC Suffield.]

but also the flow field consists primarily of the expanding detonation-product gases as distinct from air. Compared to blast at greater distance, there is proportionately much higher kinetic energy in near-field blast (Sternberg and Hurwitz 1976) as well as dramatically variant spatial and temporal energy partitions.

Near-field blast conditions have significant implications regarding consequent loading and damage processes which should be distinguished from classical understandings of static overpressure loading and damage in the far field (Baker 1973). The very strong near-field flow forces due to dynamic pressure*, especially from the impingement of the expanding detonation products, are directional and interact with structures quite differently than the static pressure component. In comparison, targets beyond the fireball expansion are subjected not only to lower blast loading because of the greater distance but also to a rapidly decreasing proportion of effects from dynamic pressure forces with distance. The energy from the dynamic pressure forces is transferred to the

---

* Dynamic pressure is a measure of the specific kinetic energy of the flow, i.e. $\frac{1}{2}\rho v^2$, where ρ and v are the density and velocity of the flow, respectively.

propagating air-blast shock wave, which is based on static rather than dynamic pressure, as the fireball expands and weakens.

Beyond the fireball, the dynamic pressure component of the blast is due to the induced flow from the passage of the air-shock wave rather than the direct effects of the expanding detonation products. This dynamic pressure or 'blast wind' decreases continuously with increasing distance such that loading of structures in the far field ($<$ 0.1 atm static pressure) is almost entirely dominated by the static pressure of the shock wave. The air-shock wave reflects* and diffracts around structures causing highly non-uniform and time-variant loading. Whether static or dynamic pressure is more important for damage is dependent on the specific vulnerability of the structure subjected to the blast.

Vehicle response to air-blast loading is dominated by the hydrodynamic forces of the blast flow-field interacting over the surfaces of a vehicle hull typically inflicting plate deformations and accelerations (refer Section 3.4) as well as imparting momentum in a global sense, thus causing global vehicle motions (refer Section 3.6).

3.2.2 Underbelly Blast

The detonation of a shallow buried charge results in a cone-shaped ejecta plume being thrown upwards by the blast, as shown in Figure 3.2; the multiphase plume consists of the soil that covered and surrounded the charge, augmented by detonation-product gases of the expanding fireball as well as casing fragmentation. Very often the 'slipstream' or edge of the ejected plume of material is quite well defined, spatially forming a cone, as shown in Figure 3.2, the cone angle being narrower with increasing depth of burial.

For buried charges, the detonation products and ejecta are channelled or vented upwards by the confinement of the surrounding ground and will inflict most of the damage experienced by a vehicle

FIGURE 3.2   Blast and ejecta from a buried charge. The ejecta plume usually forms a distinctive conical slip-stream boundary with the surrounding air. Within the ejecta plume, targets are subjected to high static pressures and dense, high-pressure, high-velocity multiphase flow of dirt overburden mixed with detonation products from the charge. Below the plume, targets would be subjected to weak blast and negligible ejecta. The contact surface is the edge of the fireball. (Courtesy of DRDC Suffield.)

---

* It is important to consider that air-shock waves exhibit distinctly different behaviour with respect to propagation and interactions at interfaces than waves considered in electromagnetics or radiology, for example. Although some aspects of acoustic (sound) wave theory can be applied to exceedingly weak air-shock waves, this concept is of little relevance for vehicle blast loading. Even when the blast flow is entirely an air-shock wave in the far field (as distinct from combined air-shock and detonation-product flow in the near field), at the surface of most structures of practical interest, the air-shock wave can be considered to be fully reflected or diffracted as distinct from being transmitted. This is because of the enormous difference in wave impedance, ρc, where ρ is the material density and c the wave propagation speed.

in its path. The blast conditions in the 'shadow zone', i.e. the zone beneath the slipstream angle of the ejecta cone, are relatively weak, with only an air-blast shock wave diffracting into this zone.

A substantial proportion of the blast energy from buried charges is transferred to the solid particulate of the soil surrounding the charge, accelerating it to form the ejecta plume; the kinetic energy of this particulate is highly directional across the span of the ejecta cone. The loading process and consequent damage from buried charges will therefore be somewhat analogous to the effect of a shotgun blast heavily laden with pellets that imparts momentum transfer directly to targets by impact. This is unlike classical air blast, where the blast acts entirely as a hydrodynamic wave, whose energy can reflect and diffract around an obstacle. UB blast is near field blast and will therefore involve a distributed zone of impinging high-density multiphase flow combined with high static pressures.

An air-shock wave will also be driven ahead of the multiphase contact surface of the ejecta plume. Over a distance, away from the vehicle, the solid and gaseous phases of the plume will begin to separate, with the momentum of the larger clumps or solid particles causing them to be thrown ahead of the decelerating gas-dynamic contact surface of the fireball. However, in the blast flow regimes of relevance for vehicle damage, the air-shock wave and the dense high-speed multiphase ejecta plume will be closely coupled. The air-shock wave will impart loading when it reflects at and diffracts around the outside surface of the vehicle hull. Although the solid particulate does not technically exhibit 'pressure', the *specific kinetic energy* (defined earlier) of the particulate phase will inflict the equivalent of a dynamic pressure loading on any obstruction in the flow.

The above aspects of the near-field blast conditions have significant implications regarding the loading and damage processes. The very strong near-field flow forces due to dynamic pressure are directional, and interact with structures quite differently to the static pressure component which dominates damage from the classical air-shock wave in the far field. The intense and localised very dense, high pressure, and for UB blast, multiphase high-speed flow of material (detonation products and ejecta) will result in stretching and bending deformation of the vehicle hull and may, in some cases, perforate it. This loading will also result in a significant momentum transfer to the vehicle, which will result in global motion of the vehicle, even without significant hull deformation or rupture. If there is coupling of the air-blast flow with ground ejecta, as would be the case with a buried explosive, the momentum imparted to a target can be nearly three times the level of an unburied explosive (Fairlie and Bergeron 2002).

UB blast loading is strongly influenced by various factors (Anderson et al. 2011; Chung Kim Yuen et al. 2012) such as explosive type, charge standoff (distance between the centre of the charge and the target), size, shape, initiation, casing and fragmentation, soil conditions and burial depth. The mechanical properties of the ground, e.g. density, porosity, granular composition, cohesion and moisture content, in which a charge is buried will also have a significant effect on the explosive loading on a target. For example, a greater target loading will be imparted by a charge laid in denser clay/ shale than a charge in dry sand (Ehrgott 2010). Soil moisture content has been shown to be of particular importance, with increased moisture content greatly enhancing the loading and momentum transferred (Fourney et al. 2005). Whilst a NATO STANAG (NATO 2006) specifies a standardised methodology for developing a landmine testbed, many variations in blast loading are possible. Soil conditions vary widely between countries attempting to standardise landmine testbeds; in particular, differences in soil density and optimum moisture content for compaction have complicated the standardisation of blast output from landmine testbeds worldwide.

3.3   SEQUENCE OF VEHICLE AND OCCUPANT RESPONSE

Analyses of UB blast effects on armoured vehicles can be subdivided into three categories as depicted in Figure 3.3: (1) localised hull response, including possible rupture, to an immediate external blast load; (2) internal vehicle dynamics, i.e. transmission of hull reaction loads to internal systems; and (3) global vehicle motions, including vehicle translation, secondary projectile impacts and rollover. The categories above are not only in approximate chronological sequence but also in general terms,

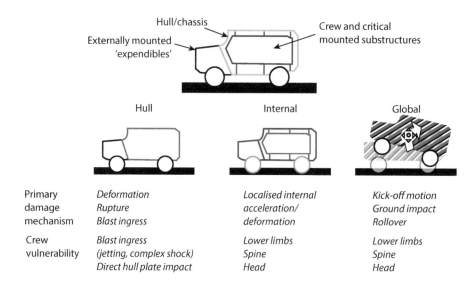

FIGURE 3.3 Concept sketch for three regimes of vehicle response to blast (hull response, internal vehicle dynamics and global vehicle motions) with typical occupant injury mechanisms.

in logical order regarding both the priority and consequence for analyses of vehicle damage and applications of protection enhancements. Indeed, Figure 3.3 is also relevant for side-blast attack of vehicles and should be viewed as a generalised depiction of vehicle response to blast attack.

Although the three phases of vehicle response are coupled and overlap in time, the distinctions are convenient for analyses since they can be associated with particular vehicle design features or components and corresponding protection strategies. The timescale and relative significance of each phase depends on the vehicle size and design as well as the magnitude of the blast threat. Vehicle occupants can be injured by the structural dynamics of all three phases as well as the blast wave itself, should the blast make ingress due to rupture of the hull.

A blast event is extremely complex, being a dynamic loading condition that is rapidly changing both spatially and temporally. Furthermore, IEDs have increasingly proliferated, and unlike standard blast threats, IEDs vary widely in charge size, shape, casing, explosive type and deployment technique, which greatly compounds the difficulty of defining their loading for protective design.

Armoured vehicles are therefore normally rated to standardised protection levels based on a level of blast loading that can be sustained without unacceptable occupant injuries (e.g. NATO 2006, 2014). If the level of blast loading exceeds the vehicle's protection level, there is a high likelihood that moderate-to-severe injuries to vehicle occupants will be sustained. This can be due to any or all three categories of vehicle response as depicted in Figure 3.3. In such circumstances, the vehicle is considered to be overmatched.

Possible causes of injury (shown in Figure 3.3) include effects from blast pressure ingress due to hull rupture, early localised internal platform accelerations or structural deformations, and inertial accelerations from late-time global vehicle motions with possible secondary impacts. Potentially serious aftermath injury modes include entrapment under debris as well as exposure to fire, toxic gases and oxygen-deficient atmospheres. A crude overview of typical injury threats to vehicle occupants from UB blast is summarised in Table 3.1 in relation to the vehicle response categories as previously described with the additional consideration of 'late-time effects' no longer related to vehicle structural response but relevant to potential occupant injuries.

Overall, there is a need to better understand the blast loading as well as the vehicle response dynamics at the level of fundamental principles and how those responses translate to the internal structures and cause occupant injuries within vehicles during the full course of a blast event. Controlled live-fire

TABLE 3.1

Overview of Typical Injury Mechanisms Correlated to Categories of Vehicle Response for Underbelly Blast

Response Category	Approximate Timescale	Typical Injury Mechanisms
Hull response	0 to milliseconds	• Explosive rupture of hull allows direct detonation product, ground ejecta and hull debris ingress (including severe late-time effects described below), with probable fatal consequences for occupants immediately above a rupture; diffracted blast pressure into the adjacent occupant compartment • Blast-induced deflection of hull can impact personnel directly causing lower-leg injuries
Internal vehicle dynamics	0 to milliseconds	• Acceleration or displacement of key fixture points such as floor joists or seat-mounting fixtures resulting in lower-leg, spinal and pelvic injuries • Diffracted blast pressure ingress via open apertures
Global vehicle motions	Milliseconds to seconds	• Inertial accelerations from the acceleration of the vehicle inflicting spinal/pelvic/head/neck/thorax injuries, secondary projectile impacts, and other types of blunt impact trauma
Late-time effects	Seconds	• Exposure to fire, toxic fumes, oxygen deficiency, pinning/incapacitation • Apnea (temporary cessation of breathing) is common after strong air-shock wave exposure

testing, i.e. full-scale destructive blast tests on actual vehicles, with appropriate scientific measurement techniques, allows analysis of the full chain of events that leads to occupants being killed or injured and informs the development of appropriate survivability systems. The discussion that follows primarily describes the effect of an underbelly blast event on vehicle and occupant loadings to better explain the vehicle response categories shown in Figure 3.3.

3.4 HULL RESPONSE

Figure 3.4 depicts a typical landmine test. The blast flow from an UB blast can interact with the underneath of a vehicle hull less than 0.25 ms after detonation. The belly of a vehicle hull is then subjected to very high strain rate loading from the blast wave and associated fragmentation and ejecta.

The hull armour of a vehicle has to withstand intense blast loading at high strain rates without rupture or excessive dynamic plate deformation into the crew survival space. Hull or plate rupture will expose any vehicle occupants immediately above a rupture to a collimated jet or flow of expanding detonation products, ground ejecta and hull debris, with probable fatal consequences. Occupants near a hull rupture are also at some risk of overpressure injuries, depending on the scale of the breach. Internal systems and occupants may also be vulnerable to direct hull plate intrusion into the occupant survivability space.

Various methods are used to locally limit dynamic hull deformations, particularly in critical areas where occupant survival spaces would be encroached. Increased armour plate strength will reduce deformation but requires significant material toughness to avoid ruptures, particularly when triaxial stress states are present. A greater plate thickness will increase the geometric stiffness and thus reduce deformation as well as reduce local plate accelerations according to F = ma. Structural reinforcement of critical areas, e.g. the use of stiffeners and even the use of fluids (Bornstein et al. 2015) are further examples of methods that may be used to reduce dynamic hull deformations.

FIGURE 3.4   A typical DST Group live-fire blast test.

    Increased standoff from the charge will reduce the blast loading on an armoured vehicle hull. This is well known in general terms, and hull shaping also has a significant effect. Flat vehicle hull designs can provide higher mobility and payload but usually provide little blast standoff or deflection. On the other hand, deep, e.g. 90° V-shaped hulls are effective at not only reducing vertical loading forces, and thus hull deformation, but also minimising the momentum transferred to the hull. This is because the flow nature of the blast allows V-shaped hulls to largely deflect the directional components of the blast wave, i.e. dynamic pressure and ejecta, and its associated loading away from the hull. This can be seen in Figure 3.5, where the deep hull angle results in greater

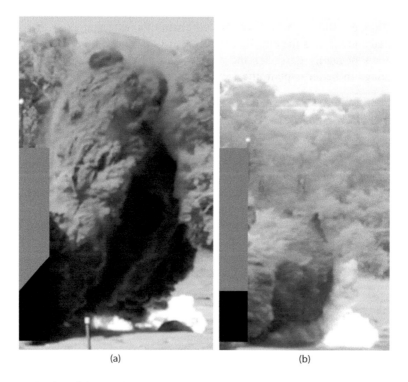

(a)   (b)

FIGURE 3.5   Deflection of ejecta and detonation products at 6 ms after detonation for a deep V-shaped, hull (a), versus a flatter hull (b). Representative hull shapes are shown in blue.

deflection of the blast ejecta and detonation products away from the hull than a flatter hull at the same point in time. By comparison with V-shaped hulls, flatter hulls are subjected to greater vertical loading and momentum transfer (Anderson et al. 2011; Chung Kim Yuen et al. 2012; Montoya and Bornstein 2012). The disadvantages of V-shaped hulls for vehicle design are generally a higher centre of gravity, and thus greater potential for rollover, a greater mass, and a smaller cabin volume, all of which may result in restrictions on vehicle payload or mobility.

'W' or 'Double V' hull shapes are now used by some military vehicle manufacturers. This, in part, allows air transport and shipping requirements to be achieved through blast-resistant designs that are more vertically space efficient than V-shaped hull geometries, although they appear to result in higher momentum transfers to the hull than V-shaped and possibly even flat vehicle hulls (Montoya and Bornstein 2012). A number of variations in 'Double V' geometry are possible.

The effectiveness of shaped hulls also depends on the charge placement relative to a specific hull geometry. This will affect not only hull deformation but also vertical and horizontal momentum transfer (Montoya and Bornstein 2012), with subsequent effects on global vehicle motion. Figure 3.6 shows how the resultant detonation flow varies for different charge locations for a 120° V-shaped hull. Off-centre charge locations can result in significant horizontal momentum transfer.

Testing and modelling of candidate vehicle hulls must take into account different placements of the threat relative to the vehicle hull and occupants to ensure a worst-case event is assessed. Adjustable height suspension systems are also being used to provide greater UB blast to hull standoff in high-threat environments while not compromising long-haul transportation requirements.

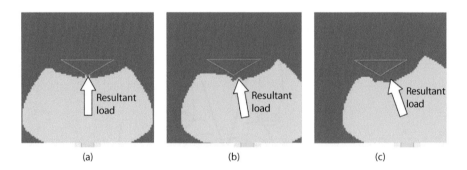

FIGURE 3.6   Comparison of hull loading vector from detonation-product flow (shown in green) 0.25 ms after detonation for a 120° V-shaped hull with the landmine location underneath its centre (a), the landmine halfway between the centre and edge of hull (b), and the landmine at the edge of the hull (c). (Reproduced from Montoya, D., and Bornstein, H., Geometric Effects of Hull Shapes on Blast Mitigation for Armoured Vehicles. In *Land Warfare Conference*, Melbourne, 29 October–2 November, pp. 355–364, 2012.)

3.4.1 CONSIDERATIONS FOR SIDE-BLAST

Side-blast loading of the hull differs from UB attack by the blast being more expanded or diffracted prior to loading the hull. The absence of ground ejecta will also result in considerably less momentum transfer.

3.5   INTERNAL VEHICLE DYNAMICS

3.5.1 LOADING

Direct interaction of a blast load with a vehicle can cause localised accelerations of the armour plates that comprise the vehicle hull. The very high rate transmission of the hull plate reaction forces and moments to internal substructures, such as the occupant seating and the floor, can occur via structural connections between such items and the outer vehicle hull. The nature and magnitude of such internal dynamic loads and their transmission will strongly depend on the exact form and location of these structural connections.

For example, a hull armour plate that experiences local blast-induced acceleration can transmit loading to a seat through any structural connection between such seats and the accelerated hull plates, with direct and more rigid connections imparting more significant loads. Another example would be the transmission of hull loading to a floor plate whose resulting deformation imparts injurious impact loads to the feet and legs of the occupants.

Post-event damage assessment of a vehicle can only be used for inspection of the final state of damage, as evidenced, for instance, by permanent hull deformation measurements. High-speed imagery in live-fire test events shows that elastic, dynamic deformations beyond those recorded by permanent hull deformation measurements are a significant damage and injury mechanism, and typically occur within the first 10 ms after detonation.

An understanding of the extent that a vehicle hull or internal substructures will dynamically deform or transmit loads is important for the effective design of the internal occupant survivability subsystems as shown in Figure 3.7. The design of such survivability subsystems should be guided by input loadings that are representative of the amplitude and duration of loads at the relevant part of the vehicle, be it seating, flooring, gunner platforms and so forth.

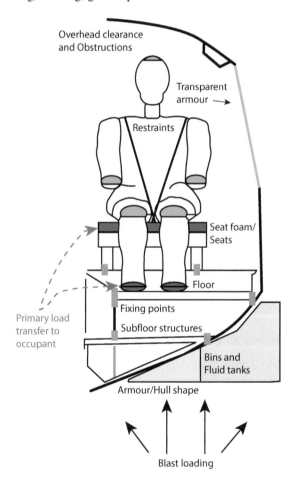

FIGURE 3.7 Schematic of underbelly blast loading, survivability subsystems and associated load paths to vehicle occupants. The primary load transfer interfaces to the human occupant are shaded in red, while orange shading denotes areas on the occupant that can also be subjected to unexpected impact loads, particularly in overmatch situations.

3.5.2 Loading on Occupants

3.5.2.1 Primary Load Transfer

The final load paths to the human occupants occur through the seat cushion, the occupant restraint system, and the floor/footrests (or soles of the boots) as shown in Figure 3.7.

Landmine-resistant seats are designed to reduce inertial loadings arising from global vehicle motions and, as such, are usually not specifically designed for mitigating short-timescale loadings from internal vehicle dynamic loads, for instance, from direct impacts of the hull or floor plates or other transmitted loads. Significant internal dynamic loadings often occur when a vehicle design has been overmatched, and any resulting hull or plate deformations or loadings that arise should not dislodge landmine-resistant seating or interfere with or damage its primary design function which is to reduce inertial loadings during global vehicle motions. Seat-mounting points on the roof or high on the side walls are often advantageous in such circumstances as those locations are usually subject to lower internal dynamic loads and deformations than floor-mounted seats. Optimum seat-mounting locations also depend on their feasibility from an engineering integration point of view.

Seat cushions (foam) are particularly pertinent to reducing spinal injuries from UB blast loadings (Wang and Bird 2001). Seat foam (if designed appropriately) can also offer a means of reducing the amplitude of internal dynamic loads in overmatch situations. While seat foams (and seats) can be optimised for specific loading rates, e.g. the work of Saunders et al. (2012), it is important that any related laboratory testing replicates the actual high-rate dynamic loadings found in an actual blast event. Laboratory drop-tower test methods can be used to replicate the final imparted vehicle velocities, but not necessarily the internal dynamic loadings from overmatch situations, being better suited to replicate inertial loadings from the global motion phase (refer Section 3.6.2.1 for a detailed description of global vehicle motions).

A similar scenario to that discussed above for seating also applies to the vehicle floor or footrests and any transfer of loading through the feet. While a number of methods are used to test systems that can reduce the short-timescale loadings discussed here (Wang et al. 2001; Golan et al. 2012), again further work needs to be conducted in order to better quantify and replicate the actual loadings transmitted within vehicles during real blast events, particularly internal dynamic loads associated with overmatch situations.

Loadings on the human occupants will also impart momentum which will cause unrestrained hand, leg and head movement, and thus potential impacts with surrounding objects and obstructions within the vehicle.

3.5.2.2 Steering Wheel Impacts

Figure 3.8 shows steering wheel deformation for a military vehicle as a result of vertical loading from a landmine event. High femur shear loads (as recorded by Hybrid III ATDs) are often caused by occupant impacts with steering wheels during such events. Potential leg injuries may be mitigated by changes in steering wheel design, greater survival space, or reducing transmitted loads. Automotive crashworthiness technologies such as airbags may also provide novel design options for military vehicles. Steering wheel airbags can take 10 ms to detect a frontal automobile crash event and trigger, and another 30 ms to deploy and inflate (Kuttenberger 2011), and the civilian automobile industry has now adapted airbags for the front, side (chest), head, knee, and more recently, seatbelt applications. Adaption and improvement of this technology may also allow this type of protection technology to be designed for the timeframe needed to prevent knee strikes against steering wheels (and other impacts) where detection, triggering and inflation are needed to mitigate impacts occurring less than 10 ms after blast detonation.

3.5.2.3 Roof Impacts

It has been observed that head strikes in a vehicle will usually be avoided in UB blast events as long as there is sufficient overhead clearance for a restrained occupant. However, sometimes excessive hull deformation in overmatch situations will reduce the survival space within a vehicle, and thus lead to

FIGURE 3.8 DST Group live-fire blast test: steering wheel after leg impact. The lower femur has initially contacted the wheel from underneath before coming to rest as shown.

head impacts. Sufficient head clearance is also required to accommodate the helmet and other add-ons such as night vision equipment. Not surprisingly, vertical space in a vehicle cabin is always at a premium; head clearance has to be traded against the vertical space required for the operation of load attenuation mechanisms associated with landmine-resistant seating (refer Section 3.6.2.1). Extra cabin height also affects the centre of gravity and adds weight to the vehicle, and thus needs to be traded against mobility requirements.

3.5.3 CONSIDERATIONS FOR SIDE-BLAST

The nature of loadings from side blast, and thus precise occupant injuries, will be affected by the specific design details of the vehicle, but more particularly the orientation of the occupant seating relative to any deforming (or impacting) side hull plates. Dynamic loadings will be transferred through the seating and then to the occupants.

Forward-facing seats (and occupants) will therefore be subjected to lateral loading (sometimes with direct hull plate impacts on the occupant), whereas occupants in side-facing seats, i.e. where the occupant is facing towards the interior of the vehicle, will be subjected to rearward loading against the parts of the body in contact with the seat. Occupants in forward-facing seats are therefore primarily at risk of injuries from loadings on the shoulder, ribs and lumbar spine/pelvis, whereas occupants in side-facing seats are primarily at risk of chest and thoracolumbar spine injuries. Neck loadings can be significant as the helmets (heads) of the occupants are not restrained and can be loaded through the seating head rests, potentially leading to excessive lateral neck flexion (for forward-facing seats) or neck flexion (for side-facing seats).

3.6 GLOBAL VEHICLE MOTIONS

3.6.1 LOADING

The blast flow-field will eventually overcome the inertia of the vehicle mass and begin to launch or lift it off the ground, imparting a kick-off velocity to its centre of mass as well as a kick-off angular

(a)

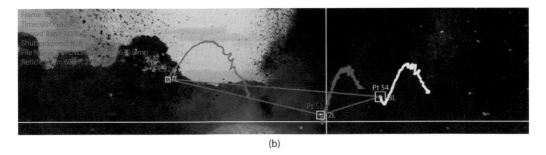

(b)

FIGURE 3.9 Surveyed ground reference plane and global motion yellow tracking balls for three-dimensional global motion tracking, (a), and typical motion tracks of tracking balls, (b). Note that image motion tracks are unrelated to the vehicle depicted.

velocity, depending on the location of the resultant blast forces. These global vehicle motions will typically not become evident until some milliseconds after the blast initiation and can last for as long as hundreds of milliseconds, depending on the vehicle mass and charge size. Injuries to the occupants in this phase are caused by either inertial loading processes in response to vehicle movement or secondary impacts of unrestrained occupants (or occupants with physical dimensions that cannot be safely accommodated by the vehicle survivability system design) with the roof, interior vehicle 'furniture' or dislodged objects.

Measurement of the initial velocity of the vehicle provides a gauge to the severity of the blast, particularly for live-fire tests. Tracking the global vehicle motions utilising high-speed three-dimensional motion tracking technology enables the overall path of the vehicle with time to be measured, as shown in Figure 3.9a and b.

3.6.2 LOADING ON OCCUPANTS

3.6.2.1 Primary Load Transfer

As discussed in Section 3.5.2, the final load paths to the human occupants are through the seat cushion, the occupant restraint system and the floor/footrests (or soles of the boots). While the discussion below is primarily related to loading on the spine and pelvis, it is also pertinent to loadings through the occupant restraint systems and floor.

When the vehicle hull (to which the blast loads are imparted) first begins to lift, the occupants are still fixed in space due to inertia. Thus the vehicle moves upwards, causing its seats to move upwards and impart a vertical load on seated occupants. This can result in potentially serious loadings on the spine and pelvis, but if the seat and restraint systems are still functional after any prior internal dynamic loading (usually in the first 20 ms), they will then operate to prevent or mitigate injuries from such accelerations.

While the vehicle and occupants will eventually reach a common velocity a specific time after the initial impulsive kick-off, any load-limiting or attenuation mechanism in a landmine-resistant seat will extend the time at which this occurs through deformation or compliance, and thus reduce spinal and pelvic loading. The vehicle will ultimately reach its maximum height and then begin to fall. Injuries may occur during this timeframe if the occupants, seats and equipment in the vehicle are not adequately restrained. This can be due to impacts of unrestrained or inadequately restrained regions of the head, legs or body with parts of the vehicle interior or even dislodged objects.

The final loading process is the impact of the vehicle on the ground. The input loading to the vehicle during vehicle set-down is analogous to that from the vehicle kick-off velocity. However, the applied force is typically of a lower magnitude due to the suspension systems on the vehicle extending the time over which the loading is applied. Hence, the loading on the spine and pelvis is usually of a lesser magnitude than that from the original explosively-imparted acceleration. The load attenuation mechanisms of some types of landmine-resistant seats may not fully recover between the initial vehicle kick-off and vehicle set-down. In certain cases, this can lead to higher spinal and pelvic loadings during vehicle set-down.

Of course, real UB blasts rarely have pure vertical loads and have loading components in all three principal directions because of either charge location or vehicle motion prior to the blast, resulting in more complex loading responses. However, the main loading component is still vertical. Blast-loading events, particularly as the vehicle usually initiates such events whilst in motion, can result in rollover of the vehicle. These motions and internal dynamic loadings can even result in occupants being dislodged from simple restraint systems. While automobile rollovers result in lateral displacement of the vehicle side pillars and/or crushing of the roof, which can cause occupant head impacts with the vehicle structure, a blast event can result in an occupant head strike to the roof. Both situations involve an impact to the head involving compression of the head and neck—landmine blast events can result in head and neck injuries similar to those sustained in automobile rollovers (McElhaney et al. 2002), and these can occur during both internal vehicle dynamics and global motion phases.

3.6.2.2 Secondary Projectiles

Secondary projectiles, such as the free-flight of the hatch handle shown in Figure 3.10, are objects that are dislodged at various times during the response of a vehicle to a blast. Other examples of

FIGURE 3.10   Free-flight of secondary projectile (hatch handle) at 155 ms into a landmine live-fire blast test.

secondary projectiles include bolts, brackets, unrestrained combat equipment, supplies and so forth. The flight of these secondary projectiles continues well into the global motion phase.

High-speed three-dimensional imaging techniques have been used within the vehicle cabin in recent live-fire trials with the aim of better understanding the potential hazards. Being able to better visualise as well as estimate the velocity vectors of secondary projectiles can help determine appropriate securement strategies. It may be that the dislodgement of secondary projectiles cannot be entirely prevented, but improved understanding through appropriate imaging measurement and visualisation can facilitate better design practices and potential mitigations, including the selection of appropriate protective eyewear.

3.6.3 CONSIDERATIONS FOR SIDE-BLAST

Depending on the precise orientation and nature of the side-blast to the specific vehicle design, the vehicle will laterally translate and/or rotate in its global motions—there will be no predominant vertical direction of motion as found in UB blast attack. When the vehicle first begins to move, the occupants are still fixed in space due to inertia, and injuries may occur due to inertial loading processes if the occupants, seats and equipment in the vehicle are not adequately restrained. Injuries can be caused by impacts of unrestrained or inadequately restrained regions of the head, legs or body with parts of the vehicle interior or even dislodged objects.

Of course, real-world side-blasts do not have pure lateral loads and have loading components in all three principal directions because of either charge location or vehicle motion prior to the blast, possibly resulting in complex vehicle motions. It is possible that significant side-blast events may also result in rollover accidents in the late time.

3.7 CONCLUSIONS

The protection of vehicle occupants against blast is a complex problem where multiple failure paths that can lead to occupant injuries must be considered. The response of a vehicle to blast from an UB or side-blast attack can be divided into three categories: (1) hull response to an immediate external blast load, (2) internal vehicle dynamics and (3) global vehicle motions. Vehicle response and survivability against blast is a broad and complex topic which has been discussed in these terms along with the associated loadings that cause injuries to vehicle occupants.

ACKNOWLEDGEMENTS

The efforts of the Australian Department of Defence, Dr Axel Bender of the Defence Science and Technology Group Land Division, and the scientific instrumentation team in Defence Science and Technology Group Maritime Division to establish and progress Australian capabilities in the conduct of high-speed data acquisition and vehicle live-fire blast tests is gratefully acknowledged. Jim Nicholls of Qinetiq Australia is also acknowledged for the establishment of the systems and procedures for high-speed three-dimensional photography. Huon Bornstein and Dr Melanie Franklyn of Defence Science and Technology Group Land Division are recognised for their insights on vehicle response and Hybrid III response/injury criteria, respectively, under blast-loading conditions.

REFERENCES

Anderson, C. E., Jr, T. Behner and C. E. Weiss. 2011. Mine blast loading experiments. *Int. J. Impact Engng*. 38: 697–706.
Baker, W. E. 1973. *Explosions in Air*, University of Texas Press, Austin, TX.
Bornstein, H., P. Phillips and C. Anderson. 2015. Evaluation of the blast mitigating effects of fluid containers. *Int. J. Impact Engng*. 75:222–228.

Chung Kim Yuen, S., G. S. Langdon, G. N. Nurick, E. G. Pickering and V. H. Balden. 2012. Response of V-shape plates to localised blast load: experiments and numerical simulation. *Int. J. Impact Engng.* 46: 97–109.

Ehrgott, J. Q., Jr. 2010. *Tactical Wheeled Vehicle Survivability: Results of Experiments to Quantify above Ground Impulse.* U.S. Army Corps of Engineers, Engineer Research and Development Centre Report ERDC/GSL TR-10-7, Vicksburg, MS, March.

Fairlie, G. and D. Bergeron. 2002. Numerical simulation of mine blast loading on structures, In *17th International Symposium on Military Aspects of Blast and Shock*, Las Vegas, NV, 10–14 June.

Fourney, W. L., U. Leiste, R. Bonenberger and D. J. Goodings. 2005. Mechanism of loading on plates due to explosive detonation. *Fragblast Int. J. Blast. Fragment.* 9:205–217.

Golan, G., Z. Asaf, E. Ran and F. Aizik. 2012. Occupant legs survivability: an assessment through the utilization of field blast test methodology. In *22nd International Symposium on Military Aspects of Blast and Shock*, Bourges, France, 4–9 November.

icasualties.org. 2014. Iraq coalition casualty count. http://icasualties.org (accessed July 20 2014).

Kuttenberger, A. 2011. Occupant-protection systems. In *Bosch Automotive Handbook*, ed. K. Reif, p. 931. 8th ed. Plochingen: Robert Bosch GmbH.

McElhaney, J. H., R. W. Nightingale, B. A. Winkelstein, V. C. Chancey and B. S. Myers. 2002. Biomechanical aspects of cervical trauma. In *Accidental Injury Biomechanics and Prevention*, ed. A. M. Nahum and J. W. Melvin, p. 332. 2nd ed. New York: Springer Verlag.

Montoya, D. and H. Bornstein. 2012. Geometric effects of hull shapes on blast mitigation for armoured vehicles. In *Land Warfare Conference*, Melbourne, 29 October–2 November, pp. 355–364.

NATO. 2006. *Allied Engineering Publication AEP-55, Vol. 2, Procedures for Evaluating the Protection Level for Logistics and Light Armoured Vehicles, Volume 2 for Mine Threat.* NATO Standardization Agency (NSA), Brussels.

NATO. 2014. *Allied Engineering Publication AEP-55, Vol. 3, (Part 1), Procedures for Evaluating the Protection Level of Armoured Vehicles – IED Threat. Edn C, Ver. 1.* NATO Standardization Agency (NSA), Brussels.

Saunders, L., J. Wang, D. Slater and J. Van den Berg. 2012. Visco-elastic polyurethane foam as an injury mitigation device in military aircraft seating. *J. Battlefield Technol.* 15:11–18.

Sternberg, H. M. and H. Hurwitz. 1976. Calculated spherical shock waves produced by condensed explosives in air and water. In *Proceedings of the 6th Symposium (International) on Detonation, ACR-221*, Office of Naval Research, Arlington VA, pp. 528–539.

Wang, J. and R. Bird. 2001. Improved seating to protect occupants of vehicles from landmines. In *Land Warfare Conference*, Sydney, November, pp. 507–516.

Wang, J., R. Bird, R. Swinton and A. Krstic. 2001. protection of lower limbs against floor impact in army vehicles experiencing landmine explosion. *J. Battlefield Technol.* 4:8–12.

4 Injury Scoring Systems and Injury Classification

Melanie Franklyn and Christine Read-Allsopp

CONTENTS

4.1 INTRODUCTION

Trauma scoring systems use a combination of the severity of an injury described anatomically and the degree of physiological derangement of the patient in order for a score to be calculated which correlates with a clinical outcome (Kondo et al. 2011). The purpose of these systems is to classify, organise and arrange types of injury and physiological disturbance following the transfer of energy to the human body. They are used to provide information and assessment of patient care and the patient journey from injury to rehabilitation and their return to pre-injury state. Scoring systems enable advanced communication between trauma surgeons, health care workers, biomechanical engineers and researchers by allowing them to discuss and interpret data in an objective and consistent manner.

In the clinical setting, injury scoring systems are used to evaluate and improve patient assessment, treatment and outcomes by using a standardised terminology to describe trauma. Patient data can be analysed within the facility, within the same region, and also internationally to compare hospital performances and patient outcomes with peer facilities. Quality performance, auditing and analysis of patient care from the time of injury can be used to identify the standards of care provided to the patient during each stage of their journey, ensuring that hospital standards are met to the highest level, and for identifying areas for improvement.

Trauma scoring systems are also important in epidemiological research, where patterns and causes of injury can be analysed in specific populations in order to monitor trends in injury causation and to develop injury prevention programmes. Additionally, epidemiological research can be used to determine the cost of injury to society; thereby identifying public health issues.

Standardised trauma scoring systems have also been used widely in biomechanical research in order to classify injury types and severities in a consistent manner, and thus determine relationships with the mechanisms of injury. Perhaps the most established of these are automotive crash databases, where injury causation due to different crash scenarios, impact directions, impact velocities and occupant characteristics can be analysed to enhance vehicle safety.

In the military, injury scoring systems are used in a number of applications analogous to civilian trauma injury data. In the clinical setting, injuries which have been coded can be analysed for casualty trauma statistics. This information can be used in conjunction with data from vehicle underbelly (UB) blast events in order to determine injury mechanisms and for the prevention of injuries in military vehicles. An alternative application of trauma scoring in defence is the integration of injury coding into computer models. These models can then be used for analysis such as predicting injury potential and probability of death outcomes from projectiles hitting a combatant from different directions.

Individual trauma scoring systems have different advantages and limitations depending on the application, for example, pre-hospital versus hospital-based scenarios, civilian versus military hospitals, and clinical versus biomechanical use. The time required to calculate a score is also significant, as this will determine its use in different types of settings; for instance, scoring systems for triage need to be quick to calculate so a rapid decision can be made on appropriate patient management and care. In addition, all scoring systems have limitations based on the population from which they were derived, which will then in turn determine their predictive power in different contexts. For example, a scoring system based on patients who sustained blunt trauma in automotive crashes may not have the same capacity to predict mortality in a population of individuals who sustained gunshot wounds in military combat situations.

In the pre-hospital setting, a Trauma Triage Tool or 'Field Triage Decision' is required in order for a rapid accurate assessment (the likely injury) and to determine the level of physiological derangement; thus enabling commencement of early resuscitation and delivery of the patient to the most appropriate facility for definitive care. The major trauma criteria (MIST) are used to ascertain patient risk for serious injury and hence transport decision and destination. The MIST criteria are defined as follows (ASNSW 2012):

M – Mechanism of injury
  I – Injuries
  S – Signs and symptoms
  T – Transport

If the patient has one of the mechanisms of injury, injuries, or signs or symptoms, they should be transported to the highest level trauma service facility within a 60-minute travel time.

Hospitals are designated by the level of trauma care they are able to provide and the availability of services. These levels are (American College of Surgeons 2014):

- Level I: Comprehensive services available including specialist amenities for acute spinal cord injury and major burns to provide total care for the patient until they are transferred to a rehabilitation centre.
- Level II: Able to initiate definitive care and provide interventions for all injured patients.
- Level III: Provides prompt assessment, resuscitation, surgery and stabilisation of the patient.
- Level IV/V: Provides initial trauma life support to the victim and prepare patients for transfer to higher levels of care.

Over the last few decades, it has been demonstrated in the literature that the '*golden hour*' concept indicates the urgency required to provide the patient with timely and effective treatment and delivery to the highest level trauma facility in the shortest time possible to improve their outcome.

4.2 TYPES OF SCORING SYSTEMS

Scoring systems can be categorised several ways, but probably most commonly: anatomical systems, physiological systems, combined systems and outcome scales.

Anatomical systems use patient anatomy to describe injuries and include parameters such as the location of the injury in the body, the organ or part of organ affected and the type of injury sustained (e.g. fracture). Physiological systems use variables such as heart rate, blood pressure, respiratory rate, motor functioning and verbal responses to classify injuries. As the name suggests, combined systems employ a combination of anatomical and physiological parameters to define injury severity. Lastly, outcome scales may involve anatomical, physiological or both of these parameters, but are used to predict the consequences or the long-term outcome resulting from trauma.

Examples of well-established anatomical scoring systems include:

- The Abbreviated Injury Scale (AIS)
- The Injury Severity Score (ISS)
- The New Injury Severity Score (NISS)

Commonly used physiological scoring systems include:

- The Glasgow Coma Scale (GCS)
- The Westmead Post-Traumatic Amnesia Scale (WPTAS)

Some combined scoring systems are:

- The Trauma and Injury Severity Score (TRISS)
- The Anatomic Profile (AP)
- A Severity Characterisation of Trauma (ASCOT)

While outcome scoring systems include:

- The Functional Capacity Index (FCI)
- The Function Independence Measure (FIM™)

These trauma scoring systems are by no means exhaustive, but are the more common ones. In the current chapter, the injury scoring systems have been classified using these three categories (refer to the table of contents for this chapter on p. 35 for all injury scales discussed).

While injury scoring systems can be classified using anatomical and physiological parameters; they can also be more generally categorised according to the method used to calculate the scores and how they are applied in trauma situations. These classifications are known as primary and secondary trauma scoring systems and are defined as follows:

1. *Primary systems*: Values are assigned directly from medical data or observations to calculate a score or value. They relate to observation and assessment, are rapid to use, and decisions can be made in the pre-hospital environment.
2. *Secondary systems*: Require scores from a primary scoring system in order to obtain a result. Relies on a validated diagnosis or patient outcome, which takes some time to obtain. However, they are generally a more reliable assessment of the patient's injury severity and predicted outcome than a primary scoring system.

4.3 ANATOMICAL TRAUMA SCORING SYSTEMS

4.3.1 THE ABBREVIATED INJURY SCALE (AIS)

Probably one of the most widely used and well known scoring systems, the AIS was initially developed through a joint initiative by the American Medical Association (AMA), the Society of Automotive Engineers (SAE), and the Association for the Advancement of Automotive Medicine (AAAM) in 1969 to address the issue of classifying injuries sustained by occupants in motor vehicle crashes. In 1971, the first AIS Dictionary was produced, and since its inception, the codes have periodically been updated with the addition of new codes, and modifications to reflect changes in injury severities due to improved diagnostic techniques and contemporary injury treatment and management. The latest revision, AIS 2005 Update 2008 (Gennarelli and Wodzin 2006), has just been superseded by a new version, AIS 2015$^{\copyright}$.

The AIS is an anatomically-based primary injury coding system, where injuries are classified according to anatomical location and nature of the injury (e.g. fracture) on a 6-point ordinal scale (the pre-dot code) combined with a severity of injury score ranging from 1 to 6 (the post-dot code). The AIS Dictionary is divided into nine chapters (Table 4.1), where each chapter represents a different body region, and the chapter numbers correspond to the first digit in the AIS Code (e.g. Chapter 1 is head; head injury codes commence with '1').

TABLE 4.1
AIS Chapters and Corresponding Body Regions

AIS Chapter	Body Region
1	Head
2	Face
3	Neck
4	Thorax
5	Abdomen
6	Spine
7	Upper extremity
8	Lower extremity
9	External and other trauma

Source: *Abbreviated Injury Scale (AIS 2005 Update 2008)*, Association for the Advancement of Automotive Medicine, Barrington, IL.

The severity of injury score, also known as the AIS score, is the most commonly used component of the system and comprises an ordinal scale ranging from 1 to 6 (Table 4.2). While the codes and injury severities are consensus derived, the injury severity values have been correlated to patient mortality in a number of studies.

Throughout the AIS Dictionary there are a number of AIS 9 codes which are used for injuries where there is insufficient information available to classify the anatomical location (specific anatomical location unknown) or the injury severity (injury severity unknown). For example, an injury to the face for which no further information was available would be coded as 200099.9: Injury to the face NFS (not further specified). These NFS codes enable non-specific trauma to be documented for epidemiological purposes; however, they cannot be used for the calculation of the patient ISS, as discussed later. Although AIS 9 codes were originally important for recording the full patient injury status, they are now less commonly employed as patient case histories will generally have comprehensive information on injuries sustained, including medical imaging reports.

Post-dot qualifiers are additional optional digits added to the AIS Code which can be used to further classify the injury as well as document the injury mechanism. Further injury classification using localisers (L1 and L2) can provide more detailed information on the injury (e.g. for a fracture to one of the thoracic vertebra, the specific vertebra injured), while cause of injury (COI) codes can be documented to identify mechanisms (e.g. driver in a passenger vehicle in an automotive crash).

Although these supplementary codes can be useful in epidemiological studies, they are less beneficial in automotive and military applications when detailed information on the vehicle damage and injury mechanisms to the occupant are required. For example, in many automotive crash databases, not only is the type of vehicle and the position of the occupant in the vehicle documented, but detailed information on the crash scenario, impact direction and other data associated with the crash from vehicle and site inspections is collected. Similarly, for military vehicle events, detailed data are usually collected pertaining to the vehicle and the circumstances of the event (e.g. see Ramasamy et al. 2011, for an overview) such as information on the blast and the deformation of the vehicle hull (see Chapter 3).

There are a number of assumptions with the AIS (AIS 2005 Update 2008), most importantly, the injury severities represent the threat-to-life in an average patient who is defined as being:

1. Between 25 and 40 years of age;
2. Sustaining no pre-existing medical conditions;
3. No treatment complications;
4. Receiving appropriate and timely medical treatment.

TABLE 4.2
The AIS Scores and Associated Injury Severities

AIS Score	Injury Severity
1	Minor
2	Moderate
3	Serious
4	Severe
5	Critical
6	Maximal or currently untreatable

Source: *Abbreviated Injury Scale (AIS 2005 Update 2008)*, Association for the Advancement of Automotive Medicine, Barrington, IL.

Patients who do not fit the above criteria may not have the same threat-to-life described by AIS, for example, an elderly motor vehicle occupant who has a pre-existing medical condition, where the chance of survival will be lower than a younger occupant who sustains the same injury but has no pre-existing medical condition.

Military combatants, however, are more likely to fit the AIS assumptions above, generally being younger and fitter than the average individual from the civilian population; therefore, the AIS severity of injury scores are likely to reflect patient mortality better for this population. Nevertheless, it has been suggested that the AIS underestimates the level of trauma in military combatants, especially in the case of penetrating injuries, for two reasons. Firstly, the AIS penetrating injury scores represent the threat-to-life based on both stab and gunshot wounds (GSWs), and penetrating injuries in military events will be predominately GSWs and penetrating fragments, where the mortality rate is significantly greater than stab wounds. Secondly, a high-velocity military weapon will have a greater mortality rate than a civilian handgun due to the greater tissue damage produced from the combat weapon. However, it has also been refuted that AIS codes underestimate the trauma in military combatants, with some military physicians and researchers stating that military combatants are more likely to access rapid medical treatment, thus increasing their chances of survival.

One of the main limitations with the AIS is that it represents threat-to-life (and other dimensions of injury severity, such as tissue damage) based on a single injury only. Consequently, a secondary injury scoring system is required to calculate threat-to-life and mortality for multiple injuries, with the most frequently used systems being the ISS and NISS. The ISS, however, is most commonly employed in hospital trauma databases. These secondary scoring systems require a primary scoring system in these cases, the AIS to calculate threat-to-life and mortality, and are discussed further in the following sections.

The AIS also cannot be applied to predict immediate or long-term functional capacity from an injury. Thus, in the case of a mounted combatant sustaining lower limb injuries in a UB blast event, the AIS cannot be used to determine how quickly that individual can affect an escape from the vehicle or be able to perform their duty after the injury has been sustained (immediate functional capacity). It also cannot be employed to predict the long-term outcomes from an injury such as a lower limb fracture; hence, another system such as the Functional Capacity Index (FCI) or the Military Functional Incapacity Scale (MFIS), which are discussed later, is needed to predict the likely long-term impairment from the injury.

Despite its limitations, the AIS has been increasingly adopted in military databases and military research. For example, the AIS is used in the US Joint Theater Trauma Registry (JTTR) to classify injury severity in military combatants and thus facilitate the improvement of patient management and care for the wounded solider.

The AIS has been used extensively in impact biomechanics to develop injury risk curves to describe the likely level of injury sustained (e.g. AIS 3 or AIS 3+), or the Maximum AIS (MAIS) to refer to the severity of the maximum injury sustained in a particular body region or in the whole body. Similarly, the AIS is likely to be a beneficial tool in developing injury risk curves for the new WIAMan (Warrior Injury Assessment Manikin) Anthropomorphic Test Device (ATD), a mannequin which is currently under development for future use in predicting military combatant injury tolerances in UB blast events.

4.3.2 The Injury Severity Score (ISS)

The ISS was developed by Baker et al. (1974) as there was a need to quantify the threat-to-life of patients with multiple traumatic injuries. In a study on 2,128 motor vehicle occupants, pedestrians and other road users who were hospitalised or whose injuries resulted in a fatality, Baker et al. (1974) found that there was a non-linear relationship between MAIS and death, and that mortality was strongly influenced by concomitant injuries to other body regions. This resulted in the development of the ISS, which is the sum of the squares of the three most severe AIS scores in three different ISS body regions, as shown by Equation 4.1:

$$ISS = a^2 + b^2 + c^2 \tag{4.1}$$

where a, b and c are the highest AIS values from three different ISS body regions, and the ISS body regions are: (1) head and neck, which includes the cervical spine; (2) face; (3) thorax, which includes the thoracic spine; (4) abdomen, which includes the lumbar spine; (5) extremities, which includes the pelvic girdle; and (6) external and other trauma. The development of the ISS highlighted some issues with the AIS version at the time, as the AIS included fatal codes for patients who died within 24 hours of sustaining an injury. Thus, patients with the same injuries could have different AIS scores (depending on whether or not death occurred) and subsequently different ISS values (Baker and O'Neill 1976).

The ISS is an ordinal scale ranging from 1 (minimal injury) to 75 (maximum) and defaults to 99 if there are any AIS 9 (unknown) injuries. However, some values in the interval are not possible to obtain due to the AIS triplet combinations used to derive the score. For instance, there is no ISS = 15, but ISS = 17 can result from several triplet combinations, e.g. (4, 1, 0) or (3, 2, 2). An example ISS calculation is shown in Table 4.3.

Since its development, the ISS has been subject to numerous validation studies, both for blunt trauma (which it was designed for) and penetrating trauma (e.g. Baker and O'Neill 1976; Bull 1975; Copes et al. 1988) and as a result, has become the most widely used scoring system in hospital trauma settings. Probably the most well known of these validation studies was a US Major Trauma Outcome Study (MTOS) conducted between 1982 and 1987 (Champion et al. 1990b; Copes et al. 1988). In this research, the authors proposed the ISS intervals in Table 4.4, correlating these ISS intervals with mortality rate for age and injury mechanism (blunt or penetrating), where the graph of ISS versus mortality for penetrating injuries is shown in Figure 4.1. Note that the horizontal axis data points are plotted near the median of each ISS band, except for ISS = 75, which is a discrete value. Of importance is that the mortality rate was shown to be higher for penetrating than for blunt injuries (blunt injuries not shown in Figure 4.1), particularly for lower ISS values. For patients < 50 years of age, the most relevant data for a military population, the mortality rate increased significantly from the ISS = 9–15 category to the ISS = 16–25 category.

The ISS is the most widely adopted scoring system for multiple injuries and provides a reasonably accurate assessment of mortality. However, the US MTOS used to assess mortality is now outdated; therefore, mortality values may not reflect patient deaths in a contemporary trauma setting where there are now improved diagnostic and treatment facilities available. This is also the case for UK MTOS data, which contains two years of data collected in the early 1990s (Yates et al. 1992). More recently, there has been debate on what constitutes major trauma, and if the ISS > 15 threshold, which was first proposed by Boyd et al. (1987) to describe >10% mortality using AIS 85, still applies.

TABLE 4.3

ISS Body Regions and Example ISS Calculation

ISS Body Region	Injury Descriptors	AIS	Square of Highest Three
Head and neck	Small cerebral subdural haematoma	4	16
Face	None		
Thorax	Two right-side rib fractures	2	4
Abdomen	Liver contusion; major	3	9
Extremity	None		
External	None		
Total			29

TABLE 4.4

Proposed ISS Intervals

ISS Interval	Most Severe Injury Combination/s
1–8	AIS 2
9–15	AIS 3
16–24	AIS 4
25–40	AIS 5 but not both AIS 5 and AIS 4
41–49	AIS 4 and AIS 5
50–66	Two AIS 5's and one AIS 4
75	One AIS 6 or three AIS 5's

Modified from Copes, W. S., et al., *J. Trauma Acute Care Surg.*, 28 (1), 69–77, 1988.

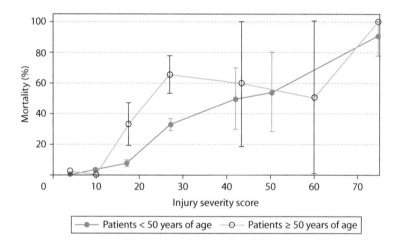

FIGURE 4.1 US Major Trauma Outcome Study (MTOS) data showing mortality rate against Injury Severity Score for defined ISS intervals. Data is for penetrating injuries only. (Adapted from Copes, W.S., et al., *J. Trauma Acute Care Surg.*, 28(1), 69–77, 1988.)

Modifications to the original AIS severity scores, which have been updated over time, in addition to changes in trauma care and management, have resulted in a shift to define major trauma as an ISS > 12 (Palmer 2007; Palmer et al. 2016; Sampalis et al. 1999), although others have used different qualifiers, such as an injury score of AIS 3+ (MacKenzie et al. 2006).

The definition of major trauma becomes significant when deciding if the patient needs immediate transport to a trauma centre with specialised trauma care, and therefore reducing the chance they are over- or undertriage. In the case of overtriage, where a patient with minor injuries is transported to major trauma centre, there is an inappropriate use of specialised resources and higher expenditure (Newgard 2012). In the case of undertriage, where a seriously injured patient is transported to a non-trauma hospital, the patient does not receive the specialised care required in a timely manner, and thus the chance of death is higher than for a patient who has been correctly triaged (Haas et al. 2010).

There are a number of significant limitations with the ISS. Firstly, while mortality generally increases with higher ISS values, this is not strictly the case; this results in anomalies such as the mortality rate for an ISS = 16 being higher than the mortality rate for an ISS = 19 (Copes et al. 1988). More importantly, the mortality for a particular ISS score depends on the AIS triplet from which it was derived (Copes et al. 1988), although this is also a limitation of the NISS

(Aharonson-Daniel et al. 2006), which is discussed later. For example, Copes et al. (1988) found that for an ISS of 17, the (4, 1, 0) triplet had a mortality rate of 18.1%, while the (3, 2, 2) triplet had a mortality rate of 2.6%, as the presence of the AIS 4 injury in the former case increases patient mortality significantly. The patient mortality also depends on the body regions from which the triplets are composed; for example, for the same ISS, a patient with a head injury will have a higher probability of death than a patient with an injury to another body region (Copes et al. 1988); however, this is not factored as each body region is weighted equally in the ISS calculation.

Another limitation with the ISS is that it cannot account for multiple injuries to the same body region, since the values in the triplet come from three different ISS body regions. In other words, in patients with multiple injuries confined to a single body region, only one of those injuries is considered in the ISS calculation (Osler et al. 1997). This has a number of important implications which reduce the predictive power of this injury measure, and is particularly important for military injuries, which often only involve one ISS body region. For example, a patient with a GSW to the abdominal region may sustain injuries to multiple abdominal organs, but no injuries to other body regions. However, in the ISS calculation, only the most severe abdominal injury will be factored. Thus, if the patient sustained three abdominal injuries, for example, two AIS 3 injuries and one AIS 2 injury, the ISS would be 9 and the patient is considered to have not sustained major trauma (while the NISS, discussed later, would be 22). However, the presence of second AIS 3 injury means the chance of death is higher than what is reflected in the ISS score; this may have implications for patient management and treatment.

Similarly, if a patient sustains multiple severe injuries to one body region, such as the abdomen, and only minor or moderate injuries to another body region, in the ISS calculation, the minor or moderate injury in the second body region is used in favour of the other, more severe, abdominal injuries. Again, this can have ramifications for military combatants, who often sustain many injuries to one body region. Thus, the aim of the ISS, which is to consider the body as a whole, conflicts with the principal of considering more severe injuries in favour of less severe injuries (Osler et al. 1997).

In summary, although the ISS has been subjected to numerous validation studies, like other trauma scoring systems, it is a better predictor of mortality for the population it was designed for; in the case of ISS, blunt injuries sustained by road trauma events. Nevertheless, it is currently employed in various military trauma databases and has been used in numerous studies for the analysis of combat data (e.g. Eastridge et al. 2012; Kelly et al. 2008).

4.3.3 THE NEW INJURY SEVERITY SCORE (NISS)

The NISS is defined as the sum of squares of the AIS values for the patient's three most severe AIS injuries, regardless of body region. Thus, the equation for NISS is the same as that for the ISS, as shown by Equation 4.2:

$$NISS = a^2 + b^2 + c^2 \qquad (4.2)$$

However, for the NISS, a, b and c are the highest AIS values from any ISS body region. The NISS was first proposed by Osler et al. (1997), where they conducted a study on 6,585 patients using two independent datasets from two Level I trauma centres. In addition to being simpler to calculate, Osler et al. (1997) found that the NISS was a better predictor of patient survival than the ISS in their analysis.

The main advantage of NISS over ISS is that multiple injuries can be included from the same body region. Thus, for patients who sustain multiple injuries to one or two body regions, the NISS is more likely to give a better estimate of survival than the ISS, particularly for severe injuries, and a more accurate prediction of short-term mortality from those injuries (e.g. Balogh et al. 2000; Brenneman et al. 1998; Lavoie et al. 2004; Osler et al. 1997).

The NISS has been shown to have superior predictive capabilities than the ISS for penetrating injuries (e.g. Frankema et al. 2005; Smith et al. 2015) and blunt injuries (e.g. Brenneman et al. 1998; Frankema et al. 2005) and mixed trauma populations (e.g. Lavoie et al. 2004; Tohira et al. 2012).

Although some research has shown the NISS does not outperform the ISS in predicting mortality, these studies have been in populations such as low-risk trauma patients (Husum and Strada 2002). Nevertheless, the NISS has yet to replace the ISS in most trauma registries. This is due to a number of reasons, including changes to historical data so the whole database can still be used for analysis, and the fact that ISS has been correlated to mortality from MTOS data. However, these issues are not as relevant as they were some years ago due to the implementation of computerised trauma registries and the automated calculation of ISS or NISS, and the fact that MTOS data is now becoming dated, and it would seem prudent to conduct new trauma outcome studies while focusing on moving towards the NISS.

Like the ISS, the NISS has been used in numerous military studies (e.g. Ramasamy et al. 2008) to analyse combat casualty data. However, analyses comparing the predictive power of the ISS and NISS for military injuries have yet to be conducted.

4.3.4 THE ORGAN INJURY SCALE (OIS)

The Organ Injury Scales (OISs) were developed by the Organ Injury Scaling Committee of the American Association for the Surgery of Trauma (AAST), which was originally formed in 1987 with the intent of providing a common nomenclature which physicians can use to describe injuries and injury severity (Moore et al. 1989, 1990, 1992, 1994, 1995). Like the AIS, the scoring systems have been modified and updated to reflect contemporary trauma management.

In the OIS system, each organ injury is graded from 1 to 6, where 1 is an injury which is not severe (or minor) and 5 is assigned to the most severe injury from which the patient can survive. Grade 6 injuries are not salvageable, i.e. the patient will not survive. Injuries are described by the organ affected, the injury severity (the grade of injury, from 1 to 6), and by anatomic description, where the injury descriptions, in general, can be categorised as follows:

- Haematoma
- Laceration
- Vascular involvement

Since their original development, the OISs have been subjected to some minor revisions.

TABLE 4.5

The 1994 Revision of the Organ Injury Scale for the Liver (Moore et al. 1994) currently used by the American Association for the Surgery of Trauma (AAST)

Grade	Type of Injury	Injury Description
I	Haematoma	Subcapsular, <10% surface area
	Laceration	Capsular tear, <1 cm parenchymal depth
II	Haematoma	Subcapsular, 10%–50% surface area
		Intraparenchymal, <10 cm diameter
	Laceration	1–3 cm parenchymal depth, <10 cm length
III	Haematoma	Subcapsular, >50% surface area or expanding. Ruptured subcapsular or parenchymal haematoma
		Intraparenchymal haematoma >10 cm or expanding
	Laceration	>3 cm parenchymal depth
IV	Laceration	Parenchymal disruption involving 25%–75% of hepatic lobe or 1–3 Coinaud's segments in a single lobe
V	Laceration	Parenchymal disruption involving >75% of hepatic lobe or >3 Coinaud's segments within a single lobe
	Vascular	Juxtahepatic venous injuries, i.e. retrohepatic vena cava/central major hepatic veins
VI	Vascular	Hepatic avulsion

An example OIS is shown in Table 4.5, which is the scale used for the liver. The liver has six grades of injury, where the lower grades are haematomas or lacerations of increasing severity, while the higher grades (Grades V and VI) also have vascular involvement. When there are multiple injuries to the same organ up to Grade III, the code advances one grade higher. For instance, a liver laceration > 3 cm parenchymal depth is a Grade III injury, but if there is also a liver haematoma 10%–50% of the surface area; the liver injury becomes a Grade IV. Consequently, the OIS has the disadvantage in that two injuries to the same organ are coded as one injury; thus, if used for injury documentation purposes, injuries can be missed. However, as the scale was designed to facilitate clinical investigation and management, the grade assigned reflects the injury severity to the organ as a whole. It is based on the anatomical disruption to a specific organ and designed to reflect the impact of injury to that organ on ultimate patient outcome (Moore et al. 1989, 1995).

Like the AIS, the OIS is a consensus-derived ordinal scale. The OISs for different organs are also used in the AIS Dictionary, with some of the AIS Code descriptors including the OIS to facilitate injury classification; however, the OIS descriptors are slightly different to the AIS ones. To date, the OISs have not been widely used for defence data.

4.3.5 The International Classification of Diseases (ICD)

The International Classification of Diseases and Health related problems, commonly abbreviated to The International Classification of Diseases (ICD), is produced by the World Health Organisation (WHO) as part of their health system development. Like the AIS, it has been updated regularly since its first inception in 1949 with ICD-6 (prior to the 6th revision, it was only a cause of death system). It has since become the standard diagnostic tool for epidemiology, health management and clinical purposes, being used to monitor the incidence and prevalence of injury, diseases and other health problems. The current version, ICD-10, was developed in 1992, with updates since this time (WHO 2011). ICD-11 is currently in the planning stages.

The ICD is used to classify injury, diseases and other health problems from medical records and death certificates for clinical, epidemiological and quality purposes and is also used for resource allocation and decision making. Although the standard ICD has been employed for applications such as health statistics, a clinical modification (ICD-CM) is also available. The ICD-CM is more precise than the standard ICD and more accurately describes the clinical status of the patient (WHO 2016). The ICD contains severity coding; however, it is generally not as specific for trauma as other classification systems, and injuries are classified as only minor, moderate or severe (unlike the AIS, which is specifically designed for trauma).

In the 1990s, Osler and colleagues established a new score for multiple injuries based on the ICD rather than the AIS. This score, denoted the ICD-derived Injury Severity Score, or ICISS, was introduced to address the limitations of the ISS, in particular, the fact that only the three highest injured body regions are considered (Osler et al. 1996) – noting that the ICISS was proposed prior to the development of the NISS in 1997. The ICISS is a survival score based on the ICD-9, where survival risk ratios (SRRs) were calculated, i.e. estimates of survival associated with each ICD code. SRRs are derived for each injury code by dividing the number of patients who survive with that code by the total number of patients who have the code. The ICISS is then defined to be the product of the SRRs of an individual patient's injuries and thus represents the likelihood that a patient will survive a particular injury.

While Osler et al. (1996) found that the ICISS was better at predicting survival than ISS in injured patients in their research, other studies have shown that the ICISS is not superior to ISS in predicting mortality (e.g. Wong and Leung 2008). Some issues which have been highlighted with ICISS include the fact that the survival risk is lower for a patient with multiple injuries, even if the injury severities themselves are moderate (Seguí-Gómez and Lopez-Valdes 2012), the SRRs depend on the database from which they were derived, and the effect of injury interaction is not factored, i.e. two injuries which, in combination, have a greater effect on mortality than either of those injuries individually

(National Center for Health Statistics 2004). Another issue is that the ICISS is based on the ICD which was not specifically developed for injury, whereas the ISS and NISS are based on the trauma-specific AIS. Thus, severity measures have been shown to be more accurate with AIS than with ICD (Cook et al. 2014). Although proponents for the ICD assert that AIS coding is too time consuming, precluding its use in some situations, the limitations of the ICISS have prevented its adoption on a widespread basis.

The ICD has been used in a large number of military studies to classify injury such as injuries due to explosions (e.g. Eskridge et al. 2012), infections due to trauma (Murray et al. 2009), treatment of wounds (e.g. Owens et al. 2008) and more general patterns of injury (e.g. Ramasamy et al. 2008; Zouris et al. 2006). However, while the ICD has been widely used in defence research, the ICISS has not, probably due to the limitations discussed above. Thus, for prediction of threat-to-life due to multiple injuries, the ISS or NISS has been adopted instead (e.g. Ramasamy et al. 2008).

4.3.6 Anatomical Military Combat Injury Scales

The AIS was initially developed from injuries sustained during automotive crashes, and although it has been extended over multiple revisions to include other injuries, such as those from penetrating trauma, it was not specifically designed for combat-related injuries. Thus, a large number of combat injuries cannot be coded (Lawnick et al. 2013), and for those which can be coded, the severities or threat-to-life of some injuries sustained in a civilian setting do not necessarily reflect the threat-to-life in combat situations. One example, which was discussed earlier in Section 4.3.1, is the case of GSWs: the AIS penetrating injury codes include both stab and (civilian) GSWs; thus, the injury severity from a higher-energy military GSW or projectile may be underestimated.

In order to address this issue, a military version of the AIS, which was denoted AIS-2005 Military, was developed by Champion et al. (2010). In order to initially identify injury patterns, the authors used a Surface Wound Mapping (SWM) database and the Surface Wound Analysis Tool (SWAT) software to develop 3D density maps of wound entry and the corresponding anatomic injury severity. Codes were assigned to the injuries using AIS 2005 as a basis, identifying combat injuries which were potentially more severe than the corresponding civilian injury, then increasing the AIS 2005 codes by one or two increments to reflect the increased severity, morbidity or mortality in a military situation. One criticism of the scale is that it had not been validated, for example, in terms of mortality prediction (Soderstrom and Seguí-Gomez 2010).

The AIS-2005 Military subsequently evolved into the Military Combat Injury Scale (MCIS), which was based on AIS 2008, and was complemented by the Military Functional Incapacity Scale (MFIS), which is designed to predict long-term outcomes (Lawnick et al. 2013) and is discussed later in Section 4.6.2. Like the AIS, the MCIS is a consensus-derived ordinal scale. However, it is a 5-point scale (as opposed to a 6-point scale like the AIS), as shown in Table 4.6. In addition, the body regions differ from the ISS in order to group the body in regions which articulate or are adjacent anatomically, with the body regions denoted as mISS or Military ISS anatomical regions (Table 4.7).

Some of the issues addressed in the development of the MCIS include the fact that many soft-tissue injuries sustained in combat are extensive, and thus well exceed the AIS 2/moderate classification in AIS in terms of injury severity. In addition, the MCIS contains more specific penetrating injury codes; descriptors appropriate for the types of complex burns sustained in military scenarios; codes for avulsion of portions of the skull, brain and mandible; more specific codes for different types of amputation; and combat stress injury. In order to verify the accuracy of the MCIS, the authors conducted a validation analysis on 992 randomly-selected injured combatants, finding that the MCIS performed better than AIS 2008 (Lawnick et al. 2013).

While Lawnick and colleagues proposed the mISS body regions for combat trauma, they did not validate the mISS scoring system (Lawnick et al. 2013). Thus, in a recent study by Le et al. (2016), the mISS was evaluated, although they used the earlier AIS-2005 Military (Champion et al. 2010) instead of the MCIS (Lawnick et al. 2013). In the study by Le et al. (2016), the mISS was calculated the same

TABLE 4.6

The Military Combat Injury Scale Levels of Severity

Severity	Severity Level	Description
1	Minor	Superficial injuries which can be treated in a theatre of combat. The casualty is likely to return to duty within 72 hours.
2	Moderate	Injuries which do not need immediate treatment. Delayed treatment in a tactical situation is not likely to result in increased morbidity or mortality.
3	Serious	Injuries not involving shock or airway compromise but should be treated within 6 hours at a medical treatment facility to avoid an increase in morbidity or mortality,
4	Severe	Injuries which may result in shock or airway compromise. Some patients will have an increased risk of death or disability if not treated at a medical treatment facility within 6 hours.
5	Likely lethal	Injuries which are not likely to be survivable in a military setting. This includes catastrophic injuries and injuries likely to be fatal within minutes of wounding.

Compiled with information from Lawnick, M. M., et al., *J Trauma Acute Care Surg.*, 75 (4), 573–581, 2013.

TABLE 4.7

Comparison of ISS and mISS Body Regions

Body Region	mISS Anatomical Regions	ISS Anatomical Regions
Head and neck	Head, neck and face	Head, neck, cervical spine
Face	N/A	Facial bones, nose, mouth, eyes and ears
Chest	N/A	Ribs, thoracic organs, thoracic spine
Torso	Injuries to the chest and abdomen, including the pelvic girdle and junctional areas such as the axilla and groin	N/A
Abdomen and pelvic contents		Abdominal organs and lumbar spine
Upper extremities	Upper extremities	Extremities and pelvic girdle
Lower extremities	Lower extremities	
Pelvic girdle	N/A	
Multiple	Injuries not confined to one specific body region	N/A
External	N/A	Soft-tissue injuries, burns and other trauma, e.g. drowning, whole-body explosion

Compiled with information from Lawnick, M. M., et al., *J Trauma Acute Care Surg.*, 75 (4), 573–581, 2013.

way as the ISS, i.e. the sum of the squares of the highest AIS-2005 Military value in three different ISS body regions. The authors then evaluated 30,364 injuries sustained in Afghanistan and Iraq by military personal from 2003 to 2014 inclusive recorded in the US JTTR, finding that the mISS predicted combat mortality better than the ISS.

As highlighted in the Editorial Critique (Polk 2016) accompanying the Le et al. (2016) paper, it has been well documented that the ISS has limitations in predicting mortality for multiple severely injured body regions (e.g. Frankema et al. 2005; Osler et al. 1997; Shin et al. 2013), particularly in high-trauma populations: the NISS was designed to overcome this limitation and therefore might be more appropriate for military populations. Nevertheless, Le et al. (2016) did not compare the mISS to the NISS in their study.

The MCIS and mISS are new concepts for injury severity coding developed specifically for use in a military setting and will require further monitoring and evaluation as their use increases.

More recently, the MCIS has also been extended to other applications such as prediction of injury from blast in a shipboard environment (Jacobs et al. 2012). Future analyses of these military-specific scales should focus on further validating the MCIS and comparing the mISS, ISS and NISS on combat data in order to determine which of these scoring systems is superior for mortality prediction.

4.3.7 THE PENETRATING ABDOMINAL TRAUMA INDEX (PATI)

The Penetrating Abdominal Trauma Index (PATI) was initially developed in 1979 (Biffl and Moore 2001) by Moore and colleagues (Moore et al. 1981) in order to quantify patients who might be at risk from postoperative complications after sustaining a penetrating injury to the abdomen. It was needed as other injury severity scales available at the time for assessing patients with multiple trauma, such as the ISS, were not specifically designed for either penetrating injuries or for abdominal injuries, and also focused on mortality rather than postoperative morbidity.

The PATI was based on 222 civilian patients (108 stab wounds and 114 GSWs) undergoing laparotomy for penetrating abdominal trauma only (i.e. they had no extra abdominal injuries), where the patient's ages ranged from 12 to 59 years and the mean age was 27 years. Patients were not analysed according to age and sex, as most of the patients were young males.

In the initial PATI score, organ risk factors, or complication risk factors, were assigned to each of the 14 organs or vessels (Table 4.8), which were then multiplied by the severity of injury score to obtain the PATI. The severity of injury score was based on a modification of AIS 1971, where 1 = minimal, 2 = minor, 3 = moderate, 4 = major and 5 = maximum, and were consensus derived using morbidity rates reported in the literature (Moore et al. 1981). Descriptors for minimal, minor and so forth to the specific organs were provided in the original 1981 Moore paper (and are not listed here).

In order to calculate the PATI, Equation 4.3 is used:

$$\text{PATI score} = (\text{Organ 1 risk factor} \times \text{injury severity}) + (\text{Organ 2 risk factor} \times \text{injury severity}) + (\text{Organ 3 risk factor} \times \text{injury severity}), \text{etc.} \quad (4.3)$$

For example, if a patient sustained a penetrating injury resulting in major debridement of the liver (injury severity = 3) and a through-and-through injury to the stomach (injury severity = 2), then the PATI is:

$$\text{PATI score} = (\text{liver risk factor} \times \text{liver injury severity}) + (\text{stomach risk factor} \times \text{stomach injury severity})$$

thus:

$$\text{PATI score} = (4 \times 3) + (2 \times 2) = 16$$

TABLE 4.8
Complication Risk Factors Assigned to the Original PATI Scores

Injured Organ or Vessel (PATI)	Organ/Complication Risk Factor
Duodenum, pancreas	5
Liver, large intestine, major vascular	4
Spleen, kidney, extrahepatic biliary	3
Small bowel, stomach, ureter	2
Bladder, bone, minor vascular	1

Source: Moore, E. E., et al., *J. Trauma.*, 21 (6), 439–445, 1981.

For both stab and GSWs, Moore et al. (1981) found that there was a dramatic increase in morbidity if the PATI > 25; thus, the authors divided the score into PATI ≤ 25 (low risk of postoperative complications) and PATI > 25 (high risk of postoperative complications). Clinically, the PATI is useful for guiding the surgeon in making operative and patient management decisions (Borlase et al. 1990).

The PATI was later modified and renamed to become the Abdominal Trauma Index (ATI) in 1989 (Biffl and Moore 2001; Borlase et al. 1990). With the change to the ATI, some of the original organ and vessel risk factors were revised (Table 4.9) based on the findings from a regression analysis of 300 trauma patients (male/female ratio was not stated) with penetrating abdominal injuries. Despite the update of the PATI to the ATI, both scoring systems are used in the literature and in clinical settings.

In numerous studies in the literature, the PATI or ATI have been used or evaluated using civilian data (e.g. Aldemir et al. 2004; Bishop et al. 1991; Chappuis et al. 1991; Croce et al. 1992; Gomez-Leon 2004), demonstrating its accuracy in predicting outcomes such as morbidity, mortality and complications after abdominal trauma. Although the PATI was developed for and validated using civilian data, there are also several studies where the PATI has been used on combat data, but it has yet to be widely adopted for military applications. For example, Sikic et al. (2001) conducted a study on 93 patients with war injuries treated at civilian hospitals between 1991 and 1995, calculating the PATI, ISS and AIS-85 from patient data. The authors found that PATI, ISS and number of injured abdominal organs (NIAO) were all good predictors of the patient developing complications, with the NIAO being the best predictor of complications and a model combining all three variables being the best predictor of death.

The study by Sikic et al. (2001) highlights a limitation of the PATI: it was designed for patients who have penetrating abdominal injuries as their only source of trauma (Moore et al. 1981), and for patients with other injuries, models using a number of predictors are likely to be needed.

TABLE 4.9
Complication Risk Factors for in the ATI

Injured Organ	Complication Risk Factor
Pancreas, colon, major vascular	5
Duodenum, liver	4
Spleen, stomach	3
Kidney, ureter	2
Bladder, bone (excluding pelvis), diaphragm, small bowel, extrahepatic biliary, minor vascular and soft tissue	1

Source: Borlase, B. C., et al., *J. Trauma Acute Care Surg.*, 30 (11), 1340–1344, 1990.

4.3.8 THE MANGLED EXTREMITY SEVERITY SCORE (MESS)

The Mangled Extremity Severity Score (MESS) was originally proposed by Johansen et al. (1990) with the aim of differentiating between salvageable and non-salvageable limbs in the case of lower extremity trauma, although since its development, it has also been used for upper extremity trauma. A mangled limb can be defined as a limb with injury to at least three out of four systems (soft tissue, bone, nerves and vessels) and are significant as they often require amputation (Prasarn et al. 2012).

The MESS is a simple rating scale for lower extremity trauma using four variables for assessment: patient age, limb ischaemia, shock and injury mechanism. The scores for each of the four categories are added to obtain the final score, where the final score ranges from 2 to 14 (Table 4.10). Using both

TABLE 4.10

Elements in the Mangled Extremity Score

Criteria	Value	Points
Patient age (years)	<30	0
	30–50	1
	>50	2
Limb ischaemia[a]	Pulse reduced or absent but normal perfusion	1
	Pulseless, paraesthesias, slow capillary refill	2
	Cool, paralysis numb/insensate	3
Shock	SBP > 90 mmHg always	0
	Hypotension transiently	1
	Persistent hypotension	2
Injury mechanism	Low energy (stab, 'civilian' gunshot, simple fracture)	1
	Medium energy (dislocation, open/multiple fractures)	2
	High energy (high-speed MVA or close-range shotgun or 'military' GSW, crush injury)	3
	Very high energy (above plus gross contamination, soft-tissue avulsion)	4

[a] Limb ischaemia score doubled for ischaemia >6 hours.

Compiled using data from Johansen, K., et al., *J. Trauma.*, 30(5), 568–573, 1990.

prospective (n = 25 patients with n = 17 limbs salvaged and n = 9 limbs requiring amputation) and retrospective analysis (n = 26 patients with n = 14 salvaged and n = 12 amputated limbs), Johansen et al. (1990) found a MESS value ≥ 7 predicted amputation with 100% accuracy. Though the score provides an early prediction of the potential outcome of an injured limb, it is designed to be used in conjunction with clinical judgement to make a decision on the most appropriate treatment option.

The MESS has also been applied to the prediction of limb salvage of the upper extremities, but the results have been conflicting, with some studies demonstrating that the MESS ≥ 7 is also predictive for upper extremity amputation (e.g. Slauterbeck et al. 1994), and other research showing it is not the right threshold value for the upper extremities (e.g. Togawa et al. 2005).

The MESS was further extended by Gregory et al. (1985) to include the effect of trauma sustained outside the extremities, with the new scoring system denoted as the Mangled Extremity Syndrome Index (MESI). Using 17 patients, Gregory et al. (1985) assigned further points for additional injuries outside the extremities and for a number of physiological parameters such as patient age. The authors subsequently defined a threshold value of 20 for the MESI: a MESI below 20 indicated that functional limb salvage could be made, whereas a MESI of above 20 meant that limb salvage was unlikely.

The MESS and MESI scores are obviously highly relevant in a military setting, and consequently have been validated and used to analyse combat data. Kjorstad et al. (2007) conducted a validation study on the use of the MESS in a combat setting, where they analysed the predictive value of the MESS for 60 limbs in 49 patients. They found that the MESS threshold of ≥ 7 was a sensitive predictor of amputation in a combat scenario; thus, it was potentially a valuable tool in combat situations where there are often mass casualties, where the MESS can be used to both assist in triage as well as manage the injuries. Conversely, Brown et al. (2009) used UK JTTR data comprised of 77 patients with 86 limbs sustaining mangled extremity injuries, 22 limbs of which required amputation. The authors found that the MESS could not be used to determine if amputation was needed in individual patients, stating that most amputations were performed when there was an ischaemic limb present. Consequently, Brown et al. (2009) suggested that the main application for the MESS is for mass casualty situations where resources might be limited and rapid decisions need to be made on which patients to treat and which limbs to salvage.

4.4 PHYSIOLOGICAL TRAUMA SCORING SYSTEMS

4.4.1 THE GLASGOW COMA SCALE (GCS) AND THE GLASGOW OUTCOME SCALE (GOS)

Initially developed by Teasdale and Jennett in 1974, the Glasgow Coma Scale (GCS) is a neurological injury scoring system developed to assess the level of consciousness in a patient who has sustained a Traumatic Brain Injury (TBI) (Teasdale and Jennett 1974). Prior to this time, the patient's state of consciousness was not quantified, but was comprised of written observations in the medical chart.

Since 1974, there have been a number of refinements to the scale (Jennett and Bond 1975; Teasdale and Jennett 1976). In the current scale, three components are assessed: eye opening (E), best verbal response (V) and best motor response (M). The GCS ranges from 3 to 15: a minimum score of 3 means the patient is in a deep coma, and the maximum score of 15 indicates that the patient is fully conscious with no neurological deficit (Table 4.11). A GCS $\leq$ 8 indicates the patient has a severe brain injury, while GCS values of 9–12 and 13–15 indicate moderate and minor brain injuries respectively. Numerous studies have shown the GCS to be accurate, especially when used by trained clinicians (e.g. Rowley and Fielding 1991), but both civilian (e.g. Heron et al. 2001) and military (e.g. Riechers et al. 2005) studies have demonstrated there can be differences in scoring, especially with untrained users. Since its inception, initially as a research tool, the GCS has been widely adopted as a method to monitor the patient's progress over time.

Jennett and Teasdale defined the comatose state as being: (1) no eye opening, (2) not obeying commands and (3) no recognisable words, and using this definition, all patients who had GCS $\leq$ 7 were in a coma, but only about half of those scoring a GCS = 8 (in their patient sample) were in a coma (Jennett 2002; Jennett and Teasdale 1977). In the clinical setting, coma is defined as a GCS $\leq$ 8, where E1, V2, M5 is the only GCS = 8 combination a patient can have and still be in a coma, and it is also the maximum score a patient can have and still be in a coma (i.e. non-eye opening, incomprehensible sounds, localises painful stimuli). Although there are other EVM combinations for GCS = 8, the patient is not in a coma for those combinations.

While there are 120 possible permutations of the GCS, not all are clinically possible: in a study of 1,390 patients with a GCS from 4 to 14 totalling 118 possible combinations, Teoh et al. (2000) found that 95 combinations were clinically possible, although their patient casemix included medical and surgical patients as well as trauma patients. A number of studies have also demonstrated that patient mortality depends on the E, V and M components; for example, all patients with a GCS = 9 will not have the same mortality. Consequently, the individual components are just as important, or more important, than the overall score. In the study by Teoh et al. (2000) as well as other research (e.g. Kung et al. 2011), it has been found that V is the best predictor of mortality for a GCS $\geq$ 9, while either M, or both V and M, are the best predictors of mortality for a GCS < 8.

TABLE 4.11

The Elements of the Glasgow Coma Scale

Eye Opening (E)	Score	Best Verbal Response (V)	Score	Best Motor Response (M)	Score
Does not open eyes	1	Makes no sounds	1	Makes no movements	1
Opens eyes in response to painful stimuli	2	Incomprehensible sounds	2	Extension to painful stimuli (decerebrate response)	2
Opens eyes in response to voice	3	Utters inappropriate words	3	Abnormal flexion to painful stimuli (decorticate response)	3
Opens eyes spontaneously	4	Confused, disoriented	4	Flexion/withdrawal to painful stimuli	4
		Oriented, converses normally	5	Localises painful stimuli	5
				Obeys commands	6

Source: Gonzalez, E., and Moore, E. E., *Encyclopedia of Intensive Care Medicine*, Springer, Berlin, 2012, pp. 982–984.

Difficulties arise in assigning the GCS when there are factors which may affect the response, for example, the patient is intubated, in a drug-induced coma or under the influence of alcohol. Opinions differ as to when to assess the GCS in these cases: options include assessing the patient pre-sedation (for a drug-induced coma) and pre-intubation, or using the motor component of the score only, e.g. in the case of intubation (Jennett 2002).

The Glasgow Outcome Score (GOS) is a scale which was designed so that patients with brain injuries can be divided into groups based on their degree of recovery (Jennett and Bond 1975) and is currently one of the most widely used measures for assessing global outcome resulting from TBI. The scale consists of five ordinal categories (Table 4.12), but, because the GOS categories are too broad to detect potentially significant differences between groups (Wright 2011), the original 5-point scale was extended in 1998 to an 8-point scale to give more specificity to the categories (Teasdale et al. 1998). The new scale, also shown in Table 4.12, was denoted the Extended Glasgow Outcome Score (GOS-E).

TABLE 4.12
The GOS and the GOS-E

GOS	Score	Description	GOS-E	Score	Description
Dead	1		Dead	1	
Vegetative State	2	Unable to interact with environment; unresponsive	Vegetative State	2	Condition of unawareness with only reflex responses. Periods of spontaneous eye opening.
Severe Disability	3	Able to follow commands; unable to live independently	Severe Disability		Dependent for daily support for mental or physical disability, usually a combination of both.
			Lower Severe Disability	3	*Lower*: Cannot be left alone for more than 8 hours at home.
			Upper Severe Disability	4	*Upper*: Can be left alone for more than 8 hours at home.
Moderate Disability	4	Able to live independently, unable to return to work or school	Moderate Disability		Have some disability such as aphasia, hemiparesis, or epilepsy, and/or deficits of memory or personality but are able to look after themselves. Independent at home but dependent outside.
			Lower Moderate Disability	5	*Lower*: Not able to return to work.
			Upper Moderate Disability	6	*Upper*: Able to return to work (even if special arrangement required).
Good Recovery	5	Able to live independently and return to work or school	Good Recovery		Resumption of normal life with the capacity to work even if pre-injury status has not been achieved. Some patients have minor neurological or psychological deficits.
			Lower Good Recovery	7	*Lower*: Deficits are disabling.
			Upper Good Recovery	8	*Upper*: Deficits are not disabling.

Compiled using data from Teasdale, G., and Jennett B., *Acta Neurochir.*, 34(1–4), 45–55, 1976; Wright, J., *Encyclopaedia of Clinical Neuropsychology*, pp. 1150–1152. Springer-Verlag, New York, 2011.

The GCS, GOS and GOS-E have been adopted for military use (Mac Donald et al. 2014; Office of The Surgeon General and Borden Institute 2013), although the scoring systems were initially developed for non-missile head injuries (Jennett and Teasdale 1977). However they appear to be reliable indicators of survival and functional outcomes generally (e.g. Kennedy et al. 1993; Mac Donald et al. 2014). For instance, Kennedy et al. (1993) found the GCS on admission to be the most important factor in predicting survival in a group of 192 civilian patients with GSWs to the brain. Other evidence on civilian GSWs presented in a review paper by Rosenfeld demonstrates that GCS is an important predictor of mortality and patient outcome (Rosenfeld 2002).

4.4.2 THE POST-TRAUMATIC AMNESIA SCORE (PTA) AND THE MILITARY ACUTE CONCUSSION EVALUATION (MACE)

Post-traumatic amnesia (PTA) is a transient sequela of TBI, defined as the period of confusion and disorientation after the traumatic event. The PTA score is used to classify the degree of retrograde, i.e. loss of memory prior to the injury, or anterograde amnesia, i.e. difficultly creating new memories after the injury (Ahmed et al. 2000; Cantu 2001). One method of classification using the PTA score alone is shown in Table 4.13, where there are six categories of TBI severity. When continuous memory returns, PTA is considered to have resolved. The severity of PTA can be classified alone or in conjunction with LOC (Loss of Consciousness) and GCS.

Evaluation of concussion and PTA in military combatants remains challenging, and instead of using traditional clinical evaluation tools to evaluate these injuries, the Military Acute Concussion Evaluation (MACE) score is employed for combat soldiers. The MACE was developed by a group of military and civilian TBI experts based on sport-concussion literature, and was first implemented for clinical use in military personnel in August 2006 (McCrea et al. 2000, 2009). It is comprised of a two-phase test with a historical component, which includes information on the nature of the event and the signs and symptoms at the time of injury, and an objective component, which is a scorable test (also known as the Standardised Assessment of Concussion, or SAC) on orientation, memory and concentration. A score of 0 to 30, commonly called the MACE score, is assigned based on patient performance (Coldren et al. 2010). A copy of the MACE pocket card can be found at: https://dvbic.dcoe.mil/files/resources/DVBIC_Military-Acute-Concussion-Evaluation_Pocket-Card_Feb2012.pdf

The MACE is not designed to be a stand-alone tool, but should be used in conjunction with a physician assessment. It has a number of significant limitations, such as being beneficial only in the acute injury period, lacking sensitivity and specificity if applied more than 12 hours post-injury

---

TABLE 4.13

Classification of TBI Using the PTA Scale Alone

Severity	PTA
Very mild	< 5 minutes
Mild	5–60 minutes
Moderate	1–24 hours
Severe	1–7 days
Very severe	1–4 weeks
Extremely severe	> 4 weeks

Source: Morris, T., *Handbook of Medical Neuropsychology. Applications of Cognitive* Neuroscience, Springer, Philadelphia, PA, 2010.

(Coldren et al. 2010; French et al. 2008). In theatre, it is often administered by emergency medical technicians who have only basic training (16 weeks). However, there is no standardised training for competence in administration of the tool (Coldren et al. 2010).

4.4.3 THE WESTMEAD POST-TRAUMATIC AMNESIA SCALE (WPTAS)

The Westmead PTA Scale, later abbreviated to the Westmead Post-Traumatic Amnesia Scale (WPTAS), was initially proposed in 1986 by Shores and colleagues from Westmead Hospital in Australia (Shores et al. 1986) and is currently the most common PTA scale used in Australia and New Zealand. The WPTAS is a brief bedside standardised test to assess how long a patient who has sustained a TBI is in PTA. The test, which was initially clinically tested in 100 patients with severe TBI, is comprised of a set of questions which the patient is asked daily in order to assess their orientation and ability to retain new information on successive days. The patient is considered to no longer have PTA after three consecutive days with a perfect score. However, if a patient has been in PTA for more than 28 days (4 weeks), on the first day they have a perfect score, they are considered to no longer have PTA.

An abbreviated version of the WPTAS, denoted as the A-WPTAS, was later developed to assess patients with mild traumatic brain injury (MTBI), which was defined as a TBI patient with a GCS 13–15 less than 24 hours post-injury (Meares et al. 2011). The A-WPTAS assessment is used in the emergency department or triaging station with the aim of testing the patient's suitability for discharge: patients with a full score can be considered for discharge and/or recovered from PTA. The scale was also designed to be used for the early identification of cognitive impairment due to an MTBI.

The A-WAPTAS is based on the GCS and a set of three picture cards (Figure 4.2), providing a maximum score of 18 (i.e. the GCS = 15 for fully conscious, plus one point for each picture card). After the GCS is administered, the patient is shown the three picture cards for about 5 seconds and is told to remember the cards. About 1 hour later, they are asked to recall the three pictures, and one point is given for each picture correctly identified. The test is repeated after another hour. For patients who cannot recall any of the pictures, or only recall one or two pictures, a set of nine images (9-object recognition chart) including the three target cards are shown, and the patient is asked to identify the three target cards. A score of one is assigned for each card correctly identified; the patient is then shown the target pictures again and then retested in 1 hour. The severity of injury is based on the time it takes for an individual to emerge from PTA; this time is calculated from the time of injury. Table 4.14 demonstrates the appropriate test method based on the duration of PTA.

The WPTAS and A-WPTAS tests are clearly relevant for military applications, and a number of authors have recommended their application for situations such as evaluating PTA after events like bomb blasts (Ford and Rosenfeld 2008; Rosenfeld and Ford 2010). However, due to the development of the MACE, which was discussed earlier, the Westmead scales are generally not used by the military. As yet, there has been no comparison between the two types of scales (Caldroney and Radike 2010).

FIGURE 4.2   The three target picture cards in the A-WPTAS test. (Reproduced from Shores 2016. Copyright permission.)

TABLE 4.14
Severity of PTA and Method of Assessment

Duration of PTA	Severity	Appropriate Measure
Less than 24 hours	Mild	A-WPTAS
1–7 days	Severe	WPTAS
1–4 weeks	Very severe	WPTAS
> 4 weeks	Extremely severe	WPTAS

Compiled using information from Meares, S., et al., *Brain Injury.*, 25 (12), 1198–1205, 2011; Shores, E. A., et al., *Med. J. Aust.*, 144 (11), 569–572, 1986.

4.4.4 THE REVISED TRAUMA SCORE (RTS)

The Revised Trauma Score (Champion et al. 1989) is derived from the earlier developed Trauma Score (Champion et al. 1981), which in turn was developed from the Triage Score (Champion et al. 1980). The Revised Trauma Score (RTS) has two versions, one for triage (T-RTS) and one for outcome evaluations (RTS), the latter of which is discussed here. The RTS is a physiological scoring system which has been correlated with the probability of survival, having been developed using data from 2,166 patients in the Washington Hospital Center database and validated on a different set of data comprising of over 26,000 patients from the US MTOS (Champion et al. 1989).

The RTS is calculated using the first set of data obtained from the patient and consists of the GCS, systolic blood pressure (SBP) and respiratory rate (RR). The raw values, i.e. the individual values of GCS, SBP and RR, are used for triage, while the coded values (values from 0 to 4) are used for outcome evaluation, i.e. the calculation of the RTS (Table 4.15). After assigning a score from 0 to 4 for each of the raw values (for example, a GCS of 14 will result in a score of 4), the codes are then used in the RTS equation (Equation 4.4) to obtain a final score:

$$RTS = 0.9368 \text{ GCS} + 0.7326 \text{ SBP} + 0.2908 \text{ RR} \qquad (4.4)$$

The RTS was modified from the Trauma Score (TS), which was developed earlier, to address limitations of the TS and also to include the GCS in order to factor in patients with major head injury who did not have major physiological changes. RTS values range from 0 to 7.8408. Champion et al. (1989) presented figures for the probability of survival based on integer values of the RTS; these have been plotted in Figure 4.3. While the RTS has been used extensively in civilian settings, its

TABLE 4.15
Raw Values of the RTS

Glasgow Coma Scale (GCS)	Systolic Blood Pressure (SBP)	Respiratory Rate (RR)	Coded Value
13–5	> 89	10–29	4
9–12	76–89	> 29	3
6–8	50–75	6–9	2
4–5	1–49	1–5	1
3	0	0	0

Compiled using data from Champion, H. R., et al., *J. Trauma Acute Care Surg.*, 29 (5), 623–629, 1989.

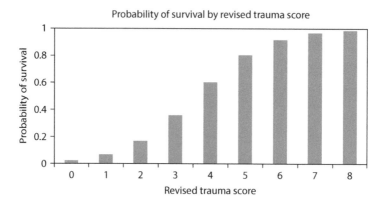

FIGURE 4.3 The probability of patient survival based on the Revised Trauma Score (using data from Champion et al., 1989).

use is more limited in combat situations, although it has been employed in several military studies, for example, Burkle et al. (1994).

4.4.5 THE CIRCULATION, RESPIRATION, ABDOMEN, MOTOR AND SPEECH (CRAMS) SCALE

The CRAMS scale is a 10-point field triage tool developed for purpose of determining which trauma patients should go to a trauma centre or high-level of care facility, and thus ensure that patients who have not sustained major trauma are not sent to a major trauma centre. The acronym CRAMS represents the five components measured as part of the scale: Circulation, Respiration, Abdomen, Motor and Speech. To calculate the score, each of the five components is assigned a value between 0 and 2 based on:

- Normal = 2 points
- Mildly abnormal = 1 point and
- Severely abnormal = 0 points

Thus, the score ranges from 0 to 10. Scores of 9 or 10 have been associated with minor trauma, while scores of 8 or less are defined as major trauma (Gormican 1982). The CRAMS score has been shown to be successful in accurately discriminating between major and minor trauma in numerous studies (e.g. Clemmer et al. 1985; Gormican 1982), although it has also been found that trauma assessment by emergency medical personnel can be almost as accurate as using the CRAMS scale (Ornato et al. 1985). When specific mechanisms of injury are factored, it has been demonstrated that the sensitivity of the CRAMS scale in predicting major/minor trauma increases significantly (Knudson et al. 1988). While not evaluated in the military literature, the potential usefulness of the CRAMS has been discussed for mass casualty events (e.g. Hoey and Schwab 2004).

4.5 COMBINED ANATOMICAL AND PHYSIOLOGICAL TRAUMA SCORING SYSTEMS

4.5.1 THE TRAUMA AND INJURY SEVERITY SCORE (TRISS)

The Trauma and Injury Severity Score (TRISS) was originally proposed by Champion and colleagues in 1983 (Champion et al. 1983) to predict the probability of patient survival, but has been revised numerous times since (Boyd et al. 1987; Champion et al. 1995), with the current modification being in 2010 (Schluter et al. 2010). The TRISS is calculated using the ISS, the RTS (and thus the GCS, systolic blood pressure and respiratory rate), the patient's age and the mechanism of injury (blunt or penetrating), where the probability of survival is calculated according to Equation 4.5:

$$Pr(s) = \frac{1}{(1 + e^{-b})} \tag{4.5}$$

and b is calculated using Equation 4.6:

$$b = b_0 + b_1 \, (\text{RTS}) + b_2 \, (\text{ISS}) + b_3 \, (\text{Age Index}) \tag{4.6}$$

The coefficients were derived using regression analysis from the US MTOS database and differ depending on whether the patient sustained blunt or penetrating trauma (Boyd et al. 1987). The Age Index is zero if the patient is < 54 years of age and one if the patient is ≥ 55 years of age. Since it was established, the TRISS has remained one of the most commonly used tools for assessing trauma fatality outcome. It can be applied for situations such as reviewing unexpected fatalities to identify opportunities for improvement in patient management and care. Alternatively, reviewing the unexpected survivors enables the clinician to determine what aspects in the patient care were performed well and correctly, and to decide if a similar patient could be managed in the same way at any time of the day, or day of the week.

Figure 4.4 shows a TRISS calculation (ISS versus RTS) on 597 patients identified as major trauma from a major trauma hospital who were aged between 15 and 55 years inclusive and

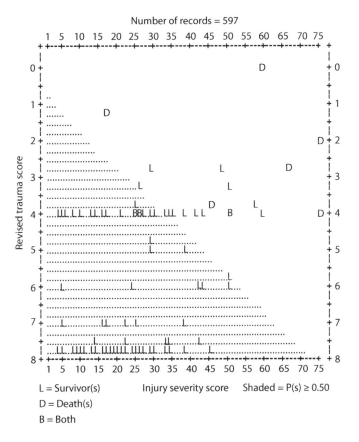

FIGURE 4.4 Calculation of the TRISS in 597 blunt trauma patients (L, D and B) 15–55 years of age inclusive from a major trauma hospital (using the Boyd et al. 1987 version). The boundary between the shaded and unshaded regions represents the 50% probability of survival line for this specific population.

sustained blunt trauma. The boundary between the shaded and unshaded regions represents the 50% probability of survival [Pr(s)] line, which is calculated for the specific population being analysed. The points marked 'L' are survivors, the points marked 'D' represent fatalities, and the points labelled 'B' contain both survivors and fatalities (the L, D and B points can represent one or many patients). Patients with a Pr(s) < 50% who survive are 'unexpected survivors' and are shown by the L or B points in the unshaded areas, while patients with a Pr(s) > 50% who die are 'unexpected deaths', and are identified by the D and B points in the shaded area. One point of interest is the region at the bottom of the graph where there is a line showing many survivors (L's) when the RTS is close to 8: patients have survived with relatively high ISS scores, up to an ISS of 45. This demonstrates that patients who are physiologically stable (high RTS value) have a high chance of survival, even if their ISS is relatively high.

Since its inception, the TRISS has been the subject of numerous studies where its validity has been analysed in different trauma populations, and new or revised coefficients have been proposed to improve its predictive power (e.g. Schluter 2011; Schluter et al. 2009, 2010). The TRISS has been used as an analysis tool in a large number of military studies to evaluate patient probability of survival (e.g. Grevitt et al. 1991; Penn-Barwell et al. 2013). In addition, the TRISS is used in both US and UK military trauma registries, and coefficients specific to military injuries have recently been calculated using the UK JTTR database (Penn-Barwell et al. 2015).

4.5.2 The Anatomic Profile (AP) and A Severity Characterisation of Trauma (ASCOT)

First introduced in 1990, the Anatomic Profile (AP) was developed in order to address the limitations of the ISS such as only considering the most severe injury in a given ISS body region and giving equal weighting to all body regions (Copes et al. 1990). The AP is essentially a modification of the ISS, but the body regions and the weightings differ. The AP body regions and associated injury severities considered are:

A: Head, brain and spinal cord (AIS 3–5)
B: Thorax and the anterior neck (AIS 3–5)
C: All other serious injuries (AIS 3–5)
D: All other non-serious injuries (AIS 1–2)

AIS 6 injuries are not included, and any region with no injury is designated a score of zero. The AP is calculated using the AIS scores (originally AIS 85) for the most severe AIS 3+ injury body regions A, B and C in addition to the most severe injury from body region D, then finding the square root of the sum of the squares using Equation 4.7:

$$AP = \sqrt{A^2 + B^2 + C^2 + D^2} \tag{4.7}$$

The AP is related to the probability of survival by logistic regression. A modified AP (mAP) was later proposed where only severe injuries are considered (Sacco et al. 1999). However, despite the increased sensitivity of both the AP and the mAP than the ISS and ICISS in predicting mortality (e.g. Frankema et al. 2005; Sacco et al. 1999), they have not gained acceptance due to the greater degree of mathematical complexity for only a modest gain in predictive performance.

The A Severity Characterisation of Trauma (ASCOT) score, first developed in 1990 (Champion et al. 1990a), involves using GCS, systolic BP, respiratory rate, patient age and AIS injury severity scores from the AP. It has been shown to be more accurate in predicting injury outcome than the TRISS (Champion et al. 1990a, 1996); however, like the AP and mAP, it is more complex to calculate than other injury scoring systems and consequently has not been widely adopted.

4.5.3   The Paediatric Trauma Score (PTS)

Civilian infants and children are often admitted to military combat support hospitals in Afghanistan and Iraq (Burnett et al. 2008; Edwards et al. 2014; Fuenfer et al. 2009), and scoring systems such as the Paediatric Trauma Score (PTS) can be used to evaluate the clinical condition of the injured child. The PTS uses both physiological and anatomical parameters to predict outcome and severity of injury in the paediatric trauma patient (0–14 years). The PTS ranges from +12 to −6, where six elements are assessed, and a lower score indicates a higher threat of mortality to the injured child (Table 4.16). In the civilian setting, the PTS is also used as a triage tool, where a score of 8 or less indicates that the patient should be taken directly to a Paediatric Major Trauma Service (Paediatric Trauma Centre).

Using two groups of paediatric trauma patients, one group from the Paediatric Major Trauma Centre in the US and the second group using data from the National Paediatric Trauma Database (US), Tepas et al. (1987) found there was a statistically significant inverse linear relationship between the PTS and the ISS. Thus, the PTS was able to not only predict injury severity but also patient mortality.

TABLE 4.16
The Paediatric Trauma Score

Paediatric Trauma Score (PTS)	+2	+1	−1
Weight	> 20 kg (44 lbs)	10–20 kg (22–44 lbs)	< 10 kg (22 lbs)
Airway	Patent	Maintainable	Unmaintainable
Systolic blood pressure	> 90 mmHg	50–90 mmHg	< 50 mmHg
CNS	Awake	+LOC	Unresponsive
Fractures	None	Closed or suspected	Multiple closed or open
Wounds	None	Minor	Major/penetrating/burns

Data obtained from Tepas, J. J., III, *J. Pediatr. Surg.*, 22 (1), 14–18, 1987.

4.5.4   Acute Physiology and Chronic Health Evaluation II (APACHE II)

First proposed by Knaus et al. (1981), the Acute Physiology and Chronic Health Evaluation (APACHE) is a severity of illness classification used for predicting mortality in patients admitted to an Intensive Care Unit (ICU). It was revised in 1985 and renamed APACHE II (Knaus et al. 1985). The APACHE II is based on the patient's age, 12 routine physiological measurements and any pre-existing disease of the patient. The score, which is calculated daily and can be used to evaluate the progress of the patient during their ICU stay, ranges from 0 to 71, with higher scores corresponding to more severe disease and a higher risk of death. The system was again revised in 1991 (Knaus et al. 1991) to become the APACHE III, with both APACHE II and III used in the literature and clinical settings.

The APACHE can be calculated using online tools such as: http://reference.medscape.com/calculator/apache-ii-scoring-system. It has been subjected to numerous validation studies, with some studies showing its value in predicting patient outcome and mortality (e.g. Rhee et al. 1990; Rutledge et al. 1993), and others highlighting its limitations, such as its assessment of acute injury in the otherwise healthy patient (e.g. McAnena et al. 1992). While the APACHE has been developed and used in civilian ICU situations, it is less relevant for military combatants, who in general have less pre-existing disease.

4.6 OUTCOME SCALES

4.6.1 THE FUNCTIONAL CAPACITY INDEX (FCI) AND FUNCTIONAL INDEPENDENCE MEASURE (FIM™)

Developed by MacKenzie et al. (1996), the Functional Capacity Index (FCI) is an outcome scale which is used to predict the patient's functional capacity 12 months after an injury was sustained, based on a single AIS injury (originally developed from AIS 90). The ranking system ranges from 1 to 5, where 1 is a poor outcome and 5 represents a full recovery at 12 months post-injury. The scale is based on 10 dimensions which are used to measure function and severity of impairment in relation to impact on the injured person's daily life, the dimensions being:

- Eating
- Excretory
- Sexual
- Ambulatory
- Hand
- Bending & lifting
- Visual
- Auditory
- Speech
- Cognition

The combination of all 10 dimensions provides a functional profile of the patient. The FCI has had several revisions since its inception, and FCI scores for AIS values are included in the currently used AIS Dictionary (AIS 2005 Update 2008). The FCI has been the subject of a number of validation studies on civilian data, with some studies demonstrating its usefulness in predicting patient functional outcome (MacKenzie et al. 2002) or improved predictive outcome with the latest version (Gotschall 2005), and other research demonstrating that predicted outcomes do not agree well with actual patient outcomes (e.g. McCarthy and MacKenzie 2001; Schluter et al. 2005).

The FCI has yet to be validated for a military population; however, it clearly has potential applications for military injury data, such as predicting long-term outcome for lower extremity injuries sustained in mounted or dismounted blast events. While it has been mentioned in the military literature, and dimensions of the FCI have been used for military research (Bacon et al. 2012), the FCI has yet to be widely implemented in defence research. Instead, a combat-specific functional capacity scoring system has been developed, which is discussed in Section 4.6.2.

The Functional Independence Measure (FIM™) scale or instrument is used to assess physical and cognitive disability, and is used to classify the ability of an individual to carry out tasks independently or the need for assistance from a device or other person. It is also employed to measure a patient's progress and assess rehabilitation outcomes (Gupta 2008). There are 18 items evaluated in FIM, 13 physical and 5 cognitive items. The 13 physical items are:

1. Eating
2. Grooming
3. Bathing
4. Dressing, upper body
5. Dressing, lower body
6. Toileting
7. Bladder management
8. Bowel management
9. Transfers – bed/chair/wheelchair
10. Transfers – toilet
11. Transfers – bath/shower

12. Walk/wheelchair
13. Stairs

The 5 cognitive items are:

1. Comprehension
2. Expression
3. Social interaction
4. Problem solving
5. Memory

The degree of need is rated on the scale from 1 to 7, where 1 is total assistance with a helper and 7 is total independence i.e. no helper required. The needs are assessed by a trained clinician at the beginning of the rehabilitation period and at intervals throughout rehabilitation to assess the individual's progress (Gupta 2008). Possible scores range from 18 to 126, with higher scores indicating more independence. Alternatively, the 13 physical items could be scored separately from the five cognitive items.

The FIM™ instrument has been used extensively in civilian applications, especially in older patients. However, it has also been applied in a number of military studies to evaluate the effect of blast and other injuries on the level of disability experienced by combatants (e.g. Clark et al. 2009; Sayer et al. 2008).

4.6.2 THE MILITARY FUNCTIONAL INCAPACITY SCALE (MFIS)

The Military Functional Incapacity Scale (MFIS) is a consensus-derived scale which was designed to be directly correlated to the MCIS (discussed earlier) and describes immediate tactical functional impairment post-injury (Lawnick et al. 2013). It is based on the predicted ability of the casualty to perform essential tasks immediately after the injury was sustained (Table 4.17). Thus it differs from the FCI, as although the FCI is based on AIS descriptors, it does not have a direct correlation with AIS severity like the MFIS.

The MFIS was developed based on the casualty's ability to shoot (load, aim or fire a weapon system), move (walk, run, crawl, enter/exit/drive a vehicle) and communicate (comprehend, receive or send verbal or non-verbal orders), which were identified as the important immediate tactical functions required by a solider in a combat situation (Lawnick et al. 2013).

The development of the MFIS highlights an important difference between civilian and military populations: the effect of the injury on the combatant's ability to resume the mission is of paramount

---

TABLE 4.17

The Military Functional Incapacity Scale and its Correlation to the Military Combat Injury Scale

MCIS	MFIS Incapacity Level	MFIS Descriptor
1	1	Able to continue mission
2	2	Able to contribute to sustaining mission
3/4	3	Lost to mission
4/5	4	Lost to military

Compiled with information from Lawnick, M. M., et al., *J Trauma Acute Care Surg.*, 75 (4), 573–581, 2013.

---

importance and could influence the success of the overall mission. Although the MFIS has not yet been used extensively, most likely as it is relatively new, it clearly has important applications for military injury research. In addition, although it is not the only functional impairment scale developed specifically for the military; for example, the Walter Reed Functional Impairment Scale was recently developed in order to assess soldier functioning in four domains: physical, occupational, social and personal; and the Walter Reed scale focuses on psychological stresses and their correlation with physical symptoms (Herrell et al. 2014) rather than traumatic injury.

4.6.3 THE INJURY IMPAIRMENT SCALE (IIS)

First proposed by States and Viano (1990), the Injury Impairment Scale (IIS) was developed to measure the degree of impairment and disability sustained by the injured patient, with a focus on permanent disability. The scale is based on a numerical score of 0 to 6 determined by the level of impairment, where 0 = No impairment, 1 = Minor, 2 = Moderate, 3 = Serious, 4 = Severe, 5 = Very severe and 6 = Total (Table 4.18). The IIS has not been widely used in civilian clinical applications or military research as the FIM™, discussed in the previous section, is generally used for rehabilitation. Thus, the IIS is only mentioned briefly here.

TABLE 4.18
IIS Codes, Level of Impairment and Example Injuries from Orthopaedics and Neurotrauma

IIS Code	Level of Impairment	Description
0	No impairment	
1	Minor	Occasional joint pain, up to 30% loss of range of motion, up to 50% loss of strength, body and extremity scarring causing loss of range of motion or strength but not affecting sensation
2	Moderate	Compression fractures of the vertebral bodies of the spine less than 20% causing occasional pain, chronic whiplash, or low back pain syndrome; repair of medial collateral or anterior cruciate ligaments; use of non-narcotic analgesics
3	Serious	Knee or hip fusion, severe traumatic arthritis, narcotic analgesics necessary for pain control
4	Severe	Paraplegia, bilateral above knee amputations, hemiplegia due to brain injury
5	Very severe	Quadriplegia; brain injury with severe decrease in cognition, speech and memory
6	Uniformly total impairment	Coma-vegetative state

Compiled using information from States, J. D., and Viano, D. C., *Accident Anal. Prev.*, 22 (2), 151–159, 1990.

4.7 CONCLUSIONS

The injury scales discussed in this chapter have been developed in order for that injuries and/or functional impairment to be classified using standard nomenclature, thereby enabling data comparison between institutions and facilitating clinical decisions. Trauma scoring systems are also invaluable as a tool in injury prevention research, where study outcomes can lead to modifications in vehicle design and body armour.

All trauma scoring systems have been designed for a specific purpose(s); thus, a suitable coding or classification system needs to be chosen based on the application and the specific population being considered. As the majority of injury coding systems have been designed for civilian populations, they may be less applicable in a military context. However, these scales are increasingly being used and validated for military data, while a number of combat-specific trauma scoring systems have also recently been developed.

This chapter was not intended to present an exhaustive list of injury scoring nomenclatures; but it includes those which are most commonly used, or those which are most relevant to combat trauma. While designed for applications such as injury prevention and assisting in clinical management, some new uses for injury coding systems have recently emerged, such as the integration of injury scoring into computational models of the combat soldier in order to predict injury potential from different projectiles, highlighting the versatility of trauma scoring systems for new applications.

ACKNOWLEDGEMENTS

The authors would like to gratefully acknowledge Sheridan Laing for drawing Figures 4.1 and 4.3, Sujanie Peiris for formatting some of the references, and Jeff Copeland for formatting references and proofreading the chapter.

REFERENCES

Aharonson-Daniel, L., A. Giveon, M. Stein, and K. Peleg. 2006. Different AIS triplets: Different mortality predictions in identical ISS and NISS. *The Journal of Trauma and Acute Care Surgery*. 61 (3):711–717.

Ahmed, S., R. Bierley, J. I. Sheikh, and E. S. Date. 2000. Post-traumatic amnesia after closed head injury: A review of the literature and some suggestions for further research. *Brain Injury* 14 (9):765–780.

Aldemir, M., I. Tacyildiz, and S. Girgin. 2004. Predicting factors for mortality in the penetrating abdominal trauma. *Acta Chirurgica Belgica*. 104 (4):429–434.

American College of Surgeons: Committee on Trauma. 2014. Introduction. In *Hospital and Prehospital Resources for Optimal Care of the Injured Patient*, ed. M. F. Rotondo, C. Cribari and R. S. Smith, pp. 1–7. Chicago, IL: American College of Surgeons.

ASNSW (Ambulance Service of New South Wales). 2012. *Protocol T1 Trauma Triage Tool: Major Trauma Criteria (MIST)*. New South Wales.

Bacon, J. R., T. J. Armstrong, and L. T. C. Brininger. 2012. The effects of functional limitations on soldier common tasks. *Work*. 41 (Suppl. 1):422–431. DOI: 10.3233/WOR-2012-0192-422.

Baker, S. P., and B. O'Neill. 1976. The injury severity score: An update. *The Journal of Trauma – Injury Infection and Critical Care*. 16 (11):882–885.

Baker, S. P., B. O'Neill, W. Haddon, Jr., and W. B. Long. 1974. The injury severity score: A method for describing patients with multiple injuries and evaluating emergency care. *The Journal of Trauma – Injury Infection and Crtitical Care*. 14 (3):187–196.

Balogh, Z., P. J. Offner, E. E. Moore, and W. L. Biffl. 2000. NISS predicts postinjury multiple organ failure better than the ISS. *The Journal of Trauma*. 48 (4):624–628.

Biffl, W. L., and E. E. Moore. 2001. Scoring systems for trauma research. In *Surgical Research*, ed. W. W. Souba and D. W. Wilmore, pp. 331–346. San Diego, CA: Academic Press.

Bishop, M., W. C. Shoemaker, S. Avakian, E. James, G. Jackson, D. Williams, P. Meade, and A. Flemming. 1991. Evaluation of a comprehensive algorithm for blunt and penetrating thoracic and abdominal trauma. *The American Surgeon*. 57 (12):737–746.

Borlase, B. C., E. E. Moore, and F. A. Moore. 1990. The Abdominal Trauma Index – A critical reassessment and validation. *The Journal of Trauma and Acute Care Surgery*. 30 (11):1340–1344.

Boyd, C. R., M. A. Tolson, and W. S. Copes. 1987. Evaluating trauma care: The TRISS method. Trauma Score and the Injury Severity Score. *The Journal of Trauma*. 27 (4):370–378.

Brenneman, F. D., B. R. Boulanger, B. A. McLellan, and D. A. Redelmeier. 1998. Measuring injury severity: Time for a change? *The Journal of Trauma*. 44 (4):580–582.

Brown, K. V., A. Ramasamy, J. McLeod, S. Stapley, and J. C. Clasper. 2009. Predicting the need for early amputation in ballistic mangled extremity injuries. *The Journal of Trauma*. 66 (4 Suppl.):S93–S98. DOI: 10.1097/TA.0b013e31819cdcb0.

Bull, J. 1975. The injury severity score of road traffic casualties in relation to mortality, time of death, hospital treatment time and disability. *Accident Analysis and Prevention*. 7 (4):249–255.

Burkle, F. M., C. Newland, S. Orebaugh, and C. G. Blood. 1994. Emergency medicine in the Persian Gulf War-Part 2: Triage methodology and lessons learned. *Annals of Emergency Medicine*. 23 (4):748–754.

Burnett, M. W., P. C. Spinella, K. S. Azarow, and C. W. Callahan. 2008. Pediatric care as part of the US Army medical mission in the global war on terrorism in Afghanistan and Iraq, December 2001 to December 2004. *Pediatrics*. 121 (2):261–265. DOI: 10.1542/peds.2006-3666.

Caldroney, R. D. and J. Radike. 2010. Experience with mild traumatic brain injuries and postconcussion syndrome at Kandahar, Afghanistan. *U.S. Army Medical Department Journal*. 22–30. https://www.ncbi. nlm.nih.gov/pubmed/21181651.

Cantu, R. C. 2001. Posttraumatic retrograde and anterograde amnesia: Pathophysiology and implications in grading and safe return to play. *Journal of Athletic Training*. 36 (3):244–248.

Champion, H. R., W. S. Copes, W. J. Sacco, C. F. Frey, J. W. Holcroft, D. B. Hoyt, and J. A. Weigelt. 1996. Improved predictions from a severity characterization of trauma (ASCOT) over Trauma and Injury Severity Score (TRISS): Results of an independent evaluation. *The Journal of Trauma*. 40 (1):42–49.

Champion, H. R., W. S. Copes, W. J. Sacco, M. M. Lawnick, L. W. Bain, D. S. Gann, T. Gennarelli, E. Mackenzie, and S. Schwaitzberg. 1990a. A new characterisation of injury severity. *The Journal of Trauma and Acute Care Surgery*. 30 (5):539–546.

Champion, H. R., W. S. Copes, W. J. Sacco, M. M. Lawnick, S. L. Keast, and C. F. Frey. 1990b. The Major Trauma Outcome Study: Establishing national norms for trauma care. *The Journal of Trauma and Acute Care Surgery*. 30 (11):1356–1365.

Champion, H. R., J. B. Holcomb, M. M. Lawnick, T. Kelliher, M. A. Spott, M. R. Galarneau, D. H. Jenkins, et al. 2010. Improved characterization of combat injury. *The Journal of Trauma*. 68 (5):1139–1150. DOI: 10.1097/TA.0b013e3181d86a0d.

Champion, H. R., W. J. Sacco, A. J. Carnazzo, W. Copes, and W. J. Fouty. 1981. Trauma score. *Critical Care Medicine*. 9 (9):672–676.

Champion, H. R., W. J. Sacco, and W. S. Copes. 1995. Injury severity scoring again. *The Journal of Trauma and Acute Care Surgery*. 38 (1):94–95.

Champion, H. R., W. J. Sacco, W. S. Copes, D. S. Gann, T. A. Gennarelli, and M. E. Flanagan. 1989. A revision of the trauma score. *The Journal of Trauma and Acute Care Surgery*. 29 (5):623–629.

Champion, H. R., W. J. Sacco, D. S. Hannan, R. L. Lepper, E. S. Atzinger, W. S. Copes, and R. H. Prall. 1980. Assessment of injury severity: The triage index. *Critical Care Medicine*. 8 (4):201–208.

Champion, H. R., W. J. Sacco, and T. K. Hunt. 1983. Trauma severity scoring to predict mortality. *World Journal of Surgery*. 7 (1):4–11.

Chappuis, C. W., D. J. Frey, C. D. Dietzen, T. P. Panetta, K. J. Buechter, and I. Cohn, Jr. 1991. Management of penetrating colon injuries. A prospective randomized trial. *Annals of Surgery*. 213 (5):492–498.

Clark, M. E., R. L. Walker, R. J. Gironda, and J. D. Scholten. 2009. Comparison of pain and emotional symptoms in soldiers with polytrauma: Unique aspects of blast exposure. *Pain Medicine*. 10 (3): 447–455.

Clemmer, T. P., J. F. Orme, Jr., F. Thomas, and K. A. Brooks. 1985. Prospective evaluation of the CRAMS Scale for triaging major trauma. *The Journal of Trauma*. 25 (3):188–191.

Coldren, R. L., M. P. Kelly, R. V. Parish, M. Dretsch and M. L. Russell. 2010. Evaluation of the Military Acute Concussion Evaluation for use in combat operations more than 12 hours after injury. *Military Medicine*. 175 (7):477–481.

Cook, A., J. Weddle, S. Baker, D. Hosmer, L. Glance, L. Friedman, and T. Osler. 2014. A comparison of the Injury Severity Score and the Trauma Mortality Prediction Model. *The Journal of Trauma and Acute Care Surgery*. 76 (1):47–52.

Copes, W. S., H. R. Champion, W. J. Sacco, M. M. Lawnick, D. S. Gann, T. Gennarelli, E. MacKenzie, and S. Schwaitzberg. 1990. Progress in characterizing anatomic injury. *The Journal of Trauma and Acute Care Surgery*. 30 (10):1200–1207.

Copes, W. S., H. R. Champion, W. J. Sacco, M. M. Lawnick, S. L. Keast, and L. W. Bain. 1988. The Injury Severity Score revisited. *The Journal of Trauma and Acute Care Surgery*. 28 (1):69–77.

Croce, M. A., T. C. Fabian, R. M. Stewart, F. E. Pritchard, G. Minard, and K. A. Kudsk. 1992. Correlation of abdominal trauma index and injury severity score with abdominal septic complications in penetrating and blunt trauma. *The Journal of Trauma and Acute Care Surgery*. 32 (3):380–388.

Eastridge, B. J., R. L. Mabry, P. Seguin, J. Cantrell, T. Tops, P. Uribe, O. Mallett, et al. 2012. Death on the battlefield (2001–2011): Implications for the future of combat casualty care. *The Journal of Trauma and Acute Care Surgery*. 73 (6):S431–S437. DOI: 10.1097/TA.0b013e3182755dcc.

Edwards, M. J., M. Lustik, M. W. Burnett, and M. Eichelberger. 2014. Pediatric inpatient humanitarian care in combat: Iraq and Afghanistan 2002 to 2012. *Journal of the American College of Surgeons.* 218 (5):1018–1023. DOI: 10.1016/j.jamcollsurg.2013.12.050.

Eskridge, S. L., C. A. Macera, M. R. Galarneau, T. L. Holbrook, S. I. Woodruff, A. J. MacGregor, D. J. Morton, and R. A. Shaffer. 2012. Injuries from combat explosions in Iraq: Injury type, location, and severity. *Injury.* 43 (10):1678–1682.

Ford, N. L., and J. V. Rosenfeld. 2008. Mild traumatic brain injury and bomb blast: Stress, injury or both. *ADF Health.* 9 (2):68–73.

Frankema, S. P., E. W. Steyerberg, M. J. Edwards, and A. B. van Vugt. 2005. Comparison of current injury scales for survival chance estimation: An evaluation comparing the predictive performance of the ISS, NISS, and AP scores in a Dutch local trauma registration. *The Journal of Trauma.* 58 (3):596–604.

Franklyn, M., and B. D. Stemper. 2008. *Abbreviated Injury Scale (AIS 2005 Update 2008).* Barrington, IL: Association for the Advancement of Automotive Medicine.

French, L., M. McCrea, and M. Baggett. 2008. The Military Acute Concussion Evaluation (MACE). *Journal of Special Operations Medicine.* 8 (1):68–77.

Fuenfer, M. M., P. C. Spinella, A. L. Naclerio, and K. M. Creamer. 2009. The U.S. Military wartime pediatric trauma mission: How surgeons and pediatricians are adapting the system to address the need. *Military Medicine.* 174 (9):887–891.

Gennarelli, T. A., and E. Wodzin. 2006. AIS 2005: A contemporary injury scale. *Injury.* 37 (12):1083–1091.

Gomez-Leon, J. F. 2004. Penetrating Abdominal Trauma Index: Sensitivity and specificity for morbidity and mortality by ROC analysis. *Indian Journal of Surgery.* 66 (6):347–351.

Gonzalez, E. and E. E. Moore. 2012. Glasgow Coma Scale. In *Encyclopedia of Intensive Care Medicine,* pp. 982–984. Berlin: Springer.

Gormican, S. P. 1982. CRAMS scale: Field triage of trauma victims. *Annals of Emergency Medicine.* 11 (3): 132–135.

Gotschall, C. S. 2005. The Functional Capacity Index, second revision: Morbidity in the first year post injury. *International Journal of Injury Control and Safety Promotion.* 12 (4):254–256.

Gregory, R. T., R. J. Gould, M. Peclet, J. S. Wagner, D. A. Gilbert, J. R. Wheeler, S. O. Snyder, R. G. Gayle, and C. W. Schwab. 1985. The Mangled Extremity Syndrome (MES): A severity grading system for multisystem injury of the extremity. *The Journal of Trauma.* 25 (12):1147–1150.

Grevitt, M. P., H. A. Muhiudeen, and C. Griffiths. 1991. Trauma care in a military hospital. *Journal of the Royal Army Medical Corps.* 137 (3):131–135.

Gupta, A. 2008. Functional independence measure and functional assessment measure. In *Measurement Scales Used in Elderly Care,* ed. A. Gupta, pp. 60–63. Oxon, UK: Radcliffe Publishing.

Haas, B., D. Gomez, B. Zagorski, T. A. Stukel, G. D. Rubenfeld, and A. B. Nathens. 2010. Survival of the fittest: The hidden cost of undertriage of major trauma. *Journal of the American College of Surgeons.* 211 (6): 804–811. DOI: 10.1016/j.jamcollsurg.2010.08.014.

Heron, R., A. Davie, R. Gillies, and M. Courtney. 2001. Interrater reliability of the Glasgow Coma Scale scoring among nurses in sub-specialties of critical care. *Australian Critical Care.* 14 (3):100–105.

Herrell, R. K., E. N. Edens, L. A. Riviere, J. L. Thomas, P. D. Bliese, and C. W. Hoge. 2014. Assessing functional impairment in a working military population: The Walter Reed Functional Impairment Scale. *Psychological Services.* 11 (3):254–264.

Hoey, B. A. and C. W. Schwab. 2004. Level I center triage and mass casualties. *Clinical Orthopaedics and Related Research.* 422:23–29.

Husum, H., and G. Strada. 2002. Injury Severity Score versus New Injury Severity Score for penetrating injuries. *Prehospital and Disaster Medicine.* 17 (1):27–32.

Jacobs, B., L. A. Young, H. Champion, M. Lawnick, M. Galarneau, V. Wing, and W. Krebs. 2012. Applying modeling and simulation to predict human injury due to a blast attack on a shipboard environment. In *Proceedings of the Human Factors and Ergonomics Society Annual Meeting,* vol. 56 (1), pp. 2359–2363. Sage.

Jennett, B. 2002. The Glasgow Coma Scale: History and current practice. *Trauma.* 4 (2):91–103. DOI: 10.1191/1460408602ta233oa.

Jennett, B. and M. Bond. 1975. Assessment of outcome after severe brain damage: A practical scale. *The Lancet.* 305 (7905): 480–484. DOI: 10.1016/S0140-6736(75)92830-5.

Jennett, B. and G. Teasdale. 1977. Aspects of coma after severe head injury. *The Lancet.* 309 (8017):878–881.

Johansen, K., M. Daines, T. Howey, D. Helfet, and S. T. Hansen, Jr. 1990. Objective criteria accurately predict amputation following lower extremity trauma. *The Journal of Trauma.* 30 (5): 568–573.

Kelly, J. F., A. E. Ritenour, D. F. McLaughlin, K. A. Bagg, A. N. Apodaca, C. T. Mallak, L. Pearse, et al. 2008. Injury severity and causes of death from Operation Iraqi Freedom and Operation Enduring Freedom: 2003–2004 versus 2006. *The Journal of Trauma and Acute Care Surgery.* 64 (2):S21–S27. DOI: 10.1097/TA.0b013e318160b9fb.

Kennedy, F., P. Gonzalez, C. Dang, A. Fleming, and R. Sterling-Scott. 1993. The Glasgow Coma Scale and prognosis in gunshot wounds to the brain. *The Journal of Trauma and Acute Care Surgery.* 35 (1): 75–77.

Kjorstad, R., B. W. Starnes, E. Arrington, J. D. Devine, C. A. Andersen, and R. M. Rush, Jr. 2007. Application of the Mangled Extremity Severity Score in a combat setting. *Military Medicine.* 172 (7): 777–781.

Knaus, W. A., E. A. Draper, D. P. Wagner, and J. E. Zimmerman. 1985. APACHE II: A severity of disease classification system. *Critical Care Medicine.* 13 (10): 818–829.

Knaus, W. A., D. P. Wagner, E. A. Draper, J. E. Zimmerman, M. Bergner, P. G. Bastos, C. A. Sirio, D. J. Murphy, T. Lotring, and A. Damiano. 1991. The APACHE III prognostic system. Risk prediction of hospital mortality for critically ill hospitalized adults. *Chest.* 100 (6):1619–1636. DOI: 10.1378/chest.100.6.1619.

Knaus, W. A., J. E. Zimmerman, D. P. Wagner, E. A. Draper, and D. E. Lawrence. 1981. APACHE-acute physiology and chronic health evaluation: A physiologically based classification system. *Critical Care Medicine.* 9 (8):591–597.

Knudson, P., C. A. Frecceri, and S. A. DeLateur. 1988. Improving the field triage of major trauma victims. *Journal of Trauma and Acute Care Surgery.* 28 (5): 602–606.

Kondo, Y., T. Abe, K. Kohshi, Y. Tokuda, E. F. Cook, and I. Kukita. 2011. Revised trauma scoring system to predict in-hospital mortality in the emergency department: Glasgow Coma Scale, Age, and Systolic Blood Pressure score. *Critical Care.* 15 (4):R191. DOI: 10.1186/cc10348.

Kung, W. M., S. H. Tsai, W. T. Chiu, K. S. Hung, S. P. Wang, J. W. Lin, and M. S. Lin. 2011. Correlation between Glasgow Coma Score components and survival in patients with traumatic brain injury. *Injury.* 42 (9):940–944.

Lavoie, A., L. Moore, N. LeSage, M. Liberman, and J. S. Sampalis. 2004. The New Injury Severity Score: A more accurate predictor of in-hospital mortality than the Injury Severity Score. *The Journal of Trauma.* 56 (6):1312–1320.

Lawnick, M. M., H. R. Champion, T. Gennarelli, M. R. Galarneau, E. D'Souza, R. R. Vickers, V. Wing, et al. 2013. Combat injury coding: A review and reconfiguration. *The Journal of Trauma and Acute Care Surgery.* 75 (4):573–581. DOI: 10.1097/TA.0b013e3182a53bc6.

Le, T. D., J. A. Orman, Z. T. Stockinger, M. A. Spott, S. A. West, E. A. Mann-Salinas, K. K. Chung, and K. R. Gross. 2016. The Military Injury Severity Score (mISS): A better predictor of combat mortality than Injury Severity Score (ISS). *The Journal of Trauma and Acute Care Surgery.* 81 (1):114–121. DOI: 10.1097/TA.0000000000001032.

Mac Donald, C. L., A. M. Johnson, L. Wierzechowski, E. Kassner, T. Stewart, E. C. Nelson, N. J. Werner, et al. 2014. Prospectively assessed clinical outcomes in concussive blast vs nonblast traumatic brain injury among evacuated US military personnel. *JAMA Neurology.* 71 (8):994–1002.

MacKenzie, E., W. Sacco, S. Luchter, J. Ditunno, C. F. Staz, G. Gruen, D. Marion, and W. Schwab. 2002. Validating the Functional Capacity Index as a measure of outcome following blunt multiple trauma. *Quality of Life Research.* 11 (8):797–808.

MacKenzie, E. J., A. Damiano, T. Miller, and S. Luchter. 1996. The development of the Functional Capacity Index. *The Journal of Trauma.* 41 (5):799–807.

MacKenzie, E. J., F. P. Rivara, G. J. Jurkovich, A. B. Nathens, K. P. Frey, B. L. Egleston, D. S. Salkever, and D.O. Scharfstein. 2006. A national evaluation of the effect of trauma-center care on mortality. *New England Journal of Medicine.* 354 (4):366–378.

McAnena, O. J., F. A. Moore, E. E. Moore, K. L. Mattox, J. A. Marx, and P. Pepe. 1992. Invalidation of the APACHE II scoring system for patients with acute trauma. *The Journal of Trauma and Acute Care Surgery.* 33 (4):504–507.

McCarthy, M. L., and E. J. MacKenzie. 2001. Predicting ambulatory function following lower extremity trauma using the functional capacity index. *Accident Analysis and Prevention.* 33 (6):821–831.

McCrea, M., M. Jaffee, K. Helmick, K. Guskiewicz, and S. Doncevic. 2009. *Validation of the Military Acute Concussion Evaluation (MACE) for In-Theater Evaluation of Combat-Related Traumatic Brain Injury.* Waukesha, WI: Waukesha Memorial Hospital.

McCrea, S., R. Bierley, J. I. Sheikh, and E. S. Date. 2000. Post-traumatic amnesia after closed head injury: A review of the literature and some suggestions for further research. *Brain Injury.* 14 (9):765–780. DOI: 10.1080/026990500421886.

Meares, S., E. A. Shores, A. J. Taylor, A. Lammél, and J. Batchelor. 2011. Validation of the abbreviated Westmead Post-traumatic Amnesia Scale: A brief measure to identify acute cognitive impairment in mild traumatic brain injury. *Brain Injury.* 25 (12):1198–1205.

Melanie Franklyn and Brian D. Stemper. 2008. *Abbreviated Injury Scale (AIS 2005 Update 2008).* Barrington, IL: Association for the Advancement of Automotive Medicine.

Moore, E. E., T. H. Cogbill, G. J. Jurkovich, J. W. McAninch, H. R. Champion, T. A. Gennarelli, M. A. Malangoni, S. R. Shackford, and P. G. Trafton. 1992. Organ Injury Scaling III: Chest wall, abdominal vascular, ureter, bladder and urethra. *The Journal of Trauma and Acute Care Surgery.* 33 (3):337–339.

Moore, E. E., T. H. Cogbill, G. J. Jurkovich, S. R. Shackford, M. A. Malangoni, and H. R. Champion. 1995. Organ Injury Scaling: Spleen and liver (1994 Revision). *The Journal of Trauma.* 38 (3) 323–324.

Moore, E. E., T. Cogbill, M. Malangoni, G. Jurkovich, H. Champion, T. Gennarelli, J. McAninch, et al. 1990. Organ injury scaling, II: Pancreas, duodenum, small bowel, colon, and rectum. *The Journal of Trauma.* 30 (11):1427–1429.

Moore, E. E., E. Dunn, J. Moore, and J. Thompson. 1981. Penetrating Abdominal Trauma Index. *The Journal of Trauma.* 21 (6):439–445.

Moore, E. E., M. A. Malangoni, T. H. Cogbill, S. R. Shackford, H. R. Champion, G. J. Jurkovich, J. W. McAninch, and P. G. Trafton. 1994. Organ Injury Scaling IV: Thoracic vascular, lung, cardiac, and diaphragm. *The Journal of Trauma and Acute Care Surgery.* 36 (3):299–300.

Moore, E. E., S. Shackford, H. Pachter, J. McAninch, B. Browner, H. Champion, L. Flint, T. Gennarelli, M. Malangoni, and M. Ramenofsky. 1989. Organ injury scaling: Spleen, liver, and kidney. *The Journal of Trauma and Acute Care Surgery.* 29 (12):1664–1666.

Morris, T. 2010. Traumatic brain injury. In *Handbook of Medical Neuropsychology. Applications of Cognitive Neuroscience*, eds. C. L. Amstrong and L. Morrow, pp. 17–32. Philadelphia, PA: Springer.

Murray, C. K., K. Wilkins, N. C. Molter, H. C. Yun, M. A. Dubick, M. A. Spott, D. Jenkins, et al. 2009. Infections in combat casualties during Operations Iraqi and Enduring Freedom. *The Journal of Trauma and Acute Care Surgery.* 66 (4):S138–S144. DOI: 10.1097/TA.0b013e31819d894c.

National Center for Health Statistics and Expert Group on Injury Severity Measurement. 2004. *Discussion Document on Injury Severity Measurement in Administrative Datasets.*

Newgard, C. 2012. Triage. In *Injury Research: Theories, Methods, and Approaches*, ed. G. Li and S. P. Baker, pp 297–315. New York: Springer.

Office of The Surgeon General and Borden Institute. 2013. Head injuries. In *Emergency War Surgery. Fourth United State Revision.* Fort Sam Houston, TX.

Ornato, J., E. J. Mlinek, Jr., E. J. Craren, and N. Nelson. 1985. Ineffectiveness of the trauma score and the CRAMS scale for accurately triaging patients to trauma centers. *Annals of Emergency Medicine.* 14 (11): 1061–1064.

Osler, T., S. P. Baker, and W. Long. 1997. A modification of the injury severity score that both improves accuracy and simplifies scoring. *The Journal of Trauma.* 43 (6):922–926.

Osler, T., R. Rutledge, J. Deis, and E. Bedrick. 1996. ICISS: An international classification of disease-9 based injury severity score. *The Journal of Trauma.* 41 (3):380–388.

Owens, B. D., J. F. Kragh, Jr., J. C. Wenke, J. Macaitis, C. E. Wade, and J. B. Holcomb. 2008. Combat wounds in operation Iraqi Freedom and operation Enduring Freedom. *The Journal of Trauma and Acute Care Surgery.* 64 (2):295–299.

Palmer, C. 2007. Major trauma and the Injury Severity Score—Where should we set the bar? *Annual Proceedings of the Association for the Advancement of Automotive Medicine.* 51: 13–29.

Palmer, C. S., B. J. Gabbe, and P. A. Cameron. 2016. Defining major trauma using the 2008. Abbreviated Injury Scale. *Injury.* 47(1):109–115.

Penn-Barwell, J. G., J. R. B. Bishop, and M. J. Midwinter. 2015. Refining the Trauma and Injury Severity Score (TRISS) to measure the performance of the UK Combat Casualty Care System. *The Bone and Joint Journal.* 97 (Suppl. 8):20–20.

Penn-Barwell, J. G., J. R. B. Bishop, S. Roberts, and M. Midwinter. 2013. Injuries and outcomes: UK military casualties from Iraq and Afghanistan 2003–2012. *The Bone and Joint Journal.* 95 (Suppl. 26): 1–1.

Polk, T. M. 2016. Editorial critique of The Military Injury Severity Score (mISS): A better predictor of combat mortality than Injury Severity Score (ISS). *The Journal of Trauma and Acute Care Surgery.* 81 (1):121. DOI: 10.1097/TA.0000000000001032.

Prasarn, M. L., D. L. Helfet, and P. Kloen. 2012. Management of the mangled extremity. *Strategies in Trauma and Limb Reconstruction.* 7 (2):57–66. DOI: 10.1007/s11751-012-0137-4.

Ramasamy, A., S. E. Harrisson, J. C. Clasper, and M. P. M. Stewart. 2008. Injuries from roadside improvised explosive devices. *The Journal of Trauma and Acute Care Surgery.* 65 (4):910–914.

Ramasamy, A., S. D. Masouros, N. Newell, A. M. Hill, W. G. Proud, K. A. Brown, A. M. J. Bull and J. C. Clasper. 2011. In-vehicle extremity injuries from improvised explosive devices: Current and future foci. *Philosophical Transactions of the Royal Society of London B: Biological Sciences.* 366 (1562): 160–170.

Rhee, K. J., W. G. Baxt, J. R. MacKenzie, N. H. Willits, R. E. Burney, R. J. O'Malley, N. Reid, D. Schwabe, D. L. Storer, and R. Weber. 1990. APACHE II scoring in the injured patient. *Critical Care Medicine.* 18 (8): 827–830.

Riechers, R. G., 2nd, A. Ramage, W. Brown, A. Kalehau, P. Rhee, J. M. Echlund, and G. S. Ling. 2005. Physician knowledge of the Glasgow Coma Scale. *Journal of Neurotrauma.* 22 (11):1327–1134. DOI: 10.1089/neu. 2005.22.1327.

Rosenfeld, J. V. 2002. Gunshot injury to the head and spine. *Journal of Clinical Neuroscience.* 9 (1):9–16.

Rosenfeld, J. V., and N. L. Ford. 2010. Bomb blast, mild traumatic brain injury and psychiatric morbidity: A review. *Injury.* 41 (5):437–443.

Rowley, G., and K. Fielding. 1991. Reliability and accuracy of the Glasgow Coma Scale with experienced and inexperienced users. *The Lancet.* 337 (8740):535–538.

Rutledge, R., S. Fakhry, E. Rutherford, F. Muakkassa and A. Meyer. 1993. Comparison of APACHE II, Trauma Score, and Injury Severity Score as predictors of outcome in critically injured trauma patients. *The American journal of Surgery.* 166 (3):244–247.

Sacco, W.J., E. J. MacKenzie, H. R. Champion, E. G. Davis, and R. F. Buckman. 1999. Comparison of alternative methods for assessing injury severity based on anatomic descriptors. *The Journal of Trauma and Acute Care Surgery.* 47 (3): 441–446.

Sampalis, J. S., R. Denis, A. Lavoie, P. Frechette, S. Boukas, A. Nikolis, D. Benoit, et al. 1999. Trauma care regionalization: A process-outcome evaluation. *The Journal of Trauma and Acute Care Surgery.* 46 (4): 565–581.

Sayer, N. A., C. E. Chiros, B. Sigford, S. Scott, B. Clothier, T. Pickett, and H. L. Lew. 2008. Characteristics and rehabilitation outcomes among patients with blast and other injuries sustained during the Global War on Terror. *Archives of Physical Medicine and Rehabilitation.* 89 (1):163–170.

Schluter, P. J. 2011. The Trauma and Injury Severity Score (TRISS) revised. *Injury.* 42 (1):90–96.

Schluter, P. J., C. M. Cameron, T. M. Davey, I. Civil, J. Orchard, R. Dansey, J. Hamill, et al.2009. Contemporary New Zealand coefficients for the trauma injury severity score: TRISS (NZ). *The New Zealand Medical Journal (Online).* 122 (1302):54–64.

Schluter, P. J., A. Nathens, M. L. Neal, S. Goble, C. M. Cameron, T. M. Davey, and R. J. McClure. Trauma and Injury Severity Score (TRISS) coefficients 2009 revision. 2010. *The Journal of Trauma and Acute Care Surgery.* 68 (4):761–770.

Schluter, P. J., R. Neale, D. Scott, S. Luchter, and R. J. McClure. 2005. Validating the functional capacity index: A comparison of predicted versus observed total body scores. *The Journal of Trauma and Acute Care Surgery.* 58 (2):259–263.

Seguí-Gómez, M., and F. J. Lopez-Valdes. 2012. Injury severity scaling. In *Injury Research: Theories, Methods, and Approaches*, ed. G. Li and S. P. Baker, pp. 281–295. New York: Springer.

Shin, E., K. N. Evans, and M. E. Fleming. 2013. Injury severity score underpredicts injury severity and resource utilization in combat-related amputations. *Journal of Orthopaedic Trauma.* 27 (7):419–423. DOI: 10.1097/ BOT.0b013e318279fa4f.

Shores, A. (2016). Copyright permission for Figure 4.2.

Shores, E. A., J. E. Marosszeky, J. Sandanam, and J. Batchelor. 1986. Preliminary validation of a clinical scale for measuring the duration of post-traumatic amnesia. *The Medical Journal of Australia.* 144 (11):569–572.

Sikic, N., Z. Korać, I. Krajacić, and J. Zunić. 2001. War abdominal trauma: Usefulness of Penetrating Abdominal Trauma Index, Injury Severity Score, and number of injured abdominal organs as predictive factors. *Military Medicine.* 166 (3):226.

Slauterbeck, J. R., C. Britton, M. S. Moneim, and F. W. Clevenge. 1994. Mangled Extremity Severity Score: An accurate guide to treatment of the severely injured upper extremity. *Journal of Orthopaedic Trauma.* 8 (4): 282–284.

Smith, B. P., A. J. Goldberg, J. P. Gaughan, and M. J. Seamon. 2015. A comparison of Injury Severity Score and New Injury Severity Score after penetrating trauma: A prospective analysis. *The Journal of Trauma and Acute Care Surgery.* 79 (2):269–274.

Soderstrom, C., and M. Seguí-Gómez. 2010. Comments and clarifications regarding improved characterization of combat injury. *The Journal of Trauma.* 69 (5):1311. DOI: 10.1097/TA.0b013e3181f2de32.

States, J. D., and D. C. Viano. 1990. Injury impairment and disability scales to assess the permanent consequences of trauma. *Accident Analysis and Prevention.* 22 (2):151–159.

Teasdale, G., and B. Jennett. 1974. Assessment of coma and impaired consciousness: A practical scale. *The Lancet.* 2 (7872):81–84. DOI: 10.1016/S0140-6736(74)91639-0.

Teasdale, G., and B. Jennett. 1976. Assessment and prognosis of coma after head injury. *Acta Neurochirurgica.* 34 (1–4):45–55.

Teasdale, G. M., L. E. Pettigrew, J. L. Wilson, G. Murray, and B. Jennett. 1998. Analyzing outcome of treatment of severe head injury: A review and update on advancing the use of the Glasgow Outcome Scale. *Journal of Neurotrauma.* 15 (8):587–597. DOI: 10.1089/neu.1998.15.587.

Teoh, L. S. G., J. R. Gowardman, P. D. Larsen, R. Green, and D. C. Galletly. 2000. Glasgow Coma Scale: Variation in mortality among permutations of specific total scores. *Intensive Care Medicine.* 26: 157–161.

Tepas, J. J., III, D. L. Mollitt, J. L. Talbert, and M. Bryant. 1987. The pediatric trauma score as a predictor of injury severity in the injured child. *Journal of Pediatric Surgery.* 22 (1):14–18.

Togawa, S., N. Yamami, H. Nakayama, Y. Mano, K. Ikegami, and S. Ozeki. 2005. The validity of the mangled extremity severity score in the assessment of upper limb injuries. *The Journal of Bone and Joint Surgery. British Volume.* 87 (11):1516–1519. DOI: 10.1302/0301-620X.87B11.16512.

Tohira, H., I. Jacobs, D. Mountain, N. Gibson, and A. Yeo. 2012. Systematic review of predictive performance of injury severity scoring tools. *Scandinavian Journal of Trauma, Resuscitation and Emergency Medicine.* 20 (1):1.

WHO (World Health Organization). 2011. *International Statistical Classification of Diseases and Related Health Problems (ICD-10).* 10th Revision. Volume 2. Instruction manual. 2010 Edition.

WHO (World Health Organization). 2016. *International Statistical Classification of Diseases and Related Health Problems (ICD-10-CM).* Preface to 2017 Edition.

Wong, S. S., and G. K. Leung. 2008. Injury Severity Score (ISS) vs. ICD-derived Injury Severity Score (ICISS) in a patient population treated in a designated Hong Kong trauma centre. *McGill Journal of Medicine.* 11 (1):9–13.

Wright, J. 2011. Glasgow outcome scale. In *Encyclopaedia of Clinical Neuropsychology*, eds. J. Kreutzer, J. DeLuca, and B. Caplan, pp. 1150–1152. New York: Springer-Verlag.

Yates, D., M. Woodford, and S. Hollis. 1992. Preliminary analysis of the care of injured patients in 33 British hospitals: First report of the United Kingdom major trauma outcome study. *British Medical Journal.* 305 (6856):737–740. DOI: 10.1136/bmj.305.6856.737.

Zouris, J. M., G. J. Walker, J. Dye, and M. Galarneau. 2006. Wounding patterns for US Marines and sailors during Operation Iraqi Freedom, major combat phase. *Military Medicine.* 171 (3):246–252.

5 The Medical Management of Military Injuries

*Damian Keene, Peter Mahoney, Johno Breeze
and Arul Ramasamy*

CONTENTS

5.1   INTRODUCTION

5.1.1   TRAUMA LOADS EXPERIENCED BY COALITION HOSPITALS

The most common mechanisms of injury during recent conflicts has been blast injury and gunshot wounds (Penn-Barwell et al. 2015). The anatomical distribution and incidence of wounds reflects the surface area, the type of conflict, the causative mechanism and any personal protective equipment (PPE) worn (Table 5.1). Conflicts also pose a significant risk to non-combatants and civilians of all ages. Military planning must take into account the immediate surgical treatment and subsequent care and repatriation of significant numbers of casualties including young children, women and expectant mothers.

TABLE 5.1

Incidences of Wounds to Body Areas of Wounded Coalition (US and UK) Soldiers Compared to Civilians and Afghan National Army (ANA) and Afghan National Police (ANP) Treated in the Coalition Field Hospital Kandahar Afghanistan 2007–2008

Body Area	Coalition Forces (%)	Local Nationals (%)	ANA/ANP (%)
Lower limb	35	26	35
Upper limb	24	34	27
Torso	10	21	17
Face and neck	24	11	16
Head	7	5	4
Eyes	0	3	1

Source: Breeze, J., et al., *Br. J. Oral Maxillofac. Surg.*, 49, 464–68, 2011.

5.1.2 TRENDS IN SURVIVAL AND MORTALITY

Improvements in combat casualty care have resulted in a sustained improvement in survival (Russell et al. 2014; Penn-Barwell et al. 2015). At the end of recent operations in Iraq and Afghanistan, casualties surviving to hospital care had an extremely high chance of survival of 97.6% (Keene et al. 2016). Fatality rates for combat injury in Afghanistan and Iraq was roughly half that of Vietnam and one-third that of World War II (Holcomb et al. 2006). The reasons for this include damage control resuscitation (DCR) (Langan et al. 2014), the re-emergence of the tourniquet, development of novel haemostatics and improved evacuation timelines (Morrison et al. 2013; Penn-Barwell et al. 2015).

Head injury remains the most common cause of death (83%) in those with non-survivable injury (Martin et al. 2009; Eastridge et al. 2011, 2012; Langan et al. 2014; Keene et al. 2016). Truncal or junctional haemorrhage is the commonest cause of death in casualties with potentially survivable injury (Eastridge et al. 2011; Morrison et al. 2013). For casualties who make it to hospital but later die from their wounds, the majority of deaths result from primary brain injury (injury occurring due to tissue damage at the time of incident) with the majority of deaths occurring within 24 hours of admission (Eastridge et al. 2012; Langan et al. 2014; Keene et al. 2016). There is currently a research focus directed towards preventative regimes (including body armour and helmet development) and pre-hospital medical treatment of head injury and haemorrhage control in an attempt to improve outcome in these casualties (Long et al. 2015).

5.1.3 THE EFFECT OF PERSONAL PROTECTIVE EQUIPMENT (PPE) ON INJURY PATTERNS

The modern soldier deploys on operations with a complex combination of PPE that is primarily designed to reduce the penetration of ballistic projectile (Breeze, Allanson-Bailey, et al. 2015). In addition, there is a requirement in the UK for the head to be protected from blunt impact (Hardaway 1978). PPE has traditionally been issued in the form of a helmet to protect the brain and body armour to protect the thorax and abdomen. The earliest design of body armour comparable to that worn today was issued to US forces in Korea, but it was in Vietnam that a clear reduction in the incidence of thoraco-abdominal injury was first demonstrated by personnel who wore it (reducing the incidence of injury from 20% to 10%). The vest covered the thorax and abdomen and consisted of a single type of ballistic protective material (originally nylon), with the aim of providing protection against fragmenting munitions. This 'soft armour' remains the primary component of body armour and is made of a flexible fabric, most commonly a para-aramid. As threats evolved, protection against high-velocity rifle bullets was desired, which continues to be achieved through the incorporation of ceramic plates, i.e. 'hard armour' (Breeze, Lewis, Fryer, et al. 2016). Despite these changes, the incidence of chest injury has remained stable, even within recent conflicts in Iraq and Afghanistan.

It was only during the Iraq conflict at the start of the 21st century that additional PPE began to be developed and issued. Initially, these were methods of increasing the area of coverage of body armour against energised fragments and included upper arm and neck protection. However, they were rarely used due to discomfort and thermal burden and, therefore, no evidence for their effectiveness in reducing injury has ever been demonstrated. In 2006, ballistic eye protection (Figure 5.1) was introduced to UK forces and represented a paradigm shift in terms of the nature of protection with the primary aim being to reduce morbidity instead of mortality. This was followed up by the introduction of pelvic protection in 2010, in an attempt to reduce the burden from genitourinary injury (Figure 5.2). Currently, no studies have quantified the exact effect of pelvic protection in terms of numbers, but surgeons operating on those wearing it have reported large reductions in the degree of injury in those wearing it (Breeze, Allanson-Bailey, et al. 2015). There is currently a drive to optimise the coverage provided by different types of protection which is likely to see further changes in the design of PPE worn in the future (Breeze, Lewis, Fryer, et al. 2016).

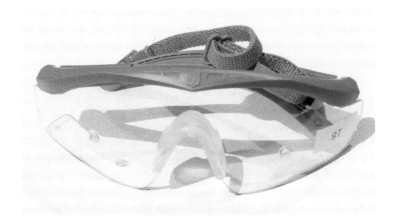

FIGURE 5.1   Low impact ballistic glasses designed to prevent the penetration of energised fragments into the eyes.

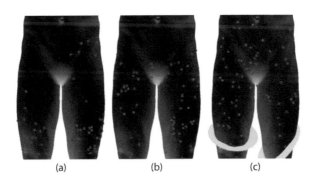

FIGURE 5.2   Surface wound mapping demonstrating lower incidence of entry wound locations in (a) coalition soldiers wearing pelvic protection (b) coalition soldiers not wearing it (c) or those in the Afghan National Army and Afghan National Police not wearing it. (Adapted from Breeze et al. 2015. With permission.)

5.2 PHYSIOLOGY OF INJURY

5.2.1 NORMAL RESPONSE TO INJURY

In order to maintain normal function, cells require oxygen for metabolic activity (aerobic respiration) and a tightly regulated environment. Both tissue pH and temperature are tightly controlled to allow optimal enzyme function and therefore cell metabolism. If oxygen consumption exceeds delivery, anaerobic respiration occurs, resulting in the production of lactic acid. If oxygen delivery increases (or the metabolic stress is reduced) aerobic respiration recommences and the lactic acid is metabolised, removing it from the blood.

5.2.2 PHYSIOLOGICAL EFFECTS OF INJURY

Traumatic injury can lead to disruption of any body tissue, resulting in abnormal function. For example, chest injury can cause ventilatory compromise leading to poor blood oxygenation, while tissue disruption can result in severe blood loss, both of which reduce oxygen delivery to the tissues resulting in anaerobic cell respiration. Initially, the body can compensate for these effects, maintaining tissue pH by buffering the effects of the lactic acid to maintain enzyme function. A fall in blood pressure leads to sympathetic stimulation, resulting in increasing cardiac output (through a rise in heart rate and contractility) and arterial and venous constriction the aim being to increase blood flow to the vital organs (Kirkman and Watts 2014). This cannot occur indefinitely if the physiological insult continues: the system will be overwhelmed, resulting in a change in body pH or 'metabolic acidosis' and eventual cardiovascular collapse (Hoyt et al. 1994).

The acidosis also affects the function of key enzymes responsible for clotting, resulting in a coagulopathy (reduction in the ability of the blood to clot), in turn this leads to further blood loss, further reducing tissue perfusion, worsening the metabolic acidosis. Poor tissue perfusion also leads to hypothermia, again worsening enzyme function. The interplay of these three factors is known as the 'lethal triad' (Figure 5.3), consisting of acidosis, coagulopathy and hypothermia that if allowed to continue results in death (Kashuk et al. 1982).

The lethal triad can be worsened by external factors (Figure 5.3), and hypothermia is worsened by removal of casualties' clothes to find injuries and the administration of cold fluids. Coagulopathy is worsened by the consumption of clotting factors and by the administration of fluids which do not contain clotting factors (dilutional coagulopathy).

In the last decade, scientific and clinical evidence has demonstrated that coagulopathy will develop independently of the lethal triad as part of the primary physiological response to trauma. The mechanism of this is unclear, but it is believed to be due to poor oxygen delivery resulting in activation of the

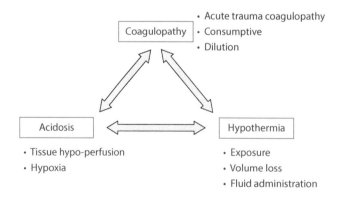

FIGURE 5.3   The Lethal Triad. (From Kashuk, J.L., et al., *J. Trauma Acute Care Surg.*, 22(8), 672, 1982.)

vascular endothelium. It is proposed that this activates protein C, a naturally occurring anticoagulant, resulting in hyperfinbrinolysis, i.e. excessive clot breakdown (Brohi et al. 2007). This is known as acute trauma coagulopathy (ATC) and is associated with significantly increased mortality (Brohi et al. 2003). ATC combined with the effects of the lethal triad on coagulation is referred to as trauma induced coagulopathy (TIC) (Brohi 2009).

5.2.3 AIMS OF TREATMENT

The key to breaking the vicious cycle of the lethal triad in severe haemorrhage is to control blood loss or 'turn off the tap'. This has led to a significant shift in the treatment priority at all stages of treatment from an airway, breathing circulation approach (ABC) to a catastrophic haemorrhage, then an ABC approach (<C>ABC) (Hodgetts et al. 2006).

Our increased understanding of the physiological effects of trauma has led to the introduction of DCR, drawing together multiple innovations in to one standardised approach. DCR is defined by the UK Defence Medical Services as 'a systemic approach to major trauma combining the <C>ABC paradigm with a series of clinical techniques from point of wounding to definitive treatment in order to minimise blood loss, maximise tissue oxygenation and optimise outcome (Hodgetts and Mahoney 2007). The key aim is to address the lethal triad and, more recently, target the ATC of traumatic injury as early as possible to attempt to mitigate its effects on mortality.

5.3 MILITARY PRE-HOSPITAL CARE – CARING UNDER THREAT

As with all medical systems, military combat casualty care follows a chain of survival (Figure 5.4). Care delivered at each stage follows an approach defined by Battlefield Advanced Trauma Life Support (BATLS) doctrine and training. This lays out the unifying principles and approach to casualty management within the UK Defence Medical Services. Care at any point is based on the <C>ABC

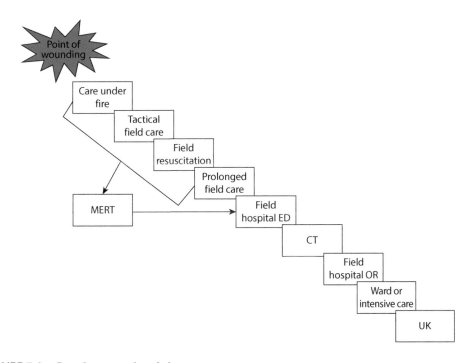

FIGURE 5.4   Casualty evacuation chain.

approach but with increasingly advanced interventions and decision making becoming available further down the chain.

5.3.1 CARE AT POINT OF WOUNDING

The first link in the chain is the delivery of care at the point of wounding. Treatment at this stage is initially limited to control of catastrophic haemorrhage and addressing airway obstruction, and is known as care under fire (CUF). If able, CUF is self-administered by the wounded soldier or else by a fellow soldier (buddy–buddy aid).

All treatments at this stage are designed to be rapidly applicable and self-sustaining as the primary role of these soldiers will be to continue their combat role. Soldiers are trained to place casualties with airway compromise in a prone position to improve airway opening and allow postural drainage of any blood. Other classically taught airway manoeuvres require sustained hands-on application to maintain airway patency such as the 'head tilt chin lift' where the neck is held flexed with the head extended. One in four soldiers is trained as a 'team medic': they receive further medical training and equipment to help address time-critical injuries.

The level of medical care delivered is inversely related to the tactical situation. As the threat decreases, the environment is deemed to become semipermissive, allowing more involved medical care to be undertaken. This next stage of care is termed tactical field care (TFC) and is performed once the immediate threat is removed: this may be by suppressing the enemy or evacuation to cover; thus providing a semipermissive environment. This is not a fully permissive environment and care is still limited to that required to ensure safe onward evacuation. The line between CUF and TFC is not distinct; they are, in fact, the two ends of a continuous spectrum that involves the delivery of clinical care with continuous tactical risk assessment as the tactical situation could deteriorate at any time.

TFC involves a rapid assessment following the <C>ABC approach; each part is addressed in turn this is known as 'vertical resuscitation'. A systematic approach is employed to reduce the risk of becoming distracted by overt injuries and ensures time-critical injuries are identified first.

Treatment not only removes the casualty from the firefight but also removes those delivering care and physically evacuating them. It is vital that they are evacuated rapidly to allow all soldiers to return to their primary function of ongoing combat operations. The casualties' injury severity is determined following a universal triage algorithm that is issued to all soldiers (Figure 5.5). Following the triage card ensures standardised casualty prioritisation for evacuation in what is likely to be a stressful situation. The triage card also allows the soldier to be able to identify if the casualty is dead.

5.3.2 HAEMORRHAGE CONTROL

As part of CUF and TFC training, soldiers are taught a sequential approach to control of external haemorrhage (Figure 5.6) with each soldier carrying pressure dressings and combat application tourniquets (CAT). In addition, team medics carry CELOX™ gauze: CELOX™ is one of many available topical hemostatic dressings that have been developed as a consequence of the recent conflicts. It contains chitosan, a structural element of shell-fish exoskeleton that increases the speed of clot formation. This is particularly useful in control of junctional bleeding where tourniquets application is not possible.

In a semipermissive environment, haemorrhage control follows this step-wise approach; however, ultimately all medical treatment will be dictated by the tactical situation. Even though an injury may be amenable to direct pressure, if the tactical situation requires, a tourniquet may be applied as it is most likely to guarantee rapid control of bleeding (Brodie et al. 2007), freeing the casualty or the treating soldier to return fire.

In order for this approach to be successful, the haemorrhage needs to be amenable to compression. Chest injuries and abdominal injuries as well as some junctional injuries (groin and axilla) may

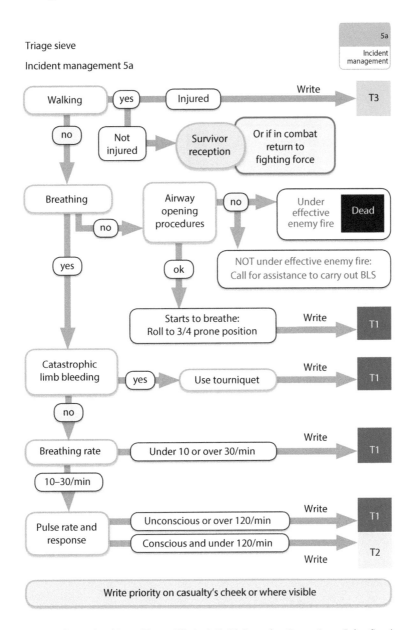

Triage sieve

Incident management 5a

FIGURE 5.5 Triage sieve algorithm. (From *Clinical Guidelines for Operations*, Joint Services Publication 999, 2012.)

not be controllable. These casualties require rapid recognition and evacuation to a surgical facility to allow surgical control.

5.3.3 Intra-Osseous Access

Rapid administration of intravenous fluids may be necessary to sustain life in the pre-hospital environment, particularly if casualty evacuation is prolonged. Intravenous access via peripheral veins maybe difficult due to peripheral shutdown and extremity injury. Intra-osseous access (IO), where a needle is inserted into the bone marrow, has been used routinely in paediatric practice for many years.

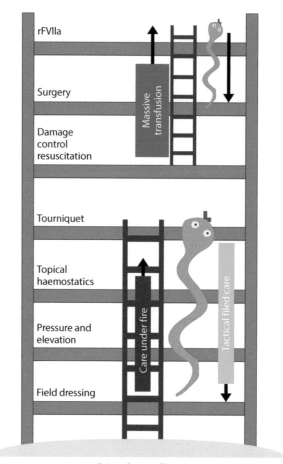

FIGURE 5.6 The haemostatic ladder. Under normal circumstances, there is progression from bottom to top of the ladder considering each intervention sequentially. However, during 'care under fire' effective direct/ indirect enemy fire, it is appropriate for catastrophic limb bleeding to immediately apply a tourniquet BUT to reassess its requirement during 'tactical field care' (firefight won), the snake takes the user back to using a field dressing, pressure and elevation at this point. (From Moorhouse, I., et al., *J. Roy. Army Med Corp.*, 153(2), 99–101, 2007.)

This offers a solution where IV insertion maybe difficult; two devices are currently used, EZ-IO® and FAST1®. EZ-IO® can be used in paediatric and adult patients and has been employed to good effect on recent operations at all stages of casualty care; it allows siting of IO access in both the upper and lower limb. FAST1® is designed for insertion into the sternum only; this is advantageous if there are multiple extremity injuries, especially given the protection afforded to the sternum by modern body armour (making the sternal injury highly unlikely).

5.3.4 FIELD RESUSCITATION AND PROLONGED FIELD CARE

The next stage of care is delivered in more secure locations referred to as regimental aid posts (RAP). As with all stages of treatment, a <C>ABC approach is employed with a further level of clinical equipment and skill being added. This includes basic monitoring and portable oxygen. This stage is known as field resuscitation and follows the BATLS system. For less severe casualties, these facilities can allow stabilisation and provide a holding capacity whilst further evacuation is coordinated. Evacuation from the point of wounding to an RAP is not always possible–direct air evacuation maybe necessary. This is discussed below.

5.4 ADVANCED PRE-HOSPITAL CARE

The distance between combat outposts and lack of secure road routes between them in Iraq and Afghanistan precluded evacuation of casualties by road. From early on in these conflicts, casualty evacuation was performed by support helicopter; the medical capability of the platform was variable and often consisted of junior personnel. Increasing timelines and casualty load/severity drove the innovation of the Medical Emergency Response Team (MERT)-enhanced concept. This provided a standardised team with a predefined medical capability operating within a set framework of standardised operating procedures. The primary principle of MERT is to project forward the capability of the hospital resuscitation teams, allowing earlier delivery of interventions usually restricted to hospital facilities with the aim of improving outcome (Morrison et al. 2013). The platform has allowed safe delivery of pre-hospital anaesthesia–the gold standard for securing patients airway and the delivery of pre-hospital blood products (Kehoe et al. 2011; O'Reilly et al. 2014). Both of these interventions form part of DCR concept of optimising tissue oxygenation as early as possible to mitigate the effects of acidosis and treat the coagulopathy.

The medical team consists of a consultant in Emergency Medicine or Anaesthesia with pre-hospital training, a registered nurse and 1–2 paramedics. This provides not only the skill set to provide the advanced interventions but also provides the higher level decision making required in the management of these casualties. The increase in the number of medical personnel present allows a shift towards 'horizontal resuscitation': this refers to the process of simultaneous and coordinated treatment rather than the sequential approach employed during the initial treatment phases.

It is important to think of MERT as a concept rather than a particular team on a particular platform. In future, operations both the team composition and vehicle used may vary, but the principle will remain the same: projection of hospital capability to allow early casualty stabilisation.

5.5 HOSPITAL RESUSCITATION

5.5.1 Patient Pathway

DCR continues with the transfer of the casualty to a field hospital; on arrival the casualty will be rapidly assessed by a pre-assembled trauma team (Table 5.2). This is similar to civilian practice, however, there are some key differences, and the most significant is that all physicians are at a consultant level (Mercer et al. 2013).

Management follows the <C>ABC paradigm using a horizontal approach of simultaneous assessment and treatment performed by multiple clinicians, significantly reducing the time needed to assess and deliver treatment. In order for this approach to be successful, a team leader is needed to coordinate efforts and maintain overall situational awareness (Mercer et al. 2010). As the treatments become more advanced, it is easy for people delivering direct clinical care to become task focused. The aim in the initial phase is twofold: (1) stabilise the casualty and (2) gain further clinical information to guide future decision making. The majority of casualties will be stabilised adequately to allow transfer to computed tomography (CT), providing detailed imaging of all injuries allowing advanced planning of any required surgical interventions (Watchorn et al. 2013).

If haemorrhage cannot be controlled, the casualty is moved directly to the operating theatre for immediate surgical control to be gained (Tai and Russell 2011). Surgery will be kept to a minimum during this phase and follows the damage control surgical principles set out in the sections below.

Once initial surgery is complete, a decision is made on the degree of physiological stability of the casualty. If necessary, they can be kept anaesthetised and moved to the intensive care unit for

TABLE 5.2
The Deployed Complex Trauma Team

Non-consultants	Consultants
Primary survey doctor (ED junior doctor)	Team leader (emergency consultant)
Operating department practitioner (ODP)	Anaesthetist 1 (airway)
Scribe (trauma nurse coordinator)	Anaesthetist 2 (central venous access)
ED nurse 1 (IV access and first blood sample)	Orthopaedic surgeon
ED nurse 2 (drugs)	General surgeon
ED nurse 3 (rapid infuser)	Plastic surgeon
ED nurse 4 (rapid infuser)	Radiologist
Runner	Deployed medical director
Radiographer	
Laboratory technician	
Theatre manager	
Interpreter	

Source: Adapted from Mercer, S.J., et al., *Anaesthesia*, 68(Suppl. 1), 49–60, 2013.
ED = emergency department, IV = intravenous.

ongoing organ support. Resuscitation continues until normal physiological parameters are achieved or until the casualty is evacuated back to the receiving hospital in the United Kingdom (for British military personnel).

Management throughout this phase follows DCR principles and has three key elements: permissive hypotension, damage control surgery and haemostatic resuscitation (Hodgetts and Mahoney 2007).

5.5.2 Permissive Hypotension

If bleeding cannot be controlled, fluid is only administered if there is loss of a radial pulse or a systolic blood pressure less than 80 mmHg (approximately 80% of normal values): this is termed permissive hypotension (Revell et al. 2002). The aim is to maintain perfusion of vital organs without causing clot disruption by raising the blood pressure and to minimise the administration of crystalloid fluids. Crystalloid fluids contain electrolytes similar to that of the plasma but no clotting factors, platelets or red blood cells. In the case of penetrating injury, evidence from civilian practice has demonstrated the effectiveness of minimising fluid administration in the pre-hospital setting (Bickell et al. 1994). Intraoperative permissive hypotension reduces blood loss, fluid administration and the degree of coagulopathy (Morrison et al. 2011).

Permissive hypotension should be thought of as a state of 'controlled' shock where a significant proportion of the bodies' tissues will deliberately be inadequately perfused, resulting in ongoing acidosis. It is vital that as soon as bleeding is controlled, a normal blood pressure should be targeted to reverse the acidosis (Morrison et al. 2013). It is important to remember that not all bleeding will require surgical intervention; application of tourniquets to extremity injuries may result in complete haemorrhage control.

Permissive hypotension in casualties with concomitant head injury is controversial as episodes of hypotension are associated with increased mortality in this group (Manley et al. 2001). Normal management in a casualty with head injury would be to target a 'normal' blood pressure. Head injury can lead to a rise in intracranial pressure due to intracerebral bleeding or tissue swelling within the closed box of the cranial vault. The use of permissive hypotension in these casualties must be constantly balanced against the risk of secondary brain injury (injury that occurs after the primary tissue injury).

The benefits of prolonged permissive hypotension in blast injury have been questioned. Animal studies have demonstrated it cannot be maintained indefinitely and a point will be reached where the physiological effects of shock cannot be reversed, resulting in death (Garner et al. 2010). A hybrid technique has been demonstrated to overcome this: permissive hypotension is allowed for the first 60 minutes, a normal blood pressure is targeted after this point by the administration of blood products irrespective of haemorrhage control (Watts et al. 2015).

5.5.3 Haemostatic Resuscitation

Hemostatic resuscitation consists of two phases: fluid administration before and after control of bleeding. The aim is to restore tissue perfusion by replacement of blood volume, ensure optimal oxygen delivery by the replacement of red blood cells, and reverse coagulopathy by the replacement of clotting factors and platelets.

During uncontrolled haemorrhage, volume loss needs to be restored quickly in order to prevent death. Since the 1980s, donated blood has been separated into four basic components: packed red blood cells (PRBC), fresh frozen plasma (FFP), platelets and cryoprecipitate. This was done to allow the administration of specific components without the risk of exposing patients to all aspects of donated blood. Prior to this, whole blood was used during trauma resuscitation. Evidence from military and civilian practice has demonstrated that resuscitating with blood products in a 1:1:1 ratio, platelets, FFP and PRBC (effectively reconstituting the donated blood), improves mortality (Borgman et al. 2007; Holcomb et al. 2015). During this initial phase, blood products are given following permissive hypotension principles in a fixed ratio to maintain adequate tissue perfusion.

Once bleeding is controlled, a more tailored or 'goal directed' approach is employed to both the volume and the type of blood products administered. At this stage, blood pressure and heart rate, the classical markers of resuscitation, may well be normal, but this is often due to extreme vaso- and venoconstriction from sympathetic stimulation. At this stage Fentynl, a potent opioid drug, is administered, this blunts the sympathetic drive of the casualty, reducing heart rate and causing vaso and venodilation (reversing the effects of the sympathetic response to injury), resulting in a drop in blood pressure. PRBC and FFP (1:1) are then given to restore the blood pressure. The aim of this process is to restore the casualties' physiological reserve by restoring their normal blood volume, allowing them to tolerate further blood loss that may occur. If this was not done, further bleeding would result in anaerobic metabolism and rapid deterioration. The degree of metabolic acidosis (due to lactic acid) is used as a marker of adequate volume resuscitation (rather than heart rate and blood pressure), the target being normalisation, this is tested at least every 30 minutes, more frequently if the casualty is unstable (Keene et al. 2013).

Coagulopathy is assessed using rotational thromboelastometry, a point of care test that provides assessment of whole blood clotting with useable results within 5–10 minutes of initiation (Woolley et al. 2013). A blood sample is placed in a fixed cup and regents added to initiate clotting, a pin rotates back and forth in the sample, as the clot forms the resistance to the pins rotation increases this produces the classic champagne glass appearance in normal samples. An increased time to clot initiation, or a reduced maximum width, indicates coagulopathy. It has been demonstrated that ROTEM® values correlate well with standard laboratory tests (Rugeri et al. 2007) and that results are available significantly faster (Doran et al. 2010). This reduces the time between the sample being taken and any intervention being delivered. In the patient with ongoing blood loss, rapid test turnaround is vital to avoid results becoming historical (Keene et al. 2013).

Haemostatic resuscitation is not limited to the administration of blood products; attention needs to be paid to all aspects of the lethal triad. Hypothermia needs to be aggressively corrected with active warming of the casualty. Calcium is a key component of the coagulation system: hypocalcaemia (low calcium levels) will lead to coagulopathy even in the presence of normal clotting factor levels. It is vital to maintain calcium at normal levels; during uncontrolled bleeding, calcium is administered at predetermined intervals. Once bleeding has been controlled, levels are checked alongside the degree of metabolic acidosis and corrected as necessary.

Tranexamic acid (TXA) is administered, if within 3 hours of injury: this is a pharmacological agent that blocks the action of plasminogen. Plasminogen is an enzyme present in the blood responsible for clot breakdown or 'fibrinolysis' – excessive fibrinolysis is associated with high mortality in trauma (Raza et al. 2013). TXA has been shown to significantly reduce mortality in the bleeding trauma casualty (CRASH-2 trial collaborators et al. 2010; J. J. Morrison 2012).

5.6 PRINCIPLES OF WAR SURGERY

Damage control surgery (DCS) is an operative strategy that sacrifices the completeness of the immediate surgical repair in an attempt to address the inevitable physiological dysfunction that accompanies significant ballistic trauma and surgery (Fries and Midwinter 2010). The philosophy of DCS was devised to counter the adverse effects of prolonged surgery on casualties already compromised by the effects of the lethal triad. Previously considered a separate phase to initial resuscitation, DCS has become an integral component within DCR of the trauma patient (Figure 5.7).

The priorities for surgery are haemorrhage control, limiting contamination and temporary closure or cover for onward evacuation where possible. Haemorrhage control is achieved by proximal vessel control followed by either ligation, suture or shunting of damaged vessels. Contamination control is achieved by closure of the ends of hollow viscus (e.g. bowel) and placement of drainage tubes. Solid organ injury can be managed by removal (e.g. spleen or single kidney injury) or therapeutic packaging, to apply direct pressure, such as in the liver. In pelvic injuries, extraperitoneal pelvic packing following external fixation of the pelvis can be performed to gain control of bleeding.

The duration of surgery is determined by the physiologic state of the patient: communication between the surgical and anaesthetic team is vital in order to ensure informed decisions are made. Attempts at revascularisation of limbs or performing complex intra-abdominal reconstruction must be deferred until the physiology of the patient has been corrected. Only when life-threatening haemorrhage has been controlled and the initial period of physiological instability overcome should wound debridement be considered. In cases of severe physiological instability, it must be recognized that this may require prolonged resuscitation on the critical care unit.

For patients being evacuated to the UK, clear communication of the operative interventions that have been performed and those that will be required is essential. Due to the absence of surgical facilities in-flight, drains should be firmly secured to the patient. Consideration should be given to pre-empt the development of abdominal or limb compartment syndrome. Compartment syndrome occurs

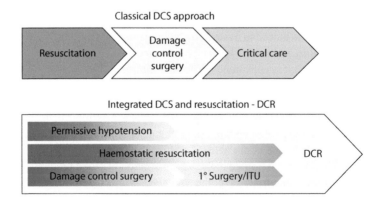

FIGURE 5.7 The paradigm shift from DCS to DCR. (Adapted from Fries, C.A., and Midwinter, M.J., *Surgery (Oxford)*, 28(11), 563–567, 2010.)

when the pressure in a sealed body compartment increases, thereby reducing blood flow to the tissues within the compartment. The resultant poor tissue perfusion leads to tissue necrosis and further swelling, generating a downward spiral of increasing compartment pressure. The management is to surgically open the compartment; in the abdomen, this is the abdominal wall (laparostomy), in the limbs, the fascia surrounding the effected muscles (fasciotomy).

5.7 ONGOING CARE: EVOLVING INJURIES AND ROYAL CENTRE FOR DEFENCE MEDICINE (RCDM) INCLUDING REHABILITATION

The reconstruction of complex ballistic trauma follows a hierarchy of management goals which aim to achieve (Ramasamy et al. 2013):

1. A clean non-contaminated wound
2. Removal of all dead tissue
3. Stabilisation of fractures
4. Timely coverage of exposed tendon and bone
5. Provision of robust, well vascularised coverage over nerve grafts and musculo-tendinous units
6. Provision of robust soft-tissue coverage of amputation stumps

It is usually not possible to reliably debride highly contaminated wounds in a single sitting, and severe complications can result from premature flap closure of open fractures in these injuries. It is important to appreciate that the wound will evolve from the time of injury. Tissue that was initially considered viable may become necrotic. Factors such as infection, particularly fungal infection, can result in serial amputation of limbs.

The timing of reconstruction must be dictated by the clinical condition of the patient. The decision to embark upon immediate complex reconstruction in an acutely sick patient, or to obtain wound closure by simpler methods, poses a challenging problem to experienced clinicians. It is worthwhile to ask three questions: What can the patient tolerate? What must be done? And what can wait? (Ramasamy et al. 2013). The development of a tailored reconstruction plan is vital to ensure there is coordinated activity between the different surgical specialties to prevent multiple trips to the operating theatre and to optimise treatment. The role of a multidisciplinary team approach involving surgeons, pain specialists, microbiologists, rehabilitation specialists and therapists cannot be underestimated and is central to this process.

In addition to their inpatient care, weekly conferences with the rehabilitation specialists allow rehabilitation plans to be formulated prior to the patients discharge and allows a comprehensive system to identify problems and to address them early. It is important to address not only the physical issues but also the psychological burden (both the patient and their families) that these life-changing injuries bring. Ensuring that appropriate psychological support is provided throughout the patient journey which is vital in improving overall functional outcome.

For these multiply injured patients, it is important to note that their care may evolve over several years, and they could require reconstructive surgery for many years after the initial injury. Where limb reconstruction has been attempted, the functional disability associated it with may necessitate late amputation. Although some studies have shown improved functional scores (Short Form, SF-36) in military patients who have undergone amputation compared to limb salvage in the short term, the long-term effects of amputation are yet to be fully understood (Doukas et al. 2013).

5.8 CONCLUSION

The recent conflicts in Afghanistan and Iraq have resulted in a severe injury burden: this has led to rapid advances in both medical techniques and systems of management that has resulted in the highest survival rates to date from combat injury (Penn-Barwell et al. 2015). The key to this has been constant adaption of the system based on continuous review of outcomes (Hodgetts et al. 2007). No one intervention can account for the improved survival, rather, it is the summed effect of the system parts, achieved through attention to detail at all stages from initial injury to discharge from hospital.

REFERENCES

Bickell, W. H., M. J. Wall, P. E. Pepe, et al. 1994. Immediate Versus Delayed Fluid Resuscitation for Hypotensive Patients with Penetrating Torso Injuries. *The New England Journal of Medicine* 331 (17): 1105–9. doi:10.1056/NEJM199410273311701.

Borgman, M. A., P. C. Spinella, J. G. Perkins, et al. 2007. The Ratio of Blood Products Transfused Affects Mortality in Patients Receiving Massive Transfusions at a Combat Support Hospital. *The Journal of Trauma: Injury, Infection, and Critical Care* 63 (4): 805–13. doi:10.1097/TA. 0b0 13e3181271ba3.

Breeze, J., L. S. Allanson-Bailey, A. E. Hepper, et al. 2015. Demonstrating the Effectiveness of Body Armour: A Pilot Prospective Computerised Surface Wound Mapping Trial Performed at the Role 3 Hospital in Afghanistan. *Journal of the Royal Army Medical Corps* 161 (1): 36–41. doi:10.1136/jramc-2014-000249.

Breeze, J., A. Gibbons, J. Combes, and A. Monaghan. 2011. Oral and Maxillofacial Surgical Contribution to 21 Months of Operating Theatre Activity in Kandahar Field Hospital: 1 February 2007–31 October 2008. *The British Journal of Oral and Maxillofacial Surgery* 49: 464–8.

Breeze, J., E. A. Lewis, R. Fryer, et al. 2016. Defining the Essential Anatomical Coverage Provided by Military Body Armour Against High Energy Projectiles. *Journal of the Royal Army Medical Corps* 162 (4): 284–90. doi:10.1136/jramc-2015-000431.

Brodie, S., T. J. Hodgetts, J. Ollerton, et al. 2007. Tournequet Use in Combat Trauma: UK Military Experience. *Journal of the Royal Army Medical Corps* 153 (4): 310–13.

Brohi, K. 2009. Trauma Induced Coagulopathy. *Journal of the Royal Army Medical Corps* 155 (4): 320–22.

Brohi, K., M. J. Cohen, M. T. Ganter, et al. 2007. Acute Traumatic Coagulopathy: Initiated by Hypoperfusion. *Annals of Surgery* 245 (5): 812–18. doi:10.1097/01.sla.0000256862.79374.31.

Brohi, K., J. Singh, M. Heron, et al. 2003. Acute Traumatic Coagulopathy. *The Journal of Trauma: Injury, Infection, and Critical Care* 54 (6): 1127–30. doi:10.1097/01.TA.0000069184.82147.06.

Clinical Guidelines for Operations. 2012. Joint Services Publication 999 October, pp. 1–244. Ministry of Defence, UK.

CRASH-2 Trial Collaborators, H. Shakur, I. Roberts, et al. 2010. Effects of Tranexamic Acid on Death, Vascular Occlusive Events, and Blood Transfusion in Trauma Patients with Significant Haemorrhage (CRASH-2): A Randomised, Placebo-Controlled Trial. *Lancet* 376 (9734): 23–32. doi:10.1016/S0140-6736(10) 60835-5.

Doran, C. M., T. Woolley, and M. J. Midwinter. 2010. Feasibility of Using Rotational Thromboelastometry to Assess Coagulation Status of Combat Casualties in a Deployed Setting. *The Journal of Trauma: Injury, Infection, and Critical Care* 69 (Suppl.): S40–8. doi:10.1097/TA.0b013e3181e4257b.

Doukas, W. C., R. A. Hayda, H. M. Frisch, et al. 2013. The Military Extremity Trauma Amputation/Limb Salvage (METALS) Study. *The Journal of Bone and Joint Surgery. American Volume* 95 (2): 138–45. doi:10.2106/JBJS.K.00734.

Eastridge, B. J., M. Hardin, J. Cantrell, et al. 2011. Died of Wounds on the Battlefield: Causation and Implications for Improving Combat Casualty Care. *The Journal of Trauma* 71 (1 Suppl.): S4–8. doi:10.1097/TA. 0b013e318221147b.

Eastridge, B. J., R. L. Mabry, P. Seguin, et al. 2012. Death on the Battlefield (2001–2011): Implications for the Future of Combat Casualty Care. *Journal of Trauma and Acute Care Surgery* 73 (6): S431–7.

Fries, C. A., and M. J. Midwinter. 2010. Trauma Resuscitation and Damage Control Surgery. *Surgery (Oxford)* 28 (11): 563–7. doi:10.1016/j.mpsur.2010.08.002.

Garner, J., S. Watts, C. Parry, et al. 2010. Prolonged Permissive Hypotensive Resuscitation Is Associated with Poor Outcome in Primary Blast Injury with Controlled Hemorrhage. *Annals of Surgery* 251 (6): 1131–9. doi: 10.1097/SLA.0b013e3181e00fcb.

Hardaway, R. M. 1978. Viet Nam Wound Analysis. *The Journal of Trauma: Injury, Infection, and Critical Care* 18 (9): 635–43.

Hodgetts, T. J., S. Davies, R. Russell, et al. 2007. Governance and Data Collection. *Journal of Royal Army Med Corps* 153 (4): 237–8.

Hodgetts, T. J., and P. F. Mahoney. 2007. Damage Control Resuscitation. *Journal of the Royal Army* 153 (4): 299–300.

Hodgetts, T. J., P. F. Mahoney, M. Q. Russell, et al. 2006. ABC to <C>ABC: Redefining the Military Trauma Paradigm. *Emergency Medicine Journal* 23 (10): 745–6. doi:10.1136/emj.2006.039610.

Holcomb, J. B., L. G. Stansbury, H. R. Champion, et al. 2006. Understanding Combat Casualty Care Statistics. *The Journal of Trauma: Injury, Infection, and Critical Care* 60 (2): 397–401. doi:10.1097/01.ta.0000203581.75241.f1.

Holcomb, J. B., B. C. Tilley, S. Baraniuk, et al. 2015. Transfusion of Plasma, Platelets, and Red Blood Cells in a 1:1:1 vs a 1:1:2 Ratio and Mortality in Patients with Severe Trauma. *Journal of the American Medical Association* 313 (5): 471–82. doi:10.1001/jama.2015.12.

Hoyt, D. B., E. M. Bulger, M. M. Knudson, et al. 1994. Death in the Operating Room: An Analysis of a Multi-Center Experience. *The Journal of Trauma: Injury, Infection, and Critical Care* 37 (3): 426–32.

Kashuk, J. L., E. E. Moore, J. S. Millikan, et al. 1982. Major Abdominal Vascular Trauma-a Unified Approach. *Journal of Trauma and Acute Care Surgery* 22 (8): 672.

Keene, D. D., G. R. Nordmann, and T. Woolley. 2013. Rotational Thromboelastometry-Guided Trauma Resuscitation. *Current Opinion in Critical Care* 19 (6): 605–12. doi:10.1097/MCC.0000000000000021.

Keene, D. D., J. G. Penn-Barwell, P. R. Wood, et al. 2016. Died of Wounds: A Mortality Review. *Journal of the Royal Army Medical Corps* 162 (5): 355–60. doi:10.1136/jramc-2015-000490.

Kehoe, A., A. Jones, S. Marcus, et al. 2011. Current Controversies in Military Pre-Hospital Critical Care. *Journal of the Royal Army Medical Corps* 157 (3 Suppl. 1): S305–9.

Kirkman, E., and S. Watts. 2014. Haemodynamic Changes in Trauma. *British Journal of Anaesthesia* 113 (2): 266–75. doi:10.1093/bja/aeu232.

Langan, N. R., M. Eckert, and M. J. Martin. 2014. Changing Patterns of in-Hospital Deaths Following Implementation of Damage Control Resuscitation Practices in US Forward Military Treatment Facilities. *Surgery* 149 (9): 904–12. doi:10.1001/jamasurg.2014.940.

Long, K. N., R. Houston, IV, J. D. B. Watson, et al. 2015. VESS Basic Science Research. Functional Outcome after Resuscitative Endovascular Balloon Occlusion of the Aorta of the Proximal and Distal Thoracic Aorta in a Swine Model of Controlled Hemorrhage. *Annals of Vascular Surgery* 29 (1): 114–21. doi:10.1016/j.avsg.2014.10.004.

Manley, G., M. M. Knudson, D. Morabito, et al. 2001. Hypotension, Hypoxia, and Head Injury: Frequency, Duration, and Consequences. *Archives of Surgery* 136 (10): 1118–23. doi:10.1001/archsurg.136.10.1118.

Martin, M., J. Oh, H. Currier, et al. 2009. An Analysis of in-Hospital Deaths at a Modern Combat Support Hospital. *The Journal of Trauma* 66 (Suppl.): S51–61. doi:10.1097/TA.0b013e31819d86ad.

Mercer, S. J., N. T. Tarmey, T. Woolley, et al. 2013. Haemorrhage and Coagulopathy in the Defence Medical Services. *Anaesthesia* 68 (Suppl. 1): 49–60. doi:10.1111/anae.12056.

Mercer, S. J., C. L. Whittle, and P. F. Mahoney. 2010. Lessons from the Battlefield: Human Factors in Defence Anaesthesia. *British Journal of Anaesthesia* 105 (1): 9–20.

Moorhouse, I., A. Thurgood, and N. Walker. 2007. A Realistic Model for Catastrophic External Haemorrhage Training. *Journal of the Royal Army medical Corp* 153 (2): 99–101.

Morrison, C. A., M. M. Carrick, M. A. Norman, et al. 2011. Hypotensive Resuscitation Strategy Reduces Transfusion Requirements and Severe Postoperative Coagulopathy in Trauma Patients with Hemorrhagic Shock: Preliminary Results of a Randomized Controlled Trial. *The Journal of Trauma* 70 (3): 652–63. doi:10.1097/TA.0b013e31820e77ea.

Morrison, J. J., J. J. Duboe, T. E. Rasmussen, et al. 2012. Military Application of Tranexamic Acid in Trauma Emergency Resuscitation (MATTERs) Study. *Archives of Surgery* 147 (2): 113. doi:10.1001/archsurg.2011.287.

Morrison, J. J., J. Oh, J. J. DuBose, et al. 2013a. En-Route Care Capability from Point of Injury Impacts Mortality after Severe Wartime Injury. *Annals of Surgery* 257 (2): 330–4. doi:10.1097/SLA.0b013e31827eefcf.

Morrison, J. J., J. D. Ross, H. Poon, et al. 2013b. Intra-Operative Correction of Acidosis, Coagulopathy and Hypothermia in Combat Casualties with Severe Haemorrhagic Shock. *Anaesthesia* 68 (8): 846–50. doi:10.1111/anae.12316.

O'Reilly, D. J., J. J. Morrison, J. O. Jansen, et al. 2014. Initial UK Experience of Prehospital Blood Transfusion in Combat Casualties. *Journal of Trauma and Acute Care Surgery* 77 (September): S66–70. doi:10.1097/TA.0000000000000342.

Penn-Barwell, J. G., S. A. G. Roberts, M. J. Midwinter, et al. 2015. Improved Survival in UK Combat Casualties From Iraq and Afghanistan: 2003–2012. *Journal of Trauma and Acute Care Surgery* 78 (5): 1014–20. doi: 10.1097/TA.0000000000000580.

Ramasamy, A., G. A. Cooper, I. D. Sargeant, et al. 2013. (I) An Overview of the Pathophysiology of Blast Injury with Management Guidelines. *Orthopaedics and Trauma* 27 (1): 1–8. doi:10.1016/j.mporth.2013.01.002.

Raza, I., R. Davenport, C. Rourke, et al. 2013. The Incidence and Magnitude of Fibrinolytic Activation in Trauma Patients. *Journal of Thrombosis and Haemostasis* 11 (2): 307–14. doi:10.1111/jth.12078.

Revell, M., K. Porter, and I. Greaves. 2002. Fluid Resuscitation in Prehospital Trauma Care: A Consensus View. *Emergency Medicine Journal: EMJ* 19 (6): 494–8.

Rugeri, L., A. Levrat, J. S. David, et al. 2007. Diagnosis of Early Coagulation Abnormalities in Trauma Patients by Rotation Thrombelastography. *Journal of Thrombosis and Haemostasis* 5 (2): 289–95. doi:10.1111/j.1538-7836.2007.02319.x.

Russell, R., N. Hunt, and R. Delaney. 2014. The Mortality Peer Review Panel: A Report on the Deaths on Operations of UK Service Personnel 2002–2013. *Journal of the Royal Army Medical Corps* 160 (2): 150–4. doi: 10.1136/jramc-2013-000215.

Tai, N. R. M., and R. Russell. 2011. Right Turn Resuscitation: Frequently Asked Questions. *Journal of the Royal Army Medical Corps* 157 (Suppl. 3): S310–14. doi:10.1136/jramc-157-03s-10.

Watchorn, J., R. Miles, and N. Moore. 2013. The Role of CT Angiography in Military Trauma. *Clinical Radiology* 68 (1): 39–46. doi:10.1016/j.crad.2012.05.013.

Watts, S., G. Nordmann, K. Brohi, et al. 2015. Evaluation of Prehospital Blood Products to Attenuate Acute Coagulopathy of Trauma in a Model of Severe Injury and Shock in Anesthetized Pigs. *Shock* 44 (August): 138–48. doi:10.1097/SHK.0000000000000409.

Woolley, T., M. Midwinter, P. Spencer, et al. 2013. Utility of Interim ROTEM® Values of Clot Strength, A5 and A10, in Predicting Final Assessment of Coagulation Status in Severely Injured Battle Patients. *Injury* 44 (5): 593–9. doi:10.1016/j.injury.2012.03.018.

6 Anthropomorphic Test Devices for Military Scenarios

Michael Kleinberger

CONTENTS

6.1 BRIEF HISTORY OF OCCUPANT PROTECTION RESEARCH

The study of impact biomechanics and occupant protection dates back almost a century to a fateful day in 1917 during World War I when a cadet pilot named Hugh DeHaven was the sole survivor of a midair collision during a training flight with the Canadian Royal Flying Corps. While recuperating from a ruptured pancreas, he started to apply the principles of engineering to his collision in an attempt to determine why he alone was able to survive the crash and subsequent fall. This began an almost 60-year personal journey to better understand occupant response to a vehicle collision and created a new field of impact biomechanics.

It took another 20 years before the field started to gain some traction when Gurdjian and Lissner founded the Bioengineering Center at Wayne State University in 1939 and started to conduct experimental testing of postmortem human subjects (PMHS) related to head injury thresholds. This centre has actively studied the principles of occupant protection for more than 75 years. DeHaven helped to establish the Cornell Aeronautical Laboratory in the early 1940s, which has since changed its name to Calspan and continues to conduct a wide array of research and testing related to occupant protection and crash survivability. These test facilities have since been joined by countless other laboratories around the world, and generations of biomechanists have been educated and trained in the field of impact biomechanics.

In 1946, Dr. John Paul Stapp became a research officer in the Army Air Force Medical Corps. He subjected live human volunteers (including himself) and subhuman primates to high levels of linear acceleration using rocket-powered sleds that could achieve speeds of over 600 mph. Like DeHaven, he wanted to better understand why some people survived vehicle collisions and others did not. He also wanted to determine the threshold for surviving high acceleration events. His research showed the importance of restraint systems and the avoidance of secondary collisions of the occupant with the vehicle interior. He also explained that the threshold of injury was increased when the subject was facing away from the crash. In 1954, he subjected himself to a deceleration from an initial speed of 632 mph to a dead stop in 1.4 seconds, resulting in a peak deceleration of 40 g's. Although this exposure resulted in a collapsed lung, retinal haemorrhage and temporary blindness, he proved that windblast and deceleration from an ejection at supersonic speeds and high altitude could be survived (Stapp 1951, 1962).

After realising that the Air Force lost nearly as many men in fatal automotive accidents as in plane crashes, Stapp began a car crash program, putting dummies into salvaged cars and running them into rigid barriers. Human volunteers tested safety belts up to 28 g's acceleration and 4,800 pounds of decelerative force. In May of 1955, Stapp invited a group of people to Holloman Air Base to witness an automotive sled demonstration and participate in discussions on automotive design and safety features. They represented the Armed Services, universities, automobile manufacturers, research laboratories, traffic and safety councils and healthcare. This meeting was repeated in successive years and became known as the Stapp Car Crash Conference, which has recently held its 59th session (http://www.stapp.org/stapp.shtml).

6.2   HUMAN SURROGATES FOR OCCUPANT PROTECTION RESEARCH

Occupant protection research typically involves subjecting a human (or surrogate) to conditions that replicate those during an impact or other high acceleration event. Test subjects can be live human volunteers, PMHS, anthropomorphic test devices (ATDs) or animals. These subjects can be placed in a vehicle and crash tested in a controlled condition or, more commonly, placed into a controlled laboratory test system that can reliably replicate the motions of a vehicle. Instrumentation and photographic targets can be attached to the anatomic regions of interest to enable post-test analysis of the occupant's response. Results from whole-body tests are typically used to (1) verify the incidence of specific types of injuries; (2) provide a better understanding of the mechanisms of injury; (3) validate the overall response of a human surrogate, such as an instrumented ATD; (4) establish a global impact threshold that could induce injury to various regions of the body; and (5) determine appropriate loading and boundary conditions to support more detailed experimentation on specific anatomic subsystems or components.

Compared to component testing, where load cells can be used to measure forces and moments at both the impact interface and restraint interface of the anatomic structure, whole-body testing is generally performed using a more limited and less invasive set of sensors to minimise their effects on the biomechanical responses of the body to the test conditions. For this purpose, kinematic markers are typically placed on anatomic landmarks to enable post-test motion analysis, while motion sensors (accelerometers and angular rate sensors) are used to directly measure the kinematics of the skeletal structures. The kinematic motions of the various body regions are a critical component of the biofidelic response corridors (BRCs) needed to verify the performance of an ATD. Strain gauges can also be attached to certain accessible anatomic surfaces of interest to provide an approximation to the forces being transmitted through the structures and the timing of any fractures. This information, combined with load and acceleration measurements made on the vehicle structures in contact with the occupant (i.e. seat, floor, belts), will provide critical information needed to verify the kinetic response of an ATD.

Each different type of test subject (volunteer, ATD, animal or PMHS) has its own benefits and limitations with respect to biomechanical testing and interpretation of results. Live human volunteers are clearly the best representatives of the operational population, but the exposures of these subjects must be kept below the injury thresholds. In addition, instrumentation must remain essentially non-invasive, typically involving the application of photographic targets, externally mounted accelerometers and occasionally EMG electrodes (Siegmund 1997, 2004). Some volunteer studies have been performed within the field of view of a high-speed x-ray system to dynamically track skeletal motions during a staged impact event (Ono 1997).

ATDs, also referred to as manikins, mannequins or dummies, offer the simplest solution from an engineering perspective for conducting biomechanical testing. By design, they are durable and repeatable and much easier to handle than other biological test subjects. Load cells, accelerometers and other types of sensors are incorporated into the ATD design and are typically connected to an onboard, in-dummy data acquisition system. Although they provide great benefits from the perspective of being a reliable engineering test device, the level of confidence in the measured responses is limited by the biofidelity of the dummy. Repeatability is generally accomplished by limiting the number of articulations or degrees of freedom, which can alter the resulting kinematics as compared with a human response. This is especially true for the spine. Overall biofidelity of the dummy will generally depend on the anatomic region under consideration, as well as the loading environment. The further the loading environment is from the conditions under which the dummy was developed and validated, the less confidence there will be in the measured response. Also, because the ATDs are designed to be durable, they do not replicate the changes in response that are associated with bony fractures and injuries to soft connective tissues. Resulting loads that greatly exceed human injury thresholds are therefore unrealistic and cannot be taken at face value. One example of using an ATD outside of its intended design environment is the use of the Hybrid III ATD (designed primarily for frontal crash loading) in the vertical underbody (UB) blast environment. Limitations and inconsistencies in the observed responses have highlighted the need for a new ATD with improved responses in the vertical direction.

Animal subjects offer the benefit of being able to measure physiological responses with active musculature during an injurious level of exposure. They also enable the study of injury responses that occur over a period of time as a result of a traumatic event. However, due to sensitivities of the public to animal testing and the difficulty in trying to scale data from animals to humans due to large anatomic differences, this type of testing has been greatly limited in recent years.

PMHS subjects probably offer the best combination of biofidelity, ease of instrumentation and interpretation of measured responses. They can be exposed to injurious levels of loading and can also be instrumented using more invasive techniques. Accelerometers and strain gauges can be attached directly to the bony structures of interest without significantly altering the structural integrity. Photographic targets can be inserted into the bones using pins. Some studies have also replaced sections of long bones with load cells to obtain direct measurements of the forces and moments acting within the anatomic region of interest (Rudd 2004). Care must be taken, however, to ensure that the insertion of the load cell does not alter the biomechanical response of the structure or create an area of stress concentration that alters the resulting failure mechanism and/or location. Subjects can also be exposed to higher levels of radiation for dynamic imaging without any health concerns. It is important to note that PMHS do not provide active muscular effects, and therefore the measured responses need to be interpreted accordingly.

Although live volunteers and human surrogates may be feasible for different types of occupant protection research, regulations and standards must be based on testing with durable and repeatable devices, such as ATDs, which enable the convenient measurement of critical occupant response parameters. In addition, it is critical that the ATD being specified for any given application is designed and certified for use under the appropriate loading environment, considering the severity, direction, rate and duration of loading applied to the occupant. For this reason, a number of different ATDs have been designed and specified over the years. Automotive dummies have been designed to measure occupant responses primarily in the horizontal plane (frontal, lateral and rear impacts). Military dummies, often called manikins, have largely been designed to measure occupant responses in the vertical direction, primarily involving ejection seats or aircraft crashes.

6.3 DEVELOPMENT OF ATDs

The design and development of ATDs has been largely an evolutionary process driven by the practical need to better understand specific mechanisms of injury or to evaluate the effectiveness of various protective systems under development. Early dummies were built on an as-needed basis to provide a relatively crude insight into occupant response during specific events of interest. These dummies were often produced in small numbers and were not durable or repeatable, which limited their use to research purposes. Dummies intended for use in certification testing or regulatory safety standards were required to be designed and manufactured to a higher standard. The development of dummies has focused on different application areas depending on the primary acceleration vectors. Automotive dummy design has focused on acceleration vectors within the horizontal (X–Y) plane and has resulted in families of dummies specifically designed for frontal, side and rear impact conditions. Early military dummies were designed to support advances in the design of aircraft ejection systems and were more concerned with providing appropriate responses in the vertical (Z) direction. More recent development of dummies for military applications has considered crash vectors for rotary and fixed wing aircraft and has largely involved modifications to existing automotive dummies. A brief history of dummy development for automotive and military applications is provided next.

6.3.1 EARLY ATD DEVELOPMENT

The use of ATDs can be traced back over a century as a means for demonstrating the performance of parachutes during their development in the early 1900s. Most of these early demonstrations were conducted using human volunteers, but one test involved a crude manikin that was dropped from the Eiffel Tower in 1911 (De Prins der Geillustreerde Bladen, February 18, 1911, pp. 88–89). Although rudimentary parachutes existed during World War I, their use was mostly limited to balloon operators who needed a means of escape in the event the balloon was shot down or set on fire. Parachutes were not being used by aviators in planes until close to the end of the war when German pilots began to wear them. Following the war, dummies began to be used for research to help improve the design and performance of parachutes and harnesses. Early dummies consisted of little more than ballast masses made of rope or sandbags connected by various linkage systems. Tests conducted using these early dummies were focused on demonstrating proper operation of the equipment and did not provide useful measurements related to human response and the likelihood of injury.

As the operating speeds of aircraft continued to increase through World War II, the military started to rely more on testing with dummies as the conditions associated with the escape from disabled planes became too dangerous for human volunteer testing. Sierra Engineering developed the Sierra Sam dummy in 1949 under contract to the US Air Force to help evaluate pilot restraint and ejection seat systems. This was a 95th-percentile male dummy with human-like shape and weight, articulated limbs and limited instrumentation to measure linear acceleration of the head. The dummy was reasonably durable and easy to repair, but its lack of repeatability limited its use.

As rocket-propelled ejection seats began to emerge in the late 1950s, it became more critical to align the centre of gravity of the occupant-seat system with the ejection thrust vector to prevent rotational instability and tumbling of the system. This required the development of a new dummy with improved biofidelity with respect to centre of mass and mass moments of inertia. These features were also critical to the space program given the relatively long duration of thrust during launch. Any misalignment could cause both the seat and astronaut to be subjected to potentially hazardous rotations and torques. Grumman and Alderson Research Laboratories worked together to develop the GARD dummy (Grumman–Alderson Research Dummy), which integrated additional sensors and telemetry systems into the dummy design to yield an overall centre of gravity and moment of inertia that closely matched properties measured in human volunteers. The GARD dummies were designed in eight different sizes ranging from 3rd to 98th percentile for Navy and Air Force aircrew populations. Instrumentation included a telemetry package and 12 transducers to measure linear and rotational accelerations and interface reaction loads between the occupant and seat. Although these dummies were quite limited in their ability to measure occupant response metrics, they were proven to be quite durable, provided appropriate ballast to the seat systems, presented anatomically correct surface dimensions for accurate wind blast effects and provided a suitable platform to mount sensors and telemetry systems (AGARD 1996).

6.4 ATDs FOR AUTOMOTIVE APPLICATIONS

In 1966, Alderson Research Laboratories introduced the VIP-50 midsized male dummy, which was the first dummy manufactured specifically for the automotive industry. Other VIP (Very Important People) dummies included a small female (VIP-5) and large male (VIP-95). Around the same time, Sierra Engineering introduced their midsized male (Sierra Stan) and small female (Sierra Susie) dummies. The VIP and Sierra dummies provided reasonable approximations for the shapes and weights of the target populations but lacked biofidelity in key anatomic regions and were also not repeatable from one dummy to the next. Features from both of these automotive dummies were combined by General Motors in the creation of the Hybrid I dummy, which offered improved biofidelity and repeatability as compared to its predecessors.

The creation of the US National Highway Traffic Safety Administration in 1970 increased the demand for the development of robust, repeatable and biofidelic test devices to assess automotive safety systems and support the implementation of new automotive safety standards. Families of dummies were designed specifically to assess occupant response and the likelihood of injury as a result of crashes in different vector directions. Specifications for these different dummies have been incorporated into the 49 CFR Part 572 standards.

6.4.1 AUTOMOTIVE FRONTAL IMPACT DUMMIES

Among the first ATDs incorporated into the standards was the family of Hybrid II and Hybrid III frontal crash dummies. Initially designed to represent a 50th percentile male occupant, these dummy designs were later expanded to include a 95th percentile male, 5th percentile female and 3- and 6-year-old children.

FIGURE 6.1 View of the automotive Hybrid III frontal dummy. (Courtesy of Humanetics Innovative Solutions, Plymouth, MI.)

The Hybrid II 50th male dummy was designed by General Motors to match the general anthropometry, mass properties and ranges of motion for the extremities. Instrumentation enabled the measurement of accelerations at the CG of the head and in the thoracic spine. Load cells were included to measure axial force along the long axis of the femurs. This dummy was incorporated into the Federal Motor Vehicle Safety Standards (FMVSS) in 1973 to help assess the efficacy of seatbelt restraint systems. Although this dummy was robust and repeatable, the biomechanical responses of the head, neck and torso were considered to be too stiff in comparison with the human response, and the instrumentation was limited.

Almost immediately following the inclusion of the Hybrid II into the FMVSS standards, General Motors began a new research effort to improve its biofidelity and enhance the instrumentation, which led to the creation of the Hybrid III. The Hybrid III dummy (shown in Figure 6.1) is currently specified in FMVSS No. 208 as the required ATD for frontal crash protection and has become the most widely used dummy in the world. Although designed specifically to provide a biofidelic response during frontal collisions, the Hybrid III has been used for research purposes well outside of its design window. This includes a wide range of applications involving vehicular collisions in lateral and rear impacts; sports applications to evaluate protective gear such as helmets and chest protectors; industrial accident reconstruction; and military applications including the assessment of protective systems for both mounted and dismounted soldiers. Dismounted applications typically involve the addition of pressure sensors to the dummy's head and/or torso in an attempt to quantify blast wave exposure. It should be noted, however, that extra caution should be taken when interpreting results from tests conducted with the Hybrid III dummy outside of the frontal collision vector.

The head of the Hybrid III dummy consists of an aluminium shell covered by a vinyl skin to provide an appropriate response to forehead impacts. The neck consists of rubber discs bonded to aluminium plates with a cable running down the centre to prevent tensile failure of the assembly. Slots are cut in the anterior portion of the discs to reduce the bending stiffness in extension. Six-axis load cells can be incorporated at both the top and bottom of the neck. The upper torso consists of six steel ribs attached anteriorly to a flexible sternal bib and posteriorly to a rigid spine box. A damping material is added to each rib to provide a biofidelic response to pendulum impacts to the chest. Chest deflection can be measured using a displacement probe located inside the torso. The lower torso consists of a curved rubber lumbar spine that attaches superiorly to the rigid spine box and inferiorly to an aluminium cast pelvis through a five-axis load cell. The lumbar spine is designed to position the dummy in a normal 'slouched' seating posture, which is appropriate for typical automotive seats with a reclined seatback. The pelvis casting is covered with urethane foam and a vinyl skin. The femur and tibia consist of steel shafts with vinyl skin that attach through a ball-and-socket joint at the hip and a clevis joint at the knee. Optional load cells are available in the femur and tibia to measure multi-axial forces and moments, and a slider mechanism is incorporated into the knee design to predict ligament injury. Bump stops are included to provide more biofidelic moment-rotation characteristics and appropriate ranges of motion for the extremities. The foot is attached to the tibia through a ball-and-socket joint with a foam pad inserted under the heel to improve the response for impact directed through the floor and into the lower leg [NHTSA website - http://www.nhtsa. gov/Research/Vehicle+Research+&+Testing+(VRTC)/Hybrid+III+50th+ Percentile+Male, accessed on March 10, 2016].

6.4.2 Automotive Side Impact Dummies

A number of different ATDs were developed to support regulations and the development of protective systems for automotive side impact collisions. Since the use of these ATDs is largely outside the scope of this chapter, they will be discussed only briefly. The original side impact dummy (SID) was developed in 1979 for the National Highway Traffic Safety Administration (NHTSA) and was eventually incorporated into 49 CFR Part 572 to support the FMVSS 214 Side Impact Standard. The SID design was focused around a unique non-symmetric torso design, which consists of a hydraulic shock absorber that links five interconnected steel ribs to the spine. There is no structural representation of the arm or shoulder, which alters the load path from lateral impacts and is one of the main criticisms of its design. Additional padding incorporated into the chest jacket is intended to represent the effects of the upper arm. Other components were initially taken from the Hybrid II, although the head and neck were eventually replaced with the Hybrid III components in the SID-HIII dummy.

The EuroSID dummy was developed by the European Experimental Vehicle Committee (EEVC) in the late 1980s. The torso consists of three separate ribs covered with foam and attached to a steel spine box through hydraulic spring-damper systems. The shoulder design allows the arm to be rotated for testing, either with or without the arm interposed between the impacting surface and the torso. The EuroSID dummy was incorporated into the European and Australian side impact regulations.

The BioSID dummy was developed by the Society of Automotive Engineers (SAE) in an attempt to improve the biofidelity of the SID and EuroSID dummies. The BioSID uses the Hybrid III head, neck and legs. The ribs are connected to the far side of the dummy, which enables greater rib deflection without permanent deformation. The BioSID dummy has never been incorporated into any regulations.

A more recent version of a side impact dummy was developed under the ISO/TC22/SC12/WG5 working group on ATDs, with its initial prototype being released in 2001 (Scherer 2001). Unlike previous side impact dummies, the WorldSID has a symmetric design and includes an in-dummy data

FIGURE 6.2 Mil-SID side impact dummy. (Courtesy of Humanetics Innovative Solutions, Plymouth, MI.)

acquisition system capable of recording over 200 channels of response data. The head consists of a polyurethane skull with a permanently attached PVC skin and a featureless face. Instrumentation at the head CG allows for the measurement of linear and rotational acceleration, with an additional two tilt sensors provided to aid initial positioning. The neck is adopted from the EuroSID dummy and allows for the adjustment of neck posture about the lateral axis. Six-axis load cells are included at the top and bottom of the neck. The torso comprises two abdominal ribs, three thoracic ribs and a shoulder rib. The ribs are made from a super-elastic alloy with damping material attached to help tune the response. Compression of the shoulder and ribs are measured using an InfraRed Telescoping Rod for Assessment of Chest Compression (IR-TRACC) system. The dummy can be used with either a full or half arm. The lumbar spine consists of an inverted U-shaped shell, which controls lateral rigidity and flexibility, with two internal vertical stiffeners. The pelvis consists of a polyurethane pelvis bone covered by vinyl flesh with ball-and-socket hip joints. It includes a 3-axis accelerometer mounted at its CoG, a 6-channel pubic symphysis load cell, an 18-channel lumbar/sacro-iliac load cell, and 6-channel femoral neck load cells. Two tilt sensors are also included to aid with initial positioning. The legs are hollow tubes covered with foam filled vinyl flesh. Each leg is instrumented with universal six-axis load cells that can be positioned in multiple locations within the femur or tibia. Potentiometers are provided to measure rotations of the ankles.

In response to observations that underbody blast exposures often involve multidirectional loading on the vehicle occupants, the Mil-SID side impact dummy (shown in Figure 6.2) was developed. This dummy uses core components from the EuroSID-2 dummy coupled with the head and neck of the Hybrid III 50th male dummy and the Mil-LX lower extremities. The Mil-SID dummy is capable of measuring both lateral and vertical forces. [Humanetics website – http://www.humaneticsatd.com/crash-test-dummies/aerospace-military/mil-sid, accessed on March 10, 2016].

6.4.3 Automotive Rear Impact Dummies

Automotive rear impact dummies were developed in the late 1990s to assess the risk of whiplash associated disorders during a relatively low-speed rear impact. The focus of these dummies was to replicate the complex motions of the human spine, which are believed to be most closely related to the mechanisms of injury under this loading scenario. As with the side impact dummies, the use of these ATDs is largely outside the scope of this chapter and will be discussed only briefly.

The BioRID dummy was developed by the University of Gothenburg and includes a complex spine design with rotational articulations within the midsagittal plane at all spine levels, with 7 cervical, 12 thoracic and 5 lumbar articulations. A torsionally controlled cable system provides proper tuning of the cervical kinematics and the position of the head relative to the neck. The RID2 dummy was developed under the European Whiplash project and is largely based on the Hybrid III design, with the ribcage of the THOR and the lumbar spine of the EuroSID. The pelvis was modified to allow adjustment of the pelvis angle and greater flexion/extension of the femurs. The neck design was changed to provide more flexibility in torsional twisting and lateral bending and also provide 15 degrees of free rotation at the nodding joint to better match observed head kinematics during volunteer testing (Cappon 2001; Philippens 2002).

6.4.4 Total Human Occupant Response Dummy

Development of the Total Human Occupant Response (THOR) dummy started in 1995 by Gesac under contract to the NHTSA. The torso was designed to help evaluate advanced safety systems that included combinations of airbags and energy-absorbing seatbelt systems. Although designed primarily for frontal and frontal oblique crash environments, the design offers multidirectional capabilities in several of its components [NHTSA website – http://www.nhtsa.gov/Research/Biomechanics+&+Trauma/THOR+50th+Male+ATD, accessed on March 10, 2016].

The head/neck system is designed to match the multidirectional kinematics from human volunteer tests, and includes a load sensing face with regional measurement capability. The neck assembly consists of stacked elliptical discs of aluminium and butyl rubber designed to provide omnidirectional kinematic biofidelity. A unique upper cervical joint allows for relatively free initial rotation of the head followed by gradually increasing resistance in flexion/extension, which is controlled by the rubber/aluminium spinal column and cables representing the anterior/posterior muscle load path. An anatomically based shoulder mechanism provides a more realistic interaction with the shoulder belt and allows for fore–aft motion, as well as elevation and depression. The thoracic spine includes a deformable element at the T7–T8 level to increase flexibility and a pitch change mechanism located just below the thoracic spine to allow the user to adjust the orientation of the spine relative to the pelvis to accommodate various initial seating postures. The lower extremity design (THOR-LX) provides more biofidelic ankle/foot motions, with a representation of the Achilles tendon (Shams 2005).

A new linkage system was incorporated into the torso to allow multidirectional displacement measurement of the rib cage relative to the spine. This system, called the CRUX, consists of gimballed telescoping rods with a string potentiometer running down the centre that are attached to Ribs 3 and 6 on either side of the centreline. The string potentiometer indicates the overall length of the rods while the angles relative to the spine are measured by two rotary potentiometers. This system allows for the measurement of rib displacement at four locations on the anterior chest surface, which enables the dummy to better distinguish asymmetric loading from various combination of seatbelt and advanced airbag systems.

6.5 RECENT DEVELOPMENTS IN AVIATION AND MILITARY MANIKINS

6.5.1 Limb Restraint Evaluator and Advanced Dynamic Anthropomorphic Manikin

As the performance of combat aircraft continued to improve, it became more difficult for crewmembers to escape from a disabled aircraft without injury. Although the development of improved ejection seat systems, such as the Advanced Concept Ejection Seat (ACES), significantly reduced the incidence of injuries and fatalities, the problem persisted as operating speeds continued to increase. High speeds introduced extreme windblast exposure, resulting in injuries to the extremities as a result of excessive motion of the joints or unrestrained impact with the seat structure. In order to evaluate the effectiveness of limb restraint devices being developed, a suitable test device was needed to measure the human response to ejection, including the response of the extremities. This led to the development of the Limb Restraint Evaluator (LRE) by Systems Research Laboratory in 1984 under contract to the US Air Force.

The LRE manikin was designed around the dimensions of the 95th percentile VIP-95 automotive dummy and used existing dummy components for the skull, upper thorax, pelvis, hands, feet and flesh. A combination of rotary, universal and ball-and-socket joints were incorporated into the manikin and designed to duplicate the appropriate degrees of freedom at the various limb articulation points. Range of motion for each joint was controlled by mechanical stops and kinematics could be measured using rotary potentiometers (White 1984).

While the LRE manikin offered major improvements in measuring the limb response of crewmembers during an ejection, it was limited to a large male size and did not provide a biofidelic response with respect to the spine. These were critical deficiencies for the Air Force, which was planning to develop improved ejection systems under their Crew Escape Technologies (CREST) program. This required the development of a new dummy with a human-like dynamic response, especially in the cervical and lumbar spine regions. The resulting dummy, called the Advanced Dynamic Anthropomorphic Manikin (ADAM), was designed to provide a human-like reactive live load into the ejection seat and possess realistic dynamics and kinematics due to windblast, impact, vibration and acceleration forces representative of those encountered during ejection (Bartol 1990).

Components of the ADAM dummies were specifically designed to match the anthropometry, mass, centre of gravity and joint centre locations, and inertial characteristics of 3rd and 97th percentile tri-service male aviators (AAMRL 1988). Several components were modified from existing automotive dummies, including the Hybrid II head, Hybrid III neck and the hands and feet from the Alderson VIP dummy. A mechanical spring-damper system was incorporated into the spinal design to provide an additional degree of freedom between the upper torso and pelvis, which greatly improved the dummy response to dynamic vertical Gz loading. The ADAM dummy includes 43 points of rotational articulation, consisting of a combination of rotational sleeve and clevis joints. The knee joint combines a clevis joint with a rotational joint that allows for full lower leg rotation when the knee is flexed 90 degrees or more and no rotation when the knee is straightened. The shoulder design includes five independent degrees of freedom to allow extension/flexion, transverse abduction/adduction, coronal abduction, elevation/depression and pronation/retraction. Soft stops were used to duplicate the increasing resistance to joint rotation as the limits of rotation are approached. An onboard 128-channel data acquisition system was incorporated into the viscera to record and store 63 channels of measured forces, accelerations and joint rotations within the dummy and an additional 56 channels of data from the seat.

6.5.2 FAA Hybrid III Dummy

The FAA Test Dummy is a modification of the Hybrid II and III frontal automotive dummies, with modifications made to the lower torso and legs. These changes were driven by new requirements for testing under emergency landing dynamic conditions, which were established by the FAA in 1988 (Gowdy 1999). These requirements are discussed further in Chapter 12. The FAA dummy design is largely based on the Hybrid III but includes a modified chest jacket, abdominal insert and upper legs from the Hybrid II. The straight lumbar spine of the Hybrid II is used, which gives the dummy an erect seated posture to accommodate positioning in upright seats, unlike the slouched automotive posture of the Hybrid III. The dummy has a similar weight distribution to the Hybrid II, but is capable of passing a Hybrid II torso flexion test, which is not possible with a Hybrid III. The head consists of an aluminium skull and cap covered by vinyl skin. A Hybrid III neck connects the head to the upper torso, which is basically that of a Hybrid III with a modified Hybrid II chest jacket made from vinyl skin over urethane foam flesh. The dummy uses Hybrid III automotive legs with the exception of the upper leg bones which are Hybrid II. Figure 6.3 shows the difference in seated posture between the FAA Hybrid III and the automotive Hybrid II, which is similar to the Hybrid III. Table 6.1 shows a comparison of weights and dimensions between the various Hybrid dummies.

6.5.3 Joint Primary Aircraft Training Dummies

In 1993, the combat exclusion law for women in the US was rescinded and the US Department of Defense redefined the anthropometric requirements for aircraft to accommodate females and smaller males in the Joint Primary Aircraft Training System (JPATS). Unlike traditional methods used for manikin sizing, where all of the body regions are representative of a single percentile subject, the JPATS program identified specific 'cases' that defined anthropometric combinations that were especially difficult to accommodate in aircraft cockpits (Plaga et al. 2005). Eight different cases were defined, including descriptions such as medium-build short limbs or largest torso. Existing Hybrid III automotive

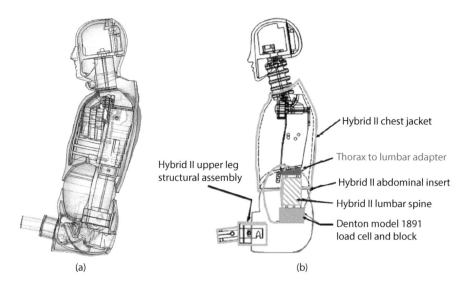

(a)   (b)

FIGURE 6.3   (a) Lateral view showing slouched posture of Hybrid II dummy and (b) upright posture of FAA Hybrid III dummy. Orange outline indicates Hybrid II parts. (Reproduced from Olivares, G., *Hybrid II and Federal Aviation Administration Hybrid III anthropomorphic test dummy dynamic evaluation test series*, Federal Aviation Administration, 2013; Taylor, A.M., et al. *Effect of passenger position on crash injury risk in transport-category aircraft*, FAA Civil Aerospace Medical Institute, 2015.)

TABLE 6.1
Comparison of Weights and Anthropometric Measurements between the Hybrid II, Hybrid III and FAA Hybrid III Dummies

Mass Distribution (kg)	Hybrid II	Hybrid III	FAA Hybrid III
Head	4.54	4.54	4.54
Neck	0.83	1.54	1.54
Upper chest	18.54	17.19	17.23
Lower chest	16.28	23.04	17.23
Upper arm (each)	2.18	2.0	2.0
Lower arm and hand (each)	2.18	2.27	2.27
Upper leg (each)	8.36	5.99	7.73
Lower leg and foot (each)	4.40	5.45	5.45
Total ATD weight	74.37	77.70	75.45
Anthropometric dimensions (cm)	**Hybrid II**	**Hybrid III**	**FAA Hybrid III**
Head circumference	57.2	59.7	57.2
Head width	15.5	15.5	15.5
Head length	19.6	20.3	19.6
Buttock to knee pivot	51.8	59.2	51.8
Knee pivot height	49.8	49.5	49.8
Sitting height	90.7	88.4	90.7

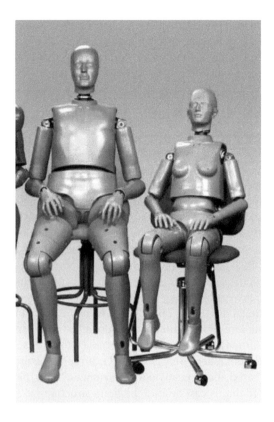

FIGURE 6.4 LARD (left) and LOIS (right) manikins from JPATS program. (From Plaga, J.A., et al., *Design and development of anthropometrically correct head forms for joint strike fighter ejection seat testing*, Air Force Research Laboratory, AFRL-HE-WP-TR-2005-0044, 2005.)

dummies were modified to meet the specifications for two of the eight cases. Modifications were made to the Hybrid III 5th-percentile female dummy to create the Case 1 LOIS (Lightest Occupant In Service) manikin. The LOIS manikin includes a straight upright spine with an approximate mass of 46.8 kg and sitting height of 833 mm. The Case 6 LARD (Large Anthropomorphic Research Dummy) manikin was created by modifying the existing Hybrid III 95th percentile male dummy. The LARD manikin also includes a straight upright spine with an approximate mass of 111.4 kg and sitting height of 965 mm. Figure 6.4 shows these two JPATS manikins.

6.5.4 Warrior Injury Assessment Manikin

The Warrior Injury Assessment Manikin (WIAMan), currently under development by the US Army, is specifically being designed for use in the UB blast environment. Anthropometry and posture are based on the Anthropometric Survey (ANSUR) II data for a 50th-percentile male soldier (Paquette 2009) and the Seated Soldier Study from the University of Michigan Transportation Research Institute (UMTRI) (Reed 2014). Dynamic response of the various anatomic regions is based on target biofidelity corridors, which have been generated from a large number of biomechanical tests performed under the program. The dummy is designed for upright military seats, with the flexibility for reclined and slouched postures. Full-body surface scans provided detailed 3D shape requirements, including facial features that allow for the proper installation of personal protective equipment (PPE), such as helmets and goggles. The ATD skeleton is heavily instrumented to include load cells, accelerometers and angular rate sensors at key locations to assess kinematic response and injury-related measurements along critical load paths. Compliant elements have been incorporated into the design to help replicate human flexibility and produce the desired response to match PMHS test data. The dummy is expected to include roughly 180 sensors for measuring the UBB relevant human response with a state-of-the-art internally distributed data acquisition system.

6.6   ATD COMPONENT DEVELOPMENT FOR MILITARY APPLICATIONS

In addition to the whole-body ATDs described earlier, several specialised components have been developed for standalone use or for use with existing ATDs to provide additional information related to human response and likelihood of injury. Several of these components have been designed specifically to assess the human response to blast and explosive events. These include the Mil-LX Lower Extremity leg, the Frangible Surrogate Leg and the FOCUS headform.

The Mil-LX, or Military Lower Extremity leg (shown in Figure 6.5) can be used as a lower leg replacement for the Hybrid III or THOR dummy legs for the evaluation of soldier protection in UB blast conditions. It incorporates design features from both dummies and includes an energy-absorbing element in the tibial shaft that provides a more biofidelic response compared to PMHS test data under high-rate vertical loading conditions. Available instrumentation includes five-axis load cells in the upper and lower tibia, along with three accelerometers in the tibia and four accelerometers in the foot. Additional discussion on the Mil-LX legs are provided in Chapter 15.

The Frangible Surrogate Leg (FSL), initially developed by the Australian Defence Science and Technology Group (DST Group), was designed to help assess the efficacy of protection systems for use against anti-personnel landmines (Bergeron 2001). It represents a 50th percentile Australian male and consists of CT-based anatomically correct synthetic bones held together by elastic connective tissue. Ballistic gelatine is cast around the bony structures to form the outer contours of the leg. Strain gauges attached to the tibia and femur enable the measurement of compressive and bending strain, as well as shock wave velocity in the bones. A modified version of the FSL lower leg (Footner 2006) included an in-line tibial load cell and shortened the length of the tibia to make it compatible for attaching to a Hybrid III ATD. Trauma assessment is achieved through visual inspection, radiographs and forensic dissection. Improvements in material selection, advanced manufacturing techniques and the inclusion of a biofidelic layered heel pad have been incorporated into the most recent design by Adelaide T&E Systems, as shown in Figure 6.6.

An advanced instrumented headform was developed by the Virginia Tech–Wake Forest Center for Injury Biomechanics in collaboration with Denton and the US Army Aeromedical Research Laboratory (Kennedy 2006). This FOCUS (Facial and Ocular CountermeasUre for Safety) headform (shown in Figure 6.7) can be installed directly on the standard Hybrid III neck and is designed to provide the capability of predicting fracture of the facial bones and injury to the eyes during various impact conditions. The external geometry of the headform is based on 50th percentile male soldier anthropometry and includes full facial features to enable proper fitting of helmets and other head-borne PPE and equipment. The headform includes eight discrete facial bones, representing the nose, mandible and bilateral frontal bone, zygoma and maxilla. Three-axis load cells are available for sensing loads applied to each facial region. In addition to the load sensitive facial bone structure, injury risk to the eyes can also be measured. This is accomplished through a modular

FIGURE   6.5   Mil-LX   Lower Extremity component. (Courtesy of Humanetics   Innovative   Solutions, Plymouth, MI.)

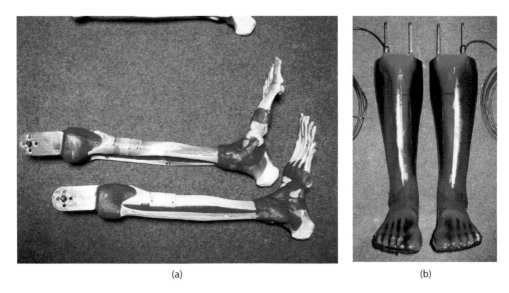

(a)   (b)

FIGURE 6.6 Views of the Enhanced Frangible Surrogate Lower Leg: (a) modified (eFSLLM) showing the internal skeletal structure with the upper tibia load cell and (b) the complete assembly incorporating the flesh material. (Courtesy of Adelaide T&E Systems, Adelaide, Australia.)

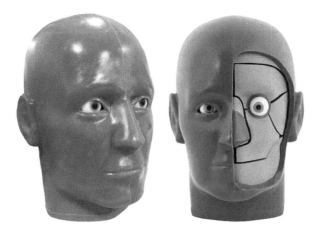

FIGURE 6.7 FOCUS (Facial and Ocular CountermeasUre for Safety) headform. (Courtesy of Humanetics Innovative Solutions, Plymouth, MI.)

design using synthetic eyes to measure blunt impact response or frangible eyes for penetrating impacts. Designs for both the facial bones and eyes were developed to match the force-deflection response from available PMHS testing.

6.7 SUMMARY AND RECOMMENDATIONS

As discussed earlier, the development of ATDs has been an evolutionary process, largely driven by specific needs of the test and evaluation community to evaluate new technologies designed to protect vehicle occupants from potentially injurious events. Various sensor suites, consisting mostly of accelerometers and load cells, have been incorporated into the ATD designs to provide quantitative measurements that can be correlated to the risk or likelihood of injury to various regions of the body.

Corresponding injury criteria are specific to a particular ATD and must be interpreted cautiously, with consideration for the inherent differences between the ATD and human responses.

For the purpose of verification or certification testing, most organisations rely on testing with the standard whole-body ATDs using established injury criteria and thresholds. For military applications, the ATDs most frequently used are the Hybrid III automotive frontal dummy and the FAA Hybrid III, the latter of which is designed for primarily vertical loading vectors. The Mil-SID is sometimes used for test conditions that are expected to result in significant lateral loading, such as with off-centre UB blast events.

As whole-body ATDs are relatively expensive and contain sensitive electronic sensors, a more practical and less expensive option is sometimes required for specific types of live-fire testing that pose a significant risk of fragmentation and penetration. In these situations, other simpler manikins may be more appropriate, even though the ability to measure biofidelic response data may be greatly limited. Various forms of simple human surrogates are commonly used in the research environment, depending on the type of event being replicated and the human response data required. In some cases, simple ballast dummies filled with water or sand may be used to provide a crude approximation of body mass. Other simple manikins, such as the Rescue Randy, have been developed to aid in First Responder training and provide a less expensive approximation of human size and weight distribution for all anatomic components of the body. These manikins can typically be fitted with accelerometers to provide a first approximation of overall kinematics resulting from a primary blast or UB blast event, but will not provide direct measurement of internal forces or moments. For events resulting in large amounts of fragmentation, various forms of ballistic or synthetic gels are often employed. These anatomic gel forms can be created using anatomic moulds for the region of interest and are intended for single-use applications. Trauma assessment is typically based on the number of fragments and the location and depth of penetration. Use of these alternative human surrogates is both acceptable and necessary for research purposes, but they lack the repeatability and durability of the standardised ATDs. They also do not offer a comparable amount of research and testing that has generally been conducted in association with the ATD development process.

Based on the current state of the art in human surrogate development, it will be necessary to continue using both standard and alternative surrogates for the evaluation of new protective technologies. This is especially true for military applications, where the surrogates are exposed to extreme conditions that are likely to damage the standardised ATDs. Some organisations involved in certification testing are left with no choice but to continue to conduct tests with fully instrumented standardised ATDs, having to repair and replace any parts that are damaged during the tests.

Finally, although outside the scope of this chapter, it is important to mention the need to conduct computational simulations in conjunction with experimental testing using ATDs. Simulations using models of the ATDs and/or humans can allow researchers to explore conditions that cannot be readily tested or to perform sensitivity or parametric analyses without the relatively high cost typically associated with experimental testing.

REFERENCES

AAMRL. *Anthropometry and mass distribution for human analogs. Volume I: Military male aviators.* Armstrong Aerospace Medical Research Laboratory, Report No. AAMRL-TR-88-010, 1988.

AGARD. *Anthropomorphic dummies for crash and escape system testing.* North Atlantic Treaty Organization, Report No. AGARD-AR-330, 1996.

Bartol, A.M., V.L. Hazen, J.F. Kowalski, B.P. Murphy, and R.P. White. *Advanced Dynamic Anthropomorphic Manikin (ADAM) Final Design Report.* Armstrong Aerospace Medical Research Laboratory, AAMRL-TR-90-023, 1990.

Bergeron, D.M., G.G. Coley, M.S. Rountree, I.B. Anderson, and R.M. Harris. *Assessment of foot protection against anti-personnel landmine blast using a frangible surrogate leg.* UXO Forum, 2001.

Cappon, H., M. Philippens, M. van Ratingen, and J. Wismans. Development and evaluation of a new rear-impact dummy: The RID2. *Stapp Car Crash Journal,* Vol. 45, pp. 225–238, 2001.

Footner, M.J., D.M. Bergeron, and R.J. Swinton. *Development and calibration of a frangible leg instrumented for compression and bending.* DSTO Technical Report No. DSTO-TR-1829, 2006.

Gowdy, V., R. DeWeese, M. Beebe, B. Wade, J. Duncan, R. Kelly, and J.L. Blaker. *A lumbar spine modification to the Hybrid III ATD for aircraft seat tests.* SAE Technical Paper No. 1999–01–1609, 1999.

Olivares, G. *Hybrid II and Federal Aviation Administration Hybrid III anthropomorphic test dummy dynamic evaluation test series.* Federal Aviation Administration, Report No. DOT/FAA/AR-11/24, 2013.

Ono, K., K. Kaneoka, A. Wittek, and J. Kajzer. Cervical injury mechanism based on the analysis of human cervical vertebral motion and head-neck-torso kinematics during low speed rear impacts. In *Proceedings of the 41st Stapp Car Crash Conference*, SAE Paper No. 973340, pp. 339–356, 1997.

Paquette, S., C. Gordon, and B. Bradtmiller. *Anthropometric survey (ANSUR) II pilot study: Methods and summary statistics.* US Army NSRDEC, Technical Report NATICK/TR-09/014, 2009.

Philippens, M., H. Cappon, M. van Ratingen, J. Wismans, M. Svensson, F. Sirey, K. Ono, N. Nishimoto, and F. Matsuoka. Comparison of the rear impact biofidelity of BioRID II and RID2. *Stapp Car Crash Journal*, Vol. 46, pp. 461–476, 2002.

Plaga, J.A., C. Albery, M. Boehmer, C. Goodyear, and G. Thomas. *Design and development of anthropometrically correct head forms for joint strike fighter ejection seat testing.* Air Force Research Laboratory, Technical Report AFRL-HE-WP-TR-2005–0044, 2005.

Reed, M. *The seated soldier study: Posture and body shape in vehicle seats.* US Army TARDEC, Report No. 24403, 2014.

Rudd, R., Y. Kitagawa, J. Crandall, and F. Poteau. Evaluation of energy-absorbing materials as a means to reduce foot/ankle axial load injury risk. *Journal of Automobile Engineering*, vol. 218(D3), pp. 279–293, 2004.

Scherer, R., D. Cesari, T. Uchimura, G. Kostyniuk, M. Page, K. Asakawa, E. Hautmann, K. Bortenschlager, M. Sakurai, and T. Harigae. Design and evaluation of the WorldSID prototype dummy. In *Proceedings of the 17th ESV Conference*, Paper No. 409, Amsterdam, June, 2001.

Shams, T., N. Rangarajan, J. McDonald, Y. Wang, G. Platten, C. Spade, P. Pope, and M. Haffner. Development of THOR NT: Enhancement of THOR Alpha—The NHTSA Advanced Frontal Dummy. In *Proceedings of the 19th ESV Conference*, Washington, DC, 2005.

Siegmund, G.P., D.J. King, J.M. Lawrence, J.B. Wheeler, J.R. Brault, and T.A. Smith. Head/neck kinematic response of human subjects in low-speed rear-end collisions. In *Proceedings 41st Stapp Car Crash Conference*, SAE Paper No. 973341, pp. 357–385, 1997.

Siegmund, G.P., D.J. Sanderson, and J.T. Inglis. Gradation of neck muscle responses and head/neck kinematics to acceleration and speed change in rear-end collisions. *Stapp Car Crash Journal*, vol. 48, pp. 419–430, 2004.

Stapp, JP. *Human exposure to linear deceleration.* Wright Air Development Center, WADC AF Technical Report 5915, 1951.

Stapp, JP. Jolt effects on man. In *Impact acceleration stress: A symposium.* Brooks Air Force Base, San Antonio, TX, National Academy of Sciences, National Research Council, Publ. 977, pp. 123–130, 1962.

Taylor, A.M., R.L. DeWeese, and D.M. Moorcroft. *Effect of passenger position on crash injury risk in transport-category aircraft.* FAA Civil Aerospace Medical Institute, Report No. DOT/FAA/AM-15/17, 2015.

White, R.P., T.W. Gustin, and M.C. Tyler. *Preliminary design of a limb restraint evaluator.* Air Force Aerospace Medical Research Laboratory, AFAMRL-TR-84-042, 1984.

7 The Mechanical Behaviour of Biological Tissues at High Strain Rates

Feng Zhu, Tal Saif, Barbara R. Presley and King H. Yang

CONTENTS

7.1   INTRODUCTION

The readiness of a military combat unit could be substantially affected when the number of troops is reduced due to injuries, which most often occur during combat missions. Prevention and adequate recovery from such injuries require precise knowledge of underlying causational factors, many of which have not been fully elucidated. Due to advances in protective equipment for military personnel, such as Kevlar body armour and advanced combat helmets, there has been an increase in survival of soldiers exposed to penetrating missiles from bullets and flying debris caused by blast winds. However, even without penetration, both bony and soft tissues can sustain damage from the high-speed pressure waves present during blasts, such as those caused by improvised explosive devices (IEDs). The higher survival rate of soldiers subjected to blasts, coupled with ruptures in unexposed soft tissues induced from high-energy blast waves, has resulted in a higher percentage of wounded soldiers in need of protection and treatment for soft tissue injuries (Mahoney et al. 2005; Okie 2005; Wolf et al. 2009). In addition to the advanced protective measures, modern soldiers are much more likely to be riding in military vehicles than soldiers who were primarily exposed to combat in jungles. Soldiers in military vehicles are additionally subjected to risks of lower extremity bone fractures due to extreme loading conditions from underbelly blasts. These extreme conditions produce intensive dynamic loads with much higher energy and strain rates than those induced in civilian settings. Impact-induced injuries in the civilian population are typically blunt trauma (Champion et al. 2003) and are induced during events such as automobile accidents, falls and contact sports. Understanding that responses are different for human tissues exposed to modern combat conditions versus combat in jungles or civilian scenarios is necessary for better mitigation of injuries, which in turn serves to better maintain combat readiness of military units.

Human tissues are viscoelastic in nature and therefore exhibit non-linear responses to different loading rates. A change in loading rate can substantially alter mechanical properties such as stiffness, strength and failure behaviour. For example, it has been found that lower limb bones exposed to vertical, high-strain loading, such as those typically seen in underbelly blasts, exhibit much greater stiffness and are more brittle than when the bones are exposed to loading that is common in civilian scenarios (Kraft et al. 2012). Because rate dependencies are usually complex and highly non-linear, a better understanding of these effects on different tissues can help in elucidating injury responses, mechanisms and thresholds, which in turn can result in improved treatments, development of anthropomorphic test devices that are more like humans, and designs for more effective and efficient protective equipment.

With the increasing use of finite element (FE) models to simulate impact responses to dynamic loading, there is a need for accurate material constitutive models. Creation of such models necessitates identification of associated material properties to describe rate dependencies of biological tissues. In the past 10 years, a number of FE models have been developed for studying combat-associated injuries. Examples include simulations of torso responses under ballistic impact (Roberts et al. 2007) and primary blast-induced traumatic brain injury (Moss et al. 2009; Taylor and Ford 2009; Chafi et al. 2010; Zhu et al. 2010a, 2013). However, the material laws used in these studies were mainly based on data obtained from material tests at low and intermediate strain rates (usually lower than 100 s^{-1}), which produce very different responses than tests performed at high strain rates. Furthermore, strain-rate effects have been disregarded in a large portion of the studies. Consequently, the accuracy of computational prediction derived from models that do not include consideration of rate effects is questionable.

Exacerbating the difficulty in identification of material properties adequate to study injuries related to current combat conditions are the limited information available regarding ballistic and blast-related strain rates and the large variations in strain rates within and among tissues that are in close proximity to one another. To the best of the authors' knowledge, strain rates experienced in different tissues of the human body from ballistic and blast events are not reported in the literature. In a numerical study of ballistic proximity impact on a long bone embedded in a gelatine block, Huang (2013) found that strain rates were not uniformly distributed. The peak strain rate at the time when the bullet just touched the femur was in the range of 10^4 s^{-1}, while the strain rates reduced to 10^3 s^{-1} just a couple of mm away. In a subsequent study, Huang (2015) used computer simulations to find the strain rate within a pig brain due to blast exposure. The maximum strain rate was found to be in the range of 10^4 s^{-1} in a very small region of the brain, but the strain rates were less than 100 s^{-1} for the rest of the brain. For these reasons, rate-dependent properties of human tissues need to be carefully investigated if numerical models are to be used to identify risks of ballistic or blast-induced injury.

To overcome limitations in current numerical models, there is a great demand to develop new test methodologies for accurate identification of material properties with the inclusion of rate dependencies. To date, studies in this important area are very much limited, and research is scattered throughout the scientific literature. This chapter brings together state-of-the-art research in this area. It begins by introducing commonly used equipment, such as the Split Hopkinson Pressure and Tension Bar systems, which can be used for characterising materials subjected to high strain-rate loading. Special considerations are given to testing biological tissues. Further in the chapter, studies of bone (cortical and trabecular) materials are outlined and investigations into soft tissues (skin, muscle, tendon, internal organ and brain) are described. For both bony and soft tissues, the rate-dependent properties reported in the different studies are listed and compared graphically. Limitations of current studies are then summarised and further work is suggested.

7.2 TECHNIQUES FOR HIGH STRAIN-RATE MATERIAL CHARACTERISATION

Testing techniques for material properties must be selected according to the specific range of strain rates that will be used. Tests making use of strain rates lower than 100 s^{-1} can be conveniently conducted with Instron or MTS (Material Testing Systems) machines or with drop towers. More details are not provided on Instron and MTS tests as this chapter focuses on high strain-rate testing. Testing with high strain rates (100–5000 s^{-1}) can be accomplished using a Split Hopkinson Bar system, also known as a Kolsky Bar (Kolsky 1963). This system has been widely accepted as an effective tool for high strain-rate characterisation of a wide range of materials (Chen et al. 2010). Based on the nature of loading, the Split Hopkinson Bar system can be classified as a Split Hopkinson Pressure Bar (SHPB) or Tension Bar (SHTB) system, as described in Sections 7.2.1 and 7.2.2, respectively.

7.2.1 SPLIT HOPKINSON PRESSURE BAR

Figure 7.1 shows a sketch of a compressive Split Hopkinson Bar system and a typical signal recording from its strain gauges. A specimen is placed between the incident and transmitted bars. An elastic pressure pulse is produced in the incident bar by an impact from a striking bar. At the interface between the incident bar and the specimen (Interface 1), the elastic stress wave is partially reflected and partially transmitted to the short specimen, deforming the specimen plastically. Similarly, at the interface between the specimen and the transmitted bar (Interface 2), the stress wave is partially reflected and partially transmitted. The specimen's material properties at high strain rates can be deduced from strain-time measurement histories obtained from the incident and transmitted bars.

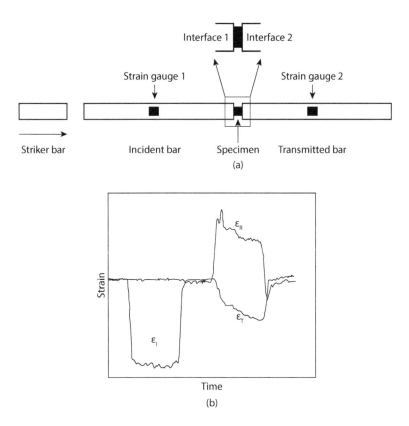

FIGURE 7.1 (a) Sketch of a Split Hopkinson Pressure Bar; (b) typical strain signals from the incident bar and transmitter bar.

One of the necessary conditions for testing with SHPB systems is dynamic stress equilibrium, which means that under dynamic loading, the stress gradient through the thickness of the specimen should remain zero. In addition, the deformation rate should be uniform within the specimen. To achieve these goals, the incident stress needs to be of sufficient duration.

According to wave propagation theory in solids, dynamic stress $\sigma_s(t)$, dynamic strain $\varepsilon_s(t)$ and the strain rate $\dot{\varepsilon}_s(t)$ at any time during loading and unloading processes can be calculated by using Equations 7.1 through 7.3,

$$\sigma_s(t) = \frac{EA}{2A_s}[\varepsilon_i(t) + \varepsilon_r(t) + \varepsilon_r(t)] \tag{7.1}$$

$$\varepsilon_s(t) = -\frac{C_0}{l_s}\int_0^t [\varepsilon_i(t) - \varepsilon_r(t) - \varepsilon_r(t)]dt \tag{7.2}$$

$$\dot{\varepsilon}_s(t) = -\frac{C_0}{l_s}[\varepsilon_i(t) - \varepsilon_r(t) - \varepsilon_r(t)] \tag{7.3}$$

where $\varepsilon_i(t)$, $\varepsilon_r(t)$ and $\varepsilon_t(t)$ are incident, reflected and transmitted strain signals, respectively; A and A_s are the cross-sectional areas of the bars and the specimen, respectively; l_s is the original length of the specimen; C_0 represents the longitudinal elastic wave velocity of the bars; and E is Young's modulus of the bars.

7.2.2 SPLIT HOPKINSON TENSION BAR

To test materials at high strain rates in tension, Split Hopkinson Tensile Bar systems have been developed with various designs (Chen et al. 2010). A schematic of a typical Split Hopkinson Tension Bar (SHTB) system is shown in Figure 7.2. The system consists of a tubular striker, an incident bar with a flange and a transmitted bar. The striker impacts the flange at the end of the input bar to produce a tensile pulse. When the wave reaches the specimen, the wave is partially transmitted and partially reflected back. By measuring the strain signals on the incident and transmitted bars, the stress and strain behaviour of the specimen can be calculated with the same equations used for compression testing: Equations 7.1 through 7.3.

7.2.3 SPECIAL CONSIDERATIONS OF SHPB AND SHTB TESTING ON BIOLOGICAL TISSUES

Although SHPB and SHTB systems have been successfully applied to test many engineering materials, there are numerous challenges in using them for the characterization of biological tissues, particularly for relatively brittle materials (e.g. cortical bone) and soft materials (e.g. soft tissues and trabecular bone). Special considerations for testing these two types of materials are discussed here.

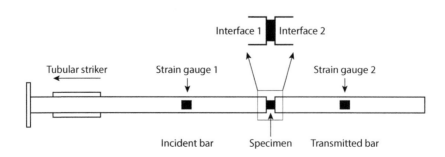

FIGURE 7.2   Schematic of a Split Hopkinson Tension Bar (SHTB) system.

Changes in the properties of pulse waves produced by the impact of a striker on an impact bar have been necessary for dealing with brittle materials. Materials with relatively brittle properties, such as cortical bone, typically exhibit linear elastic behaviour followed by sudden fracture. The failure strain is usually less than 1%. With such extremely short response durations, it is difficult to achieve dynamic stress equilibrium, which is a necessary condition in SHPB and SHTB testing. Additionally, due to the small failure strain, the deformation throughout the thickness of the specimen may not be uniform. For example, failure of the impact side of a specimen may occur before the opposite side has started to deform. In this situation, a constant deformation rate cannot be obtained. For accurate test results, a uniform deformation pattern and a constant deformation velocity must be ensured. Changes to wave transmission and reflection behaviour in the incident bar can be made by implementing a pulse shaper, which is a material layer at the interface between the striker and incident bar. Implementation of pulse shapers of appropriate materials can mitigate non-uniform patterns and varied deformation velocities in brittle materials (Christensen et al. 1972; Frantz et al. 1984; Nemat-Nasser et al. 1991). With the pulse shaper, the initial rise time and the duration of the incident pulse can be enlarged to achieve an early uniformity of the internal stress and a constant strain-rate loading within the specimen. It has been found that pulse shapers made of paper, polymers or rubbers can effectively improve the accuracy of SHPB tests on brittle materials (Chen et al. 2010).

Impedance mismatches between specimens and bar materials interfere with accurate measurements of soft materials, and the deformable nature of soft materials necessitates attention to the shapes of loading pulses. Soft materials are characterised by low strength, low stiffness and low wave impedance with values that are typically much lower than those of bar materials. This impedance mismatch can cause nearly all of the incident pulse to be reflected back, making the transmitted signal too weak to measure. Mechanical responses of soft materials are sensitive to loading rates and states of loading. The low speeds of the waves produced when testing soft specimens make stress equilibrium much more difficult to attain, since more time is needed to achieve such equilibrium. In addition, because soft specimens have low strength and are susceptible to large deformations by nature, it is difficult to ensure a uniform deformation pattern. Efforts have been made to resolve these issues. To reduce the impedance mismatch for the wave energy, incident bars made with low impedance materials (e.g. polymers, low-density metals and hollow bars) have been used. High sensitivity transducers, such as semiconductor strain gauges, have been employed to more adequately capture the weak signals. A thin specimen can clearly facilitate stress equilibrium sooner, and thus the use of a thin specimen can reduce the stress-wave attenuation. Additionally, the pulse shaping technique can generate a low loading rate but ensure a relatively constant strain-rate deformation. These modifications have been applied in testing of biological tissues making use of SHPB and SHTB systems and are reviewed in the next section.

7.3 HIGH STRAIN-RATE BEHAVIOUR OF BONY MATERIALS

Bony materials can be classified as cortical or trabecular. Cortical bone is primarily found in the shafts of long bones, and it forms the outer shell around trabecular bone at vertebrae and at both ends of long bones. The stiffness of cortical bone is usually higher than that of trabecular bone by two orders of magnitude. Representative studies on rate dependencies of cortical and trabecular bone are summarised in Sections 7.3.1 and 7.3.2, respectively.

7.3.1 STUDIES IN CHARACTERISATION OF RATE-DEPENDENT PROPERTIES OF CORTICAL BONE

Historically, traditional metallic SHPBs have been used for characterisation of cortical bone. In this section, the studies are reviewed in terms of experimental protocols, materials tested and results.

Early material tests for cortical bone at high strain rates were performed using drop towers and Instron-type machines, as summarised in Table 7.1. To conduct compressive tests on cortical bone from bovine and human femurs, McElhaney (1966) used a custom-made loading machine driven by an air gun. The tests used nearly constant strain rates of up to 4000 s^{-1}. A strong rate effect was observed from the compressive stress–strain curves. Crowninshield and Pope (1974) used an Instron machine to test cortical bone of bovine tibiae in tension. The bovine specimens were cut into dog-bone shapes and were loaded at strain rates of up to 250 s^{-1}. The modulus of elasticity, breaking stress and breaking strain were found to vary with strain rate. Similar samples made of dog-bone shaped material from bovine femurs were tested in tension by Wright and Hayes (1976) using an MTS at the highest strain rate of 237 s^{-1}, and similar strain rate effects were observed.

Use of Split Hopkinson Bar tests on cortical bone commenced in the 1970s, and this type of testing has been ongoing since that time, with a progression of new bar materials and design modifications.

TABLE 7.1

Comparison of Rate-Dependent Characterisation for Cortical Bone

Authors (year)	Sample Location	Sample Dimensions	Apparatus	Strain Rate (s^{-1})
McElhaney (1966)	Human femur	Cuboid 4.5 × 4.5 × 6.35 mm	Custom-made loading system (compression)	~0.001–1500
Tennyson et al. (1972)	Bovine femur	Cylinder Φ = 9.5 mm L = 12.7 mm	Al SHPB (compression)	~10–450
Crowninshield and Pope (1974)	Bovine tibia	Dog-bone shaped Gauge length 5 mm	Instron and drop hammer (tension)	~0.00167–250
Lewis and Goldsmith (1975)	Bovine femur	Dog-bone shaped Φ = 6.35 mm L = 19.05 mm	Al SHB (compression, tension, torsion and combined compression and torsion)	~400–7000
Wright and Hayes (1976)	Bovine femur	Dog-bone shaped Φ = 3.18 mm L = 54 mm	MTS 810 (tension)	~0.00053–237
Katsamanis and Raftopoulos (1990)	Human femur	Cuboid 216 × 10.5 × 4 mm Dog-bone shaped 54 × 10.5 × 4 mm	Al SHPB and Instron (compression and tension)	~0.00002–100
Ferreira et al. (2006)	Bovine femur	Cube 10 × 10 × 10 mm	Al SHPB (compression)	~368–1209
Adharapurapu et al. (2006)	Bovine femur	Cube 5 × 5 × 5 mm	SHPB, unknown metal (compression)	~0.001–1000
Shunmugasamy et al. (2010)	Rabbit femur	Cylinder Φ = 10 mm L = 10 mm	Inconel alloy SHPB and Instron (compression)	~0.001–775
Lee and Park (2011)	Bovine femur	Cylinder Φ = 10 mm L = 4 mm	Inconel 718 SHPB (compression)	~0.001–2700
Teja et al. (2013)	Human tibia	Cylinder Φ = 12–15 mm L = 2–4 mm	Al SHPB (compression)	~41.2–167

Φ = diameter, L = length, Al = aluminium alloy.

Tennyson et al. (1972) used an aluminium alloy SHPB to measure a wide range of strain rates (10–450 s^{-1}) on bovine cortical bone dehydrated for various lengths of time. Lewis and Goldsmith (1975) used a special Split Hopkinson Bar system made of aluminium to measure dynamic properties of compact bovine bone in compression, tension, torsion and combined torsion and compression. Fracture stress, in compression, increased with strain rate. Katsamanis and Raftopoulos (1990) tested the compressive behaviour of the human femur bone using a universal testing machine for low speeds, and an SHPB with aluminium bars for high speeds. The average dynamic magnitude of the Young's modulus was found to be 23% higher than the average static value, but the Poisson's ratio did not exhibit this effect. Ferreira et al. (2006) also used an SHPB system with aluminium bars to characterise bovine cortical bone. Cube-shaped specimens were prepared in longitudinal and transverse directions. Differences in the ultimate strengths, failure strains and elastic moduli in these directions were not statistically significant. Adharapurapu et al. (2006) used conventional material testing machines and an SHPB system to test cubic bovine bone samples under varying conditions. The results showed significant anisotropy the compressive responses, with the failure strength in the longitudinal direction greater than that in the transverse direction. Additionally, dry bone was observed to exhibit greater strength than wet bone. Strong rate effects were apparent from the test results. High strain-rate compressive responses of bone from rabbit femurs were reported by Shunmugasamy et al. (2010), who used an Inconel alloy SHPB system. Rate effects were found in both diaphyseal and epiphyseal regions. Lee and Park (2011) tested bovine femurs using an Inconel SHPB system equipped with a pulse shaper. Finally, human tibiae were tested by Teja et al. (2013) in compression with an aluminium SHPB, and a similar strain-rate effect was observed.

A summary of these studies is listed in Table 7.1. In most of the published works, the Young's moduli and strengths of bony materials were reported without entire stress–strain curves. The only curves with varieties of strain rates were reported by McElhaney (1966) on bovine femurs in compression and Crowninshield and Pope (1974) for bovine tibiae in tension. These curves are plotted in Figure 7.3. Quantitative comparisons of the Young's moduli can be seen in Figure 7.4 and yield strengths in Figure 7.5.

7.3.2 STUDIES ON THE CHARACTERISATION OF RATE-DEPENDENT PROPERTIES OF TRABECULAR BONE

As mentioned earlier in this chapter, trabecular bone has low stiffness, low strength and low impedance to propulsion of stress waves. Because of these characteristics, it is challenging to test trabecular bone with SHPB systems. In an effort to alleviate challenges in the characterisation of such materials, bars made of low impedance materials have been used in conjunction with pulse shaping techniques. Johnson (2005) used an SHPB system equipped with solid and hollow aluminium and solid polymeric pressure bars to test bovine trabecular bone at rates of up to 1300 s^{-1} and a strain magnitude of 0.07. Although the use of bars with varying properties seemed to provide comparable results in terms of density, many of the samples were too damaged to ensure accuracy of the measurements. A magnesium alloy bar was used by Shim et al. (2005) to test trabecular bone from the human cervical spine. Pilcher (2004) used an aluminium SHPB system to test bovine trabecular bone in compression. A pulse shaper was attached to better control the incident wave and to ensure dynamic stress equilibrium. A Nylon SHPB was employed by Laporte et al. (2009) for bovine trabecular bone testing. Viscoelastic wave theory was used to derive stress–strain curves at the strain rates of 1000–1500 s^{-1}. Another polymeric SHPB system made of polytetrafluoroethylene (PTFE) was employed by Teja et al. (2013) to test trabecular bone of human tibiae. A summary of representative work on trabecular bone is listed in Table 7.2. Quantitative comparisons of the Young's moduli and yield strengths are shown in Figures 7.6 and 7.7, respectively.

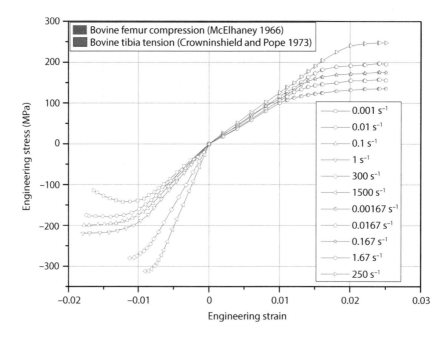

FIGURE 7.3 Typical stress–strain curves for cortical bone at various strain rates.

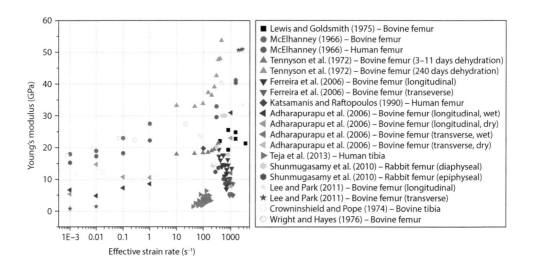

FIGURE 7.4 Comparison of Young's moduli of cortical bone at various strain rates (solid symbols: compressive tests; hollow symbols: tensile tests).

7.4 STUDIES ON THE CHARACTERISATION OF HIGH STRAIN-RATE BEHAVIOUR OF SOFT TISSUES

High strain-rate characterisation of soft tissues can be conducted with techniques similar to those used in the testing of trabecular bone. For example, the same custom-made equipment reported in the article by McElhaney (1966) was used to test the dynamic behaviours of bovine muscle in compression at

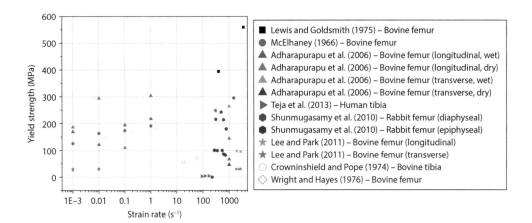

FIGURE 7.5 Comparison of yield strengths of cortical bone at various strain rates (solid symbols: compressive tests; hollow symbols: tensile tests).

TABLE 7.2
Comparison of Rate-Dependent Characterisation for Trabecular Bone

Authors (year)	Sample Location	Sample Dimensions	Apparatus	Strain Rate (s^{-1})	Theory
Johnson (2003)	Bovine femur	Cylinder $\Phi = 8$ mm $L = 6$ mm	Al SHPB (solid and hollow); PMMA SHPB + pulse shaper (compression)	~175–1300	Linear elastic; Viscoelastic
Pilcher (2004)	Bovine tibia	Cuboid $25 \times 25 \times 40$ mm	Al SHPB + pulse shaper (compression)	~154–1306	Linear elastic
Shim et al. (2005)	Human cervical spine	Cuboid $5 \times 5 \times 8$ mm	Mg SHPB and Instron (compression)	~0.001–1200	Linear elastic
Laporte et al. (2009)	Bovine femur	Cylinder $\Phi = 41$ mm $L = 14$ mm	Nylon SHPB and Instron (compression)	~0.001–1500	Viscoelastic
Teja et al. (2013)	Human tibia	Cylinder $\Phi = 12–15$ mm $L = 2–4$ mm	PTFE SHPB (compression)	~85–207.1	Viscoelastic

Φ = diameter, L = length, Al = aluminium alloy, Mg = magnesium, PTFE = polytetrafluoroethylene.

the strain rate of 1000 s^{-1}. The results exhibited significant strain-rate effects. Traditional metallic SHPB systems, with necessary modifications, were also used for soft tissue testing. Shergold et al. (2006) used a low impedance (magnesium alloy) SHPB system to conduct compressive tests on porcine skin. Semiconductor strain gauges were applied to capture the weak signals. Song et al. (2007) tested porcine muscle using an SHPB system made with an aluminium alloy together with semiconductor strain gauges. A pulse shaper was attached on the incident bar to ensure dynamic stress equilibrium and constant strain rates. In the same year, Saraf et al. (2007) made modifications on a conventional aluminium SHPB to conduct confined compressive tests on human stomach, heart, liver and lung tissues. Bulk moduli of these organs were estimated at strain rates of up to 2800 s^{-1}.

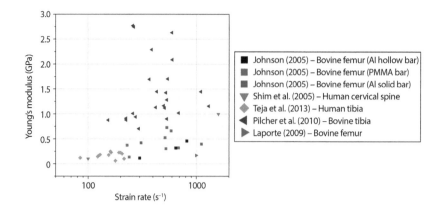

FIGURE 7.6   Comparison of Young's moduli of trabecular bone at various strain rates.

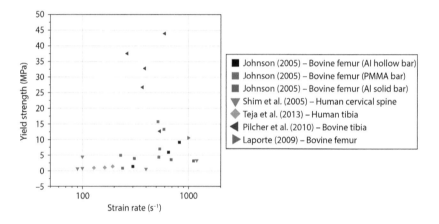

FIGURE 7.7   Comparison of yield strengths of trabecular bone at various strain rates.

And finally, Pervin et al. (2009) used a hollow aluminium alloy SHPB with a pulse shaper to test bovine liver tissue. The same SHPB was employed by the same research group to test porcine grey and white matter of the brain. The white matter was examined both along (D1) and perpendicular (D2) to the coronal section for anisotropic characterisation. It was found that white matter had higher stiffness than grey matter at all strain rates. White matter in the D1 direction was shown to be stiffer than in the D2 direction in the range of strain rates between 1000 and 3000 s^{-1}.

To resolve issues of wave impedance mismatch, viscoelastic bars made from polymeric materials have been applied in soft tissue tests. For example, a polymethylmethacrylate (PMMA) SHPB was used by Chawla et al. (2006) to characterise human muscles (including paraspinal erector spinae, hamstring and gluteus maximus muscles) in strain rates ranging from 250 to 1000 s^{-1}. Additionally, bovine muscle was tested by Van Sligtenhorst et al. (2006) using a PMMA SHPB. Polycarbonate bars were applied by Comley and Fleck (2012) to test porcine adipose at the very high strain rate of 5700 s^{-1}.

High-strain tensile behaviours of soft tissue have also been measured with SHTB systems. Most of the studies that have made use of these systems were conducted at Purdue University (West Lafayette, IN). Using an SHTB system, Cheng et al. (2009) tested bovine tendons at the

TABLE 7.3
Comparison of Rate-Dependent Characterisations for Soft Tissues

Authors (year)	Equipment (material; diameter)	Strain Rate (s^{-1})	Specimen
McElhaney (1966)	In-house loading system (compression)	~0.001–1000	Bovine muscle Cylinder $\Phi = 22.23$ mm $L = 9.53$ mm
Comley and Fleck (2012)	SHPB (Polycarbonate; $\Phi = 12$ mm), a servo hydraulic and a screw-driven test machine (compression)	~1000–5700 ~20–260 ~0.002–0.2	Porcine adipose Cylinder $\Phi = 10$ mm $L = 3\sim8$ mm
Shergold et al. (2006)	SHPB (Mg alloy; $\Phi = 12.73$ mm) + pulse shaper, Shenk Hydropuls PSA and Instron 5500R (compression)	~1500–5000 40 ~0.004–0.4	Porcine skin Cylinder $\Phi = 7.7$ mm $L = 2$ mm
Pervin et al. (2009)	SHPB (Al hollow; $\Phi_{outer} = 19.05$ mm; $\Phi_{inner} = 16.05$ mm) + pulse shaper and hydraulic-driven MTS (compression)	1000; 2000; 3000 ~1–200 ~0.01–0.1	Bovine liver Hollow cylinder $\Phi_{outer} = 10$ mm; $\Phi_{inner} = 4.7$ mm $L = 1.7$ mm
Saraf et al. (2007)	SHPB (Al alloy; $\Phi = 12.7$ mm) (compression and shear)	~300–2900 (stomach) ~60–1200 (heart) ~37–1000 (liver) ~10–7700 (lung)	Human stomach; heart; liver; lung Cylinder $\Phi = 12.7$ mm $L = 1\sim2$ mm Cuboid $9 \times 20 \times 1\sim2$ mm
Van Sligtenhorst et al. (2006)	SHPB (PMMA; $\Phi = 25.4$ mm) (compression)	1000; 1700; 2300	Bovine muscle Cylinder $\Phi = 10$ mm $L = 3.9$; 5.1; 6.3; 8.3 mm

(Continued)

TABLE 7.3 (CONTINUED)

Comparison of Rate-Dependent Characterisations for Soft Tissues

Authors (year)	Equipment (material; diameter)	Strain Rate (s^{-1})	Specimen
Song et al. (2007)	SHPB (Al alloy; Φ = 19 mm) + pulse shaper and hydraulic-driven MTS 810 (compression)	540; 1900; 3700 ~0.007-0.07	Porcine muscle Hollow cylinder Φ_{outer} = 10 mm; Φ_{inner} = 3 mm L = 3.2 mm
Chawla et al. (2006)	SHPB (Polymethylmethacrylate; Φ = 10 mm) (compression)	~250-1000	Human paraspinal erector spinae; hamstring; gluteus max Cuboid 10 × 10 × 2~8 mm
Pervin and Chen (2009)	SHPB (Al hollow; Φ_{outer} = 19 mm; Φ_{inner} = 16 mm) + pulse shaper and hydraulically-driven MTS 810 (compression)	1000; 2000; 3000 ~0.01-0.1	Porcine grey matter; white matter (D1, D2) Hollow cylinder Φ_{outer} = 10 mm; Φ_{inner} = 4.7 mm L = 1.7 mm
Lim et al. (2011)	SHTB (Steel + Al alloy; $\Phi_{Al\ alloy}$ = 12.7 mm; Φ_{Steel} = 19 mm) + pulse shaper and MTS 810 (tension)	1700; 2500; 3500 ~0.005-0.5	Porcine skin Cuboid sheet 40 × 25 × 2 mm
Nie et al. (2011)	SHTB (Steel + Al alloy; $\Phi_{Al\ alloy}$ = 12.7 mm; Φ_{Steel} = 19 mm) + pulse shaper s and MTS 810 (tension)	700; 1400; 2100 ~0.005-0.4	Porcine muscle Cuboid sheet T = 3.2 mm
Cheng et al. (2009)	SHTB (Details unknown) and MTS 810 (tension)	0.05; 2500	Bovine tendon Cuboid sheet 10 × 3 × 2 mm

Φ = diameter, L = length, T = thickness, Al = aluminium alloy, Mg = magnesium, PMMA = polymethylmethacrylate.

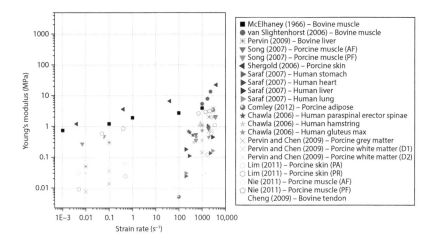

FIGURE 7.8 Comparison of the Young's moduli of soft tissues at various strain rates (solid symbols: compressive tests; hollow symbols: tensile tests).

rate of 2500 s^{-1}. The researchers found that results of tests on the tendon specimens exhibited high stiffness with evidence of rate effects. However, no details of the SHTB were provided in the literature. A modified composite SHTB with steel and aluminium bars was used to conduct tensile tests along two orthotropic directions of porcine skin (PA: parallel to the spine; PR: perpendicular to the spine). Semiconductor strain gauges and a pulse shaper were used. It was reported that for the porcine specimens, strain-rate dependencies perpendicular to the direction of the spine were more sensitive than those along the spine. Nie et al. (2011) adopted a similar composite SHTB system with a pulse shaper to characterise the rate dependency of porcine muscle along (AF) and perpendicular (PF) to the muscle fibre direction. The results suggested that at high strain rates, the sensitivity in the tensile test was less dependent on the fibre orientation relative to the loading direction. A summary of representative work on soft tissues is listed in Table 7.3. The Young's moduli are compared quantitatively and plotted in Figure 7.8.

7.5 CONCLUSIONS

Mechanical testing results at various strain rates have shown that all three types of biological tissues (i.e. cortical bone, trabecular bone and soft tissue) exhibit evidence of rate dependency. The majority of high strain-rate testing has been performed on cortical bone, and relatively less studies have been conducted on trabecular bone and soft tissue. Further, characterisation of the behaviour of each of the tissue types is fraught with difficulties for high strain-rate testing. For cortical bone, most tests are done in compression. However, tensile tests are also necessary. For trabecular bone and soft tissues, challenges are present regarding large impedance mismatches for stress-wave propulsion between bar materials and tissue samples and achievement of dynamic stress equilibrium and constant strain rates in specimens. Obstacles in high strain-rate testing of the three tissue types are now being addressed with modifications in the designs of the powerful tools: the SHPB and SHTB systems.

Cortical bone samples have an impedance to elastic stress-wave propulsion that is similar to light-weight metals. Therefore, high strain-rate characterisation is relatively straightforward with conventional SHPB and SHTB systems. However, the failure of cortical bone is dominated by tensile stress and strain, and most of the studies reported in the literature focus on compressive behaviour. To more accurately model cortical bone fractures sustained during combat missions, more tensile test data representing high strain-rate conditions is imperative. In addition, studies in the literature have predominately focused on elastic moduli and/or yield strengths, and very few stress–strain curves produced with high strain rates have been reported. These curves are crucial for derivation of material constants and failure criteria in rate-dependent constitutive equations.

Due to difficulties in sample preparation and technical issues in performing high-rate tests on trabecular bone, this type of bone has been studied less extensively than cortical bone. To date, no tensile tests on trabecular bone have been reported. Trabecular bone can be considered a soft material in regard to material property characterisation under high strain rates, and therefore it can be tested using a similar approach. Typically, impedances of trabecular bone and soft materials are very low. To reduce impedance mismatches between bars and material samples, bars have been made of low impedance materials, such as polymers and lightweight or hollow metals. Further, since this type of impedance mismatch results in most of the wave being reflected, with little transmitted to the specimen, highly sensitive strain gauges (e.g. semiconductors) are being applied to capture the resulting weak signals.

Methods to address conditions necessary for testing with SHPB systems, which are constant stress equilibrium and constant strain rates, can be achieved by controlling the incident pulses and making use of sufficiently thin material samples. Attaching a pulse shaper to the incident bar can effectively control incident pulses to provide loading rates that are more constant and have more uniform stress equilibrium. Sufficiently thin tissue samples facilitate stress equilibrium sooner and mitigate possible stress-wave attenuation and inertia effects. Preparing sufficiently thin soft tissue specimens may be challenging, since such soft materials are highly compliant. It is extremely difficult to cut such materials into thin layers with uniform thicknesses and of regular shapes.

Discrepancies in sample thicknesses and sizes may cause substantial scatter of results. This limitation can be partially resolved by the application of an optimisation method based on reverse engineering. The basic idea behind this method is to include geometric effects by using a sample-specific FE model in conjunction with a set of optimisation procedures. Using this technique, material parameters are adjusted systematically until the calculated responses match the measured ones. The advantage of this approach relies on the fact that geometric variations are eliminated due to the specificity of the samples. Consequently, optimisations are focused on matching the easily obtainable force-deflection data instead of the more difficult to determine stress–strain relationship, which is highly dependent upon accurate assessments of changing cross-sectional properties. This approach has been successfully employed to characterise the rate sensitivity of human brain tissue at intermediate strain rates (Zhu et al. 2010b).

Characterisation of soft tissues is of current interest in academia and clinical and governmental agencies. A number of studies have been conducted on different types of soft tissue, such as muscle, skin, various internal organs, brain and tendon tissues. However, the data are highly scattered due to the aforementioned challenges in testing techniques, sample preparation and discrepancies due to use of different species. While tests are mainly performed using modified SHPB and SHTB systems, there are no guidelines or standards for necessary modifications, such as the use of pulse shapers. To obtain consistent results, standard testing methodologies for soft tissues are necessary.

Injuries incurred during modern-day combat missions have resulted from underlying mechanisms that are different from those incurred in civilian scenarios and previous types of combat missions associated with jungles. With the high number of military personnel in vehicles exposed to underbelly blasts from IEDs and the rising proportion of military personnel injured but surviving combat scenarios involving blasts, much work is needed to bring research strategies in this area up-to-date. Simulations using FE models could provide essential information for such strategies. However, material properties used in current mathematical computations are based on tests that do not adequately represent conditions of modern-day combat, which involve high strain rates and all three types of human tissue. Testing to elucidate material properties associated with high strain rates is fraught with difficulties, and since there are no guidelines or standards for such specific testing, results are scattered. As a case in point, although it is known that high strain rates produce different effects than medium and low strain rates, information regarding the cut-off points among these states is not available. There is a great demand for work to develop and standardise testing techniques with which to attain the necessary material properties. Possible tools for this work include standardised testing with Split Hopkinson Bar systems and application of optimisation through reverse engineering. The work must include testing on human specimens and consideration of differences in anthropomorphic parameters, such as age and gender. Contributions to this work will result in greater protection and treatments for military personnel and will in turn provide greater readiness of military combat units.

REFERENCES

Adharapurapu, R., Jiang, F., and Vecchio, K. 2006. Dynamic fracture of bovine bone. *Materials Science and Engineering C* 26 (8):1325–1332.

Chafi, M., Karami, G., and Ziejewski, M. 2010. Biomechanical assessment of brain dynamic responses due to blast pressure waves. *Annals of Biomedical Engineering* 38 (2):490–504.

Champion, H., Bellamy, R., Roberts, P., and Leppaniemi, A. 2003. A profile of combat injury. *Journal of Trauma and Acute Care Surgery* 54 (5):S13–S19.

Chawla, A., Mukherjee, S., Marathe, R., Karthikeyan, B., and Malhotra, R. 2006. Determining strain rate dependence of human body soft tissues using a split Hopkinson pressure bar. In *International IRCOBI Conference on the Biomechanics of Impact*, pp. 173–182.

Cheng, M., Chen, W., and Weerasooriya, T. 2009. Mechanical behavior of bovine tendon with stress-softening and loading-rate effects. *Advances in Theoretical and Applied Mechanics* 2:59–74.

Chen, W., and Song, B. 2010. Split Hopkinson (Kolsky) bar: Design, testing and applications. London: Springer Science & Business Media.

Christensen, R., Swanson, S., and Brown, W. 1972. Split-Hopkinson-bar tests on rock under confining pressure. *Experimental Mechanics* 12 (11):508–513.

Comley, K., and Fleck, N. 2012. The compressive response of porcine adipose tissue from low to high strain rate. *International Journal of Impact Engineering* 46:1–10.

Crowninshield, R., and Pope, M. 1974. The response of compact bone in tension at various strain rates. *Annals of Biomedical Engineering* 2 (2):217–225.

Ferreira, F., Vaz, M., and Simoes, J. 2006. Mechanical properties of bovine cortical bone at high strain rate. *Materials Characterization* 57 (2):71–79.

Frantz, C., Follansbee, P., and Wright, W. 1984. New experimental techniques with the split Hopkinson pressure bar. In *8th International Conference on High Energy Rate Fabrication*, pp. 17–21.

Huang, Y. 2013. *ALE3D simulation of proximity hit on a long bone in a gelatin block*. Aberdeen Proving Ground, MD. US Army Research Laboratory (US): ARL report ARL-TR-6426.

Huang, Y. 2015. *Simulation of blast on porcine head*. Aberdeen Proving Ground, MD. US Army Research Laboratory (US), Weapons and Materials Research Directorate (US): ARL report ARL-TR-7340.

Johnson, T. 2005. High strain rate mechanical characterization of trabecular bone utilizing the split-Hopkinson pressure bar technique. MA thesis, Massachusetts Institute of Technology.

Katsamanis, F., and Raftopoulos, D. 1990. Determination of mechanical properties of human femoral cortical bone by the Hopkinson bar stress technique. *Journal of Biomechanics* 23 (11):1173–1184.

Kolsky, H. 1963. *Stress waves in solids*. New York: Courier Corporation.

Kraft, R., Lynch, M., and Vogel, E., III. 2012. Computational failure modeling of lower extremities. In *Proceedings of the NATO HFM-207 Symposium*, Aberdeen Proving Ground, MD. US Army Research Laboratory (US), Weapons and Materials Research Directorate (US): RTO-MP-HFM-207.

Laporte, S., David, F., Bousson, V., and Pattofatto, S. 2009. Dynamic behavior and microstructural properties of cancellous bone. Bruxelles; Belgium: *Dymat* arXiv preprint arXiv:0911.5114, pp. 895–900.

Lee, O., and Park, J. 2011. Dynamic deformation behavior of bovine femur using SHPB. *Journal of Mechanical Science and Technology* 25 (9):2211–2215.

Lewis, J., and Goldsmith, W. 1975. The dynamic fracture and prefracture response of compact bone by split Hopkinson bar methods. *Journal of Biomechanics* 8 (1):27–40.

Lim, J., Hong, J., Chen, W., and Weerasooriya, T. 2011. Mechanical response of pig skin under dynamic tensile loading. *International Journal of Impact Engineering* 38 (2):130–135.

Mahoney, P., Ryan, J., Brooks, A., and Schwab, C. 2005. *Ballistic trauma: A practical guide*. London; New York: Springer Science & Business Media.

McElhaney, J. 1966. Dynamic response of bone and muscle tissue. *Journal of Applied Physiology* 21 (4): 1231–1236.

Moss, W., King, M., and Blackman, E. 2009. Skull flexture form blast waves: A new mechanism for brain injury with implications for helmet design. *The Journal of the Acoustical Society of America* 103 (10): 108701.

Nemat-Nasser, S., Isaacs, J., and Starrett, J. 1991. Hopkinson techniques for dynamic recovery experiments. In *Proceedings of the Royal Society of London A: Mathematical, Physical and Engineering Sciences* 435 (1894):371–391.

Nie, X., Cheng, J., Chen, W., and Weerasooriya, T. 2011. Dynamic tensile response of porcine muscle. *Journal of Applied Mechanics* 78 (2):021009.

Okie, S. 2005. Traumatic brain injury in the war zone. *New England Journal of Medicine* 352 (20):2043–2047.

Pervin, F., and Chen, W. 2009. Dynamic mechanical response of bovine gray matter and white matter brain tissues under compression. *Journal of Biomechanics* 42 (6):731–735.

Pervin, F., Chen, W., and Weerasooriya, T. 2009. Dynamic compressive response of bovine liver tissues. In *Proceedings of the SEM Annual Conference*, Albuquerque, NM, Vol. 4 (1), pp. 76–84.

Pilcher, A. 2004. High strain rate testing of bovine trabecular bone. PhD dissertation, University of Notre Dame.

Roberts, J., Merkle, A., Biermann, P., Ward, E., Carkhuff, B., Cain, R., and O'Connor, J. 2007. Computational and experimental models of the human torso for non-penetrating ballistic impact. *Journal of Biomechanics* 40 (1):125–136.

Saraf, H., Ramesh, K., Lennon, A., Merkle, A., and Roberts, J. 2007. Mechanical properties of soft human tissues under dynamic loading. *Journal of Biomechanics* 40 (9):1960–1967.

Shergold, O., Fleck, N., and Radford, D. 2006. The uniaxial stress versus strain response of pig skin and silicone rubber at low and high strain rates. *International Journal of Impact Engineering* 32 (9): 1384–1402.

Shim, V., Yang, L., Liu, J., and Lee, V. 2005. Characterisation of the dynamic compressive mechanical properties of cancellous bone from the human cervical spine. *International Journal of Impact Engineering* 32 (1):525–540.

Shunmugasamy, V., Gupta, N., and Coelho, P. 2010. High strain rate response of rabbit femur bones. *Journal of Biomechanics* 43 (15):3044–3050.

Song, B., Chen, W., Ge, Y., and Weerasooriya, T. 2007. Dynamic and quasi-static compressive response of porcine muscle. *Journal of Biomechanics* 40 (13):2999–3005.

Taylor, P., and Ford, C. 2009. Simulation of blast-induced early-time intracranial wave physics leading to traumatic brain injury. *Journal of Biomechanical Engineering* 131 (6):061007.

Teja, C., Chawla, A., and Mukherjee, S. 2013. Determining the strain rate dependence of cortical and cancellous bones of human tibia using a Split Hopkinson pressure bar. *International Journal of Crashworthiness* 18 (1): 11–18.

Tennyson, R., Ewert, R., and Niranjan, V. 1972. Dynamic viscoelastic response of bone. *Experimental Mechanics* 12 (11):502–507.

Van Sligtenhorst, C., Cronin, D., and Brodland, G. 2006. High strain rate compressive properties of bovine muscle tissue determined using a split Hopkinson bar apparatus. *Journal of Biomechanics* 39 (10):1852–1858.

Wolf, S., Bebarta, M., Bonnett, C., Pons, P., and Cantrill, S. 2009. Blast injuries. *The Lancet* 374 (9687):405–415.

Wright, T., and Hayes, W. 1976. Tensile testing of bone over a wide range of strain rates: Effects of strain rate, microstructure and density. *Medical and Biological Engineering* 14 (6):671–680.

Zhu, F., Jin, X., Guan, F., Zhang, L., Mao, H., Yang, K., and King, A. 2010b. Identifying the properties of ultrasoft materials using a new methodology of combined specimen-specific finite element model and optimization techniques. *Materials & Design* 31 (10):4704–4712.

Zhu, F., Mao, H., Leonardi, A., Wagner, C., Chou, C., Jin, X., Bir, C., VandeVord, P., Yang, K., and King, A. 2010a. Development of an FE model of the rat head subjected to air shock loading. *Stapp Car Crash Journal* 54:211.

Zhu, F., Skelton, P., Chou, C., Mao, H., Yang, K., and King, A. 2013. Biomechanical responses of a pig head under blast loading: A computational simulation. *International Journal for Numerical Methods in Biomedical Engineering* 29 (3):392–407.

8 The Skull and Brain
Test Methods, Behind Armour Blunt Trauma and Helmet Design

Tom Gibson, Nicholas Shewchenko and Tom Whyte

CONTENTS

8.1 INTRODUCTION

Soldiers have worn helmets and armour for protection from enemy weapons from as early as 1015 BC. The protective equipment has made use of a variety of materials including hair-filled hides, quilted cloth, wood and metals (Houff and Delaney 1973). The design and quality of armour protection has changed significantly as the technology of the weapons and the protective equipment has progressed.

With the advent of firearms in warfare in the 15th century, the use of full armour declined. The weight required to protect from the foreseeable threat became too restricting to the mobility required on the battlefield.

World War I began with an emphasis on mobility until trench warfare developed, with a requirement for protection from shrapnel from artillery and mortar fire. Protective helmets were reintroduced as a critical piece of soldier protection. Australia and the US adopted the British Mk 1 helmet, which followed the pattern of a medieval kettle hat. It was made of 0.036 inch thick manganese steel, weighed approximately 0.9 kg and contained an integral suspension system for comfort, for standoff and to protect from impacts to the crown of the head (Houff and Delaney 1973).

The US M1 helmet became the standard helmet for the Australian army during the Vietnam War. The M1 weighed 1.3 kg and consisted of an alloy steel outer metal shell and a composite hard-hat-like inner liner with an adjustable suspension and the retention system. However, as a result of some design and use issues, this helmet was regularly not worn, or worn unfastened by troops in combat. It was regarded as cumbersome and poor fitting due to its one-size-fits-all configuration, retained significant heat, and could not be used in combination with portable communication devices (Hamouda et al. 2012).

In the early 1970s, the availability of new materials led to the development of the PASGT (Personnel Armour System Ground Troops) composite helmet for the US Army. The design of the PASGT helmet attempted to overcome the issues with the M1. The helmet was manufactured in several different sizes and weighed between 1.3 and 1.9 kg. The new design included protruding ear sections to accommodate communication equipment and a strap suspension system screwed onto the shell which, when later combined with an improved four-point retention system, was able to form a stable helmet–head interface. The stability of the helmet platform has become increasingly

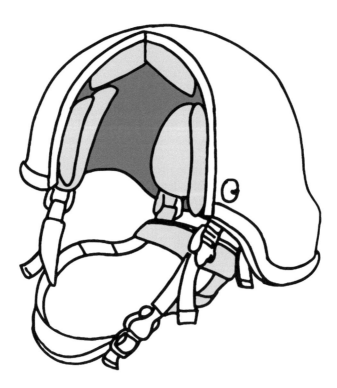

FIGURE 8.1   Sectioned view of a modern US Advanced Combat Helmet with composite ballistic shell, energy absorbing liner and four-point retention system.

important operationally with the development of night vision and advanced sighting systems. The standoff gap between the head and helmet shell of the PASGT was around 12.3 mm and provided ventilation for the transfer of heat as well as a gap for deformation under ballistic impact (Hamouda et al. 2012). The PASGT helmet was used by the US military until 2002 and provided similar levels of blunt impact protection to the crown as the M1 helmet liner.

More recent versions of the US combat helmet, such as the Advanced Combat Helmet (ACH), differ from the PASGT in the use of energy-absorbing pads within the helmet for comfort in order to attenuate blunt impacts and provide standoff for ballistic strikes (Figure 8.1). Although not improving the ability of the helmet to reject heat, the energy-absorbing liner has also been found to have a moderating effect on blast injury.

Helmet design continues to develop with the application of advances in ballistic materials and manufacturing technologies and the need to overcome changes in combat threat, such as the increased exposure to improvised explosive devices (IEDs). The US Enhanced Combat Helmet (ECH) has been in development since 2007 for the US Marine Corps and US Army. It makes use of advances in materials technology, such as the Dyneema HB80 unidirectional composite material which consists of a matrix of ultrahigh molecular weight polyethylene (UHMWPE) reinforced by carbon fibres (Kulkarni et al. 2013). This technology has allowed the design of a new generation of lightweight combat helmets with potentially equivalent or better ballistic protective capability compared to existing designs. Along with the improved ballistic performance, however, some of these new composites have been shown to have higher levels of back face deformation (BFD).

8.2 HEAD INJURY TOLERANCE

8.2.1 ANATOMY OF THE HEAD

The main components of the human head include the scalp, skull, blood vessels, meninges, cerebrospinal fluid (CSF) and the brain. The scalp (Figure 8.2) is 5–7 mm thick and consists of, from exterior to interior, the cutaneous, hair-bearing skin, a subcutaneous connective tissue layer and a thin tendinous layer. These three layers are bound together as a single unit beneath which is loose areolar tissue. The areolar tissue loosely connects the tendinous layer to the pericranium, which is the periosteum of the skull bones, and allows the superficial three layers of the scalp to move over the pericranium.

The face is covered by a relatively thin skin with muscles within it that allow for movement about the nasal, orbital and mouth openings. Deep within the facial skin are the blood vessels and nerves for the facial muscles.

The skull is made up of eight bones connected by sutures, and varies in thickness between 4 and 7 mm. The base of the skull is an irregular plate of bone containing depressions, ridges and passages for arteries, veins, nerves and the spinal cord and brainstem. The skull is a sandwich of cortical bone surrounding a diploë layer of spongy bone (Prasad et al. 1985). Fourteen bones form the face between

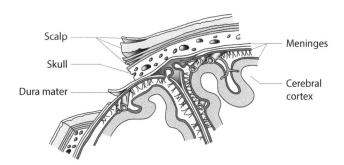

FIGURE 8.2 Cross-section of the head showing the scalp, skull bone, meninges and brain.

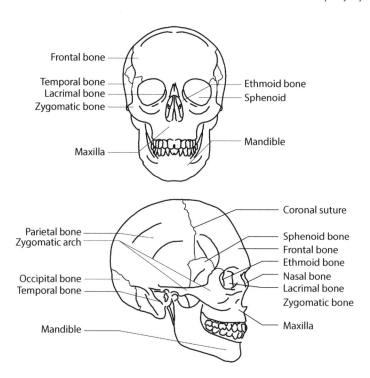

FIGURE 8.3   Main bones of the skull and face.

the forehead and the lower jaw, also fused together through complete sutures with the exception of the mandible that has two moveable attachments to the base of the skull. Facial bones are irregular in shape, frequently hollow, and some are extremely thin. The major bones of the skull and face are shown in Figure 8.3.

Beneath the skull are the meninges consisting of three layers of connective tissue: the dura mater, the arachnoid mater and the pia mater. The outermost dura mater is adherent or close to the inner surface of the bone. Beneath the dura mater is the middle covering, the thin and fibrous arachnoid. The narrow subdural space separating the dura and arachnoid is filled with a small amount of lubricant, preventing adhesion between the two membranes. A number of bridging veins cross the subdural space, draining from the underlying brain to the dura mater and the superior sagittal sinus. The third and innermost meningeal layer is the very thin, delicate and capillary-rich pia mater. The pia is intimately attached to the brain and dips down into the sulci and fissures. The gap between the arachnoid and pia is relatively large compared to the subdural space and is filled with cerebrospinal fluid.

The main parts of the brain are the cerebrum, cerebellum and brainstem. The cerebrum is the most superior part of the brain and is composed of two hemispheres accounting for 83% of the total brain mass. The surfaces of the cerebral hemispheres have ridges and furrows called gyri and sulci, respectively. A number of deeper chasms in the brain tissue are referred to as fissures, which divide the brain into lobes and separate the hemispheres. The only attachment between the hemispheres is a fibrous bundle called the corpus callosum. The cerebellum is the second largest part of the brain, accounting for around 11% of the total brain mass. It is located underneath the occipital lobes in the posterior cranial fossa of the skull. The brainstem is located at the base of the skull where the spinal cord enters the cranial cavity (Figure 8.4).

These parts of the brain are composed of grey and white matter. Grey matter is composed primarily of nerve-cell bodies concentrated in locations on the surface of the brain and deep within the brain. White matter is composed of myelinated axons that largely form tracts to connect parts of the brain to

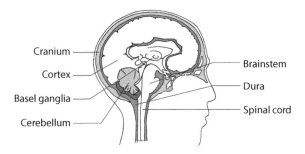

FIGURE 8.4 Brain anatomy.

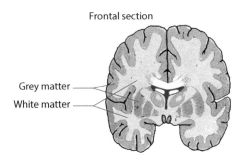

FIGURE 8.5 Frontal section of the cerebrum showing the grey and white matter in the brain.

each other. In the cerebrum and cerebellum, the grey matter is the most superficial tissue, beneath which is the white matter. The brainstem, like the spinal cord, is composed of deep grey matter surrounded by white matter fibre tracts (Marieb 1998) (Figure 8.5).

The head is supported by the neck, which consists of seven vertebrae (C1 to C7), the corresponding intervertebral discs, muscles, ligaments and flesh. The neck also includes components with less structural importance including the spinal cord, the oesophagus, the trachea and numerous blood vessels and nerves (Deng 1985).

The vertebrae consist of a thin surface layer of dense cortical bone enclosing a less dense core of porous trabecular bone. C1 interacts superiorly with the occipital condyles at the base of the skull, and C7 interacts inferiorly with the first thoracic spinal vertebra, T1. At the upper cervical spine there are two atypical vertebrae with unique anatomy called the atlas (C1) and the axis (C2). The atlas is a wide, ring-like bone which supports the skull. It has no vertebral body and consists of two lateral masses joined anteriorly and posteriorly by arches. The atlas can be thought of as a 'washer' that moves with and supports the head with the only substantial degree of relative motion between the skull and C1 in flexion–extension (or nodding) (Nelson 2011). The axis has a spindle-like protrusion called the dens or odontoid process about which the atlas rotates.

8.2.2 Physics of Head Injury Protection

The prime reason for wearing a helmet is to protect the head from injury. Injury results when the tissues of the head are unable to withstand the application of mechanical energy without anatomical or physiological alteration (Gennarelli 2005). Head injuries are multifaceted, with the injury pathophysiology and severity being dependent on the characteristics of the insult to the head in terms of location, energy and load distribution.

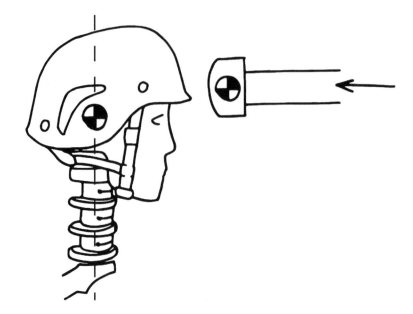

FIGURE 8.6 Impact to the forehead of a helmet mounted on a Hybrid III dummy head and neck, with the dashed line indicating the vertical axis of the head and neck and the crosshair indicating the head CoG.

Figure 8.6 illustrates loading of an upright helmeted head and neck surrogate, in this case a Hybrid III anthropomorphic test device (ATD), by a direct blunt impact to the forehead. The head contact force resulting from this direct impact to the forehead generates motion of the head centre of gravity (CoG) both in translation as well as in rotation (Figure 8.7).

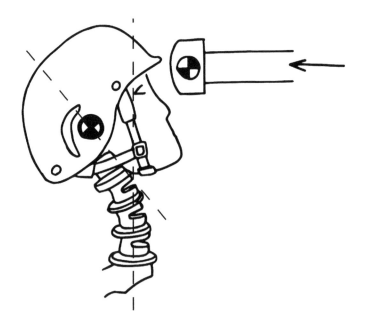

FIGURE 8.7 Combined translational and rotational motion of the helmeted head following a direct impact to the forehead, with the dashed line indicating the initial axis of the head and neck and the crosshair indicating the head CoG.

The displacement x, the velocity v and acceleration a all vary with time during an impact, and there are fundamental relations between each of them. The velocity of an object is numerically equal to the instantaneous rate of change of displacement:

$$v(t) = \frac{dx(t)}{dt} \tag{8.1}$$

Similarly, acceleration is the rate of change of velocity:

$$a(t) = \frac{dv(t)}{dt} \tag{8.2}$$

In the case of rotation, the movement of the body can be characterised in terms of the rotation kinematic terms: rotational displacement θ, rotational velocity ω and rotational acceleration α. The relationships between these terms are analogous to the linear equations:

$$\omega(t) = \frac{d\theta(t)}{dt} \tag{8.3}$$

$$\alpha(t) = \frac{d\omega(t)}{dt} \tag{8.4}$$

Basic physics, Newton's second law, dictates that a body will accelerate when a force F is applied to it. During an impact, acceleration occurs because of the forces generated by the collision of the body with another body. If the two bodies do not deform, the relation between force and acceleration is:

$$F = ma \tag{8.5}$$

where m is the mass of the body.

A rigid body will also undergo angular acceleration when a torque T is applied to it. During an impact, angular acceleration occurs because such a torque is generated. This is usually associated with impacts that have a component that induces rotational motion, such as an impact force component offset from the centre of gravity of the body or when the body is externally constrained by the neck, for example. The equivalent expression for rotational motion is:

$$T = I\alpha, \tag{8.6}$$

where T is the applied (generated) torque, I the moment of inertia and α is angular acceleration.

Since torque is a force acting about a lever arm, it is important to note that efforts to reduce force will typically reduce torque. Thus, reductions in the linear acceleration a will be accompanied by reductions in rotational acceleration α.

A body of mass m moving at a velocity V possesses translational kinetic energy as follows:

$$KE = \frac{1}{2}mV^2 \tag{8.7}$$

A body rotating will possess rotational kinetic energy as follows:

$$RKE = \frac{1}{2}I\omega^2 \tag{8.8}$$

If a body is accelerated (or decelerated), its velocity will change – it will possess more (or less) energy. This change in energy will be associated with the application of a force F, and this process of energy transfer also takes time. During the energy transfer process, for example, the head can deform under the applied load and can thus be injured.

It is fundamental that energy cannot be created or destroyed. When a body's kinetic energy changes, energy is either transferred elsewhere by changing the velocity of the impacting objects or does work by deforming. The energy change by the deformation is often considered to be absorbed. A basic principle of designing personal protective equipment is to reduce the forces likely to injure by absorbing some of the kinetic energy of the impact through the deformation of the protection offered. Thus the head is typically protected against head injury by giving it a layer of energy absorbing material – a helmet.

An impact between two or more bodies, where a large force acts on a body for a short time period, generates an impulsive force:

$$F = \frac{d(mV)}{dt} \tag{8.9}$$

The linear momentum, mV, of the system of bodies must be conserved.

If a moving object strikes the head, which is protected by a helmet with some capability to absorb the kinetic energy of the object, the forces to the head generated in the impact will be less. The extent of this reduction is a function of how much deformation is achieved and the force required to deform the helmet. The simplest relationship between the forces produced and the space required to absorb the energy is:

$$KE = Fd = \frac{1}{2}m(\Delta V)^2 \tag{8.10}$$

where d is the stopping distance, F the average force during the impact, and ΔV the change in velocity.

It can be seen that for a given kinetic energy of the impacting object, the larger the stopping distance d, the lower the force applied F. The actual force developed will be a function of the strength, stiffness, mass and shape of the helmet and impacting object and the mass, shape and stiffness of the head itself.

8.2.3 HEAD INJURY

In simple terms, the main types of head injuries include the following:

- Scalp laceration
- Skull fractures
- Focal brain injuries
  - Contusions
  - Lacerations
  - Haemorrhage
- Diffuse brain injury
  - Concussion syndrome
  - Diffuse axonal injury (DAI)

Head injury may be due to direct contact with the head, where no motion of the head is necessary, but some form of direct blow is required. The mechanisms of relatively low-speed contact injury to the head consists initially of scalp laceration and the skull bending with related skull volume changes – resulting in skull fracture, coup contusions, epidural haematoma (EDH), intracerebral haematoma (ICH), for example. Contact injury can also result from a high velocity projectile, leading to tearing and crushing of the tissues of the head.

Head injury may also be due to motion of the head. In this case, head motion in translation or rotation is required, but a direct impact is not necessary. The mechanisms of motion injury to the head are due to surface strains in the brain and associated organs – resulting in contrecoup contusion and subdural haematoma (SDH), for example, and deep strains – resulting in concussion syndromes and DAI.

It has been pointed out that the categorisation, diagnosis and treatment of severe traumatic brain injury (TBI) has been improving, and this is demonstrated in the significant reduction in severe brain-injury-related mortality in recent times (Gennarelli 2005).

The current understanding of the pathophysiology of TBI has emphasised the multifactorial nature of injury events activated by the injury, with an increasing concern that progressive, long-term consequences are a significant clinical problem (Bramlett and Dalton 2015). Histopathological and behavioural studies are indicating the progressive nature of the initial traumatic insult at all severities of TBI and the involvement of multiple pathophysiological mechanisms, including sustained injury cascades leading to prolonged motor and cognitive deficits. An increased incidence of age-dependent neurodegenerative diseases in the TBI patient population has also been observed.

8.2.4 Military Head Injury Threats

In traditional wars fought since WWI, the majority of all combat-related fatalities were caused by head wounds (Carey et al. 2000). In fact, the head has sustained up to 25% of all reported injury-causing projectile hits, despite the neck and head comprising only 12% of the body area. More recent injury surveys from the conflicts in Afghanistan and Iraq have shown that penetrating injuries account for 11% to 16% of moderate to severe TBI, with the remainder from blast (67%), direct blunt trauma due to vehicle crashes and falls (11% to 13%), and others (Wojcik et al. 2010). The types of ballistic threats associated with head injuries from World War II up to recent events have been predominantly fragments and projectiles originating from handguns, rifles, fragmenting artillery and explosions. When compared to previous conflicts in which the US was involved, the proportion of fragment head injuries from blast origins has increased relative to gunshot wounds. Penetrating injuries to the head can occur due to impacts in non-protected areas (such as the face and neck) and when the helmet shell is perforated by fragments and bullets. Ballistic related blunt trauma can occur when the defeated threat deforms the shell and subsequently impinges on the head.

Several factors suggest that the incidence of TBI currently is higher than in previous conflicts. Mortality has declined, mostly likely because of the advances in body armour worn by military personnel and improvements in the treatment of head injuries (Okie 2005). With high-quality body armour protecting the head and torso, individuals who previously may have died now survive with possible injury to the extremities, head and neck. The reduction in wounding to the thoraco-abdominal region means the proportion of head injuries compared to overall injury is also higher than in previous conflicts (Breeze et al. 2011). Furthermore, the diagnosis of mild TBI is now more frequent than in the past due to an improved understanding of this type of injury (Gennarelli 2005).

In a retrospective survey of Iraq/Afghanistan veterans, the highest prevalence of mild TBI was due to motor vehicle crashes, blasts, bullets/shrapnel and falls (Schneiderman et al. 2008). Similarly, UK military personnel returning from Afghanistan and Iraq sustained mild TBI with loss of consciousness or altered mental state most commonly in blasts, followed by falls, vehicle crashes and fragments and bullets (Rona et al. 2012). Researchers found that the majority of Type 1 TBI cases were caused by explosives (Afghanistan: 66%, Iraq: 68%) (Wojcik et al. 2010). Type 1 TBI are the most severe form and include diagnoses with recorded evidence of an intracranial injury, a moderate prolonged loss of consciousness, or injuries to the optic nerve pathways.

The characteristics of a ballistic threat can vary considerably in terms of mass, speed, material properties and construction. This variation has profound effects on the design of ballistic armour. For small arms, the AK-47 type rifle is the predominant source of ballistic threat in recent conflicts (Small Arms Survey Geneva 2012). The main ballistic threats due to small arms deployed to infantry are the 5.45 × 39 mm, 5.56 × 45 mm and 7.62 × 39 mm type rounds at muzzle velocities of between 735 m/s to 990 m/s (NRC 2014). The bullet construction varies from copper-jacketed lead cores to armour-piercing incendiary.

Fragmentation threats are more difficult to summarise due to the large range of masses, speeds, shapes, materials and even spatial distribution. Based on commonly used artillery, such as the 105 mm and 155 mm howitzer shells, the Advanced Combat Helmet (ACH) performance specification test fragment masses and speeds were found to be representative of the artillery shell threats (NRC 2014). However, some higher mass fragments in the range of 100–200 grain a were not well represented and may be a significant threat.

8.2.5 MILITARY HEAD INJURY CAUSATION

The major threats causing head injuries in military conflicts can be classified into the following three groups:

1. Ballistic threats consisting of:
   a. Penetrating projectile injuries
   b. Behind armour blunt trauma (BABT)
2. Blunt impact threats consisting of:
   a. Direct impact injuries
   b. Inertial injuries from induced head motion
3. Blast threats consisting of the following mechanisms:
   a. Primary, due to blast overpressure
   b. Secondary, due to fragments and shrapnel (ballistic threat)
   c. Tertiary, due to being thrown by blast wave into object (blunt impact threat)
   d. Quaternary, due to crushing from collapsing structures

Figure 8.8 summarises the connection between the mechanical loading related to each type of military threat to the head and the resulting mechanisms of head injury (based on Gennarelli 2005).

High-velocity ballistic penetration may be characterised by high kinetic energy deposition and stress-wave propagation resulting in three regions of tissue damage (Ryan et al. 1988):

1. The wound tract where tissues are lacerated and crushed.
2. The area adjacent to the wound track with tissue damage caused by shearing and stretching.
3. The surrounding area with blood expelled from the vessels and a lack of filling of small blood vessels.

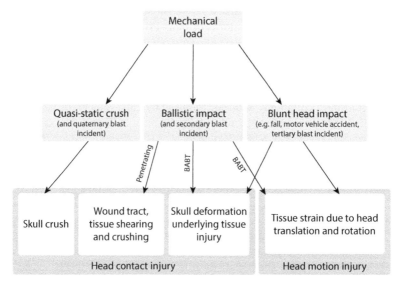

FIGURE 8.8   Type of head injury related to mechanism of mechanical load. (After Gennarelli, T.A., Head Injuries: How to Protect What. In *Snell Conference on HIC*, 2005.)

FIGURE 8.9 A composite ballistic helmet shell interior (left) and exterior (right) deformed by a ballistic impact. (Courtesy of Biokinetics and Associates Ltd., Ottawa, ON, Canada.)

A penetrating ballistic strike to the helmeted head involves direct impact by the impacting particle with the skull as a result of helmet shell perforation. Where the helmet shell has sufficient resistance to penetration, considerable deformation of the particle and dissipation of its energy may occur prior to the interaction with the head. It is this dissipation of the energy of the ballistic particle, which may reduce the head injury severity. It also brings into play another head injury mechanism due to the deformation of the helmet shell in stopping the particle: BABT. The mechanism of BABT involves energy transfer to the head when a non-penetrating ballistic projectile deforms the shell-liner system with subsequent skull contact. The composite helmet deformed by ballistic impact in Figure 8.9 demonstrates the problem.

Head injuries resulting from helmet shell impact (Bass et al. 2003; Sarron et al. 2004; Hisley et al. 2010; Rafaels et al. 2015), include the following:

- Skull fracture
- Damage to underlying tissue (brain, nerves, bloods vessels)
- Haemorrhages, abrasions and lacerations of the scalp
- Contusions of the dura
- Shear injury as a result of head and internal tissue rotation
- Cavitation caused by differential acceleration of the skull and contents
- Fractures of the cervical spine, connective tissue injury and spinal cord injury from differential head and trunk motions

Focal loading of the skull can result in simple linear fractures or complex linear-depressed fractures (Bass et al. 2003; Sarron et al. 2004; Rafaels et al. 2015). The presence of both linear and depressed fractures is unique to BABT and differentiates it from ballistic penetrating injuries. The linear fractures can be extensive and are associated with global loading of the head by the helmet, while the depressed fractures occur locally as a result of focal loading (Rafaels et al. 2015). The extent of fracture from BABT is also related to ballistic shell standoff, with the depressed fracture diameter being reduced as standoff is increased (Sarron et al. 2004).

Knowledge of the underlying brain injury related to BABT has been limited due to the lack of appropriate and available test surrogates. Postmortem human surrogate (PMHS) brains are not suitable for the analysis of post-injury physiological responses. In an investigation of BABT intracranial pressures with dry human skulls filled with silicone, Sarron et al. (2004) found peak intracranial pressures correlated well with skull fracture. The high intracranial pressures and residual forces showed significant energy transfer with the potential for brain injury. Porcine heads have also been used to provide living physiology to study anatomical and physio-pathological changes for investigation of the correlation between biomechanical responses to BABT injuries (Liu et al. 2012). The pressure onset rates were found to be similar to those seen in blast loading, suggesting that peak pressure, impulse and duration were contributors to brain injury in the blast studies.

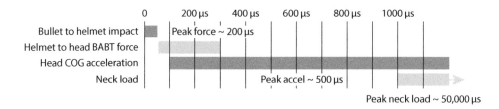

FIGURE 8.10 Typical phasing of the head–neck kinetics during non-penetrating ballistic helmet impact. (After Bass, C.R., et al., Ballistic Helmet Backface Deformation Testing with a Dummy Subject. In *Personal Armour Systems Symposium (PASS)*, Colchester, UK, 2000.)

An anatomically detailed FE head model was protected by two ballistic helmet designs, with suspension and with a padded liner, to study potential head injury metrics for BABT (Pintar et al. 2013). Defeated ballistic impacts (9 mm FMJ <360 m/s) resulted in high intracranial pressures suggesting the presence of brain contusions, while low volumetric strains indicated a low risk of DAI based on current injury thresholds. Brain trauma was also found to be less dependent on the localise loading in comparison to skull fracture, based on the estimated trauma correlating well with the impulse delivered to the head.

With the head anatomical structures having different modes of injury and associated thresholds, phasing of the loading events can be important in determining the primary cause of injury. A study carried out to establish BABT injury functions provided insight into loading to the helmeted head from a non-penetrating 9-mm FMJ bullet strike, Figure 8.10 (Bass et al. 2003). Skull loading was initiated upon shell contact with the head and reached a maximum value by 200–300 µs. The head acceleration reached a maximum much later at 500 µs. Global motion of the helmet then followed at around 1,000 µs, and it was not until maximum head translation was reached that the maximum neck forces and moments were attained, at approximately 50,000 µs. Of importance is that the primary impact event is a magnitude shorter in duration than that typically seen for head impacts in automotive or sports accidents, which typically involve head deceleration pulse durations of 3,000–50,000 µs.

Indirect head and neck injuries may also occur in the absence of head contact for oblique impacts (Hisley et al. 2010). This loading is supported by the simulation work of Aare and Halldin who observed higher brain strain responses for oblique impacts than for direct ones (Aare and Kleiven 2007; Halldin et al. 2013).

8.2.6 HEAD INJURY TOLERANCE

Much of our knowledge regarding head-injury tolerance has been developed by automotive injury researchers. As a result, the tolerance data is predominantly related to blunt impacts. A summary of the development of commonly used head injury measures is presented here. The Wayne State Tolerance Curve (WSTC), first suggested by Lissner et al. (1960), is the basis for most currently accepted indices of head injury tolerance. The WSTC made use of the observed connection between skull fracture and concussion. As originally presented, the 'curve' included six points that represented the relationship between acceleration level and impulse duration (in the range of 1–6 ms), which were found to produce a linear skull fracture in embalmed cadaver heads. In other words, short pulses of high acceleration will cause injury, while lower acceleration levels require longer duration to result in injury. The curve was later extended to pulse durations above 6 ms by the inclusion of comparative animal and cadaver impact data with human volunteer whole body acceleration test data.

Gadd (1971) found that available acceleration tolerance threshold curves, including the WSTC, NASA whole-body human tolerance literature with scaled primate experiments, when plotted on a log-log scale, gave a straight line with a $-1/2.5$ slope indicated the difference between the 'injured' and 'uninjured' regions. Gadd called this line the Severity Index (SI). An SI of 1,000 was used for the concussion tolerance level in a frontal impact. The extent to which the slope of this line was steeper than -1 indicated a greater dependence of injury on the intensity of loading, as opposed to dependence on the time of loading. Gadd also suggested that multiple impulses could be analysed for injury potential by integrating the acceleration over the entire pulse profile using the power weighting factor derived from the slope of the acceleration portion of the profile.

The relative insensitivity of the SI to high accelerations of short duration compared to lower ones with longer durations having equal SI led to the development of the Head Injury Criterion (HIC). The HIC is calculated by integrating the resultant linear acceleration of the head centre of gravity versus time curve over a time interval in which the HIC attains a maximum value:

$$HIC = \left\{ \left[\frac{1}{t_2 - t_1} \int_{t_1}^{t_2} a(t)dt \right]^{2.5} (t_2 - t_1) \right\}_{max} \tag{8.11}$$

where t_1 and t_2 are the initial and final times (in seconds) of the interval during which HIC attains a maximum value, and acceleration a is measured in g (standard acceleration due to gravity).

The maximum time duration of HIC, $t_2 - t_1$, was limited to a specific value between 3 and 36 ms, now more commonly 15 ms. HIC now often uses the 15 ms time interval to more closely match the experimental head impacts on which the measure is based – being hard contact impacts. For example, a HIC_{15} of 700 is the maximum allowed for a Hybrid III 50th percentile male occupant dummy under the provisions of the US advanced airbag regulation (National Highway Traffic Safety Administration 2000). This is estimated to represent a 5% risk of a severe injury (Mertz et al. 1997), where a severe injury is one with a score of 4+ on the Abbreviated Injury Scale (AIS) (Association for the Advancement of Automotive Medicine 1990).

The HIC has become the most widely used head injury tolerance measure, particularly in the area of automotive safety. The HIC is severely limited by being based on head kinematics using rigid body mechanics, and (skull deformation can have a major effect on its measurement); by not considering rotational acceleration; by not being directionally dependent; and, by not being based on specific head injury mechanisms. These limitations often lead to poor correlation between HIC and real-world observations.

Rotationally based injury tolerances have also been developed, with many researchers suggesting that rotation is of far greater influence than translation in causing diffuse brain injuries. Based primarily on primate experiments, bridging vein rupture leading to subdural haemorrhage (SDH) has been found to occur at levels of angular acceleration of 4,500 rad/s^2 and/or angular velocity levels of 70 rad/s. The rotational accelerations necessary to cause concussion and severe DAI for an adult have been estimated to be 4,500 and 18,000 rad/s^2, respectively (Ommaya and Hirsch 1971).

The types of blunt impacts, injury biomechanics and injury mechanisms from military threats may vary markedly from those in industrial, motor vehicle and sports accidents. These, due to availability, are commonly and possibly incorrectly often used for military applications. Figure 8.11 compares the blunt impact energy and momentum requirements from various motor vehicle and sports helmet standards with ballistic military helmet specifications. The motor vehicle and sports helmet requirements are typically associated with higher mass and lower velocity than helmets used for military ballistic projectile impacts. The available energy and momentum/impulse available from various ballistic projectiles referenced in ballistic helmet standards such as the NIJ 0106.01, and the specifications of several US combat helmets (ACH and ECH), are included in the diagram. While the ballistic projectiles may be of low mass and high velocity, the actual head injury risk will be dictated by the residual energy transmitted to the head after interacting with the ballistic shell and liner system.

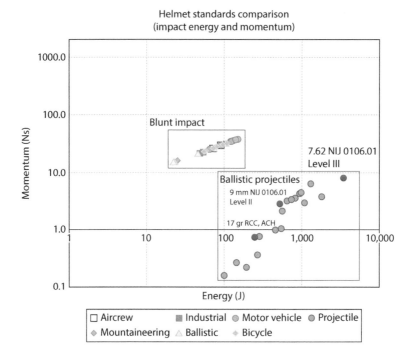

FIGURE 8.11 A comparison of energy and momentum/impulse for helmet performance standards and ballistic projectiles. (Courtesy of Biokinetics and Associates Ltd., Ottawa, ON, Canada.)

Injurious ballistic impacts are associated with peak loads occurring at about 200 µs from the time of impact in comparison to the much shorter loading events seen for blast overpressure related injuries (<100 µs) and the longer durations for blunt impact and chronic related injuries in the military environment (Table 8.1) (NRC 2014). The time duration of loading events can have a significant effect on the response and injury outcome of the human body as a result of the rate sensitivity of human tissue and bones.

In the conditions where BABT in helmeted head impacts are significant, focal skull fractures and high intracranial pressures occur (Bass et al. 2003; Sarron et al. 2004). The focal nature of the injuries suggest that injury predictions based on rigid body assumptions, such as resultant head acceleration and HIC, correlate poorly and that localise measurement of insult to the head, such as pressure or force, are more applicable (Bass et al. 2003).

TABLE 8.1

Comparison of Time to Peak Loading for Blast, Ballistic, Blunt and Chronic Injurious Events

Threat	Time until Peak Loading	Loading Type
Blast impact	3–100 µs	Peak blast overpressure
Ballistic impact	~200 µs	Peak force
Blunt impact	3,000–50,000 µs	Peak head acceleration
Chronic injuries	Hours, days, months, years	Skin irritation/cuts
		Repetitive strain
		Long-term fatigue

Source: After NRC, *Review of Department of Defense Test Protocols for Combat Helmets*, Washington, DC, 2014.

8.3 HELMET PROTECTION TO THE HEAD

8.3.1 BALLISTIC IMPACT PROTECTION

The primary purpose of a soldier's protective helmet is to prevent ballistic perforation and subsequent injury to the wearer. Using current materials technology, this requires using the helmet shell deformation to bring the ballistic projectile to a stop without perforating or contacting the head. The interaction of the projectile with the shell depends on the impact velocity, the projectile's material and the material and construction of the shell. Upon a projectile impact, the initial phase of composite damage to the helmet shell may include shearing of the layers followed by delamination and stretching of the remaining layers, as illustrated in Figure 8.12. Possible deformation of the projectile may also occur.

The manner in which the projectile is stopped involves converting all of the kinetic energy of the projectile to internal energy of the shell deformation. For helmets made from fibre reinforced composite laminates, the deformation and failure of the components contribute to the energy conversion. These may include:

- Fibre (shear and tensile failure, tensile strain)
- Matrix (tensile, compressive and shear failure)
- Laminate (delamination)

Materials used in ballistic helmet design for perforation resistance have evolved from early metal designs to those involving advanced fibre composites offering higher performance, lower mass and greater shell rigidity. Much of this can be attributed to the high strength, toughness and low density of high performance fibres such as aramids, e.g. Kevlar® and Twaron®; UHMWPE (Ultra- High Molecular Weight Polyethylenes), e.g. Dyneema® and Spectra®, PBO (Zylon®) and M5®. Materials are undergoing constant development with consideration of high tenacity UHMWPE, copolymers and nano-materials, including carbon nanotubes and graphene and boron nitride nanotubes for use in rigid armour systems. Production quantities, formability, material integration into composites, environmental degradation and cost are some of the limitations that exist with these newer technologies.

Helmets may also be designed with ceramic appliqués on the outer shell to help defeat high energy rounds, such as NIJ Level III threats, by deforming or fracturing the bullet upon impact (Freitas et al. 2014). Their relatively low density, high compressive strength and high hardness provide good performance although their low toughness, high costs and solution weight limit extensive use of ceramics. Current ceramics used in ballistic armour include boron carbides, aluminas, silicone carbides and ceramic matrix composites (CMC) to improve toughness and flexural strength. The integration of nanomaterials such as carbon nanotubes and boron nitride nanotubes into ceramics is also being investigated to improve ballistic performance (Bolduc et al. 2014).

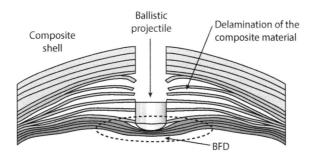

FIGURE 8.12 Partial penetration of a projectile through a composite shell with BFD.

The ballistic resistance of the shell to perforation is typically assessed by ballistic limit tests and proof tests which relate to the bullet/fragment velocity at which there is a 50% chance of penetration (V_{50}) or no chance of penetration (V_0). Most standards provide different threat categories and severity levels from fragments to bullets, each having a profound effect on armour design. Most military ballistic helmets in use today protect against fragments, tested with a fragment simulating projectile (FSP) and a right circular cylinder (RCC) from 2 to 64 grain and from the 9-mm full metal jacket (FMJ) bullet.

Ballistic helmets may provide excellent resistance to perforation but can still cause injuries to the wearer if the shell deforms sufficiently to cause contact with the head resulting in BABT. This is of concern for materials having greater ballistic efficiency, but at the expense of lower structural stiffness, such as UHMWPE. If sufficient energy is transferred to the head, skull fractures and underlying tissue injuries can occur. As a result, most ballistic helmet standards place controls on the maximum amount of deformation or load transfer to the head.

With regard to the physics of the ballistic impact event, the collision of the bullet with the helmet–head system must be considered. The collision can be treated as two bodies undergoing inelastic impact where momentum is conserved (Lucuta 2014). Kinetic energy, however, is not conserved as it is converted to strain energy during shell elastic deformation, with additional energy being converted during plastic deformation of the shell and bullet in the form of internal energy, for example, thermal energy. Any remaining energy from the impact is transferred to the head either from direct shell contact, through the suspension-liner system, or both. This results in further energy conversion during deformation of the tissue and bone and with subsequent movement of the head and neck. Both linear and rotational head kinematics have been noted to occur during ballistic impact, as predicted by numerical modelling (Aare and Kleiven 2007; Halldin et al. 2013).

Ballistic helmet performance standards now aim to control the degree of shell deformation to limit contact with the head. This shell deformation is called variously back face deformation (BFD), back face signature (BFS) or ballistic transient deformation (BTD), depending on the test standards being referenced. The BFD, BFS and BTD are often evaluated by the residual indentation left in a pliable witness medium, such as clay, placed behind the impact site. The degree of indentation, however, is difficult to directly relate to injury, and different methods are required to evaluate injury potential. Both BABT and behind helmet blunt trauma (BHBT) are common descriptors for these types of injuries. Metrics used to predict BABT/BHBT include the stresses or strains imposed on the underlying head and are related to injury risk functions or thresholds to determine their outcome. Understanding the cause and effect of BTD in terms of helmet design and projectile interactions helps to identify the critical parameters for improving performance and reduction of BABT.

The influence of shell construction on BABT outcome as measured with a load-sensing headform (Anctil et al. 2006). Two helmets were tested with similar ballistic resistance (17 gr FSP, V_{50} = 630–670 m/s) but with differing areal density (10.8 and 6.5 kg/m^3). The heavier helmet was tested with and without an impact liner. The presence of the liner resulted in a significant decrease of approximately 50% in measured impact force, a metric associated with skull fracture potential. A similar decrease was also seen between the medium and large size of the second lightweight helmet as a result of the greater standoff. In general, an impact liner and increased standoff greatly improves backface performance of existing shell technologies.

In another study, increased shell standoff was also shown to significantly reduce the risk of skull fracture in a series of 9-mm FMJ ballistic impacts with flat armour plates and dry PMHS heads. Greater standoff distances were associated with reduced size of comminuted and depressed local fractures (Sarron et al. 2004). An air gap of 12.7 mm between the inner helmet shell and the head was suggested as a good compromise between overall helmet size and level of protection.

The effect of shell materials and stiffening methods was investigated with composite panels constructed of various types of UHMWPE (Lucuta 2014). The projectiles used in the study were the 17-grain FSP and 9-mm FMJ bullet, at velocities of 560 and 434 m/s, respectively. The FSP hardness was also varied to investigate the effect of projectile deformation on perforation and BTD. Overall, the 17-grain FSP behaved differently than the 9-mm FSP, with stiffer armour configurations resulting in lower BTD for the 9-mm FMJ and higher BTD for the fragment. The hypothesis that armour with a high V_{50} rating would result in lower BTD was not confirmed due, in part, to the small sample size.

In another study of ballistic helmet performance and injury potential with refreshed PMHS skulls (i.e. initially using processed skulls, which are dehydrated due to cleaning and drying and thus making them more brittle than fresh bone, the skull is rehydrated in order to restore their ductility and strength to resemble fresh bone), Freitas noted that for 9-mm FMJ rounds shot at 427 m/s, the foam pads placed behind the impact site of the shell helped to mitigate cranial injuries and reduce intracranial pressures by attenuating the impact energy of the FMJ rounds (Freitas et al. 2014). For impacts with 7.62 calibre rounds fired at 725 and 847 m/s, the foam pads did not immediately compress, but provided a direct load path into the cranium resulting in high intracranial pressures. It is interesting to note that cranial injuries were also mitigated in the lateral regions where the shell bridged the impact area while being supported by pads on either side.

The influence of ballistic shell stiffness on injury potential was investigated with an FE model of a helmet and head (Aare and Kleiven 2007). Simulated ballistic impacts (9-mm rigid projectile, 360 m/s) were carried out on several PASGT type shells with interior suspension systems. A decrease of shell stiffness led to skull contact with subsequent high onset skull stresses local to the impact site. When shell stiffness increased to the point where no contact occurred, a small increase in brain strain levels was observed due to higher global head acceleration. Brain strains reached a minimum when the shell was compliant enough to absorb the bullet energy but not sufficiently so to cause head contact. Brain pressure did not change significantly as shell stiffness decreased until contact occurred. Oblique ballistic impacts to the shell caused a decrease in skull stresses but an increase in brain tissue strain, with a maximum strain occurring in 45° impacts due to the larger tangential load component and subsequent head rotation. With strains approaching published limits for DAI, the authors recommended that rotational kinematic requirements in ballistic helmet performance evaluations. The study was not without limitations, however, due to assumptions made in the shell material properties, bullet interactions, and simplified anatomical detail of the head model.

Reduction of head trauma can be achieved through the reduction of the energy transfer to the head and/or increase in contact area with the head. In the context of ballistic helmet design, these parameters can be related back to helmet component characteristics including the fibre tensile strength, modulus, density, laminate inter-ply adhesion, matrix type and geometric shape.

Successful helmet design is achieved when the threat is defeated in a manner that reduces loading severity to the head system to sub-injurious levels. Avenues for reducing contact and non-contact severity involve kinetic energy loss of the bullet, momentum transfer to the head, shell deformation and stiffness, behind shell load distribution, shell stand-off, and impact liner/suspension system properties. These parameters are summarised in Table 8.2. A balance between these design features is required to achieve optimal protective performance, but it must be recognise that these are not the only criteria for successful helmet design. Of equal importance are human factors considerations which need to be considered to ensure user acceptance when affording ballistic protection. Factors such as head borne weight, heat dissipation/ventilation, balance, comfort and stability need to be considered when designing the ballistic protection offered by the helmet. Many current-day combat helmet acquisitions by military organisations supplement ballistic performance requirements with human factors based rankings to ensure ultimate acceptance by the soldier and, hence, improved operational effectiveness.

8.3.2 Blunt Impact Protection

Prior to the ACH, US combat helmets such as the PASGT and the M1 were not required to provide any tested level of blunt impact protection (McIntire and Whitley 2005). Given the potential head and

TABLE 8.2

Aspects of Helmet Design in Head Loading and Trauma and Their Implications

Helmet Characteristics	Energy Implications	Contact Area Implications
Increased shell standoff	Allows for greater deformation and energy loss	Contact area may decrease and with less energy transfer
Inclusion of an impact attenuating liner	Energy loss by plastic or viscoelastic deformation of material	Load is distributed over a greater area of the head
Increasing shell structural stiffness and/or material modulus	Greater kinetic energy retained by the shell	Increased load distribution with greater material engagement during impact
Increased permanent shell damage (delamination, fibre tensile/shear failure)	Kinetic energy attenuated through permanent processes	May require greater shell standoff to reduce head contact area and loads

neck injuries resulting from motor vehicle accidents, tripping, falling and to parachutists in combat and other operational circumstances, blunt impact protection was introduced to the ACH by fitting pads with energy attenuating capabilities.

The energy-absorbing mechanism had to be robust enough to reduce the impact energy to a low injury probability level throughout a realistic range of impact velocities and environmental temperatures, regardless of the helmet impact site.

A number of different padding systems are used to achieve blunt impact protection for the US combat helmet, such as the Oregon Aero® Pads, SKYDEX, Team Wendy ZAP™ Pads, Mine Safety Appliances (MSA) and GENTEX helmet pad systems. These systems commonly use a type of foam such as polyurethane or a formed thermoplastic layer with energy absorbing properties afforded by the formed geometry (Howay 2010). The padding systems must provide adequate energy absorption across a range of temperatures and maintain wearer comfort in general use.

8.3.3 ERGONOMIC AND OPERATIONAL REQUIREMENTS

For a helmet to be effective, it must be worn. The M1 helmet was often not worn by American troops in the battlefield and high numbers of ballistic-related head injuries were observed in personnel (Carey et al. 2000). Principal to ensuring the helmet is worn are the human factors and ergonomic considerations of the helmet including head borne weight, heat dissipation/ventilation, balance, comfort and stability. The M1 design was awkward due to its one-size-fits-all configuration, significant retention of heat, lack of stability and inability to be used in combination with portable communication devices (Hamouda et al. 2012). Many military organisations supplement performance requirements with human factors rankings during helmet acquisitions to ensure ultimate acceptance by the soldier and operational effectiveness.

Weight is one of the primary concerns for the combat helmet acceptance as there is typically a trade-off between ballistic protection (thickness and coverage area) and the corresponding weight. There is a practical limit to the weight of the helmet above which it will not be worn. This weight limit has been placed at 2.5 kg for headgear which is balanced on the head.

Retention of heat is a common issue which affects the perceived comfort for all helmets, particularly in hot climates. The most efficient means of reducing the build-up of heat and moisture vapour is including an air gap between the composite shell and the head (Egglestone and Robinson 1999).

Sizing, stability and balance of the helmet are also important. An improperly-sized and unstable helmet tends to lag behind the movement of the head. Any displacement of the helmet's centre of mass from the head CoG imposes a moment on the head which can lead to fatigue of the neck muscles and discomfort.

Given the increased propensity for incorporating communication, sighting and surveillance equipment into the helmet, the helmet itself must not interfere with these devices nor interfere with the visual or acoustic fields of the wearer. Helmet mounted gear should not exert additional load on the head and neck, as this may affect balance, increase fatigue and lead to repetitive strain injury to the neck.

Finally, the helmet must be durable. The ACH is subject to multiple environmental tests including seawater immersion, flame resistance, temperature extremes and vibration, among others.

Refining the ergonomic requirements has continued through each generation of the combat helmet. The PASGT helmet addressed the primary limitations of the M1 design while also reducing the weight, thereby providing the wearer with greater comfort (Campbell 2009). In a study comparing the PASGT helmet and the ACH (Ivins et al. 2007), greater satisfaction was found with the ACH, with the majority of users satisfied with its fit, weight, overall impression and maintainability. The ACH users reported significantly less headaches after wearing the helmet for extended periods, less irritation or cuts to the skin, greater comfort, better fit and were more accepting of the weight.

8.3.4 HELMET RETENTION

Another fundamental aspect of helmet protection is that the helmet must remain on the head during likely injurious events.

In a study of aircrew helmets, Vyrnwy-Jones et al. (1989) reported that the Aviation Life Support Equipment Retrieval Program (ALSERP) statistics indicated that 21% of all aircrew helmets were ejected during a crash sequence. Peacetime statistics from the US Naval Air Development Center (NADC) indicate that 8% of helmets worn during ejections are lost during the escape sequence or during terrain contact. Improvements to the retention system in the SPH-4 helmet reduced ALSERP to 12%; however, in this later cohort study, a further 18% of all helmets were recorded as having rotated on the wearer's head during the accident sequence. Rotational displacement of the helmet during a crash sequence can expose the head of the wearer, thus allowing unprotected secondary impacts and injuries to occur.

The retention system must ensure that the helmet remains in place, not just in use, but also in extreme impact conditions. At the same time, the design of the retention system must allow the user to easily don and doff the helmet and be easily adjusted to allow proper fit.

In the ACH, the retention system consists of a four-point chinstrap with a nape pad. One standardised retention system fits all sizes and manufacturers of the ACH shell. The chinstrap uses an open cup for the chin with a side release buckle. The ACH retention system is required to meet a number of tests to ensure adequate strength, helmet stability and that the helmet remains on the head when force is applied to it.

8.4 MILITARY HELMET STANDARDS, SPECIFICATIONS AND TEST METHODS

8.4.1 EVALUATION METHODS

Performance test methods to evaluate perforation resistance have been developed since the inception of ballistic helmets. Improvements and changes to the methods have occurred over time to reflect changes in the threats, knowledge of injury mechanisms and tolerances or for improvement of the test methodology itself.

The development of suitable test methodologies and specifications to assess the protective performance of armour systems depends on the objectives to be met. Aspects for consideration may include:

- The desired measurement or response being gathered (e.g. BFD, TBI, neck loads)
- The desired level of realism of the biological system (e.g. micro or macro biofidelity)
- The desire for predictive or absolute evaluation approaches (e.g. threat input versus head impact response)

- The level of statistical confidence, repeatability and reproducibility desired
- The intended user (e.g. researcher, regulatory body, quality control)
- The costs and time for conducting the evaluations

The various evaluation methods are presented below to highlight possible approaches:

- *Guidelines and specifications*: Perhaps the simplest approach to controlling head protection is to provide design specifications that are closely related to performance. Many ballistic helmet standards, for example, stipulate a minimum shell standoff distance in an effort to control behind armour trauma in the absence of measuring the insult and related injury risks. Other requirements may include experimental procedures and measurements to assess the risk of closed head injury, e.g. peak head acceleration or applied force. The validity of the metrics or guidelines depends on the level of correlation achieved with human injury and becomes an important part of defining injury risk functions and thresholds. However, it is not uncommon to have performance specifications dictated by the performance of the existing protective technologies and not push for improvements to head protection based on the potential of new designs or technologies. The control of in-service combat helmets for new acquisitions or quality-control measures is an example of where this may apply.
- *Biological tissues*: Where greater fidelity and understanding of the injury mechanisms is required, biological tissues can be used to characterise the response and disruption of tissue on a local level. The tissue is typically isolated from the host and may not account for the bulk response as if it were *in vivo*. Tissue samples subject to direct blast loading, for example, have been used to study stress-wave propagation and cellular damage. Test methodologies based on biological tissues are better suited to research environments and not for routine combat helmet performance evaluations due to their complexity and variability during use.
- *Biological systems*: Where the response of a larger biological system is required, such as the local and global head response from bullet strike, animal models may be used where access to PMHS is limited or where a biologically similar system can be more readily tested. Such testing can provide direct insight into the response or injury mechanics. Furthermore, live animals may be used where assessment of the pathophysiological response is required, recognising that correlation with humans is needed.
- *Non-biofidelic surrogates*: Some performance evaluation methods use instruments and test equipment that provide engineering measures of the response from ballistic impact but do not respond in a human-like or biofidelic manner. The use of a witness plate to assess perforation instead of having a brain simulant or the use of clay to assess BFD are examples where the results are sufficient to determine that the necessary performance level has been achieved.
- *Biofidelic surrogates*: This is defined as a representative physical or numerical model that is human-like in response to insult. Surrogates are typically simplified in their representation of human response in order to capture the essential characteristics of interest such as global head response or skull loading. Simplicity helps reduce the number of degrees of freedom during trials, resulting in improved repeatability and reliability compared to more complex surrogates. Examples where a biofidelic surrogate may be suitable would be in the evaluation of BFD where skull forces and head accelerations may need to be measured while providing overall system response to the ballistic impact.

8.4.2 Ballistic Helmet Testing and Standards

Historically, ballistic helmet test standards have been developed to provide controls on helmet performance for perforation resistance and BFD and are typically supplemented with aspects of blunt impact attenuation, retention, human factors and build quality. Procedures for sample selection, conditioning, marking and mounting are also provided. The test standards can be provided as part of a national or institutional standard and are often imbedded into product specifications for article acceptance and quality control. Some common ballistic test standards are presented in Table 8.3, which also briefly summarises the penetration and behind armour test methodologies.

TABLE 8.3
Ballistic Helmet Test Standards and Specifications Summary

Standard	Perforation	BABT
NIJ 0106.01 'Ballistic Helmets' (NIJ 1981)	Method: Headform with witness plate Spec: No penetration of witness plate Sample: 2 (ambient, wet)	Method: Rigid DOT headform on linear rail Spec: Headform acceleration <400 G Sample: 1 (ambient)
ASTM E54.04 Subcommittee on Personal Protective Equipment (PPE)	Under development	Under development
HPW-TP-0401.01B 'Test Procedure – Bullet Resistant Helmet' (H.P. White Laboratory Inc. 1995)	Method: NIJ headform with witness plate Spec: No penetration of witness plate Sample: 2 (ambient, wet)	Method: NIJ headform with clay inserts Spec: None provided Sample: 2 (ambient, wet)
NATO STANAG 2920 'Ballistic Test Method for Personal Armour Materials and Combat Clothing', 2nd Ed. (NATO 2003)	Method: Clamped shell with witness plate Spec: No penetration of witness plate	
NATO STANAG 2920 'Ballistic Test Method for Personal Armour Materials and Combat Clothing', 4th Ed. Proposal	Method: Clamped shell with witness plate Spec: No penetration of witness headform	Method: BLSH (supplemental to Plastilina (clay) method) Spec: Peak total force <6,000 N for 25% risk fx <7,200 N for 50% risk fx
DOI-MTD-03/101 (Metropolitan Police P.P.G. 2003)	Method: Clamped shell Spec: No part of projectile penetrates inner shell	Method: Clamped shell with witness clay Spec: None provided
MPS DOI-MTD-03/010 Update (Barnes-Warden and Fenne 2014)	Method: Clamped shell Spec: No part of projectile penetrates inner shell	Method: Clamped shell with laser displacement sensor Spec: None provided
CSA Z613 Draft, 2006 (under development)	Method: Penetration headform Spec: V_{proof}. No shell perforation Sample: 3 (all ambient, per ammunition)	Method: BLSH Spec: Peak force, no limit provided Sample: 3 (all ambient, per ammunition)
Purchase Description Helmet, Advanced Combat (ACH) (NSC 2011)	Method: NIJ headform w mid-sagittal and coronal slots, witness plate Spec: V_{50}, V_{proof} (see Table 8.5). Sample: 4 sizes (V_{50}/V_{proof} = ambient, x-hot, x-cold, wet, weatherometer, aged)	Method: NIJ headform w mid-sagittal and coronal slots, clay filled Spec: <16 mm BTD top, left, right; <25.4 mm for front, rear Sample: 4 sizes (ambient, cold, hot, wet)
Purchase Description for Enhanced Combat Helmet (ECH) (MSCS 2009)	Method: NIJ headform w mid-sagittal and coronal slots, witness plate Spec: V_{50}, V_{proof} (see Table 8.5) Sample: 6 sizes (V_{50}/V_{proof} = ambient, x-hot, x-cold, wet, weatherometer, aged)	Method: NIJ headform w mid-sagittal and coronal slots, clay filled Spec: <16 mm BTD top, left, right; <25.4 mm front, rear Sample: 4 sizes (ambient, cold, hot, wet)
STANAG 2902, 2nd Ed., 2004 – Criteria for a NATO Combat Helmet	CLASSIFIED	CLASSIFIED

X = extra, Fx = fracture.

Revision of the NIJ 0106.01 standard is currently in progress by the ASTM E54.04 Subcommittee on Personal Protective Equipment (PPE) (ASTM 2015). Performance tests for perforation, behind armour trauma, and blunt force impact requirements are being reviewed, with consideration of recent developments and research studies in these areas. Improvements to calibration requirements, operational procedures and sampling methodology are also expected to improve repeatability and reproducibility.

8.4.2.1 Threat Selection

The projectiles employed in ballistic resistance and BABT evaluations need to represent the foreseeable real-world threats within the military environment. Military threats may include bullets, shrapnel and fragments with blast being a prevalent source in modern-day conflicts. The current use of 2-, 4-, 16-, 64-grain RCC fragments, the 17-grain FSP and the 9-mm FMJ in military ballistic resistance test standards, such as for the ACH (NSC 2011), need to be monitored on a regular basis for suitability since military conflicts, armour systems and their threats change, with corresponding changes to penetrating injuries or BABT.

The projectiles used for the assessment of penetration and BFD of combat helmets varies considerably between countries, with the 17-grain FSP and 9-mm FMJ ball bullets being the most common (see Figure 8.13). While the 9-mm FMJ is also referenced in law enforcement standards, including the NIJ 0106.01 Standard for Ballistic Helmets, the 17-grain FSP has good penetrating potential due to its hardness and small projected area, while the 9-mm FMJ has significant potential to cause backface deformation without penetration (Lucuta 2014). Reasons for this high deformation potential include the high mass and subsequent momentum transfer to the shell, estimated to be 2.8 Ns (Anctil et al. 2005). Large helmet deformations for defeated ballistic strikes have also been noted for the 0.30 calibre and 0.50 calibre fragments as well as the 7.62-mm rounds (Watson et al. 2008; Freitas et al. 2014).

8.4.3 Ballistic Penetration Testing

Common ballistic helmet penetration test methodologies draw on standard ballistic test practices for body armour, such as those from the US National Institute of Justice (NIJ) and NATO (STANAG).

For a test, the helmet may be rigidly fixed or placed on a support medium, such as clay, and struck by a bullet or fragment simulating projectile at various speeds measured by chronographs or Doppler radar (Figure 8.14).

The perforation of the armour is monitored by the perforation of witness plates or deformation of a support medium, such as clay, behind the target area. Definitions for perforation exist in the various standards to ensure consistent evaluation by the technician, especially in cases of partial penetration.

(a)   (b)

FIGURE 8.13 The (a) 7-grain FSP and (b) 9-mm FMJ bullet. (Courtesy of Biokinetics and Associates Ltd., Ottawa, ON, Canada.)

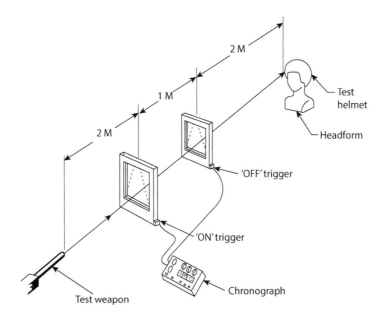

FIGURE 8.14   A ballistic helmet test setup for assessing shell perforation performance. (From NIJ, *Ballistic Helmets*, Washington, DC, 1981.)

The NATO STANAG 2920 'Ballistic Test Method for Personal Armour Materials and Combat Clothing' standard provides a methodology for assessing ballistic helmet penetration with a large range of FSPs onto a clamped shell, as pictured in Figure 8.15 (NATO 2003). The methods of V_{50} and V_0 determination are included.

The National Institute of Justice NIJ 0106.01 standard for ballistic helmets (NIJ 1981) uses a rigid headform fitted with a witness plate in the mid-coronal and mid-sagittal planes to assess the penetration resistance of ballistic helmet shells (Figure 8.16).

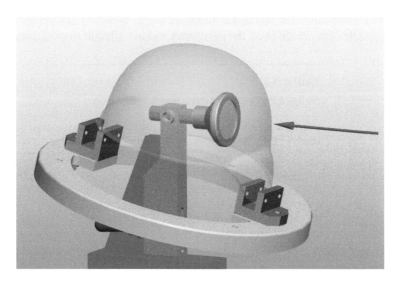

FIGURE 8.15   NATO STANAG 2920 ballistic helmet clamp test setup.

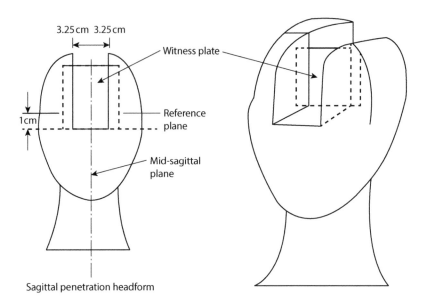

3.25 cm 3.25 cm

Witness plate

1 cm

Reference plane

Mid-sagittal plane

Sagittal penetration headform

FIGURE 8.16   The NIJ 0106.01 sagittal and coronal penetration headforms. (From NIJ, *Ballistic Helmets*, Washington, DC, 1981.)

The threats in NIJ 0106.01 cover three levels (Levels I, II-A, II), with the details summarised in Table 8.4. Both fore-aft and lateral impact sites are provided at fixed locations. Failure constitutes perforation of the witness plates for the four impact sites tested at the specified bullet velocity (V_{proof}). Some helmet manufacturers also claim ballistic protection to NIJ Level III-A (44 Magnum; Submachine Gun 9 mm), but this requirement is borrowed from the ballistic resistant protective materials standard NIJ 0108.01, as there is no such level in NIJ 0106.01.

8.4.3.1   Test Variability of Penetration Tests

Ballistic helmet performance assessment variability can stem from a number of different sources including helmet production and material variances and the experimental procedures used for measuring performance. Some production variances can be controlled through quality control processes to ensure that the materials, their processing methods, and construction fall within specifications; however, the complex interaction of all the parameters make it difficult to predict the effect on the

TABLE 8.4
Helmet Protection Levels for NIJ 0106.01

Helmet Level	Ammunition	Mass (gram)	V_{proof} (m/s)
I	22 LRHV Lead	2.6	320
	38 Special RN Lead	10.2	259
II-A	357 Magnum JSP	10.2	381
	9-mm FMJ	8.0	332
II	357 Magnum JSP	10.2	425
	9-mm FMJ	8.0	358

LRHV = long rifle high velocity, RN = round nose, JSP = jacketed soft point, FMJ = full metal jacketed.

final ballistic performance; therefore, reliance must be placed on physical testing and evaluation. Variability associated with testing is in addition to the helmet production variability; therefore, a high degree of testing accuracy and repeatability is required to differentiate true differences in helmet performance. Ideally, measurement accuracy 10 times greater than that of the differences in helmet performance to be identified is desired.

The possible sources of testing-related variability were listed by the National Research Council (NRC 2014), and these are summarised as follows (with some additions):

- *Gauge-to-gauge* (measurement) variability, which is the differences in the measurement tools, such as the tools used to quantify the degree of indentation in clay (Plastilina)
- *Operator-to-operator* variability involves either conducting the tests or interpreting the results differently
- *Lab-to-lab* variability is the reproducibility of test results from one lab to another, usually due to systematic differences
- *Lab environment* conditioning variability results from fluctuations in temperature and humidity in which the test samples and equipment are exposed to
- *Projectile speed and impact* variability may arise from variations within the speed, obliquity and yaw specifications
- *Helmet configuration* variability, involving different impact pad types and layouts, sizing and adjustments, can all change test results
- *Helmet-to-headform standoff* variability may occur when using improper headform sizes for testing or during settling of the helmet on the headform
- *Clay (Plastilina)* variability may result from material batch differences or temperature effects as it cools during testing
- *Impact location* variability from bullet trajectory inaccuracies or deviations during penetration
- *Helmet conditioning* variation, temperature, humidity and other deviations including soak times and cooling effects
- *Projectile material, hardness and geometry* variability
- *Projectile launching* variability, including spin rate and flight perturbations

It must also be recognised that reduction of variability is not always possible due to random errors or because of the time, effort and technical resources required. This does not mean, however, that methods do not exist to further reduce variability. Statistical sampling practices can reduce the perceived variability by utilising an averaged response for a sample of data that has some known distribution. From this, confidence intervals and other estimates of accuracy can be established providing the degree of certainty obtained in the test sample. Determining the number of samples to be measured, however, requires prior knowledge of the helmet and test variability. This requires repeatability and reproducibility studies to be carried out and the sources of variability identified and quantified (NRC 2014).

Ballistic helmet performance standards are not only used to establish test methods and performance requirements, but they must also deal with the practical need to test a small fraction of the whole helmet population to approximate their true performance and obtain statistically-relevant measurements. A lack of statistical confidence in the results may mean that the true helmet performance is not captured, and a risk of failure may occur – a potentially serious problem for users of the armour as well as purchasers, regulators and industry. As a result, strict controls on experimental variability and sampling methodologies need to be employed.

Experience gathered from published efforts on testing variability and sampling methods can provide some understanding of the issues. The new Canadian ballistic helmet standard in development, CSA Z613, makes use of a method to estimate the ballistic limit of a helmet by conducting a sufficient number of tests such that the V_{50} was reached ±10 m/s with a 90%–95% confidence level for statistical significance (Anctil et al. 2008). The estimated number of shots required was 20–35 per projectile, depending on shell performance variability.

In a separate study on the ACH lot (batch) acceptance for resistance to penetration, it was found that sampling methods and rejection rates could have seemingly different implications on maintaining helmet performance levels. The initial acceptance criteria in the product specification involving 20 impacts with zero allowable penetrations (the so-called 90/90 rule where a 10% chance of penetration will have a 10% chance of acceptance) and the acceptance criteria with a larger sample size involving 240 test and 17 allowable penetrations (also meeting the 90/90 rule) could allow helmets to be produced with a lower margin of safety and still be accepted (NRC 2014). In the study, sources of product and test variability were reviewed, highlighting the need for their further quantification. Specific recommendations were made to keep the larger sample sizes, but with more stringent acceptance criteria to maintain the current level of safety margin for combat helmets; to quantify systematic error sources; conduct interlab reproducibility studies; and investigate helmet configuration and impact location variability.

The above studies demonstrate the continued need to quantifying helmet and test variability to establish appropriate sampling procedures and experimental design in order to obtain confidence in the test results, whether for research or quality control applications.

8.4.4 BEHIND ARMOUR BLUNT TRAUMA MEASUREMENT

In NATO STANAG 2920 (NATO 2003), no test method exists for non-penetrating ballistic impact in the 2nd edition. However, it has recently been proposed to update the standard to include a BABT performance limit (Bolduc and Anctil 2010). The yet unreleased 4th edition recognise the Ballistic Load Sensing Headform (BLSH) as an alternative BABT assessment methodology to clay (Plastilina). It has been proposed to make use of the peak summed force value with the Blunt Criterion (BC) to evaluate the risk of skull fracture. A peak total force of 6,000 N would correspond to a 25% risk of skull fracture.

In a separate study, an alternative injury metric, the peak total impulse, was suggested to improve variability since the force measurements were susceptible to small variations (Philippens et al. 2014). However, no injury function based on impulse was provided.

The NIJ 0106.01 test methodology for ballistic impact attenuation is based on global head accelerations from ballistic impact at the rated threat level for four impact sites (Figure 8.17). The standard requires that the peak acceleration be limited to 400 G when collected in compliance with SAE J211b having a CFC 1000. This acceleration limit is similar to that of blunt impact in FMVSS 218 but without the limits on the pulse duration. Its relevance to ballistic impacts is not known.

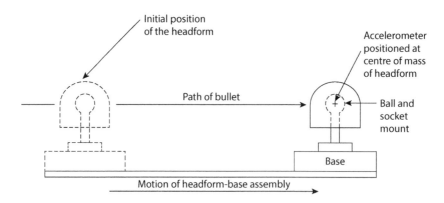

FIGURE 8.17 The NIJ 0106.01 ballistic impact headform. (From NIJ, *Ballistic Helmets*. Washington, DC, 1981.)

8.4.4.1 Clay Headform

The ballistic helmet penetration test procedure by H.P White Laboratory Inc. uses modified head-forms from NIJ 0106.01 to assess penetration and BFD of helmets intended for law enforcement use. The headform is filled with clay (Plastilina) to allow measurement of the residual indentation for non-penetrating impacts (see Figure 8.18). While this BFD measurement is not biomechanically based, it provides a comparative measure of the potential for BABT commonly used for body armour evaluations, for example, in the NIJ 0101.06 standard. The threats employed in the standard are similar to those used in the earlier version of the body armour standard, NIJ 0101.03.

The measurement of behind armour effects with clay headforms has been noted to be variable due to the difficulty in controlling its consistency, both spatially and temporally (NRC 2012). Furthermore, the degree of armour support may affect the armour performance, as a stiffer response of the clay (Plastilina) provides greater resistance to the impinging shell and so affects the peak dynamic displacement. Observations of large BFD measurement variability across helmet sizes and shot locations within the same model have been made (NRC 2014).

8.4.4.2 The Ballistic Load Sensing Headform (BLSH)

The Ballistic Load Sensing Headform (BLSH) reflects the current understanding of BABT. A brief chronological account development of the headform is presented to illustrate the recent advances in the ability to analyse behind armour deformation and its relation to injury.

Between 1998 and 2000, a NATO-sponsored BABT task group expressed interest in developing an objective test method able to measure the forces and accelerations associated with non-penetrating ballistic strikes on combat helmets, with the intent of measuring head injury risk. In initial efforts by Bolduc and Tylko (1998), the headform and neck from a standard Hybrid III automotive

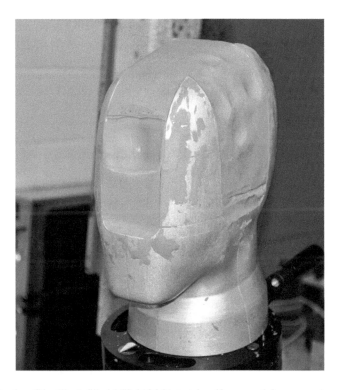

FIGURE 8.18 A clay (Plastilina) filled NIJ 0106.01 test headform used for measuring shell deformation. (Courtesy of Biokinetics, and Associates Ltd., Ottawa, ON, Canada.)

anthropomorphic test dummy (ATD) was used to measure transmitted force. In a combined Canadian and the US study, the surrogate head measurements were correlated with skull fracture (Bass et al. 2000; Waclawik et al. 2002; Bass et al. 2003). The instrumented Hybrid III headform used for the study was also fitted with a matrix of five surface-mounted piezoelectric polyvinylidenefluoride (PVDF) film sensors, having combined pressure and strain capability. A similar configuration was applied to PMHS heads. Further, the cadaveric heads were instrumented with multiple triaxial head accelerometers, skull-mounted strain gauges, upper neck force-moment transducers and intra-cranial pressure and strain sensors.

The parietal area was selected for the study, which has a fracture tolerance between the frontal and temporo-parietal areas based on skull fracture tolerances for low-rate loading studies. The selected projectile was the 9-mm FMJ, 7.7 g bullet for its ability to cause armour BFD. Bullet velocities of 400–480 m/s were used. Experiments were conducted with various headform test setups, and the response of different aramid and UHMWPE helmets investigated.

Results from the study showed that the head acceleration contained significant high-frequency content of which increased in magnitude with increasing bullet velocity. The skull pressure sensors recorded increased magnitude with higher bullet velocity but greater variability, partly attributed to off-centre impacts and local helmet variations. Both the acceleration and pressure measurements were able to differentiate responses between the aramid and UHMWPE helmets. Neck load and moment data were well below the N_{ij} automotive-based neck injury criterion, suggesting that neck injury was unlikely with the 9-mm FMJ and ballistic helmets tested. A study conducted by Bass et al. (2000) and reported on by Waclawik et al. (2002) demonstrated that peak transmitted force measured at the skull surface correlated well with skull fracture due to BFD. The head accelerations were also shown to be poor predictors of head injury when the head injury criterion (HIC) was assessed. The use of the Hybrid III head and neck with force sensors were recommended as an objective test method for assessing BABT.

Further investigation of the PVDF sensors and a miniature piezoelectric load cell was carried out to assess accuracy, repeatability and robustness in BABT applications and for inclusion into a ballistic helmet performance standard (Anctil et al. 2004). A non-lethal-type projectile (37 mm diameter, 39 gram mass, 30 m/s) was selected to replicate typical behind armour loading from a combat helmet. Results showed that best correlation with the impactor load was obtained with the miniature load cell. The work led to a prototype Miniature Load Cell (MLC) headform with loads measured at the five skull locations similar to that used in previous studies by Bass et al. (2003). The suitability of the miniature load cells was demonstrated under simulated ballistic strike.

The BLSH was further modified and used to assess the performance of various ballistic helmets (Anctil et al. 2005). The Hybrid III-shaped headform now incorporated miniature piezoelectric load cells (Kistler Type 9212, 22 kN maximum load, $f_n = 70$ kHz) with larger load plates to capture more of the behind armour loading area. Typical force responses under ballistics strike are depicted in Figure 8.19 with the headform shown in Figure 8.20. A vinyl skin covered the load cells and was similar to that of the automotive test dummy as used by Bass et al. (2003).

The researchers found that the measured peak force varied by helmet type, and the computed 50% risk of skull fracture for a specific helmet was at a force of 4,256 N. The findings of the study (Anctil et al. 2005) and a follow-on study (Anctil et al. 2006) suggested that the new load sensing test methodology was robust and sensitive to the helmet constructions, helmet/skull standoff and the type of impact liner used. Skull loading from the deformed ballistic shell was found to extend beyond the sensing area, possibly leading to incorrect force measures.

Further validation of the injury criterion at other locations of the BLSH other than the parietal area was required. New measurement locations were introduced at the front and rear of the BLSH headform. The number of load cells was increased to seven to improve the measurement accuracy (Figure 8.21) (Anctil et al. 2006). Global head acceleration and neck load instrumentation were also added for situations where significant momentum transfer to the head might result in closed head injuries.

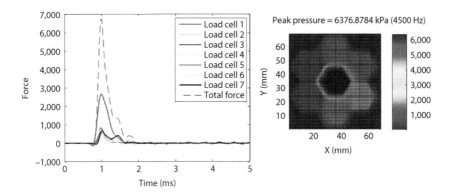

FIGURE 8.19 Typical load cell response of the improved headform under ballistic strike (Anctil et al. 2004). (Courtesy of Biokinetics and Associates Ltd, Ottawa, ON, Canada.)

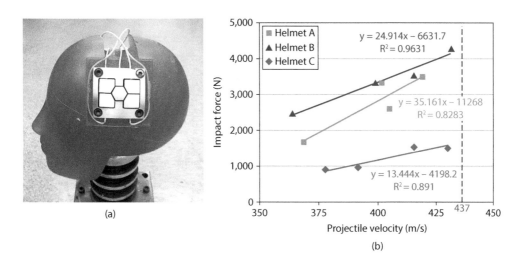

FIGURE 8.20 The modified BLSH and peak impact forces from various ballistic helmets (Anctil et al. 2005). (Courtesy of Biokinetics and Associates Ltd., Ottawa, ON, Canada.)

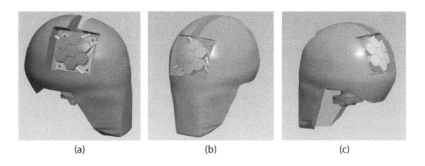

FIGURE 8.21 An improved ballistic load sensing headform with multiple impact sites shown without skin pads (Anctil et al. 2006). (Courtesy of Biokinetics and Associates Ltd., Ottawa, ON, Canada.)

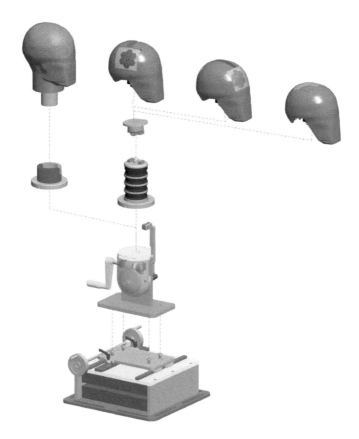

FIGURE 8.22   The latest BLSH test system. (Courtesy of Biokinetics and Associates Ltd., Ottawa, ON, Canada.)

The headform shape was also changed to that of the ISO type headform commonly used in helmet performance test standards, also providing a wider range of sizes (ISO 1983). Data processing from the sensor array was changed from using the peak force from any one load cell to the total summed force. This improved the test variability due to off-centre impacts and was able to capture more of the shell deformation. Later experimentation showed that the summed peak force and impulse were able to better discern helmet-related performance differences (Merkle et al. 2010).

The BLSH continued to undergo changes addressing issues arising from its use by several international organisations, with improvements to the data analysis methods, elastomeric skin pads, additional impact sites, and three headform sizes (ISO E, J and M) to cover the 5th percentile female to the 95th percentile male (Figure 8.22). The use of the head and neck response measurements was limited to relative comparisons due to the lack of validation with existing Hybrid III automotive-based head injury criteria.

To address the lack of an accepted comprehensive test protocol for assessing BABT with the BLSH, a test methodology was proposed by Philippens et al. (2014) and evaluated it in a round-robin test series. The findings showed some systematic variability between the BLSH systems, which limited its use to research purposes only. However, the BLSH provides greatly reduced variability compared to clay-based headforms when the BLSH is used with defined test protocols, random skin pad selection, controlled helmet placement and good experimental design.

8.4.4.3   Other BABT Test Methodologies

An alternative BABT test methodology has been developed using a spherical instrumented rigid headform to measure contact force and head/neck accelerations from non-penetrating ballistic impact (Watson et al. 2008). This methodology demonstrated correct phasing of the contact loads, but further validation with ballistic impacts was required.

A test methodology was also developed to assess behind armour effects from non-penetrating ballistic strike for the law enforcement community (Barnes-Warden and Fenne 2014). Upon review of ballistic shell deformation patterns and existing published BABT test methods, a single force sensor (PCB 200C20, f_n = 40 kHz) was used for simplicity, consistency and low cost in the context of implementing production controls for helmet acquisitions. The headform does not have a compliant skin to reduce test variability and the cost of replacement. It has anatomical features to allow accurate placement of the helmet for evaluating the coverage area. Five impact sites are currently available on a single headform, but the absence of the skin pad on shell performance has not been reported. As with other systems, a transfer function which has been correlated with PMHS data is required to assess injury risk.

Freitas et al. (2014) reported on the development of a biofidelic surrogate head for assessing BABT, with the ability to replicate skull fractures seen with PMHS heads but without the complications of handling and ethics approvals. Human craniums were commercially processed and refreshed to restore the bending strength and modulus of the bone, and subsequently filled with a Perma-gel™ brain surrogate. A silicone dura was included for pressurisation of the cerebral spinal fluid (water). A soft tissue layer of Perma-gel™ was also applied to the exterior of the skull. As a unique head is required for each test, each head had to be tuned to match the mass, size and thickness of human tissues. Sensors fitted to the skulls included surface pressure, intracranial pressure, cranial strain and triaxial head accelerations. A wide range of threats were experimented with including the 64-grain RCC (right circular cylinder), 9-mm FMJ, 7.62 × 39 (PS) and 7.62 × 51 (M80) projectiles. Different pad suspension system configurations were investigated, and the aramid shell was reinforced with a ceramic appliqué to withstand the higher energy 7.62 calibre threats. The resulting skull fracture modes were similar to those seen in other research, with PMHS heads and the resulting peak intracranial pressures also similar to those observed by Sarron et al. (2004) for protected PMHS heads. The human head surrogate (HHS) possessed good fidelity, thereby providing insight into BABT risk from different rounds and helmet configurations. No formal validations were conducted on the fracture tolerance of the craniums, but small samples of the refreshed skull bone were compared to human cranial bone in three-point bending tests to confirm strength and ductility.

8.4.4.4   Test Variability for BABT Assessment

The assessment of behind armour effects with headform-based measurement systems is a complex process involving multiple parameters and interactions. Each parameter represents a potential source of variability and source of systematic error. Identification of the major contributing parameters and their strict control is important to minimise variability. These may include:

- The bullet type and internal ballistics
- The helmet (geometric and material consistency of the shell, liner, retention, standoff, impact location)
- The headform (skin pad response if applicable, resonance, sensing area)
- The load cell bandwidth, calibration, conditioning, data collection
- The experimental design including sample size, impact sequence and number of repeats
- Data post-processing algorithms and metrics used

In addition to the test parameters that contribute to systematic errors, there are sources of error that are random in nature and cannot be typically controlled. These may include:

- Bullet interactions (deformation, failure mode, stability)
- Shell performance (deformation, delamination, shearing)
- Retention, liner performance (deformation, failure modes)
- Sensor performance (discontinuities, resonance, cross-talk)
- Data collection (artefacts, noise, bias)
- Human error (helmet placement, measurements, data collection, interpretation and transcriptions)

The number of parameters and error sources are typically defined by the particular test system and test methodology being used. Good practice entails minimising the number of parameters, controlling error sources and exercising good experimental design to minimise or overcome known sources of variability.

The BLSH system and test methodology were evaluated for variability during the development of a more comprehensive test protocol (Philippens et al. 2014). The study involved three ballistic laboratories that tested used ballistic helmets manufactured within the same year. No additional controls on production lots were implemented. Each lab employed their ballistics equipment, data conditioning and acquisition equipment. The projectile was specified as a 9-mm FMJ round to be fired from barrels with similar twist rates. The helmets were tested in two configurations with shell only and shell with suspension system. Test parameters that were controlled included the shell position, standoff, projectile obliquity, data collection rates, processing and bullet velocity (415 ± 10 m/s). The shell standoff for the shell-only test condition was controlled with physical spacers placed at the crown, sides and rear of the helmet. Three shots were used for the shell only and another nine for the helmet system. The peak force was measured for the centre load cell and the summation of all seven.

Results from the study showed that despite attempts to control standoff, considerable variation existed across the helmet samples, and was partly due to some helmet asymmetry between left and right sides. With the inclusion of bullet velocity variability, the peak total forces varied considerably within and across laboratories. Total peak force differences of 2.1–5.4 kN were noted between impact sites across the laboratories. Within laboratories, peak total force ranges of 0.2–4.6 kN were noted. No systematic differences were noted between the bare shell and full system tests.

The results suggest that differences exist between laboratories with sources of variation noted comprising the threat and its delivery, the test sample and sample positioning. Helmet standoff was identified as a major contributor to variability, with peak total force decreasing as standoff increased by approximately 260–290 N/mm, depending on the impact site. Furthermore, some variation in helmet shell laminate construction was noted along with a mounting hole located above the rear load cell array. The authors further detailed the major test parameters in Figure 8.23, but their relative contributions to test response variability were not provided.

8.4.5 DRAFT CSA Z613 BALLISTIC HELMET STANDARD

Many of the ballistic helmet standards have not been updated to reflect changes in composite material performance nor reflect the latest body of knowledge on head injury. To address these concerns, the development of a new standard was initiated under the Canadian Standards Association (CSA) Technical Committee Z613 Ballistic Helmets, elements of ballistic penetration (V_{proof}), ballistic limit (V_{50}) and BABT (Anctil et al. 2008) were considered. The evaluation of

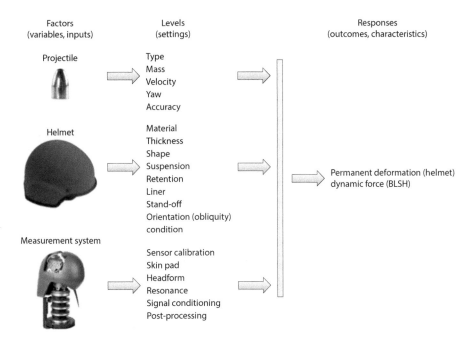

FIGURE 8.23 Test parameters related to the assessment of BABT (Philippens, Anctil, and Markwardt 2014). (Courtesy of Biokinetics and Associates Ltd., Ottawa, ON, Canada.)

the ballistic limit was considered optional and was intended for quality assurance purposes. Additional requirements for helmet retention, stability and non-ballistic blunt impact attenuation were also included. With ongoing changes to threats in the real world, a review of the ballistic threats was carried out in the context of military, law enforcement and corrections applications and resulted in four types of ballistic threats – fragments, handgun bullets, rifle bullets and shotgun slug. To further represent the severity of the various threats in different operating environments and the state of protective technologies, up to four velocity levels were defined for each projectile (Table 8.5).

BABT assessment was with the BLSH and to be evaluated at the four impact sites provided (fore-aft and lateral) (see Figure 8.24). The peak force exceeding some preset limit would constitute a failure.

8.4.6 OTHER HELMET TESTING REQUIREMENTS

Typically the specification of a military helmet will include other testing requirements other than those for the ballistic performance. For example, a test specified for the US ACH is used to measure the blunt impact energy absorption of the helmet assembly in a drop test and is based on the US Federal Motor Vehicle Safety Standard (FMVSS) 218 for motorcycle helmets. Impact attenuation is measured by determining the acceleration imparted to an instrumented rigid test headform on which a complete helmet is mounted. The helmet is fitted to a DOT headform on a monorail drop test apparatus (see Figure 8.25). It is dropped at a velocity of 10 ft/s (3 m/s) onto a hemispherical anvil, twice on each of seven locations in environmental conditions of ambient, cold and hot. Separate helmets, pads and retention systems may be used for each impact. No individual acceleration may exceed 150 G.

TABLE 8.5

Proposed Ballistic Threat Levels for the Draft CSA Z613 (Anctil et al. 2008)

Type	Protection Level	Test Projectile Description	Mass (g)	V_{50} (m/s)	V_{proof} (m/s)	BABT	Notes
Fragment	FG-A	FSP	1.10	450	390–410	–	Lightweight design with minimum protection
		Steel sphere	6.74	–	240–260	✓	
	FG-B	FSP	1.10	630	570–590	–	Current helmet design
		Steel sphere	6.74	–	390–410	✓	
	FG-C	FSP	1.10	730	670–690	–	20% increased protection over FG-B
		Steel sphere	6.74	–	490–510	✓	
Handgun	HG-A	9 mm FMJ RN	8.04	–	389–409	✓	NIJ 0101.06 Level II
	HG-B	0.357 SIG	8.10	–	439–459	✓	NIJ 0101.06 Level IIIA
	HG-C	44 Mag SJHP	15.55	–	427–447	✓	NIJ 0101.06 Level IIIA
	HG-D	9 mm penetrator	7.06	–	357–377	–	High penetrating performance handgun bullet
Rifle	RF-A	7.62 FMJ-SP	8.10	–	630–650	✓	Corresponds approximately to FG-B and HG-B, i.e. current helmet designs
	RF-B	7.62 NATO FMJ (45 degrees)	9.59	–	8.38–8.58	✓	NIJ 0101.06 Level III at 45° obliquity
	RF-C	7.62 NATO FMJ	9.59	–	8.38–8.58	✓	NIJ 0101.06 Level III at 0° obliquity
	RF-D	5.56 mm SS109	3.95	–	900–920	✓	High penetrating performance rifle bullet
Shotgun	SG-A	12-GA rifled slug	28.40	–	415–455	✓	HOSDB level SG1

Source: Courtesy of Biokinetics and Associates Ltd., Ottawa, ON, Canada.

FIGURE 8.24   The BLSH headform for assessing BABT in the draft CSA Z613 (Anctil et al. 2008). (Courtesy of Biokinetics and Associates Ltd., Ottawa, ON, Canada.)

FIGURE 8.25   Blunt impact test apparatus consisting of a monorail guided drop system with the helmet mounted on a rigid headform and a hemispherical anvil. (Courtesy of Biokinetics and Associates Ltd., Ottawa, ON, Canada.)

A retention system strength test is also specified for the ACH to measure the static and dynamic strength of the helmet retention system – the chin strap. It is also based on the US Federal Motor Vehicle Safety Standard FMVSS 218 for motorcycle helmets. The helmet is rigidly fixed by either a clamp or by being placed on a rigid headform, as shown in Figure 8.26. The retention system is attached to a grip that simulates the jaw. A vertical load of 150 pounds is applied and held for a

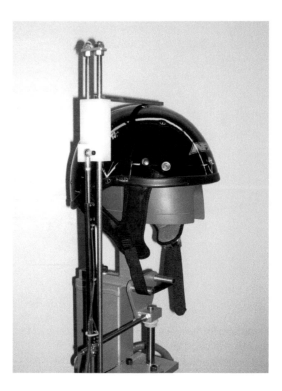

FIGURE 8.26   FMVSS 218 retention system strength test apparatus. (Courtesy of Biokinetics and Associates Ltd., Ottawa, ON, Canada.)

minimum of 1 minute. The requirements are: no component of the retention system to fail, the retention system closure device not release (open) and the webbing not slip.

8.5   DEVELOPMENT OF MILITARY HELMET DESIGN

The components of the combat helmet, including the ballistic shell and energy absorbing padding, will continue to be developed and optimised for increased performance, reduced weight and improved usability. For example, the ECH offers significantly greater perforation protection against fragments than the ACH due to the use of unidirectional UHMWPE fibre in a thermoplastic matrix. The shift to UHMWPE has been enabled by a new generation of preforms and manufacturing methods appropriate for the material (NRC 2014). At the same time, the development of these new materials has resulted in the need for more stringent BFD requirements.

Carbon nanotubes (CNTs) are another promising ballistic helmet shell material with lightweight, high strength and good energy-absorbing capacity (Kulkarni et al. 2013). Figure 8.27 shows the ballistic limit of CNTs is considerably higher than currently used ballistic helmet materials, making a polymer matrix reinforced by nanoparticles like CNTs a good candidate for the combat helmet (Kulkarni et al. 2013).

In addition to enhancing the current components, additions to the helmet are being developed to give greater protection to the currently exposed vulnerable regions of the face, eyes and neck. The increased injuries to these regions, attributable to the increased exposure to improvised explosive devices (IEDs) in the field, have prompted attention to ballistic goggles and face shields (Breeze et al. 2011; Kulkarni et al. 2013). Research into transparent materials for ballistic protection and the design of protective eyewear and face shields is ongoing and

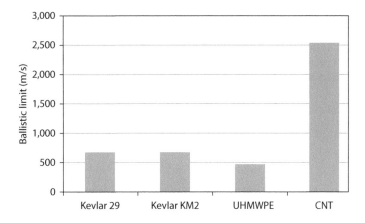

FIGURE 8.27 Comparison of the ballistic limits of various combat helmet materials with carbon nanotubes (CNT) based on molecular dynamics simulations. (After Kulkarni, S.G., et al., *US Army Res.*, 101, 313–331, 2013.)

aimed at providing more complete head protection from ballistic, blast and blunt impact threats. Currently, the Australian Army is introducing an enhanced ballistic laser ocular protection system (BLOPS) and an enhanced hearing protection system as part of the new soldier combat ensemble. These devices, as well as other maxillofacial protection systems, may remain attachments which can be easily donned and doffed (NRC 2014) but must either be accommodated by the combat helmet design or be integrated as part of the helmet, such as pictured in Figure 8.28.

Non-protective enhancements are also being made to combat helmets to improve soldier awareness of their surroundings in the field. In the largest US Army head-protection research and development project, the 'Helmet Electronics and Display System – Upgradable Protection' (HEaDS-UP) program explicitly acknowledges that the helmet is no longer simply a device to prevent injury from fragments and blunt impact. The helmet has become a platform to provide the soldier with new capabilities to enhance their survivability (NRC 2014). Components that may be included in future helmets include high-resolution miniature displays and sensors, along with a body-worn computer that could provide digital maps, tactical overlays and other battlefield data (Carey et al. 2000; NRC 2014). Auditory communication features such as a radio headset and microphone can be incorporated into the helmet, allowing

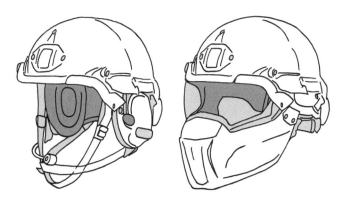

FIGURE 8.28 US Advanced Combat Helmet soldier ensembles with hearing protection (left) and integrated facial protection (right).

communication with squad members and others. Current attachments such as night vision and other devices, such as remote weapons sighting, global position devices and laser range finder have been suggested additions for integrating into the combat helmet (Carey et al. 2000).

The overriding ergonomic limitation for the soldier in the field remains the maximum weight that can be carried by the soldier. This factor is repeatedly identified in discussion of future battlefield helmet designs, along with comfort and the need for more efficient cooling systems (Carey et al. 2000; Breeze et al. 2011; Hamouda et al. 2012; NRC 2014).

REFERENCES

Aare, M., and S. Kleiven. 2007. Evaluation of Head Response to Ballistic Helmet Impacts Using the Finite Element Method. *International Journal of Impact Engineering* 34 (3): 596–608. doi:10.1016/j.ijimpeng. 2005.08.001.

Anctil, B., D. Bourget, G. Pageau, J.-P. Dionne, M. Wonnacott, K. Rice, and A. Toman. 2008. The Development of a Ballistic Helmet Test Standard. In *Personal Armour Systems Symposium (PASS)*, Brussels, Belgium.

Anctil, B., D. Bourget, G. Pageau, K. Rice, and J. Lesko. 2004. Evaluation of Impact Force Measurement Systems for Assessing Behind Armour Blunt Trauma for Undefeated Ballistic Helmets. In *Personal Armour Systems Symposium (PASS)*, The Hague, The Netherlands.

Anctil, B., M. Keown, D. Bourget, and G. Pageau. 2005. A Novel Test Methodology to Assess the Performance Ballistic Helmets. In *22nd International Symposium on Ballistics*, Vancouver, BC.

Anctil, B., M. Keown, D. Bourget, G. Pageau, K. Rice, and G. Davis. 2006. Performance Evaluation of Ballistic Helmet Technologies. In *Personal Armour Systems Symposium (PASS)*, Leeds, United Kingdom.

ASTM. 2015. *Subcommittee E54.04 on Personal Protective Equipment (PPE)*.

Barnes-Warden, J., and P. Fenne. 2014. Ballistic Helmet Evaluation Developments. In *Personal Armour Systems Symposium (PASS)*, Cambridge, United Kingdom.

Bass, C. R., B. Bogges, B. Bush, M. Davis, R. Harris, S. Rountree, S. Campman, et al. 2003. Helmet behind Armour Blunt Trauma. In *Proceedings of the RTO AVT/HFM Specialists' Meeting on Equipment for Personal Protection (AVT-097) and Personal Protection: Bio-Mechanical Issues and Associated Physio-Pathological Risks (HFM-102)*, Koblenz, Germany.

Bass, C. R., B. Boggess, and B. Bush. 2000. Ballistic Helmet Backface Deformation Testing with a Dummy Subject. In *Personal Armour Systems Symposium (PASS)*, Colchester, United Kingdom.

Bolduc, M., and B. Anctil. 2010. Improved Test Methods for Better Protection, a Proposal for STANAG 2920. In *Personal Armour Systems Symposium (PASS)*, Quebec City, Canada.

Bolduc, M., J. Lo, R. Zhang, D. Walsh, S. Lin, B. Simard, K. Bosnick, M. O'Toole, M. Bielawski, and A. Merati. 2014. Toward Better Personal Ballistic Protection. In *Personal Armour Systems Symposium (PASS)*, Cambridge, United Kingdom.

Bolduc, M., and S. Tylko. 1998. Hybrid III Head Response to Undefeated Combat Helmet in a Ballistic Environment. In *Personal Armour Systems Symposium (PASS)*, Colchester, United Kingdom.

Bramlett, H. M., and D. W. Dalton. 2015. Long-Term Consequences of Traumatic Brain Injury: Current Status of Potential Mechanisms of Injury and Neurological Outcomes. *Journal of Neurotrauma* 32 (23): 1834–1848.

Breeze, J., I. Horsfall, A. Hepper, and J. Clasper. 2011. Face, Neck, and Eye Protection: Adapting Body Armour to Counter the Changing Patterns of Injuries on the Battlefield. *British Journal of Oral and Maxillofacial Surgery* 49 (8): 602–606. doi:10.1016/j.bjoms.2010.10.001.

Campbell, S. 2009. PASGT Helmet Test: An Example of Effective Intra-Government Testing Collaboration. *International Test and Evaluation Association Journal* 30: 557–561.

Carey, M. E., M. Herz, B. Corner, J. McEntire, D. Malabarba, S. Paquette, and J. B. Sampson. 2000. Ballistic Helmets and Aspects of Their Design. *Neurosurgery* 47 (3): 678–689. doi:10.1227/00006123-200009000-00031.

Deng, Y. C. 1985. *Human Head/Neck/Upper-Torso Model Response to Dynamic Loading*. University of California, Berkley, CA.

Egglestone, G. T., and D. J. Robinson. 1999. *Venting of a Ballistic Helmet in an Attempt to Reduce Thermal Loading*. Melbourne, VIC: Defence Science and Technology Organisation.

Freitas, C. J., J. T Mathis, N. Scott, R. P. Bigger, and J. Mackiewicz. 2014. Dynamic Response Due to Behind Helmet Blunt Trauma Measured with a Human Head Surrogate. *International Journal of Medical Sciences* 11 (5): 409–425. doi:10.7150/ijms.8079.

Gadd, C. W. 1971. Tolerable Severity Index in Whole-Head Non-Mechanical Impact. In *15th Stapp Car Crash Conference*, pp. 809–816, New York.

Gennarelli, T., Abbreviated Injury Scale 1990. Barrington, IL: Association for the Advancement of Automotive Medicine 1990.

Gennarelli, T. A. 2005. Head Injuries: How to Protect What. In *Snell Conference on HIC*.

Halldin, P., D. Lanner, R. Coomber, and S. Kleiven. 2013. Evaluation of Blunt Impact Protection in a Military Helmet Designed to Offer Blunt & Ballistic Impact Protection. In *1st International Conference on Helmet Performance and Design*, pp. 1–12, Imperial College London.

Hamouda, A. M. S., R. M. Sohaimi, A. M. A. Zaidi, and S. Abdullah. 2012. Materials and Design Issues for Military Helmets. In *Advances in Military Textiles and Personal Equipment*, edited by E. Sparks, pp. 103–138. Oxford, UK: WoodHead.

Hisley, D., J. Gurganus, J. Lee, S. Williams, and A. Drysdale. 2010. *Experimental Methodology Using Digital Image Correlation (DIC) to Assess Ballistic Helmet Blunt Trauma*. Army Research Laboratory, Aberdeen Proving Ground, MD.

Houff, C. W., and J. P. Delaney. 1973. *Historical Documentation of the Infantry Helmet Research and Development*. Human Engineering Laboratory, Aberdeen Proving Ground, MD.

Howay, K. 2010. *Characterization and Modification of Helmet Padding System to Improve Shockwave Dissipation*. Clemson University, SC.

H.P. 1995. *White Laboratory Inc. Test Procedure – Bullet Resistant Helmet*. Maryland, USA.

ISO. 1983. *ISO/DIS 6220—Headforms for Use in the Testing of Protective Helmets*. International Organization for Standardization, Switzerland.

Ivins, B. J., K. A. Schwab, J. S. Crowley, B. J. McEntire, C. C. Trumble, F. H. Brown, and D. L. Warden. 2007. How Satisfied Are Soldiers with Their Ballistic Helmets? A Comparison of Soldiers' Opinions about the Advanced Combat Helmet and the Personal Armor System for Ground Troops Helmet. *Military Medicine* 172 (6): 586–91.

Kulkarni, S. G., X.-L. Gao, S. E. Horner, J. Q. Zheng, and N. V. David. 2013. Ballistic Helmets—Their Design, Materials, and Performance against Traumatic Brain Injury. US *Army Research* 101: 313–331. http://digital commons.unl.edu/usarmyresearch/201.

Lissner, H. R., M. Lebow, and F. G. Evans. 1960. Experimental Studies on the Relation between Acceleration and Intracranial Pressure Changes in Man. *Surgery Gynecology, and Obstetrics* 3: 329–338.

Liu, H., J. Kang, J. Chen, G. Li, X. Li, and J. Wang. 2012. Intracranial Pressure Response to Non-Penetrating Ballistic Impact: An Experimental Study Using a Pig Physical Head Model and Live Pigs. *International Journal of Medical Sciences* 9 (8): 655–664. doi:10.7150/ijms.5004.

Lucuta, V. 2014. Ballistic Limit and Dynamic Back-Face Deformation of Helmet Shell Composite Materials and Corresponding Deformation of the Projectiles. In *Personal Armour Systems Symposium (PASS)*, Cambridge, United Kingdom.

Marieb, E. N. 1998. *Human Anatomy and Physiology*. 4th Ed. CA: Benjamin/Cummings Science Menlo Park, CA.

McIntire, B. J., and P. Whitley. 2005. *Blunt Impact Performance Characteristics of the Advanced Combat Helmet and the Paratrooper and Infantry Personnel Armor System for Ground Troops Helmet*. Fort Rucker, AL: U.S. Army Aeromedical Research Laboratory.

Merkle, A. C., K. Carneal, A. Wickwire, K. Ott, C. J. Freitas, and S. Tankard. 2010. Evaluation of Threat Conditions and Suspension Pad Configurations in Determining Potential for Behind Helmet Blunt Trauma. In *Personal Armour Systems Symposium (PASS)*, Quebec City, Canada.

Mertz, H. J., P. Prasad, and N. L. Irwin. 1997. Injury Risk Curves for Children and Adults in Frontal and Rear Collisions. In *Proceedings of the 41st Stapp Car Crash Conference*, pp. 13–30, Warrendale, PA.

Metropolitan Police P.P.G. 2003. *Ballistic Helmet Test Method*. UK.

MSCS. 2009. *Purchase Description Helmet, Enhanced Combat Helmet*. Quantico, VA: Marine Corps Systems Command, CESS: 57.

National Highway Traffic Safety Administration. 2000. *Title 49 Code of Federal Regulations (CFR) Part 571 Section 208, Occupant Crash Protection*. National Highway Traffic Safety Administration, Washington, DC.

NATO. 2003. *STANAG 2920 2nd Edition—Ballistic Test Method for Personal Amour Materials and Combat Clothing.* NATO Standardization Agency, Belgium.

Nelson, T. S. 2011. *Towards a Neck Injury Prevention Helmet for Head-First Impacts: A Mechanical Investigation.* The University of British Columbia, Vancouver, Canada.

NIJ. 1981. *Ballistic Helmets.* National Law Enforcement and Corrections Technology Centre, Washington, DC.

NRC. 2012. *Testing of Body Armor Materials—Phase III.* Washington, DC.

NRC. 2014. *Review of Department of Defense Test Protocols for Combat Helmets.* The National Academies Press, Washington, DC.

NSC. 2011. *Purchase Descritpion Helmet, Advanced Combat Helmet (ACH).* U.S. Army Natick Soldier Systems Center, Natick, MA.

Okie, S. 2005. Traumatic Brain Injury in the War Zone. *The New England Journal of Medicine* 353 (20): 2043–2047. doi:10.1056/NEJM200508113530621.

Ommaya, A. K., and A. E. Hirsch. 1971. Tolerances for Cerebral Concussion from Head Impact and Whiplash in Primates. *Journal of Biomechanics* 4: 13–21.

Philippens, M. M. G. M., B. Anctil, and K. C. Markwardt. 2014. Results of a Round Robin Ballistic Load Sensing Headform Test Series. In *Personal Armour Systems Symposium (PASS)*, Cambridge, United Kingdom.

Pintar, F. A., M. M. G. M. Philippens, J. Y. Zhang, and N. Yoganandan. 2013. Methodology to Determine Skull Bone and Brain Responses from Ballistic Helmet-to-Head Contact Loading Using Experiments and Finite Element Analysis. *Medical Engineering & Physics* 35 (11): 1682–1687. doi:10.1016/j.medengphy.2013.04.015.

Prasad, P., J. W. Melvin, D. F. Huelke, A. I. King, and G. W. Nyquist. 1985. Head. In *Review of Biomechanical Impact Response and Injury in the Automotive Environment. Task B Final Report*, edited by J. W. Melvin and K. Weber. Ann Arbor, MI: Department of Transport, National Highway Traffic Safety Administration. pp. 1–44.

Rafaels, K. A., H. C. Cutcliffe, R. S. Salzar, M. Davis, B. Boggess, B. Bush, R. Harris, et al. 2015. Injuries of the Head from Backface Deformation of Ballistic Protective Helmets under Ballistic Impact. *Journal of Forensic Sciences* 60 (1): 219–225. doi:10.1111/1556-4029.12570.

Rona, R. J., M. Jones, N. T. Fear, D. Oxon, L. Hull, N. Jones, A. S. David, N. Greenberg, M. Hotopf, and S. Wessely. 2012. Mild Traumatic Brain Injury in UK Military Personnel Returning From Afghanistan and Iraq: Cohort and Cross-Sectional Analyses. *The Journal of Head Trauma Rehabilitation* 27 (1): 33–44. doi:10.1097/HTR.0b013e318212f814.

Ryan, J. M., G. J. Cooper, and R. L. Maynard. 1988. Wound Ballistics: Contemporary and Future Research. *Journal of the Royal Army Medical Corps* 134 (3): 119–125.

Sarron, J.-C., M. Dannawi, A. Faure, J.-P. Caillou, J. D. Cunha, and R. Robert. 2004. Dynamic Effects of a 9 mm Missile on Cadaveric Skull Protected by Aramid, Polyethylene or Aluminum Plate: An Experimental Study. *The Journal of Trauma* 57 (2): 236–242; discussion 243.

Schneiderman, A. I., E. R. Braver, and H. K. Kang. 2008. Understanding Sequelae of Injury Mechanisms and Mild Traumatic Brain Injury Incurred during the Conflicts in Iraq and Afghanistan: Persistent Postconcussive Symptoms and Posttraumatic Stress Disorder. *American Journal of Epidemiology* 167 (12): 1446–1452. doi:10.1093/aje/kwn068.

Small Arms Survey Geneva. 2012. *Small Arms Survey 2012.* Cambridge, United Kingdom: Cambridge University Press..

Vyrnwy-Jones, P., C. R. Paschal, and R. W. Palmer. 1989. *Evaluation of Helmet Retention Systems Using a Pendulum Device.* Fort Rucker, AL: U.S. Army Aeromedical Research Laboratory.

Waclawik, S., M. Bolduc, and C. R. Bass. 2002. Development of a Non-Penetrating, 9mm, Ballistic Helmet, Test Method. In *Personal Armour Systems Symposium (PASS)*, pp. 61–68, The Hague, The Netherlands.

Watson, C., A. Webb, and I. Horsfall. 2008. Assessment of Potential Blunt Trauma Under Ballistic Helmets. In *Personal Armour Systems Symposium (PASS)*, Brussels, Belgium.

Wojcik, B. E., C. R. Stein, K. Bagg, R. J. Humphrey, and J. Orosco. 2010. Traumatic Brain Injury Hospitalizations of U.S. Army Soldiers Deployed to Afghanistan and Iraq. *American Journal of Preventive Medicine* 38 (1 Suppl): S108–S116. doi:10.1016/j.amepre.2009.10.006.

9 The Skull and Brain
Mechanical Properties of the Brain under Impact

Lynne Bilston

CONTENTS

9.1 INTRODUCTION

9.1.1 BRIEF HISTORY

Brain tissue mechanical properties have been of interest to researchers for approximately half a century, with an increasing focus from the late 1960s onwards. Early studies were largely motivated by a desire to understand the mechanisms of traumatic brain injury. They included assessments of creep properties (e.g. Dodgson 1962), oscillatory measurements of viscoelasticity (Koeneman 1966) and large deformation viscoelastic properties (e.g. Fallenstein et al. 1969; Engin and Wang 1970; Estes and McElhaney 1970; Galford and McElhaney 1970; Metz et al. 1970; Shuck and Advani 1972; Wang and Wineman 1972) of *ex vivo* tissue samples. Unfortunately, methodological issues that have only become apparent in retrospect limit the utility of much of this early data. These issues will be touched on later in the chapter; however, other publications have reviewed these issues in more depth (Bilston et al. 2008; Cheng et al. 2008; Hrapko et al. 2008). Subsequently, more rigorous rheological studies of *ex vivo* brain mechanics have been performed across different species, and while considerable variability remains in the literature, there is now a solid consensus that brain tissue is a highly non-linear viscoelastic material, whose properties vary considerably with the loading rate and deformation magnitude. More recently, the focus has shifted to measuring brain tissue properties *in vivo*, with the emergence of non-invasive imaging techniques such as magnetic resonance elastography that can measure brain properties in the living human (e.g. Green et al. 2008).

9.1.2 WHY STUDY THE MATERIAL PROPERTIES OF THE BRAIN?

The mechanical properties of the brain underpin the response of the brain tissue to any applied mechanical loading. Such loading ranges from rapid loads that occur during traumatic events that give rise to brain injury, both in civilian and military contexts, to quasi-static loading that occurs during growth and development of the brain (Budday et al. 2014, 2015), and some chronic brain disorders. An example of the latter is chronic hydrocephalus, where slow dilation of the fluid-filled ventricles in the centre of the brain compresses the brain parenchyma, leading to neurological symptoms and cognitive dysfunction. In the military context, the primary focus has been on trauma, where understanding the mechanical response of the brain to blast or other loading scenarios is essential for designing protective equipment for military staff. Experimental data on brain tissue mechanics is also essential for developing mathematical models that describe this response, called 'constitutive models', used in computational simulations of the brain under load. Such simulations are now the primary tools used in the design of protective equipment to mitigate injury in both civilian and military applications. Computational simulations are also widely used in studying brain disorders and are now becoming widely accepted for simulating the brain's response during neurosurgery, both for training surgeons (e.g. Chan et al. 2013) and surgical planning and intraoperative navigation (e.g. Garlapati et al. 2014).

9.2 BRAIN TISSUE STRUCTURE

The brain is an anatomically and microstructurally complex organ. Its high metabolic demands require rich vascularisation, and there is considerable local specialisation in microstructure arising from local functional arrangements of brain nuclei. In this section, we briefly review the anatomical and microstructural features of brain tissue, with relevance to its mechanical behaviour.

9.2.1 BRAIN STRUCTURE

The human brain has a mass of approximately 1.2 kg in adult females and 1.3 kg in males (Hartmann et al. 1994). It sits within the rigid skull, surrounded by the fibrous layers of the meninges and a thin layer of cerebrospinal fluid. The skull, meninges and CSF provide protection and cushioning for the brain during the activities of normal daily life, at least those that have dominated the evolutionary timescale over which these structures developed. Figure 9.1 depicts the gross anatomy of the brain in a sagittal plane. The brain is divided into the cerebrum, the brainstem (including the midbrain, pons and medulla) and the cerebellum. The cerebrum is divided into four lobes–the frontal, parietal, temporal and occipital, all of which have an undulating surface made up of sulci and gyri. The two hemispheres of the brain are joined by a thick band of white matter called the corpus callosum. Underneath sit the ventricles, a series of fluid-filled spaces in the centre of the brain, and the cerebellum, pons and medulla, which connects to the spinal cord.

Overlying the surface of the brain are three fibrous layers that make up the meninges, with the outer layer, the dura mater, being the thickest. Underneath the dura mater is the arachnoid mater, and apposed to the surface of the brain is the pia mater. The space between the arachnoid and pia mater is known as the subarachnoid space and is filled with cerebrospinal fluid. The two layers are bridged by a fine network of trabeculae, reminiscent of a spider's web (hence the name of the arachnoid mater). These layers are shown schematically in Figure 9.2.

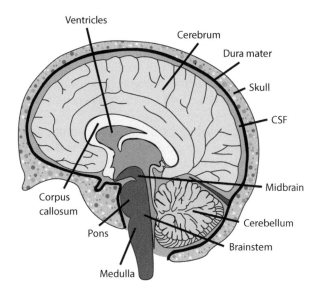

FIGURE 9.1 The human brain in the midsagittal plane, showing the major structures. (© L. Bilston, used by permission.)

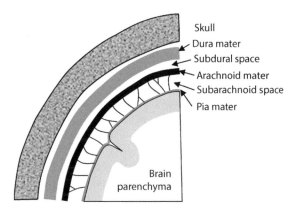

FIGURE 9.2 Schematic of the meningeal layers between the skull and brain. (© L. Bilston, used by permission.)

9.2.2 BRAIN TISSUE COMPOSITION AND MICROSTRUCTURE

The brain parenchyma is made up of white and grey matter. Grey matter consists primarily of unmyelinated nerve fibres (axons) and neural cell bodies, and white matter consists of myelinated axons. The white colour is due to the presence of a myelin sheath that surrounds these nerve fibres, created by oligodendrocytes wrapped around them. The latter are one of a number of types of glial cell that provide support to neural cells, as well as other non-neural functions in the brain.

9.3 MECHANICAL PROPERTIES OF BRAIN TISSUE

9.3.1 OVERVIEW OF MECHANICAL BEHAVIOUR

The brain's mechanical behaviour is complex–its response to applied loading (both mechanically and physiologically) varies with both the amount of deformation and the rate at which this deformation is applied. The brain is thus a viscoelastic material, including both fluid and solid characteristics. If brain tissue is stretched and held at its stretched length, the stress in the tissue will decrease over time–a process known as 'relaxation', and quantified by the relaxation modulus. If a constant force is applied, then the tissue will continue to deform over time–this is known as creep and is quantified by the creep modulus. Other commonly measured quantities are the storage and loss moduli that describe the elastic and viscous components of response under oscillatory loading.

To make things even more challenging, its mechanical response is non-linear. In practice, this means that one cannot measure the properties of the brain under a narrow range of loading conditions and use these to predict the brain's mechanical behaviour over a wider range of loading conditions. As an example, we cannot measure brain properties at small strains and moderate loading rates and use these to predict how the brain responds to large deformations at very high loading rates, such as those that occur during blast loading. This mechanical complexity also means that simple 'linear viscoelastic' constitutive models cannot be validly used to simulate the mechanical behaviour of the brain. Indeed, brain tissue behaves as a linear material only at very small strains, typically in the range of up to 0.1%–0.5% (Bilston et al. 1997), requiring the use of more complex equations to describe its non-linear behaviour for most physiologically interesting strain levels.

Brain tissue is also heterogeneous. While differences between the white and grey matter are relatively obvious, there is considerable variation between different brain regions, depending on the local neural structure. The corpus callosum, for example, is a very dense, highly ordered fibre bundle, consisting primarily of aligned myelinated axons. This leads to anisotropic mechanical behaviour, where the mechanical properties in the fibre direction are different to directions perpendicular to the fibres. Other white matter regions have less aligned axonal structures, including crossing fibres, and are less structurally anisotropic. Grey matter regions also display considerable structural inhomogeneity, which may influence their local mechanical response.

Brain tissue is often assumed to be incompressible at moderate and high loading rates, due to its high water content. In some applications, where very slow loading is applied, fluid can be redistributed within the tissue, and porous models are used to describe this phenomenon (e.g. Cheng and Bilston 2007).

Other systemic factors can influence the mechanical response of the brain, although these are less well studied. Brain tissue has been suggested to soften during the ageing process (Sack et al. 2009), and there are a small number of studies suggesting changes during early development, although a consistent picture has yet to emerge (Thibault and Margulies 1998; Prange and Margulies 2002; Gefen et al. 2003). Factors such as tissue perfusion may also be important (Bilston 2002; Gefen and Margulies 2004; Hatt et al. 2015), but the precise nature of these effects also remains to be established. Neurodegeneration has also been shown to alter brain tissue properties (Wuerfel et al. 2010; Murphy et al. 2011; Murphy et al. 2012).

In the following subsections, the current state of knowledge of the mechanical behaviour of brain tissue under different loading types is described.

9.3.2 SHEAR PROPERTIES OF BRAIN TISSUE

Shear loading is depicted schematically in Figure 9.3. Pure shear does not alter the volume of the loaded material. Shear loading is thought to be the primary deformation pattern involved in a particularly severe form of traumatic brain injury known as diffuse axonal injury (see Chapter 8 for the different types of brain injury), which occurs when the head is brought to a rapid stop, resulting in rotational motion of the brain within the skull (Gennarelli et al. 1982).

There have been numerous studies of *ex vivo* brain tissue samples subjected to shear loading, typically by sandwiching samples between parallel plates, where one of the plates is held

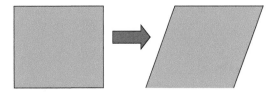

FIGURE 9.3   Schematic of shear loading.

fixed while the other is either moved parallel to it, either in a straight line or rotated around an axis perpendicular to the plates (see Figure 9.4). The latter is the most common in commercial rheometers.

The shear properties of soft tissues are quantified in terms of the shear moduli, which are usually given the symbol, G, often with subscripts or superscripts to denote whether creep, relaxation, or storage and loss moduli are being reported. The relaxation modulus, $G(t)$, is the ratio of the shear stress recorded in response to an applied 'step' shear strain. In a linear viscoelastic material, this would be a consistent function of time only. However, in brain tissue, the shear relaxation modulus varies with the applied strain beyond the linear viscoelastic limit. The shear creep modulus is the ratio of the strain response over time to an applied 'step' shear stress to the magnitude of that shear stress. There are very limited creep data for brain tissue. More common is data reporting the storage and loss modulus (G' and G'', respectively) from oscillatory tests, where a sinusoidal shear loading is applied and the resulting shear stress recorded. The resulting shear stress in the linear viscoelastic regime is also sinusoidal but lags the input strain by a factor that indicates the relative contributions of the viscous and elastic components. This situation arises from the fact that a purely elastic response to a sinusoidal input shear strain, $\gamma(t) = \sin \omega t$, will be in phase with the applied shear, $\sigma = G'\gamma$, where σ is the stress and γ is the shear strain, whereas the viscous response will be $\pi/2$ out of phase, since in a fluid, $\sigma = G'' d\gamma/dt$. These two moduli are often written using complex notation, where $G^* = G' + iG''$, to indicate this.

9.3.2.1   Linear Viscoelastic Behaviour

At *very* small strains, the brain can be characterised using just the linear viscoelastic shear moduli, as its response is independent of the strain applied. This limit is approximately 0.1%–0.5% strain (Bilston et al. 1997; Brands et al. 2000; Nicolle et al. 2005; Hrapko et al. 2006;

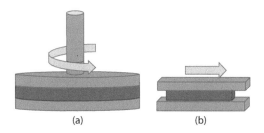

(a)   (b)

FIGURE 9.4   Schematic of two common setups for measuring shear properties of excised brain samples. The tissue sample (indicated in red) is sandwiched between parallel plates (blue). One plate is fixed (the lower plate in these examples), while the other is moved parallel to the fixed plate. Rotational shear (a) and linear shear (b).

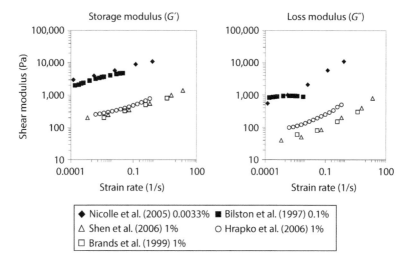

FIGURE 9.5 Storage and loss moduli for *ex vivo* brain tissue samples reported in the literature. Note that data collected at higher strains tends to measure lower shear moduli, consistent with shear thinning observed in strain sweep experiments, suggesting these data may be outside the linear viscoelastic regime. Other methodological issues may also influence data quality and values, such as post-mortem time, tissue handling and the frequency used. Data from Bilston, L.E., et al., *Biorheology*, 34(6), 377–385, 1997; Brands, D.W., et al., Stapp Car Crash Conference, San Diego, CA., 1999; Hrapko, M., et al., *Biorheology*, 43, 623–636, 2006; Nicolle, S., et al., *Biorheology*, 42, 209–223, 2005; Shen, F., et al., *J. Biomech. Engin.*, 128, 797–801, 2006.

Shen et al. 2006), as determined by conducting tests while slowly increasing the magnitude of the strain until the shear moduli recorded begin to change. This is called a 'strain sweep' experiment. Thus, for most problems of interest to the injury research and development community, this data is of limited interest.

Figure 9.5 shows the results of oscillatory tests conducted on *ex vivo* brain tissue samples over a range of strain rates, which were reported to have been conducted be within the linear viscoelastic regime (Bilston et al. 1997; Brands et al. 1999; Nicolle et al. 2005; Hrapko et al. 2006; Shen et al. 2006). This data highlights that brain tissue is a very soft solid, with considerable viscoelasticity. Note that it exhibits a 'power-law' behaviour, with increasing shear storage and loss moduli with increasing strain rate. There is, however, considerable difference in the magnitude of the shear moduli reports. A few of these studies are likely to have been conducted at strains that exceed the linear viscoelastic limit, and this likely accounts for the lower values reported in the studies conducted at 1% strain, since the shear properties of brain decrease beyond the linear regime (Bilston et al. 1997). This is called 'shear thinning'.

The relaxation properties of brain tissue have been studied less frequently, but the limited data available in the linear viscoelastic regime shows a rapid initial relaxation, followed by slower relaxation that continues beyond the length of the experiment (see Figure 9.6). Plotted on a log-log scale, this appears as power-law relaxation.

In the last decade, there have been a large number of studies reporting *in vivo* brain shear moduli measured using magnetic resonance elastography (MRE). This technique, which can be conducted on living humans, relies on that fact that the propagation of acoustic waves through a material is dependent on the inherent viscoelastic properties of the material (Muthupillai et al. 1995). Magnetic resonance imaging is used to visualise propagating displacement waves, making use of special motion-sensitive imaging sequences that are time-locked to an

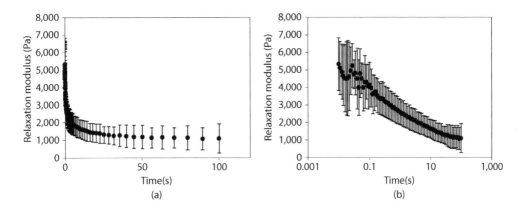

FIGURE 9.6   Linear viscoelastic relaxation modulus for brain in shear, plotted on (a) linear and (b) logarithmic time scales. (Data from Bilston, L.E., et al., *Biorheology*, 34, 377–385, 1997.)

externally applied vibration. The equations for wave propagation through a linear viscoelastic material are then solved numerically to estimate the shear moduli of the tissue:

$$\rho \partial_t^2 \vec{u} = \mu \nabla^2 \vec{u} + (\lambda + \mu)\nabla(\nabla \vec{u}) + \eta \partial_t \nabla^2 \vec{u} + (\xi + \eta)\partial_t \nabla(\nabla \vec{u}), \tag{9.1}$$

where $\vec{u}(x,t)$ is the displacement field measured using MRI, ρ is the density of the medium, μ is the shear modulus, λ the second Lamé coefficient, η the shear viscosity and ξ is the longitudinal viscosity. As with rheometry experiments mentioned above, the shear moduli reported with MRE varies with frequency, but most human studies have been conducted between approximately 30–200Hz (Hamhaber et al. 2007; Green et al. 2008; Kruse et al. 2008; Sack et al. 2008; Liu et al. 2009; Murphy et al. 2011; Streitberger et al. 2011; Zhang et al. 2011a; Guo et al. 2013; Hatt et al. 2015; Jamin et al. 2015), while some animal studies use higher frequencies (e.g. Atay et al. 2008; Schregel et al. 2012). A few earlier studies reported much higher values, but the current consensus is that brain shear moduli in this frequency range are of the order of 1–4 kPa, with lower values being reported at lower frequencies, and higher values at higher frequencies. Some research groups are currently using this technique to create standardised 'atlases' of regional brain properties (Guo et al. 2013), but this is not yet well established. A typical image of visualised brain displacements, along with the matching map of brain shear properties is shown in Figure 9.7.

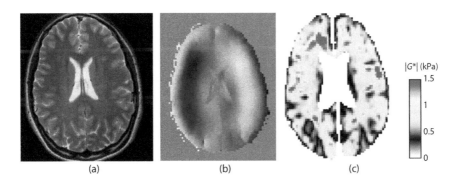

FIGURE 9.7   (a) Axial anatomical image of the human brain, (b) sample brain displacement image and (c) matching shear modulus map from a human volunteer undergoing a magnetic resonance elastography study. (© L. Bilston, used by permission.)

9.3.2.2 Non-Linear Behaviour

At strains larger than the linear viscoelastic limit, the measured mechanical properties of brain tissue, including the shear moduli, depend on the strain at which they are measured. This is demonstrated in Figure 9.8, which shows the shear moduli at increasing strains for both oscillatory and relaxation tests. Note the shear thinning behaviour, where the apparent shear modulus decreases with increasing strain.

Large strain (non-linear) behaviour can be measured using a range of different test paradigms, including large amplitude oscillatory shear (Wilhelm et al. 1998; Hyun et al. 2002), constant strain-rate shear and relaxation. Large amplitude oscillatory shear (LAOS) has been used for other tissues (Lamers et al. 2013; Tan et al. 2013) but not for brain to date. Data analysis for LAOS involves decomposing the non-sinusoidal (non-linear) shear stress developed in response to a sinusoidal input shear strain into its harmonic components in a technique known as Fourier Transform rheology (Wilhelm et al. 1998). Large strain oscillatory shear data does appear in the literature (e.g. Fallenstein et al. 1969; Arbogast et al. 1995; Brands et al. 2000; Darvish and Crandall 2001); however, most of these studies do not use appropriate non-linear methods to analyse the data, and thus the results are not generalisable beyond the specific test conditions under which they were conducted.

Relaxation has been used commonly to characterise brain tissue samples, and like other studies, there is considerable variability in the magnitude of the reported relaxation modulus (Arbogast et al. 1995; Brands et al. 2000; Bilston et al. 2001; Shen et al. 2006; Takhounts et al. 2003), as shown in Figure 9.9. However, there is a broadly consistent decrease in $G(t)$ as strain increases, both within studies and across studies (Figure 9.9).

Constant shear rate tests show the effect of increasing strain and strain rate on shear response of brain tissue (Donnelly and Medige 1997; Bilston et al. 2001; Hrapko et al. 2006). They also can be used to identify the onset of tissue damage, which indicates failure strains (i.e. the strains at which the tissue begins to rupture), exceeding 50% strain. The available data show that brain tissue can sustain quite high strains prior to failure in shear in comparison to traditional engineering materials. It should be noted, however, that these strains of smaller magnitudes are likely to cause functional failure of neural tissue, even in the absence of structural failure, since function is typically affected at lower strains than those that cause physical damage (Galbraith et al. 1993). Selected constant shear rate test data in the literature is shown in Figure 9.10.

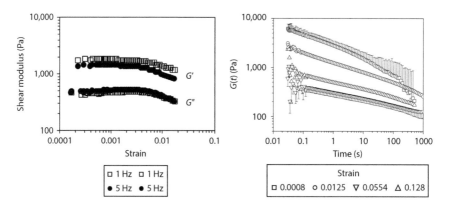

FIGURE 9.8 At strains outside the linear viscoelastic regime (i.e. greater than ~0.5% strain), brain tissue properties vary with the strain applied, demonstrating shear thinning behaviour. (data from Bilston, L.E., et al., *Biorheology*, 38, 335–345, 2001.)

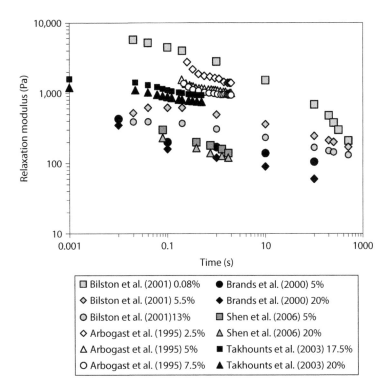

FIGURE 9.9   Relaxation behaviour of brain tissue at large strains. There is a broadly consistent decrease in relaxation modulus with increasing strain, although the magnitudes of the relaxation modulus are quite variable for a given shear strain. (Data from Arbogast, K.B., et al., Proceedings of the 39th Stapp Car Crash Conference, Coronado, CA., 1995; Bilston, L.E., et al., *Biorheology*, 38, 335–345, 2001; Brands, D.W.A., et al., Stapp Car Crash Conference, 2000; Shen, F., et al., *J. Biomech. Engin.*, 128, 797–801, 2006; Takhounts, E., et al., *Stapp Car Crash J.*, 47, 79–92, 2003.)

9.3.3   COMPRESSIVE PROPERTIES OF BRAIN TISSUE

Brain tissue can be compressed due to external loading that is sufficient to cause very large skull deformations or fracture, but it is also common in neurological disorders, such as hydrocephalus or intracranial tumours, that compress the brain tissue from within. In the case of hydrocephalus, enlarging ventricles that arise from changes in cerebrospinal fluid flow pathways compress the brain 'from the inside out', and can cause serious neurological dysfunction. In military contexts, entrapment of the head between rigid materials, or external loading of the skull, can cause compression and fracture of the skull, thus allowing compression of the underlying brain tissue, which is normally well protected by the rigid skull.

Compressive brain tissue properties have been measured using traditional engineering techniques, such as compression of *ex vivo* tissue samples between two parallel platens, direct indentation of the brain in live animals, and more recently using impact loading using Split Hopkinson Pressure Bar techniques. The latter method involves placing the specimen between two rigid bars. One of these bars is struck rapidly (e.g. by a high pressure), thus sending a high impact stress wave into the sample and onto the second bar where it is detected (see schematic in Figure 9.11). The sample stress and strain are estimated from the incident and transmitted waves, from which the compressive modulus can be estimated.

Estes and McElhaney (1970) were among the first groups to conduct compression tests on brain tissue, on rhesus and human brain samples at large strains over a broad a range of loading rates, including

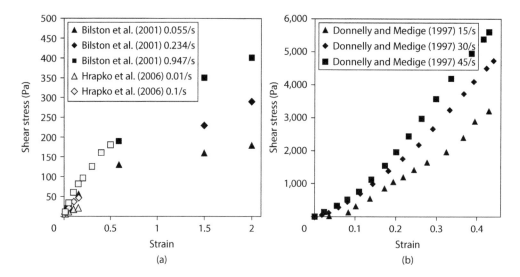

FIGURE 9.10 Selected constant shear rate tests of brain tissue at different strain rates: (a) lower shear rate data and (b) higher shear rate data. Note the consistent trend for increasing shear stress for a given strain at higher strain rates. (From Bilston, L.E., et al., *Biorheology*, 38, 335–345, 2001; Donnelly, B.R., and J. Medige., *J. Biomech. Engin.*, 119, 423–32, 1997; Hrapko, M., et al., *Biorheology*, 43, 623–636, 2006).

FIGURE 9.11 Schematic of Split Hopkinson Pressure Bar experiment.

high loading rate tests. They observed substantial increases in stiffness with increasing strain rate, consistent with the shear properties mentioned in the previous section, and non-linear stress–strain responses in compression. At lower strain rates, Miller and Chinzei (1997) and Cheng and Bilston (2007) reported compression data for brain tissue, confirming its non-linear viscoelastic behaviour. Data collected on compression by Tamura et al. (2007) at moderate- to high-rate loading rates display similar strain-rate sensitivity but less clear-cut non-linear stress–strain behaviour than other studies. Their numerical results are somewhat lower than those of Estes and McElhaney for similar loading conditions, possibly due to the samples in the earlier studies being tested long after the animal's death. This is known to affect mechanical properties due to post-mortem tissue changes. At higher strain rates, Pervin and Chen (2009) used a modified Hopkinson split bar technique, again confirming the brain's strong strain-rate sensitivity. Their data, collected at 1,000–3,000/s from very fresh samples reported much higher stiffness than Estes and McElhaney at 40/s, indicating that for high loading rate applications, the properties cannot be estimated from lower strain-rate data. This is particular importance for modelling blast loading in military contexts. Zhang et al. (2011b) used a similar technique, also showing a power-law stress–strain behaviour. A selection of this data is shown in Figure 9.12.

While compressive oscillatory tests, analogous to those often conducted in shear, can be conducted on neural tissues, this has rarely been done. As a result, formal assessment of the linear viscoelastic limit for brain tissue in compression (or in tension) has not been done, although it is likely to be broadly similar to the strain limit for shear (<1% strain).

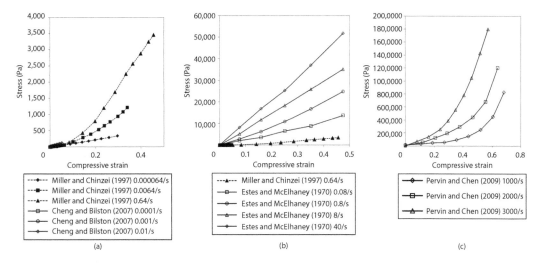

FIGURE 9.12 Compressive properties of brain tissue at low to moderate loading rates (a) and moderate (b) and high loading rates (c).

Relaxation properties of brain in compression have been measured at small and large strains, albeit only in low strain-rate conditions. Cheng and Bilston (2007) reported relaxation tests of fresh *ex vivo* brain tissue samples and showed that the relaxation response after a given compressive strain application at three different compression rates was consistent (see Figure 9.13a). Tamura et al. (2007) compressed brain tissue at three different large strains (20%–70%), and also reported that the shape of the relaxation was consistent across these conditions (see Figure 9.13b), although there was some strain dependence, consistent with the non-linear behaviour of brain tissue observed in other test protocols.

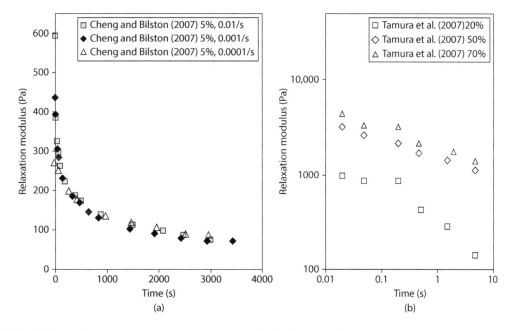

FIGURE 9.13 Selected compressive relaxation data for brain tissue. [Data in (b) has been calculated from Tamura, A., et al., *J. Biomech. Sci. Eng.*, 2, 115–126, 2007.]

Indentation tests have been performed in both non-human primates and rodents (e.g. Fallenstein et al. 1969; Gefen and Margulies 2004; Shulyakov et al. 2009) and on tissue slices (e.g. Elkin et al. 2007; Finan et al. 2012). Because indentation induces a complex non-uniform strain pattern under the indenter, interpretation of the data is more difficult than for other test paradigms. This is partly because of the non-linear viscoelastic properties of brain tissue, which means that traditional engineering theoretical analyses of indentation tests cannot be used, since they assume linear behaviour. Finite element modelling has often been used to extract mechanical parameters from indentation tests. Indentation of the live non-human primate brain conducted by Fallenstein et al. (1969) showed linear viscoelastic behaviour (reflected as a purely sinusoidal force recorded in response to a sinusoidal indentation) at very small amplitudes of indentation and non-linear behaviour (i.e. non-sinusoidal force response) at larger amplitudes.

9.3.4 Tensile Properties of Brain Tissue

The brain is rarely loaded in *pure* tension *in vivo*, due to its relatively protected location in the cranium, floating in cerebrospinal fluid. The lower parts of the brain, including the brainstem, can be subject to tension via the spinal cord. This can occur during spinal flexion or when there is rapid inertial deceleration of the head in the sagittal plane. In the latter case, the cranial contents continue to rotate after the skull stops, thus creating a tensile load on the brainstem.

In traditional engineering contexts, tension tests are the most commonly conducted tests, as it is relatively straightforward to cut a uniform material specimen and stretch it between two sets of clamps. From a tension test, Young's modulus (defined as the slope of the stress/strain curve in the linear region) and other parameters, such as yield and failure stress, can be obtained. However, this is extremely challenging with fragile brain tissue, due to the difficulty in cutting specimens and gripping them effectively. Some researchers have used tissue glue to load specimens in tension with some degree of success. The analysis is more complex than for traditional engineering specimens, as long, narrow brain tissue specimens are nearly impossible to prepare and test. This means that the non-uniform deformation of the specimens must be taken into account when analysing data. Moreover, there is often no appreciable linear region, so the behaviour must be analysed using non-linear elasticity models.

The data that exists for tensile properties of brain tissue has all been collected at low strain rates (see Figure 9.14 for a selection of available data). These data indicate that brain is both strain-rate

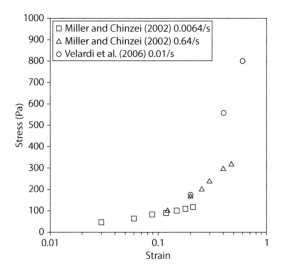

FIGURE 9.14   Selected tensile brain properties data. (data from Miller, K., and K. Chinzei., *J. Biomech.* 35, 483–90, 2002; Velardi, F., et al. *Biomech. Model. Mechanobiol.*, 5, 53–61, 2006.)

dependent and non-linear in tension, as has been observed for shear and compression. While the stress–strain curves reported are non-linear, it appears that brain tissue exhibits strain softening rather than the more common strain stiffening observed in many other soft tissues.

While impact loading rate data for tension is not available to date, it may be possible to use newer methods, such as a tensile modification to the Split Hopkinson Pressure Bar technique, to obtain impact-relevant data.

9.4 LIMITATIONS IN OUR CURRENT KNOWLEDGE OF BRAIN TISSUE MECHANICS FOR MILITARY APPLICATIONS

One of the main limitations in current brain tissue mechanics knowledge for military applications is that much of the data has been collected at moderate or small strains and at low or moderate loading rates. For the most part, these methods have been used due to their convenience as well-established protocols and equipment are more readily available. Very high strain-rate data that would provide insight into brain tissue responses at loading rates that would be expected to occur during explosive events, primarily to two studies using the Split Hopkinson Pressure Bar technique (Pervin and Chen 2009; Zhang et al. 2011b). Moreover, the data that is available is typically from studies using only a single loading type (e.g. compression or shear) rather than direct comparison between responses under different loading types applied to specimens in the same experimental series under similar experimental conditions. Data on failure of brain tissue relevant to penetrating injuries either from projectiles or explosive fragments is not available. This is another major gap for military applications. While there are a small number of studies comparing living and dead tissues in animals, the effects of perfusion, post-mortem time, tissue hydration and other details of *ex vivo* test protocols have only received limited attention (e.g. Hrapko et al. 2008; Zhang et al. 2011b). Also, until recently, most data has been collected on animal tissues. This limits our understanding of human brain mechanics as it is not known how animal neural tissue mechanics differs from human tissue mechanics. While recent MR elastography studies have begun to measure *in vivo* human properties at very small strains and moderate loading rates, there is still much to be done in understanding how living brain tissue differs from *ex vivo* specimens.

The combination of these factors, together with the strong non-linearity of brain tissue mechanics and the complexity of the brain's physiology means that it is not straightforward to extrapolate from current data to either mechanical or physiological responses to loading conditions that occur in military settings. There is clearly scope to develop new experimental methods for measuring a very high strain-rate loading of neural tissues in shear, tension and compression and in combination, particularly *in vivo*. Such methods could then be used to obtain robust measures of the complex non-linear mechanical behaviour of brain tissue that can form the basis for accurate constitutive modelling of brain mechanics and development of computational models and injury criteria.

ACKNOWLEDGEMENTS

Lynne Bilston is supported by an NHMRC senior research fellowship. She wishes to acknowledge support from the Australian Research Council for some of the work presented in this chapter.

REFERENCES

Arbogast, K.B., D.F. Meaney, and L.E. Thibault. 1995. Biomechanical characterization of the constitutive relationship for the brainstem. *Proceedings of the 39th Stapp Car Crash Conference*, Coronado, CA.
Atay, S.M., C.D. Kroenke, A. Sabet, P.V. Bayly, S.M. Atay, C.D. Kroenke, A. Sabet, and P.V. Bayly. 2008. Measurement of the dynamic shear modulus of mouse brain tissue in vivo by magnetic resonance elastography. *Journal of Biomechanical Engineering* 130 (2):021013.

Bilston, L.E. 2002. The effect of perfusion on soft tissue mechanical properties: A computational model. *Comput Methods Biomech Biomed Engin* 5 (4):283–90.

Bilston, L.E., E.C. Clarke, and S. Cheng. 2008. Brain tissue mechanical properties—Making sense of 5 decades of test data. In *The Pathomechanics of Tissue Injury and Disease, and the Mechanophysiology of Healing*, edited by A. Gefen, pp. 1–18. Kerala, India: Research Signpost.

Bilston, L.E., Z. Liu, and N. Phan-Thien. 1997. Linear viscoelastic properties of bovine brain tissue in shear. *Biorheology* 34 (6):377–85.

Bilston, L.E., Z. Liu, and N. Phan-Thien. 2001. Large strain behaviour of brain tissue in shear: Some experimental data and differential constitutive model. *Biorheology* 38 (4):335–45.

Brands, D.W., P. Bovendeerd, G. Peters, J. Wismans, M. Paas, and J. Van-Bree. 1999. Comparison of the dynamic behaviour of brain tissue and two model materials. Stapp Car Crash Conference, San Diego, CA.

Brands, D.W.A., P.H.M. Bovendeerd, G.W.M. Peters, and J. Wismans. 2000. The large shear strain dynamic behaviour of in vitro porcine brain tissue and a silicone gel model material. *Stapp Car Crash Conference*.

Budday, S., P. Steinmann, and E. Kuhl. 2014. The role of mechanics during brain development. *Journal of the Mechanics and Physics of Solids* 72:75–92.

Budday, S., P. Steinmann, and E. Kuhl. 2015. Physical biology of human brain development. *Frontiers in Cellular Neuroscience* 9:257.

Chan, S., F. Conti, K. Salisbury, and N.H. Blevins. 2013. Virtual reality simulation in neurosurgery: Technologies and evolution. *Neurosurgery* 72:A154–64.

Cheng, S., and L.E. Bilston. 2007. Unconfined compression of white matter. *Journal of Biomechanics* 40 (1): 117–24.

Cheng, S., E.C. Clarke, and L.E. Bilston. 2008. Rheological properties of the tissues of the central nervous system: A review. *Medical Engineering and Physics* 30 (10):1318–37.

Darvish, K.K., and J.R. Crandall. 2001. Nonlinear viscoelastic effects in oscillatory shear deformation of brain tissue. *Medical Engineering & Physics* 23 (9):633–45.

Dodgson, M.C.H. 1962. Colloidal structure of brain. *Biorheology* 1 (1):21–30.

Donnelly, B.R., and J. Medige. 1997. Shear properties of human brain tissue. *Journal of Biomechanical Engineering* 119 (4):423–32.

Elkin, B.S., E.U. Azeloglu, K.D. Costa, and B. Morrison. 2007. Mechanical heterogeneity of the rat hippocampus measured by atomic force microscope indentation. *Journal of Neurotrauma* 24 (5): 812–22.

Engin, A.E., and H.C. Wang. 1970. A mathematical model to determine viscoelastic behavior of in vivo primate brain. *Journal of Biomechanics* 3 (3):283–96.

Estes, M.S., and J.H. McElhaney. 1970. Response of brain tissue to compressive loading. ASME Paper 70-BHF-13. New York, NY: American Society of Mechanical Engineers.

Fallenstein, G.T., V.D. Hulce, and J.W. Melvin. 1969. Dynamic mechanical properties of human brain tissue. *Journal of Biomechanics* 2 (3):217–26.

Finan, J.D., B.S. Elkin, E.M. Pearson, I.L. Kalbian, and B. Morrison, 3rd. 2012. Viscoelastic properties of the rat brain in the sagittal plane: Effects of anatomical structure and age. *Annals of Biomedical Engineering* 40 (1):70–8.

Galbraith, J.A., L.E. Thibault, and D.R. Matteson. 1993. Mechanical and electrical responses of the squid giant axon to simple elongation. *Journal of Biomechanical Engineering* 115:13–22.

Galford, J.E., and J.H. McElhaney. 1970. A viscoelastic study of scalp, brain, and dura. *Journal of Biomechanics* 3:211–21.

Garlapati, R.R., A. Roy, G.R. Joldes, A. Wittek, A. Mostayed, B. Doyle, S.K. Warfield, et al. 2014. More accurate neuronavigation data provided by biomechanical modeling instead of rigid registration. *Journal of Neurosurgry* 120 (6):1477–83.

Gefen, A., N. Gefen, Q. Zhu, R. Raghupathi, and S.S. Margulies. 2003. Age-dependent changes in material properties of the brain and braincase of the rat. *Journal of Neurotrauma* 20 (11):1163–77.

Gefen, A., and S.S. Margulies. 2004. Are in vivo and in situ brain tissues mechanically similar? *Journal of Biomechanics* 37 (9):1339–52.

Gennarelli, T.A., L.E. Thibault, J.H. Adams, D.I. Graham, C.J. Thompson, and R.P. Marcincin. 1982. Diffuse axonal injury and traumatic coma in the primate. *Annals of Neurology* 12 (6):564–74.

Green, M.A., L.E. Bilston, and R. Sinkus. 2008. In vivo brain viscoelastic properties measured by magnetic resonance elastography. *NMR in Biomedicine* 21 (7):755–64.

Guo, J., S. Hirsch, A. Fehlner, S. Papazoglou, M. Scheel, J. Braun, and I. Sack. 2013. Towards an elastographic atlas of brain anatomy. *PLoS One* 8 (8):e71807.

Hamhaber, U., I. Sack, S. Papazoglou, J. Rump, D. Klatt, and J. Braun. 2007. Three-dimensional analysis of shear wave propagation observed by in vivo magnetic resonance elastography of the brain. *Acta Biomaterialia* 3 (1):127–37.

Hartmann, P., A. Ramseier, F. Gudat, M.J. Mihatsch, and W. Polasek. 1994. Normal weight of the brain in adults in relation to age, sex, body height and weight. *Pathologe* 15 (3):165–70.

Hatt, A., S. Cheng, K. Tan, R. Sinkus, and L.E. Bilston. 2015. MR elastography can be used to measure brain stiffness changes as a result of altered cranial venous drainage during jugular compression. *American Journal of Neuroradiology* 36 (10):1971–7.

Hrapko, M., J.A.W. van Dommelen, G.W.M. Peters, and J.S.H.M. Wismans. 2006. The mechanical behaviour of brain tissue: Large strain response and constitutive modelling. *Biorheology* 43 (5): 623–36.

Hrapko, M., J.A.W. van Dommelen, G.W.M. Peters, and J.S.H.M. Wismans. 2008. The influence of test conditions on characterization of the mechanical properties of brain tissue. *Journal of Biomechanical Engineering* 130 (3):031003.

Hyun, K., S.H. Kim, K.H. Ahn, and S.J. Lee. 2002. Large amplitude oscillatory shear as a way to classify the complex fluids. *Journal of Non-Newtonian Fluid Mechanics* 107 (1–3):51–65.

Jamin, Y., J.K. Boult, J. Li, S. Popov, P. Garteiser, J.L. Ulloa, C. Cummings, G. Box, et al. 2015. Exploring the biomechanical properties of brain malignancies and their pathological determinants in vivo with magnetic resonance elastography. *Cancer Research* 75(7):1216–22.

Koeneman, J.-B. 1966. Viscoelastic properties of brain tissue. M.Sc thesis, Case Institute of Technology.

Kruse, S.A., G.H. Rose, K.J. Glaser, A. Manduca, J.P. Felmlee, C.R. Jack, Jr., and R.L. Ehman. 2008. Magnetic resonance elastography of the brain. *NeuroImage* 39 (1):231–37.

Lamers, E., T.H.S. van Kempen, F.P.T. Baaijens, G.W.M. Peters, and C.W.J. Oomens. 2013. Large amplitude oscillatory shear properties of human skin. *Journal of the Mechanical Behavior of Biomedical Materials* 28:462–70.

Liu, G.R., P.Y. Gao, Y. Lin, J. Xue, X.C. Wang, B.B. Sui, L. Ma, et al. 2009. Brain magnetic resonance elastography on healthy volunteers: A safety study. *Acta Radiologica* 50 (4):423–9.

Metz, H., J. McElhaney, and A.K. Ommaya. 1970. A comparison of the elasticity of live, dead, and fixed brain tissue. *Journal of Biomechanics* 3:453–8.

Miller, K., and K. Chinzei. 1997. Constitutive modelling of brain tissue: Experiment and theory. *Journal of Biomechanics* 30 (11–12):1115–21.

Miller, K., and K. Chinzei. 2002. Mechanical properties of brain tissue in tension. *Journal of Biomechanics* 35 (4):483–90.

Murphy, M.C., G.L. Curran, K.J. Glaser, P.J. Rossman, J. Huston, J.F. Poduslo, C.R. Jack, J.P. Felmlee, and R.L. Ehman. 2012. Magnetic resonance elastography of the brain in a mouse model of Alzheimer's disease: Initial results. *Magnetic Resonance Imaging* 30 (4):535–9.

Murphy, M.C., J. Huston, 3rd, C.R. Jack, Jr., K.J. Glaser, A. Manduca, J.P. Felmlee, and R.L. Ehman. 2011. Decreased brain stiffness in Alzheimer's disease determined by magnetic resonance elastography. *Journal of Magnetic Resonance Imaging* 34(3):494–8.

Muthupillai, R., D.J. Lomas, P.J. Rossman, J.F. Greenleaf, A. Manduca, and R.L. Ehman. 1995. Magnetic resonance elastography by direct visualization of propagating acoustic strain waves. *Science* 269 (5232):1854–7.

Nicolle, S., M. Lounis, R. Willinger, and J.F. Palierne. 2005. Shear linear behavior of brain tissue over a large frequency range. *Biorheology* 42 (3):209–23.

Pervin, F., and W.W. Chen. 2009. Dynamic mechanical response of bovine gray matter and white matter brain tissues under compression. *Journal of Biomechanics* 42 (6):731–5.

Prange, M.T., and S.S. Margulies. 2002. Regional, directional, and age-dependent properties of the brain undergoing large deformation. *Journal of Biomechanical Engineering* 124 (2):244–52.

Sack, I., B. Beierbach, U. Hamhaber, D. Klatt, J. Braun, I. Sack, B. Beierbach, U. Hamhaber, D. Klatt, and J. Braun. 2008. Non-invasive measurement of brain viscoelasticity using magnetic resonance elastography. *NMR in Biomedicine* 21 (3):265–71.

Sack, I., B. Beierbach, J. Wuerfel, D. Klatt, U. Hamhaber, S. Papazoglou, P. Martus, et al. 2009. The impact of aging and gender on brain viscoelasticity. *Neuroimage* 46 (3):652–7.

Schregel, K., E. Wuerfel née Tysiak, P. Garteiser, I. Gemeinhardt, T. Prozorovski, O. Aktas, H. Merz, D. Petersen, J. Wuerfel, and R. Sinkus. 2012. Demyelination reduces brain parenchymal stiffness quantified in vivo by magnetic resonance elastography. *Proceedings of the National Academy of Sciences* 109 (17):6650–5.

Shen, F., T.E. Tay, J.Z. Li, S. Nigen, P.V.S. Lee, and H.K. Chan. 2006. Modified Bilston nonlinear viscoelastic model for finite element head injury studies. *Journal of Biomechanical Engineering* 128 (5):797–801.

Shuck, L.Z., and S.H. Advani. 1972. Rheological response of human brain tissue in shear. *Journal of Basic Engineering* 94:905–11.

Shulyakov, A., F. Fernando, S. Cenkowski, and M. Del Bigio. 2009. Simultaneous determination of mechanical properties and physiologic parameters in living rat brain. *Biomechanics and Modeling in Mechanobiology* 8 (5):415–25.

Streitberger, K.-J., E. Wiener, J. Hoffmann, F.B. Freimann, D. Klatt, J. Braun, K. Lin, et al. 2011. In vivo viscoelastic properties of the brain in normal pressure hydrocephalus. *NMR in Biomedicine* 24 (4):385–92.

Takhounts, E., J.R. Crandall, and K. Darvish. 2003. On the importance of nonlinearity of brain tissue under large deformations. *Stapp Car Crash Journal* 47:79–92.

Tamura, A., S. Hayashi, I. Watanabe, K. Nagayama, and T. Matsumoto. 2007. Mechanical characterization of brain tissue in high-rate compression. *Journal of Biomechanical Science and Engineering* 2 (3):115–26.

Tan, K., S. Cheng, L. Jugé, and L.E. Bilston. 2013. Characterising soft tissues under large amplitude oscillatory shear and combined loading. *Journal of Biomechanics* 46 (6):1060–6.

Thibault, K.L., and S.S. Margulies. 1998. Age-dependent material properties of the porcine cerebrum: Effect on pediatric inertial head injury criteria. *Journal of Biomechanics* 31 (12):1119–26.

Velardi, F., F. Fraternali, and M. Angelillo. 2006. Anisotropic constitutive equations and experimental tensile behavior of brain tissue. *Biomechanics and Modeling in Mechanobiology* 5 (1):53–61.

Wang, H.C., and A.S. Wineman. 1972. A mathematical model for the determination of viscoelastic behavior of brain in vivo. II. Relaxation response. *Journal of Biomechanics* 5 (6):571–80.

Wilhelm, M., D. Maring, and H.W. Spiess. 1998. Fourier-transform rheology. *Rheologica Acta* 37 (4):399–405.

Wuerfel, J., F. Paul, B. Beierbach, U. Hamhaber, D. Klatt, S. Papazoglou, F. Zipp, P. Martus, J. Braun, and I. Sack. 2010. MR-elastography reveals degradation of tissue integrity in multiple sclerosis. *NeuroImage* 49 (3): 2520–5.

Zhang, J., M.A. Green, R. Sinkus, and L.E. Bilston. 2011a. Viscoelastic properties of human cerebellum using magnetic resonance elastography. *Journal of Biomechanics* 44 (10):1909–13.

Zhang, J., N. Yoganandan, F.A. Pintar, Y. Guan, B. Shender, G. Paskoff, and P. Laud. 2011b. Effects of tissue preservation temperature on high strain-rate material properties of brain. *Journal of Biomechanics* 44 (3): 391–6.

10 The Skull and Brain
Computer Models for the Head and Its Protection

Kwong Ming Tse, Long Bin Tan and Heow Pueh Lee

CONTENTS

10.1 INTRODUCTION

10.1.1 DEFINITION OF HEAD INJURY AND BRAIN INJURY

Head injury is one of the most common and the most severe form of trauma-related injury (Mayer et al. 1981). In the context of this chapter, head injury refers to any physical injury to the head, which may or may not involve the brain. The phrases 'head injury' and 'brain injury' are often used interchangeably in medical literature (Field 1976; Anderson and McLaurin 1980; Kraus and

McArthur 2000). In fact, there is no agreement in the definition of head injury for epidemiological purposes, and the definitions in each of the studies differ. Depending on the specialty of surgeons and the surgical pathology over time, head injury demands a broad definition. Head injury, in anatomical terms, refers to any physical trauma to the body above the lower border of the mandible (Fearnside and Simpson 1997). On the other hand, brain injury is more properly defined as 'physical damage to, or functional impairment of, the cranial contents from acute mechanical exchange, excluding birth trauma' (Jagger et al. 1984). In the context of this chapter, head injury refers to any physical damage to both the extracranial and intracranial contents, which may or may not involve the brain, while brain injury is termed as injury to intracranial tissues.

10.1.2 MECHANISMS OF BRAIN INJURY

Head injury mechanisms are introduced in Chapter 8 and will be discussed in more detail in this section, particularly on brain injury mechanisms. Currently, there are various theories regarding mechanisms for brain injury, namely positive pressure, negative pressure, pressure gradient and rotational effects.

At the instant impact loading occurs, the head is subjected to an external force, either a static or dynamic force. Upon an impact to the head, contact phenomena resulting from the impact, for example, bending, fracture and penetration of the skull, produce local effects such as scalp laceration, skull fracture, extradural haematoma (EDH), cerebral contusion and intracerebral haemorrhage (ICH) (Bullock and Graham 1997). At the same time, the scalp absorbs and distributes external energy from the mechanical load. This external energy subsequently gives rise to either linear (translational) or angular (rotational) movement of the head. More often, the injury is of a combined translational and rotational acceleration type as the impact is generally directed eccentrically to the head.

Although the cranium is obviously an important protective feature, ridges along its base can cause physical damage in the event of acceleration or deceleration types of injuries when the brain collides with the rough bony cranial compartment (Mansfield et al. 1995). Pure translational acceleration creates intracranial pressure gradients which may result in regional variation in cerebral blood flow for long duration impacts (Johnston and Rowan 1974). Pure rotational acceleration also causes rotation of the skull relative to the brain, which is likely to tear parasagittal bridging veins (Bullock and Graham 1997). These movements, which give rise to tissue deformation, twisting and bending, result in extracranial and intracranial injuries. The skull-deformation-angular acceleration theory proposed by Holbourn (1943), suggests two main causes of brain injury: skull deformation (bending and fracture) causing damage because of local brain distortion, and angular acceleration, generating shear strains and diffuse brain injury (Lollis et al. 2008). The severity of head injuries increases with acceleration levels. Early animal experiments demonstrated that cerebral concussion, with or without contusion, was readily produced at low- and moderate-acceleration levels, while subdural haematomas were found in high-acceleration cases (Adams et al. 1982; Gennarelli and Thibault 1982).

Various components in the head respond differently in motion due to the differences in their material properties and their anatomical locations. The rigid skull, when struck by blunt object, moves faster than the viscoelastic brain, which is still stationary at the instant of impact. This produces a compressive wave at the site of impact which propagates within the brain. Consequently, positive pressure arises at the impact site, causing coup injury at the impact site. Positive pressure usually occurs in coup injuries when a stationary head is struck by a blunt object; however, it can occur in contrecoup injuries, particularly during a fall when the head is striking a stationary object (Hardy et al. 1994).

Negative pressure develops on the contralateral side of the head as a result of the tension generated by the moving skull that is leaving behind the response-delayed brain. The negative pressure could be also due to a tensile wave formed by the reflection of the original compressive stress wave from the impact site. Negative pressure at the contrecoup site is formed upon impact, while that at the coup site, it arises due to the rapid recovery of the deformed skull to its original geometry. As suggested by Denny-Brown and Russell (1940) as well as Gross (1958), contrecoup injury is not only caused by negative pressure alone; cavitation collapse, which refers to violent collapse of vacuum cavity formed under brain's membranes, could also be responsible. Brain damage may also occur as a result of cavitation (formation of bubbles) because of short-duration reductions in intracranial pressure (Bullock and Graham 1997).

Angular acceleration was proposed as a mechanism of brain injury by Holbourn (1943). It was hypothesised that 'shear strain and strain generated by rotation alone could cause cerebral concussion as well as contrecoup contusion' (King et al. 2003). Löwenhielm (1975) also stated that the 'deep brain could be injured while the surface was not injured and that the zone of maximum shear became deeper as the angular acceleration pulse duration increased' (King et al. 2003). Major proponents of rotation as cause of head injuries are Gennarelli et al. 1981, 1982) in their experimental works using live subhuman primates and physical models. They hypothesised that brain injuries such as concussion, diffuse axonal injury (DAI) and subdural haematomas (SDHs) were induced by shear strain generated by angular acceleration.

However, aside from these individual hypotheses, combinations of theories have been formed. Gurdjian and Lissner (1945) as well as Thomas et al. (1967) found that impact-induced acceleration and pressure are responsible for the development of pressure gradients. The intracranial tissue attempts to move from areas of high pressure to areas of low pressure. Upon impact, these pressure gradients can also cause traumatic injury to the brain, as intracranial shear stresses produced by pressure gradients result in local deformation of brain tissue. As acceleration and pressure increase, the pressure gradients become higher, and these elevated intracranial pressure gradients give rise to shear effects in grey matter–white matter junctions, resulting in DAI (Thomas et al. 1967).

Contact forces resulting from an impact on the layered structure such as the brain produce local effects such as scalp laceration, skull fracture, EDH, cerebral contusion and ICH (Bullock and Graham 1997). In contrast, acceleration, which results in movement of the brain and leads to development of intracranial pressure gradients as well as shear, tensile and compressive strains, can cause both focal and diffuse injuries to the brain. These inertial mechanisms are responsible for the most important and potentially fatal injuries, such as acute SDHs (resulting from tearing of subdural bridging veins) and DAI (Bullock and Graham 1997).

10.1.3 Finite Element Modelling – An Economical and Ethically Acceptable Alternative

To investigate different head injury mechanisms, animals, cadavers and volunteer living subjects have been used extensively in experimental tests for several decades to monitor the head's physiologic response over time. Despite the amount of invaluable information provided by these experiments, these tests not only raised issues in morality and ethics but also raised many research concerns such as inflexibility and unreliability of these potentially biased experimental data due to the limited number of subjects and non-standardised experimental procedures. Alternatively, crash test dummies, which were developed using information gleaned from animal testing and cadaver research, may provide better control over experimental configurations. Human surrogates, such as the Hybrid III anthropomorphic test device (ATD), the National Operating Committee on Standards for Athletic Equipment (NOCSAE) headform and Facial and Ocular CountermeasUre for Safety (FOCUS) headform, which are discussed further in Chapter 6, have been used to obtain head accelerations and impact forces for various low to high velocity loading scenarios. However, the exact deleterious effect of these induced head accelerations and impact forces on any particular brain region have yet to be fully understood, and more research is required to bridge the gap between clinical observations and biomechanical parameters.

Moreover, because of the poor biofidelity of these physical models in terms of representing the interior anatomical components of the head, they cannot be used to delineate the actual mechanism of head injury. As some of the internal biomechanical responses of the brain can neither be measured easily, nor measured *in vivo* by experimental techniques, numerical modelling of the head offers a cost-effective alternative to experimental methods in estimating the internal biomechanical responses of the human head, and they can also more accurately represent the real-life scenario than experimental laboratory tests. There have been a number of attempts to model the human head analytically since 1960s (Hodgson et al. 1967; Hodgson and Patrick 1968). However, due to the mathematical difficulties in formulation and obtaining the analytical solution, these mathematical or analytical head models could only be represented by simplified geometries, or modelled with lumped-mass-spring-damper techniques with limited degrees-of-freedom (DOF). On the other hand, numerical approaches like the finite element (FE) method greatly simplify the analytical problem by using approximations to the analytical formulation or solution, thus making the FE method more adaptable to digital computation techniques, while the accuracy of the solution can be controlled by specifying different parameters such as the geometrical accuracy and the level of convergence. The ability of the FE approach to handle irregular shapes and inhomogeneous materials make it particularly suitable for modelling the mechanical structures and responses of the human head. In the last few decades, advancements in technology (cheaper computing tools and faster codes) led to the rapid development of more complex and anatomically realistic FE models of the human head and brain. Biomechanical tissue responses such as skull stresses, intracranial pressure (ICP) and strains, which can neither be measured easily nor measured *in vivo* by experimental techniques, can be determined from FE simulations and validated with available experimental data in order to model alternative scenarios and establish new injury criteria. With these promising techniques, numerous FE models have been developed by researchers in different countries and used in assessing head injuries sustained in military scenarios. In this chapter, the most widely used FE head and helmet models are reviewed, and suggestions for future military research are discussed.

10.2 FINITE ELEMENT MODELLING OF HUMAN HEAD

10.2.1 EARLY FE HEAD MODELS (70–80s)

Early FE head models were generally simplified using regular geometry (such as spherical or ellipsoidal shells or solids) (Hardy and Marcal 1971; Chan 1974; Shugar 1975; Khalil and Hubbard 1977). To the authors' best knowledge, in 1971, Hardy and Marcal (1971) made the first attempt to model the behaviour of human head responses using a three-dimensional (3D) FE skull model, which consisted of triangular shell elements. Several years later, Shugar (1975) improved the model by adding an elastic fluid-filled brain, which was firmly attached to the skull. Thereafter, Chan (1974) modelled the human head as a spherical shell and a prolate ellipsoid, while Khalil and Hubbard (1977) used an ellipsoidal shell for the skull and an inviscid fluid core as the intracranial content. A spherical scalp-skull-dura-brain model was constructed by Saczalski and Richardson (1978) to examine the biomechanical responses of a protected and unprotected human headform when subjected to impact. These geometrically simplified models considered the anatomic and material features of only one of the two major structures, namely the skull or the brain. In 1981, Hosey and Liu (1981) developed a 3D homomorphic model representing the basic anatomy of skull, brain and neck. This model was later considered to be the most comprehensive FE head and neck model in the 80s (Yang et al. 2009). Their model not only included the basic anatomy of head, brain and neck, but also incorporated the inertial and material properties of the head and neck (Figure 10.1). Despite being simplified and considered anatomically inaccurate, these earlier attempts in modelling the human head and brain represented the first step in the numerical analysis of head injury biomechanics.

10.2.2 ESTABLISHED FE HEAD MODELS (90s–PRESENT)

It was only in the 1990s, when computing capability had advanced considerably, that the development of more realistic and comprehensive 3D FE models of human head, based on the specific

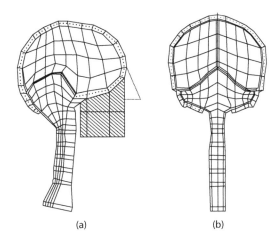

(a)   (b)

FIGURE 10.1 Finite element head model developed by Hosey and Liu in 1981 (Hosey and Liu 1981). (a) Sagittal cross-section and (b) coronal cross-section. (Reprinted with kind permission from Springer Science+Business Media: *Med. Biol. Eng. Comput.*, Finite-element models of the human head, 34(5), 1996, 375–381, Voo, L., et al., Copyright © 2016).

geometry of the head, was possible. Among the existing FE head models, the principal ones commonly reported in the literature are the Wayne State University Head Injury Model (WSUHIM), led by King and his team (Ruan et al. 1993; Zhou et al. 1995; Zhang et al. 2001); the University of Louis Pasteur (ULP) FE Head Model by Kang and Willinger (Kang et al. 1997); the Kungliga Tekniska Högskolan (KTH) FE Head Model by Kleiven (Kleiven and Hardy 2002); the Simulated Injury Monitor (SIMon) FE Head Model by Takhounts and Eppinger (2003); and the University College Dublin Brain Trauma Model (UCDBTM) by Horgan and Gilchrist (2003) (Table 10.1).

10.2.2.1 Wayne State University Head Injury Model (WSUHIM)

In the early 1990s, King and his team from Wayne State University (WSU) spent their efforts to build a 3D FE model of human skull and brain, which was regarded as the first version of the Wayne State University Head Injury Model (WSUHIM) (Figure 10.2a). The developer, Ruan and colleagues (Ruan et al. 1993), then used the WSUHIM to investigate the effect of lateral impacts on head injury. This preliminary version of the WSUHIM consisted of 6,080 nodes and 7,351 elements, with the skull, brain and cerebrospinal fluid (CSF) modelled as eight-noded hexahedron elements while the scalp, dura mater and falx cerebri comprised four-noded shell elements. The outputs of this model have only been compared with the time histories of the impact force, head acceleration and ICP obtained from a single cadaveric experiment from the study by Nahum and colleagues (Nahum et al. 1977).

The WSUHIM was later improved by Zhou et al. (1995) by using a more refined mesh (17,656 nodes and 22,995 elements). In addition, Zhou et al. (1995) differentiated between the white and grey matters of the brain, used different material properties for these two types of tissue, and included ventricles as well as parasagittal bridging veins. Zhou et al. (1995) found that the differentiation of white and grey matter in the cerebrum resulted in larger variations in shear stress distribution patterns without affecting ICP distribution, while the inclusion of the ventricles in the brain was necessary to match regions of higher shear stress to locations of DAI. This finding is qualitatively similar to porcine experimental results in which DAI was found in the white matter at its interfacial surface adjacent to the grey matter and ventricles (Ross et al. 1994).

The model was later revised and improved by Al-Bsharat et al. (1999), where various numerical definitions of the sliding interface between the skull, CSF and brain were studied (Figure 10.2b). The totally revised model included most of the essential anatomical features of a 50th percentile male head,

TABLE 10.1

Literature Survey of Some Non-Subject-Specific FE Head Models

Authors	Year	Dimension	No. of Elements	Model Descriptions	Skull–Brain Interface	Findings
Hardy and Marcal (1973)	1971	3D		Linear elastic skull	Nil	In compression, the skull is stronger in frontal loading than lateral loading.
Kenner and Goldsmith (1972)	1972	3D		Compressible, inviscid fluid in an axisymmetrical, thin, spherical shell (elastic for the skull shell, viscoelastic for the brain fluid)		Predicted 20% higher strains in their skull model than those measured in their experimental impact on a spherical aluminium shell filled with water.
Nickell and Marcal (1974)	1974	2D		Linear elastic skull for vibration response study	Nil	Four modal shapes and frequencies for each end condition; Modes 1, 2 are flexural and support rotation while Mode 3 supports the cavitation theory.
Chan (1974)	1974	3D		Modelled the human head with 2 configurations: one as a linearly viscoelastic spherical shell and the other as a linearly viscoelastic prolate ellipsoid shell, both with a linearly viscoelastic core bonded to it		Both shear strain and reduced cerebral pressure are equally important in brain injury (cavitation and rotation mechanisms of head injury).
Shugar (1975)	1975	2D		Three-layered fluid-filled skull with brain modelled as homogeneous, nearly incompressible fluid	Common nodes	Higher brain stresses predicted in a free skull than in a fixed skull; lower stress/strain levels and maximum shear at brain surface for non-impact loading.
Shugar and Kahona (1975)	1977	3D		Thin layer modelled as subarachnoid space between the skull and the brain	Common nodes	Rotational acceleration contributes more to the rupturing of bridging veins than translational acceleration does. As such, shear strain on the brain surface may be a better predictor for contusions.
Ward and Thompson (1975)	1975	3D		Rigid skull shell with linearly elastic core and CSF	Common nodes	

(Continued)

TABLE 10.1 (CONTINUED)
Literature Survey of Some Non-Subject-Specific FE Head Models

Authors	Year	Dimension	No. of Elements	Model Descriptions	Skull–Brain Interface	Validation	Findings
Khalil and Hubbard (1977)	1977	2D		Fluid-filled, single-layered or multilayered circular and ellipsoidal shells (elastic for the scalp and the skull layers; viscoelastic for the brain fluid)	Common nodes		Load spatial distribution strongly influenced skull strains and thus skull fracture initiation.
Nahum et al. (1977)	1977	3D	189	Linear elastic solid brain	Common nodes	Pressure (validated by two of the cadaveric tests performed by the same study)	
Saczalski and Richardson (1978)	1978	3D		Spherical brain of nearly incompressible material, a covering of linear elastic dura, a linear elastic spherical skull and a layer of non-linear scalp material			
Hosey and Liu (1981)	1981	3D	786	Homomorphic head and neck model including scalp, skull, falx, dura, CSF, brain and spinal cord and cervical column.	Common nodes with CSF	Initial inertial characteristics of the brain	Linear distribution of pressure gradient along the axis of impact with zero pressure over the anterior of foramen magnum; negative ICP at contrecoup region occurred 1.6 ms after impact
Ueno et al. 1989, 1991)	1989	2D		2D model including a layered rigid skull and linear elastic brain	Tied nodes	Pressure	Translational acceleration gives rise to ICP while rotational acceleration contributes to intracranial shear; combined translational and rotational acceleration gives larger values than the individual components of acceleration
	1991	2D		Advanced 2D model by including layered rigid skull and linear elastic brain	Tied nodes	Pressure	

(Continued)

TABLE 10.1 (CONTINUED)
Literature Survey of Some Non-Subject-Specific FE Head Models

Authors	Year	Dimension	No. of Elements	Model Descriptions	Skull–Brain Interface	Validation	Findings
Ruan et al. (1993, 1994)	1993	3D	7351	Developed WSUHIM Version I, which included the scalp, the three-layered skull, CSF, dura mater, falx cerebri and brain	CSF modelled as solid element	Pressure [Nahum et al. (1977)'s frontal tests]	In good agreement against the experimental data from human cadavers by Nahum et al. (1997)
	1994	3D	9146		CSF modelled as solid element	Pressure	Viscoelasticity of brain has insignificant effect on ICP in frontal impact; impactor velocity affects brain responses more than impactor's mass does; HIC is proportional to the impact force, coup pressure and maximum brain shear stress
Zhou et al. (1995)	1995	3D	22995	Improved on WSUHIM Version I by using finer elements	Combination of tied nodes and sliding without separation	Pressure [Nahum et al. (1977)'s frontal tests]; relative motion magnitude	Differentiation of CNS and inclusion of ventricles are necessary to match regions of higher shear stress DAI locations; sagittal rotation causes higher strain in bridging veins and higher stress in corpus callosum than coronal rotation.
Kang and Willinger (Kang et al. 1997; Willinger et al. 1999)	1997	3D	13208	Developed the University of Louis Pasteur, Strasbourg Finite Element Head Model (ULP FE Head Model)	CSF modelled as solid element	Stress	A good correlation between numerical head response, in terms of intracranial stress, and observation of brain injuries in autopsy in the motorcycle accident replication
	1999	3D	25300		CSF modelled as solid element	Pressure [Nahum, Smith, & Ward (1977)'s and Trosseille, Tarriere, & Lavaste (1992)'s frontal tests]	CSF transfer during head impact cannot be successfully modelled simply by introducing compressibility to the intracranial space; Brain viscoelasticity has no significant effect on ICP

(Continued)

TABLE 10.1 (CONTINUED)

Literature Survey of Some Non-Subject-Specific FE Head Models

Authors	Year	Dimension	No. of Elements	Model Descriptions	Skull–Brain Interface	Validation	Findings
Kumaresan and Radhakrishnan (1996)	1996	3D		Homeomorphic head and neck model which consisted of the skull, CSF, the brain, its partitioning membranes and the neck	CSF modelled as solid element		The partitioning membranes of the brain can affect the ICP and maximum shear stress of the brain
Zhang et al. (2001)	2001	3D	245000	Improved and constructed WSUHIM Version II by improving the facial features and introducing a sliding interface between linearly elastic skull and viscoelastic brain	Sliding w/o separation; tied interface	Pressure [Nahum et al. (1977)'s frontal tests]	ICP is largely a function of the translational acceleration of the head, while the maximum shear stress is more sensitive to rotational acceleration.
Kleiven and Hardy (2002)	2002	3D	18400	Developed the KTH FE head model which consists of the scalp, skull, brain, meninges, CSF, 11 pairs of parasagittal bridging veins and a simplified neck and included sliding condition between skull and brain	Tied interface between skull and dura; sliding interface between meninges and brain	Pressure [Nahum et al. (1977)'s and Trosseille et al. (1992)'s frontal tests]; relative motion (Hardy et al. 2001)	Found 50% probability of concussion for a maximum principal strain of 0.21 in corpus callosum and 0.26 in the grey matter
King et al. (2003)	2003	3D	314500	Latest and completely revised WSUHIM with viscoelastic brain and elastic-plastic bones.	Tied interface		Strain rate and the product of strain and strain rate in the midbrain region appeared to be the best injury predictors for concussion
Horgan and Gilchrist (2003)	2003	3D	50000	Developed the University College Dublin Brain Trauma Model (UCDBTM),which includes a scalp, a three-layered skull, dura, CSF, pia, falx, tentorium, cerebral hemispheres, cerebellum and brainstem			Found that the short-term brain shear modulus had significant effect on frontal ICP and von Mises stress; CSF bulk modulus had significant effect on contrecoup pressure; a coarsely meshed model is adequate for pressure response

(Continued)

TABLE 10.1 (CONTINUED)

Literature Survey of Some Non-Subject-Specific FE Head Models

Authors	Year	Dimension	No. of Elements	Model Descriptions	Skull–Brain Interface	Validation	Findings
Takhounts & Eppinger (Takhounts and Eppinger 2003; Takhounts et al. 2008)	2003	3D	7852	A simplified model built for fast computation (known as the SIMon FE Head Model) which did not account for both cerebellum and midbrain	Tie-break interface	Relative motion (Hardy et al. (2001)'s 3 neutral density targets tests for frontal, occipital, lateral impacts in 2001); Pressure (Nahum et al. (1977)'s & Trosseille et al. (1992)'s PMHS tests in 1977 & 1992, respectively)	Found three injury mechanisms: (a) cumulative strain damage measure (CSDM) as a correlate for DAI; (b) dilatation damage measure (DDM) as a correlate for contusions; (c) relative motion damage measure (RMDM) as a correlate for acute SDH
	2008	3D	45875	Advanced model including skull, dura-CSF and brain based on the outer brain surfaces.	Tied interface (due to numerical difficulties)		Cumulative strain damage measure (CSDM) and especially relative motion damage measure (RMDM) correlated well with rotational acceleration and rotational velocity; maximum principal strain correlated well with RMDM, and angular head kinematic measures; maximum principal stress did not correlate with any kinematic measure
Belingardi et al. (2005)	2005	3D	55264	Developed a numerical model generated from CT scan data which was composed of scalp, three-layered skull, facial bones, dura mater, CSF, brain tissues, ventricles, falx and tentorium membranes		Pressure [Nahum et al. (1977)'s frontal tests]	The absence of membranes and ventricles raised peak ICP by 17% in frontal region and 18% in posterior region

(Continued)

TABLE 10.1 (CONTINUED)
Literature Survey of Some Non-Subject-Specific FE Head Models

Authors	Year	Dimension	No. of Elements	Model Descriptions	Skull–Brain Interface	Validation	Findings
Zong et al. (2006)	2006	3D		Simplified model consisting of a three-layered non-homogeneous skull, impressible CSF, homogeneous brain	CSF modelled as solid element (with and without fluid option to damp the pressure oscillation)		

CSF = cerebrospinal fluid, ICP = intracranial pressure, CNS = central nervous system, DAI = diffuse axonal injury, CSDM = cumulative strain damage measure, DDM = dilatation damage measure, RMDM = relative motion damage measure, SDH = subdural haematoma, CT = computed tomography.

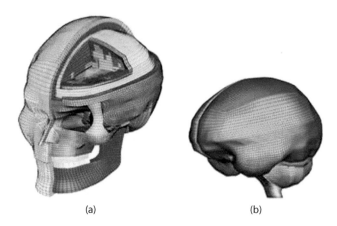

(a)   (b)

FIGURE 10.2   The Wayne State University Head Injury Model (WSUHIM). (Reprinted from Makwana, R., *Development and Validation of a Three-Dimensional Finite Element Model of Advanced Combat Helmet and Biomechanical Analysis of Human Head and Helmet Response to Primary Blast Insult*, M.S., Wayne State University, Ann Arbor, MI, 2012. Copyright © 2012 by Rahul Makwana. With permission.)

including the scalp, a three-layered skull, the dura, falx cerebri, tentorium, pia, CSF hemispheres of the cerebrum with distinct white and grey matter, the cerebellum, the ventricles, the brainstem and the facial bones. It consisted of 281,800 nodes and 314,500 elements, with detailed facial features. Al-Bsharat et al. (1999) found that the inclusion of sliding skull-CSF-brain interface would improve the correlation of the simulated relative skull–brain displacement with the measured data reported by King et al. (1999) using high-speed radiography. To the authors' knowledge, this latest revised version of the WSUHIM has been validated against 35 experimental cases in terms of brain pressure, brain displacements, skull bone responses and facial bone responses (Mao et al. 2013). It has also been validated with real-world data including the National Football League (NFL)'s 28 experimental reconstructions of NFL impacts (Viano et al. 2005) as well as 4 automotive crashes (Franklyn et al. 2005). The WSUHIM has become one of the most well-known FE models of human head and brain. This version of the WSUHIM has been used extensively by many of the researchers in the field and is still in active use today.

10.2.2.2   Simulated Injury Monitor Head Model

At about the same time as the WSUHIM was being established, DiMasi et al. (1991) from the National Highway Traffic Safety Administration (NHTSA) developed another head model known as Simulated Injury Monitor I (SIMon I) model. The SIMon I was a simplified model with a non-deformable skull, developed mainly to minimise run time so that it could be used by original equipment manufacturers who routinely conducted a large number of simulations representing crash tests. This first generation of SIMon model (SIMon I) consisted of the skull, the meninges (the dura) and the cerebri hemispheres separated by a relatively thick falx, where the average thickness was 7 mm, which aimed to avoid poor quality meshes associated with a low aspect ratio (Figure 10.3). The brain was modelled as a viscoelastic material (Flugge) model with static and dynamic shear moduli of 0.0172 MPa and 0.0345 MPa, respectively, while the remaining components were modelled as linear elastic materials. A slip interface with a low coefficient of friction was introduced at the dura-cerebral cortex interface to facilitate brain motion. The authors hypothesised that the risk of sustaining a DAI was proportional to the fraction of the brain volume which exceeded a preset injury threshold, a parameter they called the cumulative strain damage measure (CSDM) (Eppinger et al. 1994; DiMasi et al. 1995). It was shown from the simulation results that the HIC was closely related to translational

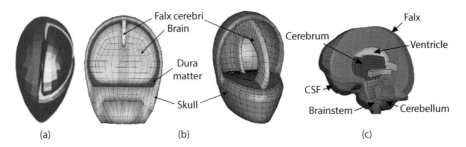

FIGURE 10.3 (a) The Simulated Injury Monitor (SIMon) I head model developed by DiMasi et al. (1991). (Reprinted with kind permission from Springer Science+Business Media: DiMasi, F., et al., *The Finite Element Method in the 1990's*, Simulated head impacts with upper interior structures using rigid and anatomic brainmodels, 1991, pp. 333–345, Copyright © 1991.) (b) The meshed view of the SIMon I model (Takhounts and Eppinger 2003, Takhounts et al. 2008) showing various components of the model. (Reprinted from *Accident Analysis & Prevention*, 40(3), Marjoux, D., et al., Head injury prediction capability of the HIC, HIP, SIMon and ULP criteria, 1135–1148, Copyright © 2008, with permission from Elsevier.) (c) The revised version of SIMon FE Head Model (Takhounts et al. 2008). (Modified with kind permission from Springer Science+Business Media: Takhounts, E. G., et al., *Stapp Car Crash Journal*, Investigation of traumatic brain injuries using the next generation of simulated injury monitor (SIMon) finite element head model, 2008, 52, 1–31, Copyright © 2008.)

kinematics only, whereas the CSDM responded to rotation or combined translational and rotational conditions (Eppinger et al. 1994; DiMasi et al. 1995).

The SIMon I model was further refined and improved in 2003 by Takhounts and Eppinger (2003), who had chosen the linear viscoelastic material model to describe the brain tissues after they performed a parametric study using several non-linear material models. This SIMon head model was of relatively simple geometry due to the run-time constraint of 2 hours in order to run large batches of crash test simulations. The model consisted of a rigid skull, a combined dura-and-CSF layer, a brain, the falx and bridging veins, but did not take the cerebellum and the midbrain in account (Figure 10.3a and b). The model represented the head of a 50th percentile male and weighed 4.7 kg. Based on a logistic regression of the model-predicted responses and scaled animal injury outcomes, the authors derived three injury measures to estimate the risk associated with three intracranial injury types: (1) the CSDM, which exceeds 55% of the brain volume, represents a serious risk of sustaining DAI; (2) the dilatation damage measure (DDM), which when lower than −100 kPa (tension), represents an occurrence of brain contusion and focal lesions; and (3) the relative motion damage measure (RMDM), which when equal to one corresponds to the occurrence of an acute SDH. In their study, it was also found that side impact resulted in more severe rotational accelerations and could potentially be more injurious than frontal impact.

In 2008, Takhounts and his co-workers (Takhounts et al. 2008) developed a new geometrically-detailed FE head model comprising the cerebrum, cerebellum, falx, tentorium, pia-arachnoid complex with CSF, ventricles, brainstem and parasagittal blood vessels (Figure 10.3c). The new model could be used to simulate combined translational and rotational accelerations of up to 400 g and 24,000 rad/s^2.

10.2.2.3 University of Louis Pasteur (ULP) Head Model

The research group led by Willinger from the University of Louis Pasteur (ULP) of Strasbourg spent enormous efforts to build another FE model of the human head. The initial version was a simplified 3D model developed based on 18 slices of magnetic resonance imaging (MRI) data by Willinger (Willinger et al. 1995), with the initial intention of determining the natural frequencies of the head in order to design protective devices that can prevent the head from being exposed to these resonant frequencies (Willinger et al. 1995). This model was composed of six elements of the head: the skull, CSF, cerebral hemispheres, cerebellum, the cerebral trunk and membranes.

In 1996, Turquier et al. (1996) used the ULP head model to replicate the cadaveric experiments conducted by Trosseille et al. (1992) in which seated cadaveric specimens were impacted at the face by impactor, simulating head impact onto a steering wheel. The trend of the simulated responses in terms of intracranial accelerations (at the cular nucleus, frontal and occipital lobes), epidural pressures (at the frontal, occipital and temporal regions) as well as ICP (at the third and lateral ventricles) agreed well with that of the cadaveric experiments. However, there were significant numerical oscillations in both acceleration and pressure pulses. These numerical oscillations, which were believed to be due with the modelling assumption of the subarachnoid space, could be minimised slightly by including damping to the model. In the following year, Kang et al. (1997) revised the ULP model and validated it against Nahum et al. (1977)'s experimental cadaveric ICP data. Further validation with the experimental cadaveric ICP data of both Nahum et al. (1977)'s short-duration impacts (<15 ms) and Trosseille et al. (1992)'s long duration impacts (>15 ms) have been reported in Willinger et al.'s (1999) work. The predicted responses correlated reasonably well with the experimental data for one of the short-duration test but not for the longer duration impact. The authors had investigated the effect of subarachnoid space's compressibility on pressure responses and found that the correlation with Trosseille et al. (1992)'s long duration impact test was much better with lower compressibility (bulk modulus of around 2 GPa).

The group had introduced a Lagrangian formulation for the CSF and an elastic-brittle constitutive law for the skull–brain interface to describe the mechanical behaviour of the bone to simulate fractures. The ULP FE head model was modelled with 13,208 elements. The material properties of scalp, skull, CSF, tentorium and falx were all homogeneous and isotropic, except for the brain, which was viscoelastic (Khalil and Viano 1982) (Figure 10.4). This ULP FE head model has been used extensively in various applications such as automobile crash tests, preventive protection and forensic medicine (Willinger and Baumgartner 2003; Baumgartner and Willinger 2003; Pinnoji et al. 2006; Pinnoji and Mahajan 2007; Raul et al. 2008; Milne et al. 2012). Moreover, a validated FE cervical neck model was built by the group (Meyer et al. 2004a, 2004b; Meyer et al. 2012), and the neck model was then incorporated into ULP FE head model.

10.2.2.4   Kungliga Tekniska Högskolan (KTH) Head Model

In 2002, the Swedish Kungliga Tekniska Högskolan (KTH) group consisting of Kleiven, von Holst and Hardy proposed a detailed and parameterised FE model of an adult head and a simplified neck,

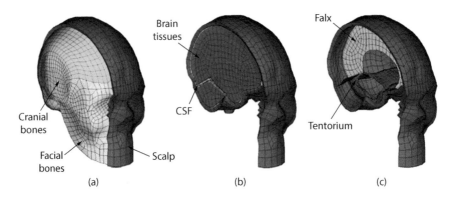

(a)   (b)   (c)

FIGURE 10.4   The Université Louis Pasteur (ULP) FE Head Model. (a) Cranial and facial Bones; (b) the intracranial contents; (c) falx and tentorium. (Marjoux et al. 2008). (Reprinted from Accident Analysis & Prevention, 40(3), Marjoux, D., et al., Head injury prediction capability of the HIC, HIP, SIMon and ULP criteria, 1135–1148, Copyright © 2008, with permission from Elsevier.)

known as the Kungliga Tekniska Högskolan (KTH) head model (Kleiven and Hardy 2002). They constructed six head models of different sizes to study the effect of head size in terms of von Mises stresses and HIC for a series of frontal impacts and inertial loading. From their findings, they stated that the head injury criterion (HIC) does not include the effect of head size and proposed that any new head injury criteria should take the variation in head size into account. Later on, Kleiven and his co-workers (Kleiven 2003, 2006) revised their KTH model and found that the shortest bridging veins displaced the most, specifically those bridging veins that were oriented in the plane of motion and were angled in the direction of motion during rotation of the head in the sagittal plane. The KTH head model included the scalp, skull, brain, meninges, CSF, 11 pairs of parasagittal bridging veins and a simplified neck (Figure 10.5). The KTH head model consisted of 18,400 elements and was modelled with isotropic, homogeneous, non-linear and viscoelastic constitutive material properties based on the work of Mendis et al. (1995). Moreover, dissipative effects were also taken into account through linear viscoelasticity by introducing viscous stress that is linearly related to the elastic stress. Besides the ICP data in the impact tests of Nahum et al. (1977) and Trosseille et al. (1992), the KTH model had been validated against the only brain strain data available, which was from Hardy et al. (2001)'s cadaveric experiments (Kleiven 2006). In 2007, Ho and Kleiven (2007) reconstructed a brain vasculature model based on 3D CT angiographs and incorporated it into the KTH head model by merging the nodes of the vasculature model with those neighbouring brain nodes. It was found that the inclusion of vasculature has minimal effect on the strain responses. Recently, in 2014, Giordano and Kleiven (2014) used the validated KTH FE model of the human head to evaluate the hypothesis that strain in the direction of the neuronal fibres (axonal strain) is a better predictor of TBI than maximum principal strain (MPS), anisotropic equivalent strain (AESM), the cumulative strain damage measure (CSDM), the HIC and the brain injury criterion (BrIC). Using reconstructions of the mild TBI accident dataset obtained from the American NFL (Newman et al. 2000), axonal strain was shown to best predict mild TBI in these 58 accident reconstructions.

10.2.2.5 University College Dublin Brain Trauma Model (UCDBTM)

In 2003, Horgan and Gilchrist (2003) developed the University College Dublin Brain Trauma Model (UCDBTM), with the initial purpose of simulating the simple automobile-pedestrian accidents. Horgan and Gilchrist (2003)'s skull–brain complex included a scalp, a three-layered skull, dura, the CSF, pia, falx, tentorium, cerebral hemispheres, cerebellum and brainstem, comprising

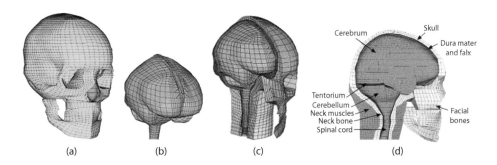

(a)   (b)   (c)   (d)

FIGURE 10.5 Kungliga Tekniska Högskolan (KTH) FEHM. (a) The skull model; (b) the brain model with 11 pairs of bridging veins; (c) sagittal cross-section of the head and neck model. (Reprinted from KTH, *Numerical Modeling of the Human Head*, 2008. Copyright © 2008 by Kleiven S. With permission.) (d) Anatomical structures modelled in KTH model. (Reprinted from *Journal of Biomechanics*, 35(2), Kleiven, S., and von Holst, H., Consequences of head size following trauma to the human head, 153–160, Copyright © 2002, with permission from Elsevier.)

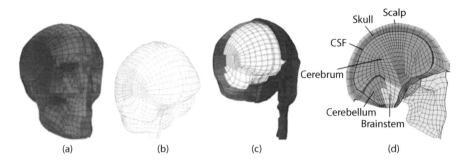

FIGURE 10.6 The University College Dublin Brain Trauma Model (UCDBTM). (a) The skull, (b) the brain and (c) sagittal cross-section. (Reprinted with kind permission from Springer Science+Business Media: *International Journal of Crashworthiness*, The creation of three-dimensional finite element models for simulating head impact biomechanics, 8(4), 2003, 353–366, Horgan, T.J., and Gilchrist, M. Copyright © 2003.) (d) The midsagittal section illustrating the intracranial contents. (Reprinted from Forero Rueda, M A., et al., *Comput. Methods Biomechanics Biomed. Eng.*, 14(12), 1021–1031, 2011. Copyright (2011) by Taylor & Francis Group. With permission.)

50,000 elements in total (Horgan and Gilchrist 2003) (Figure 10.6). A parametric study was then performed to investigate the effects of both the mesh size and material properties on the model-predicted responses. The authors found that a coarsely meshed model would be adequate for predicting ICP responses, while a finer meshed model would be required for more detailed investigations. The short-term shear modulus (G_0) of the brain tissue was also found to have the most significant influence on frontal ICP and von Mises stress, while the contrecoup ICP was significantly influenced by the CSF's bulk modulus. The model was validated against both the ICP data of Nahum et al. (1977)'s and Trosseille, Tarriere, and Lavaste (1992)'s cadaveric impact tests as well as Hardy et al. (2001)'s experimental relative displacement data (Horgan and Gilchrist 2003; Horgan and Gilchrist 2004). The UCDBTM is available on the Internet through the BEL repository managed by the Istituti Ortopedici Rizzoli, Bologna, Italy (available from: http://www.biomedtown.org/ biomed_town, 1st Aug 2013) and has been used widely by other researchers, for example, Yan and Pangestu (2011).

10.2.2.6 Singapore Institute of High Performance Computing (IHPC) Head Model

It was much later when this in South-East Asia began to focus on head injury. In 2006, Lee and his co-workers (Zong et al. 2006) constructed a 3D FE head and brain model to study the structural intensity inside the head, and the model was loaded dynamically in the front, rear and lateral directions (Figure 10.7). The geometrical information was taken from Koenig (1998), and the model was relatively simplified with an average mesh size of several centimetres. Their structural intensity field distribution for the three different impact directions revealed that the power flow paths through the skull medium propagated to the spinal cord region with significant structural intensity values, thus suggesting a high possibility of spinal cord injury. The model was validated with the ICP data from both Nahum et al. (1977)'s and Trosseille et al. (1992)'s cadaveric experiments.

10.3 CURRENT RESEARCH AND METHODOLOGY ON FE MODELLING

In the recent years, the structural detail in FE models of the human head has been greatly enhanced through the use of medical imaging data. The current modelling and simulation methodology revolves around the use of either computed tomography (CT) or magnetic resonance imaging

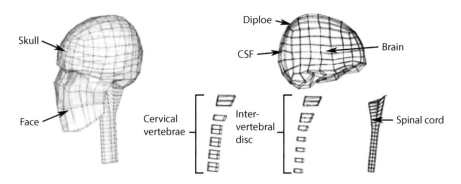

FIGURE 10.7 The simplified FE model of head and brain created by Zong, Lee, and Lu (2006). (Modified from *Journal of Biomechanics*, 39(2), Zong, Z., et al., A three-dimensional human head finite element model and power flow in a human head subject to impact loading, 284–292, Copyright 2006, with permission from Elsevier.)

(MRI) scanning machines to obtain high resolution slice images of the human head. Medical image processing software, such as Mimics (Materialise, Leuven, Belgium) or Simpleware (Simpleware, Exeter, UK), is then used to segment the various components of the scanned object to obtain the individual solid models. Segmentation is a semiautomatic process of classifying different components in the head or grouping elements with the same properties. It essentially requires skilled professionals to identify the various components of the human multilayered head, such as the scalp, skull, meninges, falx, CSF and the white and grey matter of cerebrum and cerebellum from the grey-scale differences of the image slices. Various components are then segmented and grouped into different colour-coded masks using the intensity based algorithms. The imaging processing software will then piece up all the image slices and build the 3D surface mesh accordingly. The surface mesh can usually be exported as stereolithography (STL) file and imported into meshing pre-processors such as 3-matic (Materialise, Leuven, Belgium), Hypermesh (Altair HyperWorks, Troy, MI, USA) or TrueGrid (XYZ Scientific Applications, Inc., Pleasant Hill, CA, USA), for mesh reparation and/or 3D mesh generation. The 3D FE mesh of the highly anatomical realistic human head can then be exported into file formats compatible with various commercial FE software packages, such as Abaqus (Dassault Systemes SIMU-LIA, RI, USA), Ansys (Ansys, Inc., Canonsburg, PA, USA), LS-DYNA (Livermore Software Technology Corp., Livermore, CA, USA) or NASTRAN (MSC Software Corp., Newport, CA, USA) before it is used for simulation purposes. However, constructing 3D FE head models that contain both a high level of anatomical detail and sophisticated constitutive models remains a challenge and there is often a tradeoff between the computational efficiency and the sophistication of the FE models.

10.3.1 RECENT SUBJECT OR PATIENT-SPECIFIC HEAD MODEL

10.3.1.1 Johns Hopkins University Head Model

In 2009, Robert and his co-workers (Roberts et al. 2009) from Johns Hopkins University developed a human head FE model based on the geometry of a human surrogate head, with a Hybrid III ATD neck attached to it. This FE model of human head was created with LS-DYNA based on the initial mesh obtained from VOLPE National Transportation Research Center. It had the identical exterior geometry of the human ATD head, but with a human interior skull and brain (Figure 10.8a) and was modelled with 130,000 hexahedral, tetrahedral, shell and beam elements. The aim of their study was to conduct both shock tube experiments and FE simulations using the head model with a Hybrid III

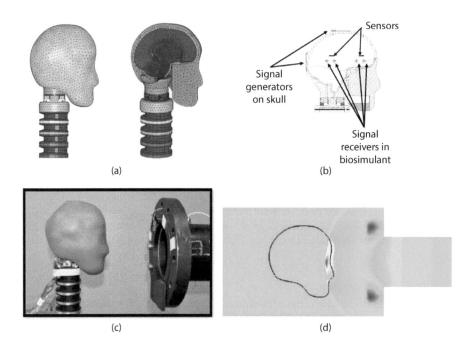

FIGURE 10.8   (a) The FE head model with Hybrid III anthropomorphic test device (ATD) neck attached to it, (b) location of the electromotive force (EMF) displacement sensors and pressure sensors in the surrogate head, (c) the surrogate head placed in front of the shock tube in the experimental environment and (d) the simulated shock wavefront impacting on the FE head model. (Reprinted from *Journal of Biomechanics*, 45(16), Roberts, J.C., et al., Human head–neck computational model for assessing blast injury, 2899–2906, Copyright 2012, with permission from Elsevier.)

neck to establish the governing head parameters that affected the global head rotation, ICP and head displacements (Figure 10.8c and d). The Maxwell-Kelvin viscoelastic material model for the brain tissues and a low friction surface-to-surface interface at skull–CSF–brain boundaries were found to best replicate conditions seen in shock tube experiments. The difference between experimental and numerical-predicted global head rotation and peak ICP was 15% (Roberts et al. 2009) and 25% (Roberts et al. 2012), respectively, while the predicted relative skull–brain displacement was only about 30%–40% of those measured in the human surrogate experiments (Roberts et al. 2012). Coincidently, the peak von Mises effective strain, tensile shear strain and global head rotation were observed to be at about 10 ms after the initial impact of the blast wave, implying that global head rotation might be the primary contributor to the peak intracranial strain in blast-induced TBI events (Harrigan et al. 2010).

10.3.1.2   The Navy Research Lab Simpleware Head Model by Brown University

Brown University has a long history of research on head injury; as mentioned earlier, Hardy and Marcal (1971) made the first attempt in 1971 to construct a 3D FE elastic skull model (Figure 10.9). Since then, to the authors' best knowledge, there is a paucity of published works on head injury using FEM. More recently, Brown University and Rhode Island Hospital collaborated with Exeter Simpleware (Exeter, UK) and Dassault Systemes SIMULIA Group to reconstruct a subject-specific FE model of the human head from the MRI data of a healthy adult male using Simpleware (Bar-Kochba et al. 2012; Morse et al. 2014). This model is known as the Navy Research Lab (NRL)-Simpleware model. It is a 3D model comprising 1.5 million four-node, linear tetrahedral elements and is divided into 12 anatomical regions, namely the white matter,

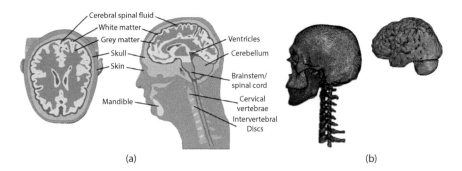

(a) (b)

FIGURE 10.9 (a) A midsagittal cut showing all of the anatomical sections of the head with the exception of the eyes. (b) The FE mesh, which consists of the white matter, grey matter, ventricles and cerebellum, skull, skin, mandible, cerebrospinal fluid, spinal cord, eyes, vertebrae and intervertebral discs. (Modified from Bar-Kochba, E., et al., Finite element analysis of head impact in contact sports, *SIMULIA Community Conference (SCC)*, SIMULIA, Providence, RI, 2012. Copyright © 2012 by Christian Franck. With permission.)

grey matter, ventricles and cerebellum, skull, skin, mandible, cerebrospinal fluid, spinal cord, eyes, vertebrae and intervertebral discs.

The NRL-Simpleware FE head model was created with the initial intention to simulate traumatic brain injuries and to assess protective headgear performance in contact sports. Bar-Kochba et al. (2012) simulated frontal impacts, commonly encountered in the NFL, and investigated on two different boundary conditions, namely bonded and free boundaries, for the skull-CSF interaction. It was found that boundary condition plays a significant role in determining the ICP distribution. More sport-related injuries, such as stick impacts in girls' and women's lacrosse, were simulated using the model to investigate the intracranial biomechanical responses (Figure 10.10a). In recent work by Morse et al. (2014), reduced anatomical sets comprising the skull, brain, ventricles and skin were used, with the skull being modelled as a linear elastic material, while the soft tissues were modelled by a hyperelastic material model. This reconstruction of unprotected lacrosse stick-head impacts provides the foundation for a quantitative methodology of injury prediction in girls' and women's lacrosse. It should be noted that the NRL-Simpleware FE head model had been validated against Nahum, Smith, and Ward (1977)'s ICP data in cadaveric tests as well as Morse et al. (2014)'s kinematic data obtained in experiments with the actual players striking on the head surrogate (Figure 10.10b).

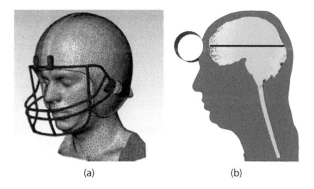

(a) (b)

FIGURE 10.10 (a) The NRL-Simpleware FE head model with a football helmet, (b) midsagittal view showing the ICP distibution when the head is striked by a rigid cylinder of 80kg at frontal region. (Modified from Bar-Kochba, E., et al., Finite element analysis of head impact in contact sports, *SIMULIA Community Conference (SCC)*, SIMULIA, Providence, RI, 2012. Copyright © 2012 by Christian Franck. With permission.)

10.3.1.3 The National University of Singapore (NUS) Head Model

In 2012, Tse (Tse et al. 2014a) from the National University of Singapore (NUS) developed a subject-specific FE model of a human head which consisted of the skull, cerebrum, cerebellum, brainstem, CSF, as well as the nasal cartilages. The model was reconstructed based on the CT and MRI data of healthy male adults and was referred as the NUS FE Head Model I (Figure 10.11a–c). A modal analysis was conducted using this model to determine its modal responses in terms of resonant frequencies and mode shapes under vibration (Tse et al. 2013). These modal responses are found to be reasonably in agreement with the literature (Grandjean 1980; Chaffin and Andersson 1991; O'Brien 2002; Meyer et al. 2004a; Genta 2012). Additional and rarely reported modal responses such as 'mastication' modes of the mandible and flipping of nasal lateral cartilages were identified (Figure 10.11d). This modal validation in terms of modal shapes suggested a need for detailed modelling to identify all the additional frequencies of each individual part.

The NUS FE Head Model I has also been used to simulate nine common impact scenarios of facial injuries to correlate traumatic brain injuries (TBI) with facial injuries (Tse et al. 2015a) (Figure 10.11e). From the study, it was concluded that the severity of brain injury is highly associated with the location of impact to the brain, whereas facial injury is not necessarily closely related to brain injury, despite the proximity of the facial skeleton and the skull. The authors hypothesised that the midface was capable of absorbing considerable energy and protecting the brain from impact. The nasal cartilages also helped to dissipate a portion of the impact energy in the form of large-scale

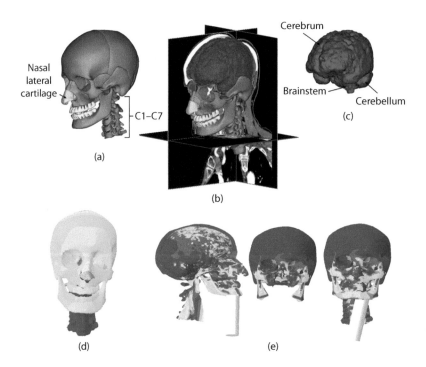

FIGURE 10.11 The NUS FE model I of the skull–brain. (a) The skull and nasal cartilages, (b) the components which were segmented from the CT data and (c) the brain tissues (Tse et al. 2014a). (Reprinted from Tse, K.M., et al.: Development and validation of two subject specific finite element models of human head against three cadaveric experiments. *International Journal for Numerical Methods in Biomedical Engineering* 2014. 30(3). 397–415. Copyright Wiley-VCH Verlag GmbH & Co. KGaA. With permission.) Its applications in (d) modal analysis (Tse et al. 2013) and (e) blunt impact (Tse et al. 2015a). (Reprinted from *Accident Analysis & Prevention*, 79, Tse, K.M., et al., Investigation of the relationship between facial injuries and traumatic brain injuries using a realistic subject-specific finite element head model, 13–32, Copyright © 2015, with permission from Elsevier.)

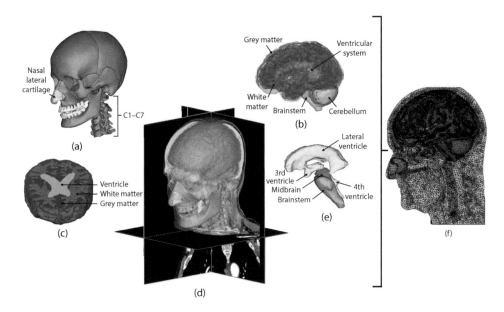

FIGURE 10.12 Various components of the NUS FE model II of the human skull and brain segmented by Mimics, from CT and MRI data. (a) The skull; (b) the brain tissues; (c) the transverse view of the brain tissues showing the ventricle, white and grey matters; (d) The segmented components in CT data; and (e) the ventricles and brainstem. (f) The midsagittal view of the final meshed model. (Modified from Tse, K.M., et al.: Development and validation of two subject-specific finite element models of human head against three cadaveric experiments. *International Journal for Numerical Methods in Biomedical Engineering*. 2014. 30(3). 397–415. Copyright Wiley-VCH Verlag GmbH & Co. KGaA. With permission.)

deformation and fracture, with the vomer-ethmoid diverging stress to the 'crumpling zone' of air-filled sphenoid and ethmoidal sinuses; in its most natural manner, the face protects the brain. However, with anatomic proximity of the facial skeleton and cranium, any larger impact force would be sufficient to cause TBI.

Similar to the NRL-Simpleware FE head model of Brown University, another detailed subject-specific FE model of human head, comprises the skull and nasal cartilages with the overlying soft tissue as well as the intracranial contents which were further separated into white matter, grey matter, the cerebral peduncle (midbrain) and the ventricles, was developed by Tse et al. (2014a) from the NUS. The improved NUS FE Head Model II also included some of the interior details such as the airway as well as air-containing sinuses namely the frontal sinus, sphenoidal sinuses and maxillary sinuses, which have often been ignored in earlier models. The NUS FE Head Model II weighed 4.73 kg and consisted of 1,337,903 linear tetrahedral elements (Figure 10.12). Both of Version I and II of the models had been validated with the ICP data of Nahum, Smith, and Ward (1977) and Trosseille, Tarriere, and Lavaste (1992)'s cadaveric impact tests, as well as the brain strain data of Hardy et al.'s cadaveric experiments (Tse et al. 2014a).

10.4 PRIMARY PERSONAL PROTECTIVE EQUIPMENT–COMBAT HELMETS

Helmets, being the oldest form of personal protective equipment (PPE), are known to have been worn in ancient Greece and Rome and also through the Middle Ages. These ancient helmets were made of bronze and iron which offered protection for the combatant's head from cutting blows with swords and flying arrows (Figure 10.13a). Since the early 20th century, steel helmets (e.g. M1 steel helmets) were actively used by the warring nations in the world wars to protect

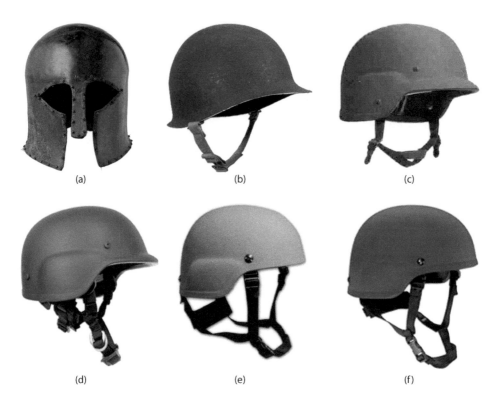

(a)   (b)   (c)

(d)   (e)   (f)

FIGURE 10.13   (a) Ancient Greek bronze helmet (Reprinted from The Bronze Helmet of Corinthian Type, The Metropolitan Museum of Art 2003. With permission.); (b) M1 steel helmet during world wars (Reprinted from Medals Gone Missing 2015, *Uniform & Kit Issued to the US Army & Marine Corps during WWII*. With permission) as well as modern combat helmets, e.g. (c) PASGT helmet; (d) LWH; (e) MICH; and (f) ACH (reprinted from Helmets (2015). Copyright © 2014 by Operation Helmet, with permission).

soldiers from ballistic shrapnel and fragments (Figure 10.13b). Modern helmet design initiated in 1972 after it was realised that adequate protection to the head would save soldiers from serious head injuries or concussion, and this was vital to winning battles. The Personnel Armour System Ground Troops (PASGT) helmet was officially adopted in 1984 as the first polymer helmet used by the US military to replace the M1 steel helmet that had been used since 1942 (Figure 10.13c). It was constructed of Kevlar® Aramid 29, a resilient fireproof plastic that is five times stronger than steel by weight and is backed by phenolic resin. The PASGT helmet is known to provide ballistic protection for the head from fragmenting munitions and handgun bullets from a distance and is capable of deflecting rifle rounds fired at an angle. The US Department of Justice rated the PASGT helmet at Level II, that can effectively stop 9-mm NATO rounds and 0.44 Magnum rounds. The PASGT helmet was replaced by its look-alike successor used the Light Weight Helmet (LWH). The LWH had a four-point chinstrap built-in and weighed about 170 g (or 6 ounces), which is lighter than the PASGT (Figure 10.13d). In early 21st century, the US Army Special Operations Command had replaced the PASGT with the more superior Modular Integrated Communications Helmet (MICH), which was incorporated with both excellent ballistic protection and the ability to interface with most tactical headsets and mikes (Figure 10.13e).

In 2002, the PASGT was completely phrased out by the US military in place of the Advanced Combat Helmet (ACH) which is based on the design of MICH (Figure 10.13f). It is made of newer Aramids eight times stronger than steel by weight (McEntire and Whitley 2005) and is lighter than the PASGT helmets. Moreover, the ACH was also proven to have improved ballistic performance, with the ACH being able to withstand a ballistic impact from a 9-mm round at close range while the PASGT would fail, as shown in the National Institute of Justice (NIJ) tests (McEntire and Whitley 2005). The function of these modern combat helmets is to attenuate shock by absorbing the energy generated during impact through deformation and dissipating the energy rapidly via delamination and failure of its shell without injuring the wearer. Further information on these helmets can be found in Chapter 8.

Despite the fact that modern combat helmets can prevent bullets of handguns and even some rifles from penetrating them, traumatic injuries to both the skull and brain can still occur due to the excessive mechanical responses of the helmet and the head. A ballistic impact on a helmet can generate peak head acceleration which may exceed the tolerance level of what the tissues can bear, thus causing irreversible damage to these tissues. Injuries to head can also occur when the projectile has sufficient energy to cause the interior helmet shell to come in contact with the underlying tissue. and this is known as 'rear effect' (Carroll and Soderstrom 1978). In order to minimise the morbidity and mortality resulting from ballistic head injuries, it is particularly important to ensure that combat helmets are capable and effective in providing protection from ballistic impact.

10.4.1 EARLIER RESEARCH ON THE HELMETED HEAD

Earlier research on head protection could be traced back to the 1960s when experiments, as reported in Alley (1964), Patrick (1966), Ewing and Irving (1969), and Rayne and Maslen (1969) were performed extensively to investigate the required head protection in an event of traumatic head injury. Due to ethical issues, these tests could only be performed using a simplified experimental representation of the human head, for example, spherical metal shells filled with liquid. Later on, biofidelic headforms such as Side Impact Dummies (SIDs) and Hybrids were used for head injury prediction (Haug et al. 2004). However, different and non-standardised test procedures used by helmet manufacturers made comparison of helmets difficult. Moreover, maintenance of experimental equipment and replacement of damaged components incur costs. FE simulations, which serve as a cost-effective alternative to these experimental tests, were then used to understand the mechanism of head injury during impact and how the use of helmets attenuate injuries.

10.4.2 FINITE ELEMENT MODELLING OF COMBAT HELMETS

10.4.1.1 Earlier FE Helmet Models

The first few numerical studies by Khalil and his co-workers (Goldsmith and Khalil 1973; Khalil 1973) used a simplified axisymmetric elastic head-helmet FE model to investigate the dynamic response of head-helmet in a localised short-duration impact. In 1974, Khalil, Goldsmith, and Sackman (1974) conducted a low-velocity ballistic impact study using a simplified axisymmetric elastic head-helmet FE model and validated the model with corresponding experiments (Figure 10.14).

10.4.1.2 FE Helmet Models

Thereafter, tremendous efforts had been spent on research of head protective helmets using the FE method (Van Hoof et al. 1999; Van Hoof et al. 2001; Baumgartner and Willinger 2003; Aare and Kleiven 2007; Tham, Tan, and Lee 2008; Lee and Gong 2010; Yang and Dai 2010). Since 1999, van Hoof et al. (Van Hoof et al. 1999; Van Hoof et al. 2001; Van Hoof and Worswick 2001) from the University of Waterloo developed a composite PASGT helmet model and combined it with a French Delegation Generale pour l'armement (DGA) FE head model (modified from the WSUHIM Version I), which was originally developed for automotive crash studies, in order to investigate the responses of a helmeted head subjected to a ballistic impact (Figure 10.15). The PASGT helmet model was reconstructed from coordinates measured from about 16,000 points on the outer surface

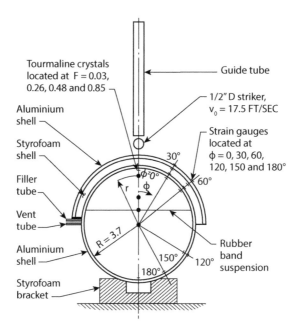

FIGURE 10.14 FE model of the ballistic impact on helmeted head of Khalil's study (Khalil, Goldsmith, and Sackman 1974). (Reprinted from *International Journal of Mechanical Sciences*, 16(9), Khalil, T.B., et al., Impact on a model head-helmet system, 609–625, Copyright 1974, with permission from Elsevier.)

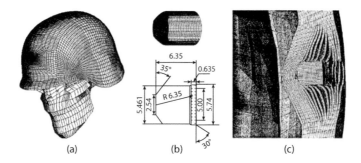

FIGURE 10.15 (a) The FE model of the helmeted head subjected to frontal ballistic impact (Van Hoof and Worswick 2001); (b) the FE model of the FSP bullet with its dimension; and (c) deformed mesh plots obtained from simulation of a 1.1 g FSP frontally impacting a helmeted head (Van Hoof and Worswick 2001). (Reprinted from Van Hoof, J., and M.J. Worswick, Combining head models with composite helmet models to simulate ballistic impacts, 2001, Copyright Defence Research and Development Canada (DRDC). With permission.)

of the helmet and then meshed with brick elements. The 9.5 mm PASGT helmet was modelled with the impacted region using refined meshes. The helmet model comprises 19 layers across its thickness, with two elements for each layer. It should be noted that no helmet liner or interior cushioning system was modelled as van Hoof assumed the effects of low-mass projectiles (like the 1.1 g fragment-simulating projectile, FSP, used in their study) were localised in impacted region and rigid body motion was negligible (Van Hoof and Worswick 2001). The University of Waterloo composite damage model was based on Continuum Damage Mechanics (CDM) theory (Voyiadjis 1987; Matzenmiller et al. 1995). It employs a strain-based failure initiation criterion in addition to property degradation which is used to model intralamina failure of the helmet and interlamina delamination. The University of Waterloo composite damage model, which was designed to predict the penetration and delamination of the laminated helmet shell, had been implemented using VEC-DYNA (Livermore Software Technology Corp., Livermore, CA, USA) FE hydrocode (Figure 10.15c). Van Hoof and his team performed both experimental and FE studies on the response of the woven composite helmet materials to ballistic impact and found that the helmet interior exhibited large deformation which could exceed the gap between inner helmet shell and head (Van Hoof et al. 2001).

It should be noted that these preliminary studies by van Hoof had ignored the modelling of the helmet liner or interior padding as it was found that for a low-mass projectile, the effects of the impact were localised in the impacted region and the rigid body motion of the helmet was negligible (Van Hoof and Worswick 2001). Since then, research has also focused on the helmet shell materials (Tan and Ching 2006; Rao et al. 2009; Mamivand and Liaghat 2010; Feli, Yas, and Asgari 2011; Ha-Minh et al. 2011a, 2011; Abu Talib et al. 2012) and helmet liners or cushion system (Mills 2007; Goel 2010). In 2007, Aare and Kleiven (2007) simulated a series of ballistic impacts of higher mass projectile (a rigid 8 g lead bullet which is heavier than Van Hoof and Worswick (2001)'s 1.1 g FSP) travelling at 360 m/s impacting a validated FE model of a PASGT helmet (validated against data from shooting tests, (Test Protokoll-Nr (2001)) combined with the detailed KTH FE head model (Figure 10.16). The helmet mechanical material properties used in their study are based on the material properties presented in Van Hoof et al. (2001). The failure of the woven fabric-reinforced aramid laminates of the helmet shell were modelled with a Chang-Chang composite failure model (Chang and Chang 1987) available in LS-DYNA (Livermore Software Technology Corp., Livermore, CA, USA) which includes matrix cracking, matrix compressive failure (Hashin Failure Criterion), fibre-matrix shearing and fibre breakage (Yamada Sun failure criterion). Aare and Kleiven (2007) studied the effects of helmet shell stiffness in frontal and oblique impact directions (0°, 22.5°, 45°, and 67.5° to the frontal impact direction) on the load levels in a human head during a ballistic impact, and they suggested that the helmet shell deflections should not exceed the initial gap between the helmet shell and the head in order to prevent the rear effect. From their parametric study, it seemed that the stiffness of the helmet shell had substantial influence on skull stresses: skull stresses increased considerably when there was contact between helmet shell and the head. However, slightly lower brain strains were seen in the non-contact impact case than in the contact case, and this may have been because of different injury mechanisms in which the non-contact impacts give rise to inertial-induced intracranial strain, while the contact impacts result in localised high skull stresses. Aare and Kleiven (2007) also observed that the helmets used in both their study and van Hoof et al.'s (Van Hoof et al. 1999; Van Hoof et al. 2001; Van Hoof and Worswick 2001) studies were too stiff compared to the experimental observations used for validation of the model.

Tham, Tan, and Lee (2008) reconstructed the FE model of a Kevlar® helmet by measuring the geometrical coordinates of the helmet using a coordinate measuring machine (CMM) (Figure 10.17). The helmet was meshed with hexahedral elements and was assigned with strain-based failure criteria,

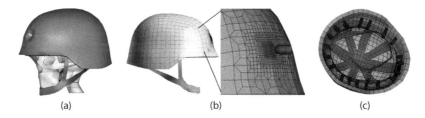

(a) (b) (c)

FIGURE 10.16 (a) Frontal impact of an 8 g lead bullet impacting the FE model of a helmeted head (Aare and Kleiven 2007), (b) the FE model of a PASGT helmet with the meshes at the impact region refined and (c) the FE model of the strap-netting system in the PASGT helmet (Aare and Kleiven 2007). (Reprinted from *International Journal of Impact Engineering*, 34(3), Aare, M., and Kleiven, S., Evaluation of head response to ballistic helmet impacts using the finite element method, 596–608, Copyright 2007, with permission from Elsevier.)

FIGURE 10.17 Geometrical coordinates of the Kevlar® helmet being picked up by the CMM (Tham et al. 2008) (Reprinted from *International Journal of Impact Engineering*, 35(5), Tham, C.Y., et al., Ballistic impact of a Kevlar® helmet: Experiment and simulations, 304–318, Copyright 2008, with permission from Elsevier.)

with subsequent property degradation under ballistic impact. However, the modelling of delamination of the composite laminates was not included in the study.

To determine the responses of the Kevlar® helmet, Tham et al. (2008) conducted both experimental and FE simulations of ballistic test of a 11.9 g spherical projectile launched from a gas gun at a velocity of 205 m/s (Figure 10.18a). The helmet responses from the simulations were found to be in good agreement with the experimental results. Once validated with the experimental results, an additional two ballistic tests were performed to determine whether the Kevlar® helmet could pass the ballistic test standards for combat helmets, namely the MIL-H-44099A and NIJ-STD-0106.01 Type II. In the MIL-H-44099A test, the top of the Kevlar® helmet was struck by a 1.1 g FSP with an impact velocity of 610 m/s (Figure 10.18b). It was revealed that an FSP with impact velocity higher than 610 m/s is required to perforate the helmet. As for the NIJ-STD-0106.01 Type II test, an 8 g, 9-mm full-metal-jacketed (FMJ) bullet struck the helmet at 358 m/s and the simulation result showed that the Kevlar® helmet was capable of deflecting the FMJ bullet travelling at 358 m/s (Figure 10.18c). However, no head model was incorporated in Tham, Tan, and Lee's (2008) simulation; hence, the analyses were restricted to mechanical responses of the helmet without any insights to head or brain injuries.

In 2010, Lee and Gong (2010) incorporated Zong et al.'s (2006) IHPC FE head model with Tham et al.'s (2008) Kevlar® helmet in a series of ballistic simulations of the frontal and lateral impacts of the FSP and FMJ (Figure 10.19a and b). Models of the helmet liner or strap-netting suspension

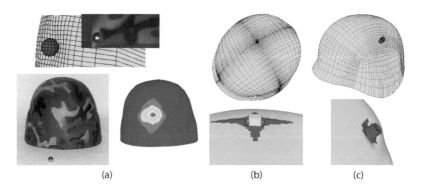

(a)   (b)   (c)

FIGURE 10.18 The FE simulations of ballistic impact tests, (a) the gas gun ballistic test in the study (a 11.9 g spherical steel projectile striking the rear of the helmet at 205 m/s); (b) the MIL-H-44099A (a 1.1 g FSP striking the top of the helmet at 610 m/s); and (c) the NIJ-STD-0106.01 Type II (an 8 g, 9 mm FMJ striking the side of the helmet at 358 m/s) (Tham, Tan, and Lee 2008). (Reprinted from *International Journal of Impact Engineering*, 35(5), Tham, C.Y., et al., Ballistic impact of a KEVLAR® helmet: Experiment and simulations, 304–318, Copyright 2008, with permission from Elsevier.)

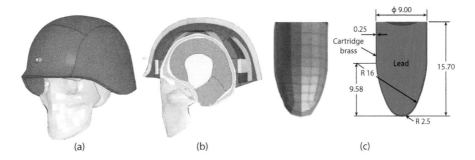

FIGURE 10.19 Frontal impact of a FMJ projectile on the helmeted IHPC FE head model. (a) Isometric view and (b) midsagittal view (Lee and Gong 2010). (c) Dimensions and mesh of the FMJ projectile (Lee and Gong 2010). (Reprinted from *Computer Methods in Biomechanics and Biomedical Engineering*, 13(5), Lee, H.P., and S.W. Gong, Finite element analysis for the evaluation of protective functions of helmets against ballistic impact, 537–550, Copyright 2010, with permission from Taylor & Francis.)

systems were also included (Figure 10.19c). It was found that the helmet with an interior strap-netting system was able to deflect both the FSP and FMJ without penetration of the helmet shell, with more superior protection against smaller and lighter FSPs than FMJ. However, this did not indicate that no severe head injuries would be sustained. In fact, head injuries can still be severe in the helmeted regions where the interior net straps were of higher stiffness or/and where the helmet–head gap was small, as there would be less room for impact energy dissipation.

Yang and Dai (2010) from Texas Tech University focused on evaluating the rear effect by having the helmeted FE head model impacted by an 8 g parabellum handgun bullet at different impact angles and impact positions (Figure 10.20). The helmet was modelled using geometry similar to the PASGT, with a shell thickness of 8 mm. The helmet liner or interior cushioning system was not modelled as its influence on the helmet–skull interaction was presumed to be minimal (Yang and Dai 2010). The Chang-Chang composite failure material model was used to model the failure of the helmet shell. The HIC and other biomechanical parameters were used for the assessment of head injury. It was found that rear impact would give the highest skull stresses and HIC values, while top impacts

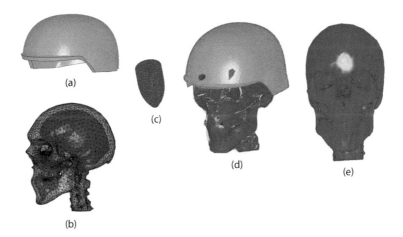

FIGURE 10.20 Yang and Dai (2010)'s FE models of (a) a PASGT helmet; (b) a skull–brain head model; (c) an 8 g handgun bullet (Yang and Dai 2010). (d) Frontal ballistic impact scenario and (e) its corresponding skull stress (Yang and Dai 2010). (Reprinted from *Computer-Aided Design and Applications*, 7(1), Yang, J., and J. Dai, Simulation-based assessment of rear effect to ballistic helmet impact, 59–73, Copyright 2010, with permission from Taylor & Francis.)

would result in the highest ICP and maximum brain strains. As such, Yang and Dai (2010) concluded that the HIC values did not necessarily have the same trend as the values of von Mises stress, ICP and brain principal strain.

Makwana and Zhang (Makwana 2012; Zhang et al. 2013) from Wayne State University (WSU) reconstructed an FE model of the ACH using 3D laser scanner, and meshed the 10-mm helmet shell using 8-noded brick elements. The helmet liner comprised Team Wendy® 7-pads, and the pads were modelled by outlining the surface contours on the interior helmet shell and extruding them to the desired thickness (9–10 mm). The entire helmet assembly was meshed with about 70,000 elements, with average element size of about 3 mm. Makwana (2012)'s ACH model adopted the transversely isotropic material properties of Van Hoof et al. (2001) and used the in-built composite failure material model, MAT_22 [an orthotropic material model adopted within LS-DYNA (Livermore Software Technology Corp., Livermore, CA, USA)], with optional brittle failure for composites (based on Chang-Chang composite failure model) (Makwana 2012). In-house uniaxial compressive tests were performed to obtain the Team Wendy®'s stress–strain relationships (Makwana 2012). In order to validate the ACH model, simulations of a series of drop tests based on the US Army impact test standard for the ACH helmet were performed with a Department of Transport (DOT) headform (Figure 10.21a). The helmeted headform was impacted at seven impact locations (front, left, crown, right, rear, left nape and right nape) at 3.0 m/s (10 fps) and 4.3 m/s (14.14 fps), and the headform's mean and peak accelerations were then compared with McEntire and Whitley (2005)'s experimental values. The predicted accelerations were found to be within the experimental range for almost all the impact directions except for two lateral impacts. After validation of the ACH model, it was then integrated with the WSUHIM to simulate the same impact scenarios used in the ACH helmet validation study (Figure 10.21b). Biomechanical parameters such as ICP, intracranial strain, and strain rates were monitored and compared between the different impact cases to assess brain responses during standard helmet impact tests. From the simulation, it was observed that an increase in head accelerations would give rise to an increase in ICP, suggesting that ICP might be predicted by the resulting head accelerations in the current blunt impact-loading condition. The observed ICP patterns were in accordance with coup and contrecoup phenomena which had been postulated as mechanism of cerebral contusion. It was

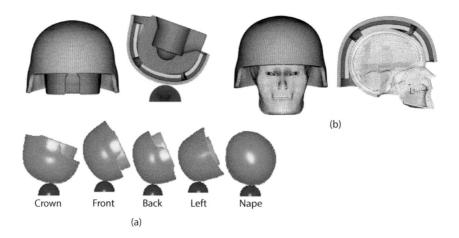

(b)

Crown   Front   Back   Left   Nape

(a)

FIGURE 10.21 The blunt impact scenario simulated by Makwana (2012) using the FE models of (a) DOT headform and (b) WSUHIM FE. (Reprinted from Makwana, R., 2012, Development and validation of a three-dimensional finite element model of advanced combat helmet and biomechanical analysis of human head and helmet response to primary blast insult, M.S., Wayne State University, Ann Arbor, MI, Copyright 2012, with permission from Rahul Makwana.)

also found that the impact velocity of 4.31 m/s (or 14.14 fps) would result in ICP ranging from 200 to 240 kPa, which may exceed Ward et al. (1980)'s ICP threshold for serious brain injury at 235 kPa, suggesting that the soldiers were at greater risk of sustaining moderate/severe brain injuries (Makwana 2012), while the intracranial strains were within the subconcussive range according to the thresholds proposed by Zhang and King (King et al. 2003; Zhang et al. 2003).

Tan, Tse and Lee, from National University of Singapore (NUS), performed both experimental tests and FE simulations of frontal and lateral ballistic impacts of a 11.9 g spherical steel projectile launched from a gas gun and struck on the helmeted Hybrid III headform with two different interior helmet padding systems (strap-netting type and Oregon Aero (OA) foam types), at speeds of approximately 200 m/s (Figure 10.22) (Tan et al. 2012).

Both the FE models of the Hybrid III headform and the ACH helmet were reconstructed from CT scans (Figure 10.23a and b). The helmet liners, namely the strap-netting configuration and the 7-pad foam configuration, were modelled using computer-aided design (CAD) and performed by applying a helmet-fitting step in which the helmet and head were made rigid, with the helmet and its liners being fitted onto the head. Despite that the actual physical ACH contains more than 20 laminate layers, the FE model of the ACH in Tan et al. (2012)'s study was modelled with just four layers in order to reduce computational cost and increase computational efficiency. The headform was built with 4-noded tetrahedral elements, while the helmet model was meshed with 8-noded hexahedral elements. It should be noted that the ACH FE model had been validated against ballistic impacts (Tan et al. 2013). Tan et al. (2012) adopted the elastic material properties of Van Hoof et al. (2001) for the ACH, while the material properties of the straps and OA foams were obtained from literature as well as in-house uniaxial tensile and compressive tests, respectively (Tan et al. 2012). The failure of the fabric-reinforced aramid laminates of the helmet shell were modelled using the Hashin-Fabric Criterion, which takes into account the bidirectional strength of the fibres which are woven into a fabric laminate. Tables 10.2 and 10.3 summarise all the elastic material properties and failure properties of the various models discussed in this chapter.

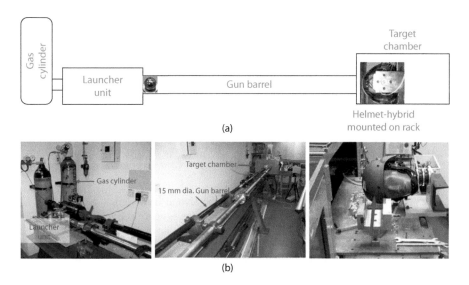

FIGURE 10.22 (a) Schematic diagram and (b) the experimental setup of ballistic impact on helmeted Hybrid III headform and neck using gas gun (Tan et al. 2012). (Reprinted from *International Journal of Impact Engineering*, 50, Tan, L.B., et al., Performance of an advanced combat helmet with different interior cushioning systems in ballistic impact: Experiments and finite element simulations, 99–112, Copyright 2012, with permission from Elsevier.)

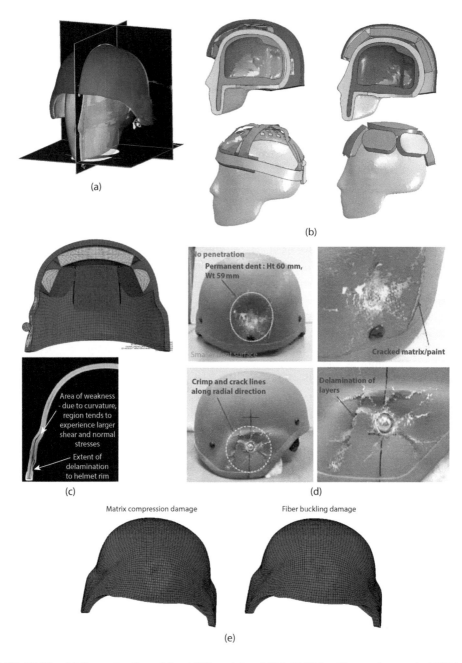

FIGURE 10.23 (a) Reconstruction of the ACH model and Hybrid III headform using pre-test CT images. (b) The FE model with both the strap-netting and OA foam padding system. (c) Comparison of the FE simulated helmet delamination with post-test CT images. (d) Post-test observation of the ACH (with strap-netting configuration) for both front and side impact. (e) Simulated matrix compression and fiber buckling damage on the ACH (with OA foam padding) (Tan et al. 2012). (Reprinted from *International Journal of Impact Engineering*, 50, Tan, L.B., et al., Performance of an advanced combat helmet with different interior cushioning systems in ballistic impact: Experiments and finite element simulations, 99–112, Copyright 2012, with permission from Elsevier.)

TABLE 10.2
Helmet Elastic Material Properties Used in Various Helmet Models

Authors	E11 (GPa)	E22 (GPa)	E33 (GPa)	υ12	υ13/υ32	G12 (GPa)	G23/G13 (GPa)	ρ (g/cm^3)	Additional Description
Van Hoof et al. (2001)	18.5	18.5	6	0.25	0.33	0.77	2.715	1.23	1. With modelling of interply delamination of the helmet (19 layers/32 elements) 2. Modelled degradation of helmet properties with failure initiation criteria
Aare and Kleiven (2007)	18.5	18.5	6	0.25	0.33	0.77	2.715	1.23	1. No modelling of helmet property degradation upon failure initiation 2. No modelling of interply delamination
Tham et al. (2008)	17.989	17.989	1.948	0.08	0.698/0.0756	1.857	0.2235	1.65	1. Uses strain-based failure criteria; includes property degradation modelling 2. No modelling of interply delamination
Lee and Gong (2010)	18.5	18.5	6	0.25	0.33	0.77	2.715	1.23	1. No modelling of helmet property degradation upon failure initiation 2. No modelling of interply delamination
Yang & Dai (2010)	164	3.28	3.28	0	0	3.28	3.28	1.44	1. No modelling of interply delamination
Makwana (2012)	18.5	18.5	6	0.25	0.33	0.77	2.72	1.23	Nil
Tan, Tse & Lee (Tan et al. 2012; Tse 2013; Tse et al. 2015b)	18	18	4.5	0.25	0.33	0.77	2.6	1.23	1. With modelling of interply delamination of Helmet (four layers of elements) through thickness 2. Modelled degradation of helmet properties with failure initiation criteria

TABLE 10.3
Helmet Failure Properties Used in Various Helmet Models

Authors	X_{1t} (MPa)	X_{1c} (MPa)	X_{2t} (MPa)	X_{2c} (MPa)	X_{3t} (MPa)	X_{3c} (MPa)	S_{12} (MPa)	S_{13} (MPa)	S_{23} (MPa)	S_n (MPa)	S_s (MPa)	Additional Description
Van Hoof et al. (2001)	555	555	555	555	1200	1200	77	1086	1086	34.5 (19 layers)	9 (19 layers)	1. Tensile/compressive strength in through ply direction is stronger than along fibre direction 2. Shear strength along fibre axes are lower than that along thickness of laminate 3. Strain-based failure criteria for intralaminar failures; models property degradation
Aare and Kleiven (2007)	555	555	555	555	1200	1200	77	1086	1086	–	–	1. Did not use S_n & S_s * values for interlaminar delamination modelling 2. No modelling of property degradation 3. Uses Hashin, Chang-Chang & Yamada Sun Criteria
Tham et al. (2008)	1080	1080	1080	1080	39	39	92.9*	11.18*	11.18*	–	–	1. *Shear strengths are not available in publication, assumes 0.05 failure strain to derive obtained strengths 2. Interlaminar failure not physically modelled, but taken into account by a constant a 3. Model for property degradation
Lee and Gong (2010)	555	555	555	555	1200	1200	77	1086	1086	–	–	1. Did not use S_n & S_s values for interlaminar delamination modelling 2. No modelling of property degradation 3. Uses Hashin, Chang-Chang & Yamada Sun Criteria

(Continued)

TABLE 10.3 (CONTINUED)

Helmet Failure Properties Used in Various Helmet Models

Authors	X_{1t} (MPa)	X_{1c} (MPa)	X_{2t} (MPa)	X_{2c} (MPa)	X_{3t} (MPa)	X_{3c} (MPa)	S_{12} (MPa)	S_{13} (MPa)	S_{23} (MPa)	S_n (MPa)	S_s (MPa)	Additional Description
Yang and Dai (2010)	2886	2886	1486	1700	1486	1486	1886	1586	1886	-	-	1. Tensile/Compressive strength along fibre direction is stronger than through ply direction 2. Shear strength along fibre axes are similar to those along thickness of laminate 3. Uses Chang–Chang criterion, no interlaminar delamination modelling
Tan, Tse & Lee (Tan et al. 2012; Tse 2013; Tse et al. 2015b)	555	555	555	555	1050	1050	77	1060	1086	823.2 (4 layers)	588 (4 layers)	1. Models for Intralaminar failure using Hashin-Fabric Criteria 2. Models for interlaminar failure using surface traction criteria 3. Model for property degradation

* S_n and S_s are the normal strength and shear strength of the helmet surface, respectively.

Tan et al. (2012) found that there is a reasonably good correlation between the simulation results and experimental data in terms of headform acceleration time histories and post-test observation of the helmet failure modes (via a post-test CT scan and optical images).

Good correlation of the type and spatial extent of helmet damage between experiment and FE results further authenticated the quality of the FE model in predicting the correct transmitted impact force to the head (Figure 10.23c and d). The authors concluded that the OA foam paddings helped to reduce peak acceleration experienced by the headform for frontal impacts, mainly because of the superior capability in absorbing impact shock through foam deformation (which allows distribution of impact forces to larger surface areas) and plateau characteristics of the foam stress–strain curves (which deforms the foam without transmitting high contact impact forces to the head). A similar conclusion on the effect of foam stiffness in mitigating ballistic head impacts has been reported by Lee and Gong (2010). For lateral impacts, the impact energy was primarily absorbed through the large deflection of the helmet shell, while the maximum deflection of the helmet shell in frontal impacts was found to be lesser than that observed for lateral impact.

In order to determine the severity of the intracranial injuries sustained from ballistic impacts when equipped with the combat helmet with the same interior straps and cushions, the same group, Tse, Tan and Lee (Tse 2013; Tse et al. 2014b), performed a series of ballistic impact simulations (frontal, lateral, rear and crown) of the FMJ bullet striking the validated NUS FE Head Model II at impact velocity of 358 m/s, with an impact duration of 4 ms (Figure 10.24). The same two helmet liners or cushioning system used in their earlier work (Tan et al. 2012) were employed in these simulations, where the aim was to determine the effectiveness of the helmet liners in terms of intracranial parameters. An additional case of a crown impact, which resembles the military personnel in a prone position with his or her face facing the ground while being subjected to a gunshot at the helmet, was also included in the analysis for a more complete and comprehensive understanding of head injuries from ballistic impacts. Besides the failure modelling of the composite helmet and the FMJ bullet, Tse (2013) and Tse et al. (2014) implemented a helmet-fitting step of 4 ms prior to the actual ballistic step, in which the head and cushions were pre-stressed similar to that of the real-life scenario.

Similar to Aare and Kleiven (2007), Tse (2013) and Tse et al. (2014b) benchmarked their simulation results with the experimental pressure values and ranges of Sarron et al. (2004)'s cadaveric experimental work, since this work is the only available experimental pressure data where helmeted heads were impacted by the same FMJ projectile used in the experiments with similar impact energy and a similar helmet–head gap. It was found that crown impacts were the most severe with the maximum skull stress, while both the lateral and rear impacts gave the highest ICP, principal strain, and shear strain. Similar to Yang and Dai (2010), Tse (2013) and Tse et al. (2014b) also stated that individual biomechanical metrics, such as a high HIC values, do not necessarily have the same trend as other metrics, such as high shear strain. It was also observed that compared to the strap-netting configuration, the OA cellular foams are more superior in limiting the transmission of force by being better able to absorb more energy prior to the plateau phase and densification occurring.

Pintar et al. (2013) conducted a similar study of ballistic impacts (frontal, left, right and rear) on a helmeted magnesium ISO headform attached to Hybrid III dummy neck. The helmet was equipped with two helmet liners, a foam inner liner and a belt inner liner. Twenty-four ballistic tests were performed with three tests in each direction for each of the two helmets. The type of helmet was not identified, except that both helmets tested were new helmets. A series of FE ballistic simulations were then performed on the helmeted UCDBTM, which was developed by Horgan and Gilchrist (2003, 2004) using the mean experimental helmet-to-head contact force histories as boundary conditions. No helmet or interior liner was modelled or included in the simulation. The skull and brain responses in terms of percentage of the elemental volume exceeding the selected thresholds to the entire volume were compared with the cumulated strain damage measure (CSDM) brain injury criteria for injury assessment.

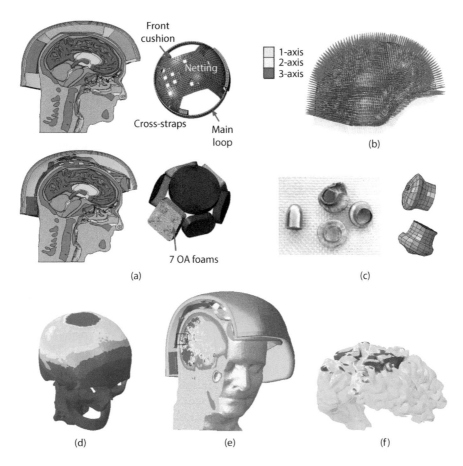

FIGURE 10.24 (a) The midsagittal view of the FE model of the helmet-liner-head assembly. (b) Material orientation for the composite helmet. (c) Comparison of the simulated and experimental deformed FMJ bullets. Contour plots of (d) skull stress; (e) ICP as well as (f) shear strain of the white matter in the ballistic impacts. (Adapted from Tse, K.M., *Development of a Realistic Finite Element Model of Human Head and Its Applications to Head Injuries*, Department of Mechanical Engineering, National University of Singapore, Singapore, 2013.)

Most recently, Salimi Jazi et al. (2014) reconstructed an FE model of the human head using geometrical information from MRI. Similar to the work of Tham et al. (2008), Salimi Jazi et al. (2014) reconstructed the FE model of an ACH using a CMM to obtain the geometrical coordinates of the exterior helmet shell. The ACH was modelled with brick elements with 4 layers and its material properties were obtained from Van Hoof et al. (2001)'s study. In Salimi Jazi et al. (2014)'s work, four different types of the low-density foam, expanded polystyrene (EPS) and expanded polypropylene (EPP) were examined to determine the effectiveness of padding material in protecting the head from sustaining head injury in ballistic events. In addition, the effect of the padding system against different angles (0°, 30° and 60° to the direction of impact in sagittal plane) and directions of ballistic impacts (frontal, lateral and rear impacts) were examined. Frontal impacts were found to be the most severe among all the three impact directions, with the highest ICP and shear stress. Consistent with the findings by Tse (2013) and Tse et al. (2014b), Salimi Jazi et al. (2014) also highlighted that the elasticity of the padding material and location of ballistic impact plays an important role in absorbing and transferring the impact energy within the helmet-interior liner-head system.

10.5 FUTURE DIRECTIONS AND SOME RECOMMENDATIONS

Beyond crash and ballistic impact simulations, there has also been a trend towards modelling the effect of blast loading on the human head, and understanding the resulting brain injury outcomes. Modelling the effect of blast waves arose due to wars in Afghanistan and Iraq, where soldiers had been wounded by improvised explosives devices (IEDs) used in the conflict zones. Death and injuries resulting from IEDs have accounted for more than 60% of the US casualties in Afghanistan (Larry 2012). IEDs have caused a high incidence of TBI rather than death due to the Kevlar® body armour and helmet that helps shield soldiers from blast shrapnel, thereby preventing severe penetrating injuries (Okie 2005). Since 2009, a number of FE studies related to blast-induced TBI (bTBI) using FE models of human head integrated with a combat helmet have been performed, in which the mechanisms of TBI from exposure to blast waves were investigated (Moore et al. 2009; Moss et al. 2009; Taylor and Ford 2009; Grujicic 2011; Grujicic et al. 2011, 2012; Salimi Jazi 2015). However, the helmet alone was found to be insufficient to provide adequate cover to the head region and prevent closed brain injuries caused by blast waves. This has led to some studies on the addition of a face shield in front of the helmet to help mitigating the effects of blast waves on the head (Nyein et al. 2010; Courtney et al. 2012; Tan et al. 2014; Sarvghad-Moghaddam et al. 2015; Rodríguez-Millán et al. 2016). Most notably, a study by researchers at Massachusetts Institute of Technology (MIT) concluded that combining the helmet with a face shield made of polycarbonate impeded the direct transmission of the blast waves to the face region, hence lowering intracranial stresses (Nyein et al. 2010; Bettex 2013).

Another possible alternative would be the design of a face guard (Figure 10.25a), complete with protection of the eyes, ears, mandible and neck against frontal blast waves on the face and wave ingress into gap spaces between the helmet and the head. Research has shown that shockwaves pass through the eyes, nose and mouth, and can still damage the brain. The face shield is responsible for mitigating the effect of the blast wave by absorbing and diffracting most of the wave energy away from the vital parts of the head, especially the face. The FE study by the Defense and Veterans Brain Injury Center (DVBIC) and MIT showed that face shields reduced intracranial stresses (Bettex 2013), while another study conducted by Courtney et al. (2012) demonstrated that different materials such as polycarbonate and laminated glass absorb and reflect the blast wave; thereby reducing transmission to the face. The study of face-shield dimensions (thickness and width coverage) and of their aerodynamic contours, as well as the incorporation of advanced material such as aerogel onto the shield laminate, has also been performed (Tan 2014; Rodríguez-Millán et al. 2016).

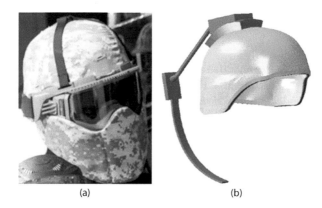

(a)   (b)

FIGURE 10.25   (a) Complete face and head protection with the incorporation of face shield or ballistic eye wear and mandible guard. (Reprinted from Defence Update 2007, Personal Gear at the Modern Day Marine Expo, 2015. Copyright Defence Update. With permission.) (b) 3D model of neck support with advanced combat helmet (ACH). (Reprinted from Cheong, S. Y. C., 2015, *Design of Protective Helmet in Mitigating Ballistic Impacts*, Department of Mineral Engineering, National University of Singapore, Singapore. Copyright Heow Pueh Lee. With permission.)

Traditional helmets lack protection and support for the neck. Blasts often force soldiers off their feet, and this can jerk and twist the neck and head that can cause injury. Shoulder harnesses or neck supports in the form of exoskeleton have also been proposed (Figure 10.25b) and modelled to analyse the potential to mitigate head and neck injury while retaining robustness of head movement in the military scenarios. If these designs are proven to be practical and useful, it is likely that the concept could be taken up by helmet manufacturers for the motorcycle and bicycle industries to minimise mortality rates in traffic accidents.

Some ongoing efforts have also been made by using head models to investigate neck injuries by including the cervical spine, where the spinal component begins at the base of the skull and extends down to the thoracic spine. Realistic modelling of the passive and active properties of cervical musculature tissues has also be performed, for example, the realistic cervical musculature tissue model developed by Hedenstierna (2008) which was implemented to the KTH FE head model. The aim of these type of models is to re-examine the injury mechanisms of head impacts, where the neck could be a precursor or conduit for brain injury or could be a primary injury by itself, such as a ruptured carotid artery, which supplies the brain with blood, or a collapsed or damaged trachea that is caused by a compressive blast wave impact.

With regard to material advancement, the research of stronger and lighter composite combat helmets and of better interior foam inserts and paddings that can cushion the head against impact is ongoing in some manufacturing companies and research institutions. The use of shear-thickening fluid pads and pads with auxetic or aerogel materials to mitigate impact response are also being investigated (Cheong 2015). More advanced and realistic material models of intracranial contents are needed to simulate the motion of the intracranial contents. The current approach for modelling the CSF is to use linear elastic solid elements with low shear modulus and high bulk modulus or a 'fluid' option in which the element's ability to support shear stress is lost and only compressive hydrostatic stress states are possible (Zhou et al. 1995). It is worth mentioning that the CSF in the central nervous system (CNS) is not stationary and is often described as the 'third circulation' (Cushing 1969). Therefore ideally, fluid elements with appropriate boundary condition should be used to simulate the moving CSF.

On simulation aspects, there are also efforts to enhance existing head models to incorporate oral and nasal cavities to look into the effects of such cavities on wave impact transmission. The effects of more complex blast scenarios such as the biological head model being subjected to multiple consecutive blasts from different directions or the effect of wave reflections due to the subject being in an enclosed space or taking cover behind a rigid structure are also some of the trends observed in more recent publications (Tan et al. 2014). Although the existing head models may predict the response accrued to blunt impact of relatively smaller impulse relatively accurately, impacts of greater intensity requiring the incorporation of helmet penetration damage and skull fracture modelling due to highly non-linear or large deformations of the head has yet to be reported.

The traditional macroscale acceleration-based injury criteria for TBI can easily be measured experimentally and had become the basis for TBI protocols, including automobile safety standards and helmet testing standards. However, these cannot be correlated with microscale injury criterion. On the other hand, injury criteria generated by FE replication of real-life scenarios with correlation to clinical data may vary from model to model. Alternatively, there are ongoing attempts to establish a cellular axonal injury criterion for TBI using *in vitro* and *in vivo* studies (Morrison et al. 2011; Wright and Ramesh 2012). However, it is a difficult task to bridge the injury mechanism and its relation between the tissue and the cellular level using an FE head model. As proposed by Wright and Ramesh (Wright et al. 2013) from John Hopkins University, another possible direction would be multiscale modelling to bridge or couple the cellular mechanism of axonal injury to that of tissue level, and eventually to macroscale level. There is still a lot of potential future research to work on head injury mechanisms, and the research community should continue to turn current research limitations into future research topics.

10.6 CONCLUDING REMARKS

In this chapter, numerous FE models of human head and combat helmets for ballistic impacts have been presented and future directions towards the investigation of TBI are discussed. The combination of a detailed FE model of the human head and combat helmet as well as realistic failure modelling of the composite helmet may be crucial in order to obtain an accurate assessment of traumatic head injuries caused by ballistic impacts. With rapid advance in computing facilities and commercial software, more head models with detailed anatomical features have been developed to show they can better predict head injuries than the previous models. Despite the invaluable insight and knowledge these models have provided in understanding head injury biomechanics and their usefulness in the development of future ballistic helmet designs, any continued development of newer models is limited by the lack of experimental data in which to validate them. Thus, it is clear that more detailed and comprehensive high-quality experiments should be conducted, especially for ballistic impacts, which are designed to acquire head biomechanics and injury data for validating the existing models and deriving new injury thresholds.

REFERENCES

Aare, M., and S. Kleiven. 2007. Evaluation of head response to ballistic helmet impacts using the finite element method. *International Journal of Impact Engineering* 34 (3):596–608. doi: 10.1016/j.ijimpeng.2005.08. 001.

Abu Talib, A. R., L. H. Abbud, A. Ali, and F. Mustapha. 2012. Ballistic impact performance of Kevlar-29 and Al2O3 powder/epoxy targets under high velocity impact. *Materials & Design* 35 (0):12–19. doi: 10.1016/ j.matdes.2011.08.045.

Adams, J. H., G. I. Graham, and T. A. Gennarelli. 1982. Neuropathology of acceleration-induced head injury in sub-human primate. In *Head injury: Basic and clinical aspects*, edited by R. G. Grossman and P. L. Gildenberg, pp. 141–150. New York: Raven Press.

Al-Bsharat, A. S., W. N. Hardy, K. H. Yang, T. B. Khalil, S. Tshman, and A. I. King. 1999. Brain/skull reactive displacement magnitude due to blunt head impact: New experimental data and model. In *Paper read at 43rd Stapp Car Crash Conference*, Warrendale, PA.

Alley, R. H., Jr. 1964. Head and neck injuries in high school football. *JAMA* 188:418–422.

Anderson, D., and R. McLaurin. 1980. Report on the national head and spinal cord injury survey. *Journal of Neurosurgery* 53 (Suppl):S1–S43.

Bar-Kochba, E., M. Guttag, S. Sett, J. A. Franck, K. McNamara, J. J. Crisco, J. Blume, and C. Franck. 2012. Finite element analysis of head impact in contact sports. In *SIMULIA Community Conference (SCC)*, SIMULIA, Providence, RI.

Baumgartner, D., and R. Willinger. 2003. Finite element modelling of human head injuries caused by ballistic projectiles. In *Paper read at NATO Research & Technology Organisation (RTO) Specialist Meeting*, Koblenz, Germany.

Belingardi, G., G. Chiandussi, and I. Gaviglio. 2005. Development and validation of a new finite element model of human head. In *Paper read at 19th International Technical Conference on the Enhanced Safety of Vehicles*, Washington, DC.

Bettex, M. 2013. *Heading off trauma: Study suggests adding a face shield to military helmets would help more soldiers avoid blast-induced brain injuries*. [cited 25 June 2015]. Available from: http://newsoffice.mit. edu/2010/brain-injuries-1123.

Bronze helmet of Corinthian type. 2015. The Metropolitan Museum of Art 2003 [cited 27 November 2015]. Available from: http://www.metmuseum.org/collection/the-collection-online/search/257634.

Bullock, R., and D. I. Graham. 1997. Non-penetrating injuries of the head. In *Scientific foundations of trauma*, edited by G. J. Cooper and H. Dudley, pp. 101–126. Oxford: Butterworth-Heinemann.

Carroll, A. W., and C. A. Soderstrom. 1978. A new nonpenetrating ballistic injury. *Annals of Surgery* 188 (6): 753–757.

Chaffin, D. B., and G. Andersson. 1991. *Occupational biomechanics*. 2nd ed. New York: Wiley.

Chan, H. S. 1974. Mathematical model for closed head impact. In *Paper read at 18th Stapp Car Crash Conference*, Ann Arbor, MI.

Chang, F. K., and K. Y. Chang. 1987. A progressive damage model for laminated composites containing stress concentrations. *Journal of Composite Materials* 21:834–855.

Cheong, S. Y. C. 2015. *Design of protective helmet in mitigating ballistic impacts.* Department of Mineral Engineering, National University of Singapore, Singapore.

Courtney, E. D., A. C. Courtney, and M. W. Courtney. 2012. Blast wave transmission through transparent armour materials. *Journal of Battlefield Technology* 15 (2):19–22.

Cushing, H. 1969. The third circulation and its channels. In *Selected papers on neurosurgery*, edited by D. D. Matson, pp. 289–319. New Haven, CT: Yale University Press.

Denny-Brown, D., and W. R. Russell. 1940. Experimental cerebral concussion. *The Journal of Physiology* 99 (1):153.

DiMasi, F., R. Eppinger, and F. Bandak. 1995. Computational analysis of head impact response under car crash loadings. In *Paper read at 39th Stapp Car Crash Conference*, San Diego, CA.

DiMasi, F., P. Tong, J. H. Marcus, H. C. Gabler, III, and R. H. Eppinger. 1991. Simulated head impacts with upper interior structures using rigid and anatomic brain models. In *The finite element method in the 1990's*, edited by E. Oñate, J. Periaux, and A. Samuelsson, pp. 333–345. Berlin: Springer.

Eppinger, R., M. Kleinberger, R. Morgan, N. Khaewpong, F. A. Bandak, and M. Haffner. 1994. Advanced injury criteria and crash test evaluation techniques. Paper 90-S1-O-11. In *Paper read at the NHTSA 14th International Technical Conference on Experimental Safety Vehicles*, Munich, Germany.

Ewing, C. L., and A. M. Irving. 1969. Evaluation of head protection in aircraft. *Aerospace Medicine* 40 (6): 596–599.

Fearnside, M. R., and D. A. Simpson. 1997. Epidemiology. In *Head injury: Pathophysiology and management of severe closed injury*, edited by P. Reilly and R. Bullock, pp. 1–23. London: Chapman & Hall Medical.

Feli, S., M. H. Yas, and M. R. Asgari. 2011. An analytical model for perforation of ceramic/multi-layered planar woven fabric targets by blunt projectiles. *Composite Structures* 93 (2):548–556. doi: 10.1016/j.compstruct. 2010.08.025.

Field, J. H. 1976. *Epidemiology of head injuries in England and wales: With particular application to rehabilitation.* London: HMSO.

Forero Rueda, M. A., L. Cui, and M. D. Gilchrist. 2011. Finite element modelling of equestrian helmet impacts exposes the need to address rotational kinematics in future helmet designs. *Computer Methods in Biomechanics and Biomedical Engineering* 14 (12):1021–1031. doi: 10.1080/10255842.2010.504922.

Franklyn, M., B. Fildes, L. Zhang, Y. King, and L. Sparke. 2005. Analysis of finite element models for head injury investigation: Reconstruction of four real-world impacts. *Stapp Car Crash Journal* 49:1–32.

Gennarelli, T. A., J. H. Adams, and D. I. Graham. 1981. Acceleration induced head injury in the Monkey. I. The model, its mechanical and physiological correlates. In *Experimental and Clinical Neuropathology*, edited by K. Jellinger, F. Gullotta, and M. Mossakowski, pp. 23–25. Berlin: Springer.

Gennarelli, T. A., and L. E. Thibault. 1982. Biomechanics of acute subdural hematoma. *The Journal of trauma* 22 (8):680–686.

Gennarelli, T. A., L. E. Thibault, J. H. Adams, D. I. Graham, C. J. Thompson, and R. P. Marcincin. 1982. Diffuse axonal injury and traumatic coma in the primate. *Annals of neurology* 12 (6):564–574. doi: 10.1002/ana. 410120611.

Genta, G. 2012. Wheeled vehicles and rovers. In *Introduction to the mechanics of space robots*, p. 597. Springer.

Giordano, C., and S. Kleiven. 2014. Evaluation of axonal strain as a predictor for mild traumatic brain injuries using finite element modeling. *Stapp Car Crash Journal* 58:29–61.

Goel, R. 2010. *Study of an advanced helmet liner concept to reduce TBI: Experiments & simulation using sandwich structures.* Berkeley, CA: Department of Aeronautics and Astronautics, Massachusetts Institute of Technology (MIT).

Goldsmith, W., and T. B. Khalil. 1973. Effect of a protective device in the reduction of head injury. In *Paper read at International Conference on Biokinetics of Impacts*, Amsterdam, the Netherlands.

Grandjean, É. 1980. *Fitting the task to the man: An ergonomic approach.* 3rd ed. Taylor & Francis.

Gross, A. G. 1958. A new theory on the dynamics of brain concussion and brain injury. *Journal of Neurosurgery* 15 (5):548–561. doi: 10.3171/jns.1958.15.5.0548.

Grujicic, A. 2011. *A computational investigation for potential improvements in shock mitigation efficacy of polyurea augmentations to the current advanced combat helmet.* M.S., Clemson University, Ann Arbor, MI.

Grujicic, A., M. LaBerge, M. Grujicic, B. Pandurangan, J. Runt, J. Tarter, and G. Dillon. 2012. Potential improvements in shock-mitigation efficacy of a polyurea-augmented advanced combat helmet. *Journal of Materials Engineering and Performance* 21 (8):1562–1579. doi: 10.1007/s11665-011-0065-3.

Grujicic, M., W. C. Bell, B. Pandurangan, and P. S. Glomski. 2011. Fluid/structure interaction computational investigation of blast-wave mitigation efficacy of the advanced combat helmet. *Journal of Materials Engineering and Performance* 20 (6):877–893. doi: 10.1007/s11665-010-9724-z.

Gurdjian, E. S., and H. R. Lissner. 1945. Mechanism of head injury as studied by the Cathode Ray Oscilloscope. *The Journal of Nervous and Mental Disease* 102 (4):425.

Ha-Minh, C., F. Boussu, T. Kanit, D. Crépin, and A. Imad. 2011a. Analysis on failure mechanisms of an interlock woven fabric under ballistic impact. *Engineering Failure Analysis* 18 (8):2179–2187. doi: 10.1016/j.engfailanal.2011.07.011.

Ha-Minh, C., T. Kanit, F. Boussu, and A. Imad. 2011b. Numerical multi-scale modeling for textile woven fabric against ballistic impact. *Computational Materials Science* 50 (7):2172–2184. doi: 10.1016/j.commatsci.2011.02.029.

Hardy, C. H., and P. V. Marcal. 1971. *Elastic analysis of a skull.* Office of Naval Research, Contract No. N00014-67-A-0191–0007, Division of Engineering Brown University.

Hardy, C. H., and P. V. Marcal. 1973. Elastic analysis of a skull. *Journal of Applied Mechanics* 40 (4):838–842. doi: 10.1115/1.3423172.

Hardy, W. N., C. Foster, M. Mason, K. Yang, A. King, and S. Tashman. 2001. Investigation of head injury mechanisms using neutral density technology and high-speed biplanar X-ray. In *Paper read at 45th Stapp Car Crash Conference*, San Antonio, TX.

Hardy, W. N., T. B. Khalil, and A. I. King. 1994. Literature review of head injury biomechanics. *International Journal of Impact Engineering* 15 (4):561–586. doi: 10.1016/0734-743x(94)80034-7.

Harrigan, T. P., J. C. Roberts, E. E. Ward, and A. C. Merkle. 2010. Correlating tissue response with anatomical location of mTBI using a human head finite element model under simulated blast conditions. In *26th Southern Biomedical Engineering Conference SBEC 2010, April 30–May 2, 2010, College Park, Maryland, USA*, edited by K. E. Herold, J. Vossoughi, and W. E. Bentley, pp. 18–21. Berlin: Springer.

Haug, E., H.-Y. Choi, S. Robin, and M. Beaugonin. 2004. Human models for crash and impact simulation. In *Handbook of numerical analysis*, edited by N. Ayache, pp. 231–452. Elsevier.

Hedenstierna, S. 2008. *3D Finite element modeling of cervical musculature and its effect on neck injury prevention.* Huddinge, Sweden: Division of Neuronic Engineering, Royal Institute of Technology.

Helmets. 2015. Operation-Helmet.Org 2014 [cited 27 November 2015]. Available from: http://operation-helmet.org/helmet/.

Ho, J., and S. Kleiven. 2007. Dynamic response of the brain with vasculature: A three-dimensional computational study. *Journal of Biomechanics* 40 (13):3006–3012. doi: 10.1016/j.jbiomech.2007.02.011.

Hodgson, V. R., E. S. Gurdjian, and L. M. Thomas. 1967. Development of a model for the study of head injury during impact tests. SAE 670923. In *Paper read at the 11th Stapp Car Crash Conference*, Anaheim, CA.

Hodgson, V. R., and L. M. Patrick. 1968. Dynamic response of the human cadaver head compared to a simple mathematical model. SAE 680784. In *Paper read at the 12th Stapp Car Crash Conference*, Detroit, MI.

Holbourn, A. H. S. 1943. Mechanics of head injuries. *Lancet* 2:438–441.

Horgan, T. J., and M. Gilchrist. 2003. The creation of three-dimensional finite element models for simulating head impact biomechanics. *International Journal of Crashworthiness* 8 (4):353–366.

Horgan, T. J., and M. D. Gilchrist. 2004. Influence of FE model variability in predicting brain motion and intracranial pressure changes in head impact simulations. *International Journal of Crashworthiness* 9 (4):401–408.

Hosey, R. R., and Y. K. Liu. 1981. A homeomorphic finite element model of the human head and neck. In *Finite elements in biomechanics*, edited by B. R. Simon, R. H. Gallagher, P. C. Johnson, and J. F. Gross, pp. 379–401. Wiley.

Jagger, J., J. Levine, and J. Jane. 1984. Epidemiologic features of head injury in a predominantly rural population. *Journal of Trauma* 24:40–44.

Johnston, I. H., and J. O. Rowan. 1974. Raised intracranial pressure and cerebral blood flow. *Journal of neurology, neurosurgery, and psychiatry* 37 (5):585–592.

Kang, H. S., R. Willinger, B. Diaw, and B. Chinn. 1997. Validation of a 3D anatomic human head model and replication of head impact in motorcycle accident by finite element modeling. In *Paper read at 41st Stapp Car Crash Conference*, Lake Buena Vista, FL.

Kenner, V. H., and W. Goldsmith. 1972. Dynamic loading of a fluid-filled spherical shell. *International Journal of Mechanical Sciences* 14 (9):557–568. doi: 10.1016/0020-7403(72)90056-2.

Khalil, T. B. 1973. *Impact on a model head-helmet system.* Berkeley, CA: University of California.

Khalil, T. B., W. Goldsmith, and J. L. Sackman. 1974. Impact on a model head-helmet system. *International Journal of Mechanical Sciences* 16 (9):609–625. doi: 10.1016/0020-7403(74)90061-7.

Khalil, T. B., and R. P. Hubbard. 1977. Parametric study of head response by finite element modeling. *Journal of Biomechanics* 10 (2):119–132. doi: 10.1016/0021-9290(77)90075-6.

Khalil, T. B., and D. C. Viano. 1982. Critical issues in finite element modelling of head impact. In *Paper read at 26th Stapp Car Crash Conference*, Ann Arbor, MI.

King, A. I., K. H. Yang, W. N. Hardy, A. S. Al-Bsharat, B. Deng, P. C. Begeman, and S. Tashman. 1999. Challenging problems and opportunities in impact biomechanics. In *Paper read at ASME Bioengineering Conference*.

King, A. I., K. H. Yang, L. Zhang, W. Hardy, and D. Viano. 2003. Is head injury caused by linear or angular acceleration? In *Paper read at International IRCOBI Conference on the Biomechanics of Impacts*, Lisbon, Portugal.

Kleiven, S. 2003. Influence of impact direction on the human head in prediction of subdural hematoma. *Journal of neurotrauma* 20 (4):365–379. doi: 10.1089/089771503765172327.

Kleiven, S. 2006. Evaluation of head injury criteria using a finite element model validated against experiments on localized brain motion, intracerebral acceleration, and intracranial pressure. *International Journal of Crashworthiness* 11 (1):65–79.

Kleiven, S., and W. N. Hardy. 2002. Correlation of an FE model of the human head with local brain motion–consequences for injury prediction. In *Paper read at 46th Stapp Car Crash Conference*, November, Ponte Vedra, FL.

Kleiven, S., and H. von Holst. 2002. Consequences of head size following trauma to the human head. *Journal of Biomechanics* 35 (2):153–160. doi: 10.1016/s0021-9290(01)00202-0.

Koenig, H. A. 1998. *Modern computational methods*. Taylor & Francis, Philadelphia, PA.

Kraus, J. F., and D. L. McArthur. 2000. Epidemiology of brain injury. In *Head injury*, edited by P. R. Cooper and J. G. Golfinos, pp. 1–26. New York: McGraw-Hill.

Kumaresan, S., and S. Radhakrishnan. 1996. Importance of partitioning membranes of the brain and the influence of the neck in head injury modelling. *Medical and Biological Engineering and Computing* 34 (1): 27–32. doi: 10.1007/bf02637019.

Larry, S. 2012. *Top general expects IED problem to rise in Afghanistan*. CNN.

Lee, H. P., and S. W. Gong. 2010. Finite element analysis for the evaluation of protective functions of helmets against ballistic impact. *Computer Methods in Biomechanics and Biomedical Engineering* 13 (5): 537–550. doi: 10.1080/10255840903337848.

Lollis, S. S., P. B. Quebada, and J. A. Friedman. 2008. Traumatic brain injuries. In *Handbook of clinical neurology*, edited by G. B. Young and E. F. M. Wijdicks, pp. 17–41. Edinburgh: Elsevier Science Publishers B.V.

Löwenhielm, P. 1975. Mathematical simulation of gliding contusions. *Journal of Biomechanics* 8 (6):351–356. doi: 10.1016/0021-9290(75)90069-x.

Makwana, R. 2012. *Development and validation of a three-dimensional finite element model of advanced combat helmet and biomechanical analysis of human head and helmet response to primary blast insult*. M.S., Wayne State University, Ann Arbor, MI.

Mamivand, M., and G. H. Liaghat. 2010. A model for ballistic impact on multi-layer fabric targets. *International Journal of Impact Engineering* 37 (7):806–812. doi: 10.1016/j.ijimpeng.2010.01.003.

Mansfield, R. T., D. W. Marion, and P. M. Kochanek. 1995. Traumatic head or spinal injury. In *Pediatric transport medicine*, edited by K. McCloskey and R. A. Orr, pp. 285–297. St. Louis, MO: Mosby.

Mao, H., L. Zhang, B. Jiang, V. V. Genthikatti, X. Jin, F. Zhu, R. Makwana, et al. 2013. Development of a finite element human head model partially validated with thirty five experimental cases. *Journal of biomechanical engineering* 135 (11):111002. doi: 10.1115/1.4025101.

Marjoux, D., D. Baumgartner, C. Deck, and R. Willinger. 2008. Head injury prediction capability of the HIC, HIP, SIMon and ULP criteria. *Accident Analysis & Prevention* 40 (3):1135–1148. doi: 10.1016/j.aap.2007.12.006.

Matzenmiller, A., J. Lubliner, and R. L. Taylor. 1995. A constitutive model for anisotropic damage in fiber-composites. *Mechanics of Materials* 20 (2):125–152. doi: 10.1016/0167-6636(94)00053-0.

Mayer, T., M. L. Walker, D. G. Johnson, and M. E. Matlak. 1981. Causes of morbidity and mortality in severe pediatric trauma. *JAMA* 245 (7):719–721.

McEntire, B. J., and P. Whitley. 2005. *Blunt impact performance characteristics of the advanced combat helmet and the paratrooper and infantry personnel armor system for ground troops helmet*. USAARL Report 2005-12. United States Army Aeromedical Research Laboratory.

Mendis, K. K., R. L. Stalnaker, and S. H. Advani. 1995. A constitutive relationship for large deformation finite element modeling of brain tissue. *Journal of Biomechanical Engineering* 117 (3):279–285.

Meyer, F., N. Bourdet, K. Gunzel, and R. Willinger. 2012. Development and validation of a coupled head-neck FEM—Application to whiplash injury criteria investigation. *International Journal of Crashworthiness* 18 (1):40–63. doi: 10.1080/13588265.2012.732293.

Meyer, F., N. Bourdet, R. Willinger, F. Legall, and C. Deck. 2004a. Finite element modelling of the human head-neck: Modal analysis and validation in the frequency domain. *International Journal of Crashworthiness* 9 (5):535–545. doi: 10.1533/ijcr.2004.0309.

Meyer, F., R. Willinger, and F. Legall. 2004b. The importance of modal validation for biomechanical models, demonstrated by application to the cervical spine. *Finite Elements in Analysis and Design* 40 (13–14): 1835–1855. doi: 10.1016/j.finel.2003.11.005.

Mills, N. J. 2007. Finite element modelling of foam deformation. In *Polymer foams handbook: Engineering and biomechanics applications and design guide*, pp. 115–145. Oxford: Butterworth-Heinemann.

Milne, G., C. Deck, R. P. Carreira, Q. Allinne, and R. Willinger. 2012. Development and validation of a bicycle helmet: Assessment of head injury risk under standard impact conditions. *Computer Methods in Biomechanics and Biomedical Engineering* 15 (Suppl. 1):309–310. doi: 10.1080/10255842.2012.713623.

Moore, D. F., A. Jerusalem, M. Nyein, L. Noels, M. S. Jaffee, and R. A. Radovitzky. 2009. Computational biology—Modeling of primary blast effects on the central nervous system. *Neuroimage* 47 (Suppl. 2): T10–T20. doi: 10.1016/j.neuroimage.2009.02.019.

Morrison, B., B. S. Elkin, J.-P. Dollé, and M. L. Yarmush. 2011. In vitro models of traumatic brain injury. *Annual Review of Biomedical Engineering* 13 (1):91–126. doi: 10.1146/annurev-bioeng-071910-124706.

Morse, J. D., J. A. Franck, B. J. Wilcox, J. J. Crisco, and C. Franck. 2014. An experimental and numerical investigation of head dynamics due to stick impacts in girls' lacrosse. *Annals of Biomedical Engineering* 42 (12):2501–2511. doi: 10.1007/s10439-014-1091-8.

Moss, W. C., M. J. King, and E. G. Blackman. 2009. Skull flexure from blast waves: A mechanism for brain injury with implications for helmet design. *Physical review letters* 103 (10):108702.

Nahum, A. M., R. Smith, and C. C. Ward. 1977. Intracranial pressure dynamics during head impact. In *Paper read at 21st Stapp Car Crash Conference*, San Diego, CA.

Newman, J. A., N. Shewchenko, and E. Welbourne. 2000. A proposed new biomechanical head injury assessment function—The maximum power index. In *Paper read at 44th Stapp Car Crash Conference*, Atlanta, GA.

Nickell, R., and P. Marcal. 1974. In vacuo model dynamic response of the human skull. *Journal of Engineering Industry* 4:490–494.

Numerical Modeling of the Human Head. 2013. KTH, 2008 [cited 21 July 2013]. Available from: http://www.kth.se/en/sth/forskning/forskningsomraden/2.3593/projekt/2.3790/numerisk-modellering-av-det-manskliga-huvudet-1.14227.

Nyein, M. K., A. M. Jason, L. Yu, C. M. Pita, J. D. Joannopoulos, D. F. Moore, and R. A. Radovitzky. 2010. In silico investigation of intracranial blast mitigation with relevance to military traumatic brain injury. *Proceedings of the National Academy of Sciences* 107 (48):20703–20708. doi: 10.1073/pnas.1014786107.

O'Brien, T. G. 2002. Testing the workplace environment. In *Handbook of human factors testing and evaluation*, edited by S. G. Charlton and T. G. O'Brien, pp. 568. Mahwah, NJ: Lawrence Erlbaum Associates, Inc.

Okie, S. 2005. Traumatic brain injury in the war zone. *The New England Journal of Medicine* 352 (20): 2043–2047. doi: 10.1056/NEJMp058102.

Patrick, L. M. 1966. Head Impact Protection. In *Paper read at Head Injury Conference*.

Personal Gear at the Modern Day Marine Expo. 2015. Defence Update 2007 [cited 27 November 2015]. Available from: http://defense-update.com/20071101_mdm07_soldier.html.

Pinnoji, P., and P. Mahajan. 2007. Finite element modelling of helmeted head impact under frontal loading. *Sadhana* 32 (4):445–458. doi: 10.1007/s12046-007-0034-6.

Pinnoji, P. K., and P. Mahajan. 2006. Impact analysis of helmets for improved ventilation with deformable head model. In *International IRCOBI Conference on the Biomechanics of Impact*, Madrid, Spain.

Pintar, F. A., M. Mat, M. G. M. Philippens, J. Y. Zhang, and N. Yoganandan. 2013. Methodology to determine skull bone and brain responses from ballistic helmet-to-head contact loading using experiments and finite element analysis. *Medical Engineering & Physics* 35 (11):1682–1687. doi: http://dx.doi.org/10.1016/j.medengphy.2013.04.015.

Rao, M. P., Y. Duan, M. Keefe, B. M. Powers, and T. A. Bogetti. 2009. Modeling the effects of yarn material properties and friction on the ballistic impact of a plain-weave fabric. *Composite Structures* 89 (4): 556–566. doi: 10.1016/j.compstruct.2008.11.012.

Raul, J.-S., C. Deck, R. Willinger, and B. Ludes. 2008. Finite-element models of the human head and their applications in forensic practice. *International Journal of Legal Medicine* 122 (5):359–366. doi: 10.1007/s00414-008-0248-0.

Rayne, J. M., and K. R. Maslen. 1969. Factors in the design of protective helmets. *Aerospace Medicine* 40 (6): 631–637.

Roberts, J. C., T. P. Harrigan, E. E. Ward, T. M. Taylor, M. S. Annett, and A. C. Merkle. 2012. Human head–neck computational model for assessing blast injury. *Journal of Biomechanics* 45 (16):2899–2906. doi: 10.1016/j.jbiomech.2012.07.027.

Roberts, J. C., E. E. Ward, T. P. Harrigan, T. M. Taylor, M. A. Annett, and A. C. Merkle. 2009. Development of a human head-neck computational model for assessing blast injury. Paper IMECE2009-11813. In *Paper read at ASME International Mechanical Engineering Congress and Exposition (IMECE)*, 13th–19th November 2009, Lake Buena Vista, FL.

Rodríguez-Millán, M., L. B. Tan, K. M. Tse, H. P. Lee, and M. H. Miguélez. 2016. Effect of full helmet systems on human head responses under blast loading. *Materials & Design* 117:58–71.

Ross, D. T., D. F. Meaney, M. K. Sabol, D. H. Smith, and T. A. Gennarelli. 1994. Distribution of forebrain diffuse axonal injury following inertial closed head injury in miniature swine. *Experimental Neurology* 126 (2):291–298. doi: 10.1006/exnr.1994.1067.

Ruan, J. S., T. Khalil, and A. I. King. 1994. Dynamic response of the human head to impact by three-dimensional finite element analysis. *Journal of Biomechanical Engineering* 116 (1):44–50.

Ruan, J. S., T. B. Khatil, and A. I. King. 1993. Finite element modeling of direct head impact. In *Paper read at 37th Stapp Car Crash Conference*, San Antonio, TX.

Saczalski, K. J., and E. Q. Richardson. 1978. Nonlinear numerical prediction of human head/helmet crash impact response. *Aviation, Space, and Environmental Medicine* 49 (1 Pt. 2):114–119.

Salimi Jazi, M. 2015. *Examination of the impact of helmets on the level of transferred loads to the head under ballistic and blast loads*. Ph.D., North Dakota State University, Ann Arbor, MI.

Salimi Jazi, M., A. Rezaei, G. Karami, F. Azarmi, and M. Ziejewski. 2014. A computational study of influence of helmet padding materials on the human brain under ballistic impacts. *Computer Methods in Biomechanics and Biomedical Engineering* 17 (12):1368–1382. doi: 10.1080/10255842.2012.748755.

Sarron, J.-C., M. Dannawi, A. Faure, J.-P. Caillou, J. Da Cunha, and R. Robert. 2004. Dynamic effects of a 9 mm missile on cadaveric skull protected by aramid, polyethylene or aluminum plate: An experimental study. *Journal of Trauma-Injury, Infection, and Critical Care* 57 (2):236–243.

Sarvghad-Moghaddam, H., M. S. Jazi, A. Rezaei, G. Karami, and M. Ziejewski. 2015. Examination of the protective roles of helmet/faceshield and directionality for human head under blast waves. *Computer Methods in Biomechanics and Biomedical Engineering* 18 (16):1846–1855. doi: 10.1080/10255842.2014.977878.

Shugar, T. A. 1975. Transient structural response of the linear skull brain system. In *Paper read at 19th Stapp Car Crash Conference*, San Diego, CA.

Shugar, T. A., and M. C. Kahona. 1975. Development of finite element head injury model. *Journal of American Society of Civil Engineers* 101:223–239.

Takhounts, E., and R. Eppinger. 2003. On the development of the SIMon finite element head model. In *Paper read at 47th Stapp Car Crash Conference*, San Diego, CA.

Takhounts, E. G., S. A. Ridella, V. Hasija, R. E. Tannous, J. Q. Campbell, D. Malone, K. Danelson, J. Stitzel, S. Rowson, and S. Duma. 2008. Investigation of traumatic brain injuries using the next generation of simulated injury monitor (SIMon) finite element head model. *Stapp Car Crash Journal* 52:1–31.

Tan, L. B., F. S. Chew, K. M. Tse, V. B. C. Tan, and H. P. Lee. 2014. Impact of complex blast waves on the human head—A computational study. *International Journal for Numerical Methods in Biomedical Engineering* 30 (12):1476–1505. doi: 10.1002/cnm.2668.

Tan, L. B., K. M. Tse, H. P. Lee, and V. B. C. Tan. 2013. Ballistic impact analysis on the Advanced Combat Helmet (ACH)-testing and simulation. In *9th International Conference on Composite Science and Technology (ICCST): 2020—Scientific and Industrial Challenges*, edited by M. Meo. Sorrento, Italy: Destech Publications, Inc.

Tan, L. B., K. M. Tse, H. P. Lee, V. B. C. Tan, and S. P. Lim. 2012. Performance of an advanced combat helmet with different interior cushioning systems in ballistic impact: Experiments and finite element simulations. *International Journal of Impact Engineering* 50 (0):99–112. doi: 10.1016/j.ijimpeng.2012.06.003.

Tan, V. B. C., and T. W. Ching. 2006. Computational simulation of fabric armour subjected to ballistic impacts. *International Journal of Impact Engineering* 32 (11):1737–1751. doi: 10.1016/j.ijimpeng.2005.05.006.

Tan, Y. H. 2014. *Design of transparent composite face shields for shock wave mitigation*. Singapore: Department of Mineral Engineering, National University of Singapore.

Taylor, P. A., and C. C. Ford. 2009. Simulation of blast-induced early-time intracranial wave physics leading to traumatic brain injury. *Journal of Biomechanical Engineering* 131 (6):061007.

Test Protokoll-Nr: 01M072-01, Date: 21.03.2001; 01M073A01, 21.03.2001; 01M079-01, 29.03.2001; 01M085A01, 30.03.2001; 01M085B01, 30.03.2001, Bericht-No: 140–143, Date: 19.03.2001. Lohstrasse 5, 97638 Mellrichstadt, Germany: Beschussamt Mellrichstadt. 2001.

Tham, C. Y., V. B. C. Tan, and H. P. Lee. 2008. Ballistic impact of a KEVLAR® helmet: Experiment and simulations. *International Journal of Impact Engineering* 35 (5):304–318. doi: 10.1016/j.ijimpeng.2007.03.008.

Thomas, L. M., V. L. Roberts, and E. S. Gurdjian. 1967. Impact-induced pressure gradients along three orthogonal axes in the human skull. *Journal of neurosurgery* 26 (3):316–321. doi: 10.3171/jns.1967.26.3.0316.

Trosseille, X., C. Tarriere, and F. Lavaste. 1992. Development of a FEM of the human head according to a specific test protocol. In *Paper read at 30th Stapp Car Crash Conference*, Warrendale, PA.

Tse, K. M. 2013. *Development of a realistic finite element model of human head and its applications to head injuries*. Singapore: Department of Mechanical Engineering, National University of Singapore.

Tse, K. M., L. B. Tan, S. J. Lee, S. P. Lim, and H. P. Lee. 2014a. Development and validation of two subject-specific finite element models of human head against three cadaveric experiments. *International Journal for Numerical Methods in Biomedical Engineering* 30 (3):397–415. doi: 10.1002/cnm.2609.

Tse, K. M., L. B. Tan, S. J. Lee, S. P. Lim, and H. P. Lee. 2015a. Investigation of the relationship between facial injuries and traumatic brain injuries using a realistic subject-specific finite element head model. *Accident Analysis & Prevention* 79 (0):13–32. doi: http://dx.doi.org/10.1016/j.aap.2015.03.012.

Tse, K. M., L. B. Tan, S. P. Lim, and H. P. Lee. 2013. Conventional and complex modal analyses of a finite element model of human head and neck. *Comput Methods Biomech Biomed Engin* 18 (9):961–973. doi: 10.1080/10255842.2013.864641.

Tse, K. M., L. B. Tan, V. B. C. Tan, B. Yang, V. B. C. Tan, and H. P. Lee. 2015b. Effect of helmet liner systems and impact directions on severity of head injuries sustained in ballistic impacts: A finite element (FE) study. *Medical & Biological Engineering & Computing*. July 13 [Epub ahead of print]. doi: 10.1007/s11517-016-1536-3.

Tse, K. M., L. B. Tan, B. Yang, V. B. C. Tan, S. P. Lim, and H. P. Lee. 2014b. Ballistic impacts of a full-metal jacketed (FMJ) bullet on a validated finite element (FE) model of helmet-cushion-head. In *Paper read at The 5th International Conference on Computational Methods (ICCM)*, 28th–30th July 2014, Cambridge, UK.

Turquier, F., H. S. Kang, X. Trosseille, R. Willinger, X. Trosseille, F. Lavaste, C. Tarriere, and A. Domont. 1996. Validation study of a 3D finite element head model against experimental data. In *Paper read at 40th Stapp Car Crash Conference*, Albuquerque, NM.

Ueno, K., J. W. Melvin, E. Lundquist, and M. C. Lee. 1989. Two-dimensional finite element analysis of human brain impact responses: Application of a scaling law. *American Society of Mechanical Engineers (ASME), Applied Mechanics Division (AMD)* 106:123–124.

Ueno, K., J. W. Melvin, M. E. Rouhana, and J. W. Lighthall. 1991. Two-dimensional finite element model of the cortical impact method for mechanical brain injury. In *112th American Society of Mechanical Engineers (ASME) Winter Annual Meeting, Applied Mechanics Division (AMD)*.

Uniform & kit issued to the US Army & Marine Corps during WWII. 2015. Medals Gone Missing 2015 [cited 27th November 2015]. Available from: http://www.medalsgonemissing.com/Uniform-Kit-issued-to-the-US-Army–Marine-Corps-During-WW2/2.html.

Van Hoof, J., D. S. Cronin, M. J. Worswick, K. V. Williams, and D. Nandlall. 2001. Numerical head and composite helmet models to predict blunt trauma. In *Paper read at 19th International Symposium on Ballistics*, 7–11 May 2001, Interlaken, Switzerland.

Van Hoof, J., M. J. Deutekom, M. J. Worswick, and M. Bolduc. 1999. Experimental and numerical analysis of the ballistic response of composite helmet materials. In *Paper read at 18th International Symposium on Ballistics*, 15–19 November 1999, San Antonio, TX.

Van Hoof, J., and M. J. Worswick. 2001. *Combining head models with composite helmet models to simulate ballistic impacts*. Waterloo, ON: Waterloo University.

Viano, D. C., I. R. Casson, E. J. Pellman, L. Zhang, A. I. King, and K. H. Yang. 2005. Concussion in professional football: Brain responses by finite element analysis: Part 9. *Neurosurgery* 57 (5):891–916; discussion 891–916.

Voo, L., S. Kumaresan, F. Pintar, N. Yoganandan, and A. Sances. 1996. Finite-element models of the human head. *Medical and Biological Engineering and Computing* 34 (5):375–381. doi: 10.1007/bf02520009.

Voyiadjis, G. Z. 1987. Introduction to continuum damage mechanics, L. M. Kachanov, Martinus Nijhoff Publishers, 1986. No. of pages: 135. Price U.S. $39.50. ISBN: 90-247-3319-7. *International Journal for Numerical and Analytical Methods in Geomechanics* 11 (5):547–547. doi: 10.1002/nag.1610110509.

Ward, C. C., M. Chan, and A. M. Nahum. 1980. Intracranial pressure-a brain injury criterion. In *Paper read at 24th Stapp Car Crash Conference*, Warrendale, PA.

Ward, C. C., and R. B. Thompson. 1975. The development of a detailed finite element brain model. In *Paper read at 19th Stapp Car Crash Conference*, New York.

Willinger, R., and D. Baumgartner. 2003. Human head tolerance limits to specific injury mechanisms. *International Journal of Crashworthiness* 8 (6):605–617. doi: 10.1533/ijcr.2003.0264.

Willinger, R., H. S. Kang, and B. Diaw. 1999. Three-dimensional human head finite-element model validation against two experimental impacts. *Annals of Biomedical Engineering* 27 (3):403–410. doi: 10.1114/1.165.

Willinger, R., L. Taleb, and C. M. Kopp. 1995. Modal and temporal analysis of head mathematical models. *Journal of Neurotrauma* 12 (4):743–754. doi: 10.1089/neu.1995.12.743.

Willinger, R., L. Taled, and P. Pradoura. 1995. Head biomechanics: From the finite element model to the physical model. In *Paper read at International IRCOBI Conference on the Biomechanics of Impacts*, Brunnen, Switzerland.

Wright, R. M., A. Post, B. Hoshizaki, and K. T. Ramesh. 2013. A multiscale computational approach to estimating axonal damage under inertial loading of the head. *Journal of neurotrauma* 30 (2):102–118. doi: 10.1089/neu.2012.2418.

Wright, R. M., and K. T. Ramesh. 2012. An axonal strain injury criterion for traumatic brain injury. *Biomechanics and Modeling in Mechanobiology* 11 (1–2):245–260. doi: 10.1007/s10237-011-0307-1.

Yan, W., and O. D. Pangestu. 2011. A modified human head model for the study of impact head injury. *Computer Methods in Biomechanics and Biomedical Engineering* 14 (12):1049–1057. doi: 10.1080/10255842.2010.506435.

Yang, J., and J. Dai. 2010. Simulation-based assessment of rear effect to ballistic helmet impact. *Computer-Aided Design and Applications* 7 (1):59–73. doi: citeulike-article-id:9593024.

Yang, J., J. Dai, and Z. Zhuang. 2009. Human head modeling and personal head protective equipment: A literature review. In *Digital human modeling*, edited by V. Duffy, pp. 661–670. Berlin: Springer.

Zhang, L., R. Makwana, and S. Sharma. 2013. Brain response to primary blast wave using validated finite element models of human head and advanced combat helmet. *Frontiers in Neurology* 4:88. doi: 10.3389/fneur.2013.00088.

Zhang, L., K. H. Yang, R. Dwarampudi, K. Omori, T. Li, K. Chang, W. N. Hardy, T. B. Khalil, and A. I. King. 2001. Recent advances in brain injury research: A new human head model development and validation. In *Paper read at 45th Stapp Car Crash Conference*, November, San Antonio, TX.

Zhang, L., K. H. Yang, A. I. King, and D. Viano. 2003. A new biomechanical predictor for mild traumatic brain injury—A preliminary finding. In *Paper read at ASME Summer Bioengineering Conference*.

Zhou, C., C. T. B. Khalil, and A. I. King. 1995. A new model comparing impact responses of the homogeneous and inhomogeneous human brain. In *Paper read at 39th Stapp Car Crash Conference*, San Diego, CA.

Zong, Z., H. P. Lee, and C. Lu. 2006. A three-dimensional human head finite element model and power flow in a human head subject to impact loading. *Journal of Biomechanics* 39 (2):284–292. doi: 10.1016/j.jbiomech.2004.11.015.

11 The Neck
Neck Injuries in Military Scenarios

Kwong Ming Tse, Jianfei Liu, Victor P.W. Shim,
Ee Chong Teo and Peter Vee Sin Lee

CONTENTS

11.1 EPIDEMIOLOGY OF NECK INJURY

Cervical spine injury, often referred as neck injury, is a common problem with a wide range of severities from minor ligamentous injury to most severe spinal cord injuries that are sustained in impact-related scenarios in both civilian and military contexts (Rhee et al. 2006; Torretti and Sengupta 2007; Mahoney et al. 2016). The human neck is vulnerable to injury as it connects to the head superiorly and to the torso inferiorly, and both the head and torso are much heavier than the neck. Upon impact on either the head or torso or both, inertial forces of the head or torso can load or bend the neck tissues beyond their tolerable limits, thus resulting in cervical spine injury.

Consequently, head injury is often seen in conjunction with cervical spinal injury.

Cervical spine injury occurs more frequently in motor vehicle accidents than in the military events, with whiplash injury predominating in motor vehicle accidents (Yadla et al. 2008). Cervical spine injury can have long-lasting devastating effects such as neck pain and stiffness, occipital headache, cognitive function loss, numbness of the upper limbs and even permanent disability (sensory and motor loss in arms, body and legs) to the injured individuals. In its most severe form, cervical spine injury is life threatening or fatal, especially when the injury occurs in the upper cervical vertebrae. It was estimated in 2011 that there are about 12,000 new cases of cervical spine injury in the United States (US) each year, 42% of which are due to motor vehicle accidents (*Cervical spine injury*, 2016).

Cervical spine injury does not only account for a high mortality rate, but the chance of a permanent disability is also high. This poses a serious public health problem with significant social and economic costs. The World Health Organization (WHO) estimated that the lifetime cost (exclusive of the consequent loss of earning capacity) for a person injured with a cervical spine injury at the age of 25 in the US is approximately 4.6 million US dollars (WHO 2013). Consequently, due to the high injury incidence rate in the civilian population, there has been tremendous research effort, especially by the automotive industry, in analysing cervical spine injury through retrospective, clinical, experimental and numerical investigations. All these studies have been successful in providing a better understanding of cervical spine injury mechanisms, providing insight in design of anthropomorphic test devices (ATDs) for cervical spine injury assessment and promulgating safety regulations, in particular in the automotive environment. On the other hand, cervical spine injury research in military events is comparably sparse and has received lesser dissemination. With the recent rise in international conflicts, there has been an increasing use of improvised explosive devices (IEDs), and the number of individuals from both civilian and military populations afflicted with blast-related traumatic injury has been increasing. Based on data for the period from 2000 to 2009 (Schoenfeld et al. 2012), the incidence of cervical fractures and spinal cord injuries with fractures are estimated to be 29 and 70 per million in the US military, respectively. Despite the lower injury incidence of cervical spine fractures in the military than in automotive crashes, the fatality rate of the cervical spine injury sustained by military personnel in the battlefield is extremely high as their cervical spines are exposed to more severe G forces loaded in a vertical compressive manner over a very short span of time. For instance, the fatality rate for fighter pilot ejection at heights below 500 feet is as high as 53.7% (*Literature review on biomechanics of injury causation*, 2012) with the most common sites of cervical spine fracture during high-speed pilot ejection being C2, C5 and C6. Recent military data for underbelly (UB) blast injury demonstrates that blast-related spinal injury contributed to only 13.5% of casualties, while lower extremity injuries account for the majority of casualties (Vogel and Dootz 2007). However, it should be noted that spinal injuries, especially in the upper cervical vertebrae, are more likely to be fatal and are potentially more detrimental than lower extremity injuries, as cervical injuries can lead to direct impingement of the spinal cord and rupturing of major arteries and veins by the fractured bones.

11.2 ANATOMY OF HUMAN NECK

11.2.1 CERVICAL SKELETON

The human cervical spine, as shown in Figure 11.1, is the most superior portion of the vertebral column, located between the skull (C0) and the thoracic vertebrae (T1–T12). It consists of 7 distinct vertebrae, referred to as C1–C7 (see Chapter 12 for further discussion on the anatomy of other regions of the spine). The cervical spine can be divided into two components, namely the upper cervical spine and lower cervical spine, whereas the upper cervical spine consists of the atlas (C1) and the axis (C2) while the lower cervical spine is made up of C3–C7, each with similar shape.

11.2.1.1 Upper Cervical Spine

The first two cervical vertebrae (C1 and C2) are quite different in structure from the rest of the cervical vertebrae as they permit articulating movements of the head and neck. The atlas (C1) articulates superiorly with the occiput and inferiorly with the axis (C2). Unlike the other cervical vertebrae, the

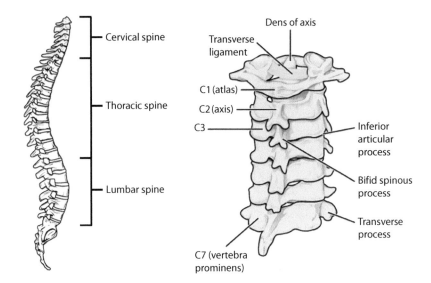

FIGURE 11.1 Human spinal anatomy. (Modified from *Anatomy & physiology: The vertical column*, OpenStax CNX 2014, Licensed under Creative Common (CC) BY 3.0. Available from: http://cnx.org/contents/ FPtK1zmh@6.27:4CMef3D9@4/The-Vertebral-Column. With permission.)

ring-shaped atlas (C1) does not have a vertebral body but consists of two lateral masses which are joined together by the anterior and posterior arches. On each of the lateral masses, there is a superior and inferior facet joint. These superior facets articulate with the occipital condyles, which face downward and outward laterally. The relatively flat inferior articular facets face downward and inward to articulate with the superior facets of the axis (C2). On the other hand, the axis (C2) has a large vertebral body, which contains the odontoid process (or den). The den articulates with the anterior arch of the atlas (C1) via its anterior articular facet and is held in place by the transverse ligament (TL). The axis (C2) articulates with the atlas (C1) via its superior articular facets, providing a pivot around which the atlas and head rotate. A vertical pillar of bone or dens projects upwards from the superior surface of the body of the axis (C2) (Oliver and Middleditch 1991). Figure 11.2 shows an anatomical diagram of the atlas (C1) and axis (C2) in their superior-posterior view.

11.2.1.2 Lower Cervical Spine

The C3 vertebra is the most superior vertebra in the lower cervical spine, and it is commonly acknowledged as the typical cervical column (Bogduk and Mercer 2000). Thus, the structure of the C3 vertebra closely resembles that of the rest of the vertebrae making up the cervical column (C4–C7). In general, the vertebra in the lower cervical spine is divided into two basic regions, a vertebral body and a posterior arch. The bone in both regions is composed of an outer layer of cortical bone and a core of cancellous bone. The shell of the cortical bone is thin on the discal surfaces of the vertebral body and is thicker in the posterior arch and its processes [25]. The C3 vertebral body is approximately rectangular in shape when viewed superiorly, while the C7 vertebral body is more of diamond shaped (Figure 11.3). The vertebral foramen in the cervical column is generally triangular when viewed superiorly. The spinous processes of the cervical column are generally short but increase in length for the inferior cervical vertebrae (Figure 11.3). They are bifid or branched into two at their posterior ends except for the vertebra prominens (C7) (Figure 11.3). The cervical transverse processes are generally wide and contain foramina through which the vertebral arteries pass superiorly as they go towards the brain.

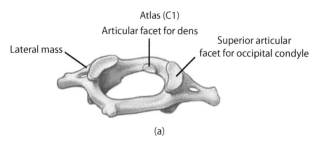

(a)

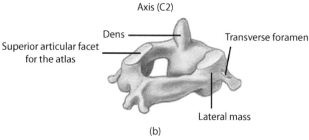

(b)

FIGURE 11.2   (a) The atlas (C1) and (b) the axis (C2) in superior-posterior view. (Reprinted from Bridwell, K., *Vertical column*, 2016, Copyright © by Dr. Keith Bridwell. Available from: http://www.spineuniverse.com/anatomy/vertebral-column. With permission.)

11.2.2 Cervical Ligaments

Ligaments resist loads most effectively along the direction in which their fibres are aligned. They readily resist tensile forces but buckle when subjected to compression. When a spinal functional unit is subjected to complex force and torque vectors, the individual ligaments provide tensile resistance to external loads by developing tension to stabilise its motion. The ligaments associated with the upper and lower cervical spine segments are generally different to each other in structure and properties. The upper ligaments are those associated with the occiput, atlas, and the anterior and lateral aspect of the axis (C2). The lower cervical ligaments encompass all other ligaments of the C3 vertebra and the typical cervical column (Oliver and Middleditch 1991).

11.2.2.1 The Upper Cervical Spine Ligaments

Figure 11.4 shows a diagram of the upper cervical ligaments. The two alar ligaments are symmetrically placed, arising from the posterior part of the tip of the dens. A greater portion of the fibres are inserted onto the occipital condyles, while some fibres are attached onto the lateral masses of the atlas (Oliver and Middleditch 1991). The apical ligament is short and thick geometrically, and attaches the tip of the dens to the anterior margin of the foramen magnum. The thick TL holds the odontoid in place and passes between the tubercules on the medial side of the lateral masses of the atlas. The TL is thicker centrally, where some fibres extend up to the occiput and the others pass downwards to the body of the axis (C2) (Oliver and Middleditch 1991).

11.2.2.2 The Lower Cervical Spine Ligaments

Figure 11.5 displays the lower cervical ligament group which comprises of: the anterior longitudinal ligament (ALL), posterior longitudinal ligament (PLL), capsular ligament (CL), ligamentum flavum (LF), ligamentum nuchae (LN), interspinous (ISL) and supraspinous (SSL) ligaments. The ALL lies anterior to the vertebral bodies and is attached to the basilar part of the occipital bone from which it

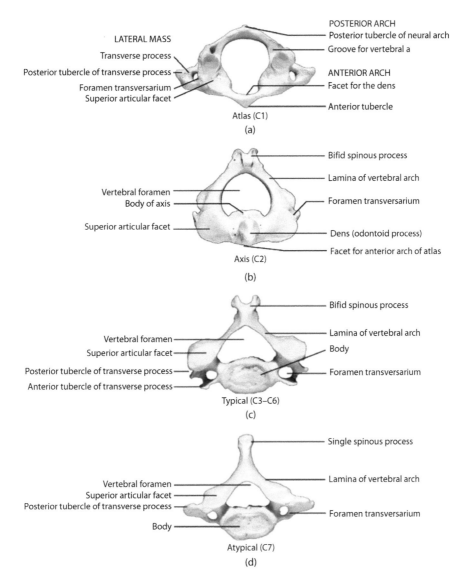

LATERAL MASS
Transverse process
Posterior tubercle of transverse process
Foramen transversarium
Superior articular facet

POSTERIOR ARCH
Posterior tubercle of neural arch
Groove for vertebral a

ANTERIOR ARCH
Facet for the dens
Anterior tubercle

Atlas (C1)
(a)

Bifid spinous process
Lamina of vertebral arch
Vertebral foramen
Body of axis
Foramen transversarium
Superior articular facet
Dens (odontoid process)
Facet for anterior arch of atlas

Axis (C2)
(b)

Bifid spinous process
Lamina of vertebral arch
Vertebral foramen
Superior articular facet
Body
Posterior tubercle of transverse process
Anterior tubercle of transverse process
Foramen transversarium

Typical (C3–C6)
(c)

Single spinous process
Lamina of vertebral arch
Vertebral foramen
Superior articular facet
Posterior tubercle of transverse process
Foramen transversarium
Body

Atypical (C7)
(d)

FIGURE 11.3 The superior view of (a) C1; (b) C2, (c) C3–C6 and (d) C7. (Netter 1997). (Reprinted from *The back*, Pocket Dentistry 2015. Copyright © by Pocket Dentistry. Available from: http://pocketdentistry. com/2-the-back/. With permission.)

extends to the tubercle of the atlas and attaches to the front of the vertebral bodies. It consists of several layers of fibres fixed to the superior and inferior margins of the vertebral bodies and the intervertebral disc. The superficial fibres are long and extend over three or four vertebrae. The middle fibres extend over two or three vertebrae, while the deepest fibres are attached to adjacent vertebrae (Oliver and Middleditch 1991). The PLL is posterior to the vertebral body and lies inside the vertebral canal. It is attached to the body of the axis (C2) and also to the margins of the vertebral bodies and the intervertebral discs. It consists of superficial fibres extending over three or four vertebrae and deep fibres which extend between adjacent vertebrae. This ligament is broad and uniform in width in the cervical spine. It is stretched on neck flexion and relaxed in extension. The CL is attached to the margins of the articular facets. They are loose in the cervical spine and, therefore, allow considerable mobility (Oliver and Middleditch 1991). The LF is a predominantly yellow elastic tissue that connects the

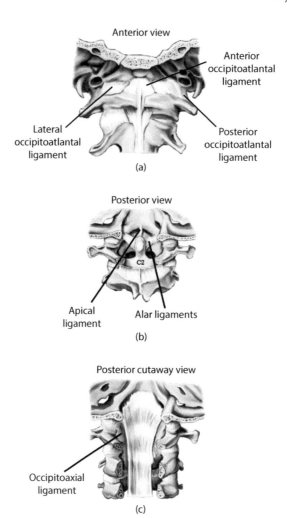

FIGURE 11.4   The upper cervical spine ligaments. (a) Anterior, lateral and posterior occipitoatlantal ligam-nets; (b) apical and alar ligaments; (c) occipitoaxial ligament. (Modified from Surange, P. N., *Anatomy of spine*, Slideshare, 2010, Copyright © 2010 by Dr. Surange. With permission.)

laminae of adjacent vertebrae. These ligaments, which are broad and long in the neck, allow flexion to occur, but prevent hyperflexion by breaking the movement so that the end of range is not large (Oliver and Middleditch 1991). The LN is a flat membranous structure that extends from the region between the cervical spinous processes anteriorly to the back of the neck posteriorly and spans the region between the occiput superiorly to the spinous process of C7 inferiorly. The ISL are poorly developed in the cervical region, consisting of a think, membranous, translucent septum. The SSL originates in the ligamentum nuchae and continues along the tips of the spinous processes as a round strand down to the sacrum (Cramer and Darby 2013).

11.2.3   CERVICAL FACET JOINTS

The cervical facet joints are synovial joints, with each consisting of two matching bony articular sur-faces covered with a 1- to 2-mm thick layer of articular or hyaline cartilage (Figure 11.6). The hyaline-lined portion of a superior and inferior articular process is known as the articular facet. The junction between the superior and inferior articular facets on one side of two adjacent vertebrae is known as a

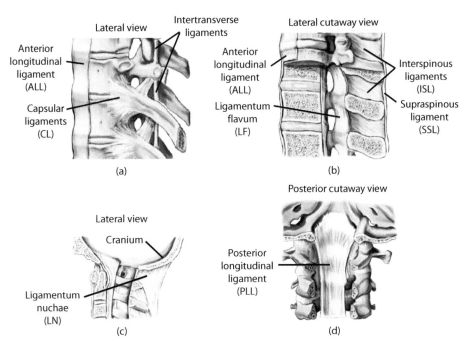

FIGURE 11.5   The lower cervical spine ligaments. (a) ALL, CL and Intertransverse ligaments; (b) ALL, LF, ISL and SSL; (c) LN; (d) PLL. (Modified from Surange, P. N., *Anatomy of spine*, Slideshare, 2010, Copyright © 2010 by Dr. Surange. With permission.)

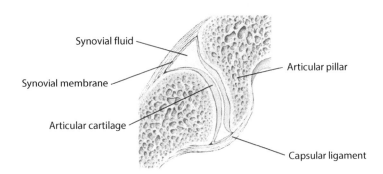

FIGURE 11.6   Cross-sectional view of the synovial facet joint of the cervical spine. (Modified from *Anatomy and function*, University of Maryland Medical Center, 2003, Copyright © 2003 by DePuy Acromed. Available from: http://umm.edu/programs/spine/health/guides/anatomy-and-function. With permission.)

facet joint (Cramer and Darby 2013). There is a gap filled with synovial fluid between the articular cartilages, which is viscous and lubricates the cartilage for ultra-low friction sliding (Figure 11.6). Containing the synovial fluid within the synovial joint is the flexible synovial membrane (Figure 11.6). In the cervical facet joint, the joint capsular (JC) ligament encases the facet joint to provide joint strength in tension.

The facet articulations of the upper cervical spine are extremely complex due to the convexity and concavity of the vertebral body; with the kidney-shaped facets and their higher internal margins compared to the medial margins (Middleditch and Oliver 2005). The atlanto-occipital joints are formed by the articulation of the concave articular facets on the lateral masses of the atlas with the convex facets on the occipital condyles (Oliver and Middleditch 1991). The lower cervical vertebrae (C3–C7) have similar or typical convexity and concavity, with their superior surface concave transversely and convex anteroposteriorly (Middleditch and Oliver 2005). The articular facets on the inferior surface of the vertebral body articulate with the uncinated processes of the subjacent vertebra (Middleditch and Oliver 2005). The functional joint biomechanics in the cervical spine are determined by the size and orientation of the articular facets (Pal et al. 2001). These allow much of the flexion and extension that occurs in the head–neck junction and at least one half of the axial rotation of the cervical spine which is inevitably coupled with lateral flexion (Bogduk and Mercer 2000). The superior articular surface of the facet joint undergoing an axial rotation slides up the inferior surface, causing a lateral bending motion between the vertebrae. Conversely, when undergoing a lateral bend, the superior articular surface of the compressed facet joint will slide downwards and posteriorly, causing a rotation between the vertebrae

11.2.4 Intervertebral Discs

Each intervertebral disc, which acts as a fibrocartilaginous joint or a symphysis between adjacent vertebrae in the vertebral column, cushions the vertebral inertial compressive load, allows slight vertebral movement (vertical displacement and angular displacement) without having the bony vertebrae rubbing against each other (Figure 11.7). The three basic components of the disc are the annulus fibrosus, the nucleus pulposus and the vertebral end plate (Oliver and Middleditch 1991) (Figure 11.7). The annulus fibrosus is a composite structure consisting of concentric layers or lamellae of collagen fibres that forms the peripheral portion or the outer ring of the intervertebral disc. It functions to absorb pressure from the central jellylike shock absorber, the nucleus pulposus (Oliver and Middleditch 1991). The cervical annulus is well developed and thick anteriorly; but it tapers laterally and posteriorly towards the anterior edge of the uncinate process on each side. The nucleus is a semi-fluid-gel-filled domain centred in each disc, and it takes up approximately 40%–60% of the disc. Being an incompressible fluid, the nucleus can be deformed under pressure without a reduction in volume. This property makes it an effective shock absorber, enabling it to both accommodate to movement and to transmit some of the compressive load from one vertebra to the next. The superior and inferior vertebral body is composed of think plate of bone and a thin layer of cartilage. The vertebral end plate forms an interface between the vertebral bone and the disc. The vertebral end plates enable even and effective stress distribution from the bone to the disc and allow diffusion across the plate between the vertebral body and the disc. Under high compressive loading, the end plates are the most common

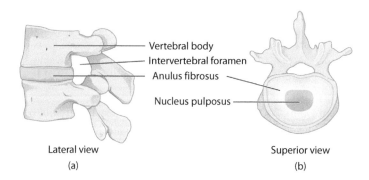

Vertebral body
Intervertebral foramen
Anulus fibrosus
Nucleus pulposus

Lateral view   Superior view
(a)   (b)

FIGURE 11.7   (a) Intervertebral discs (blue) in between adjacent vertebrae and (b) a gel-filled centre called the nucleus of the interverebral disc and a tough fibrous outer ring called the annulus. (Reprinted from *Anatomy & physiology: The vertical column*, 2014, Licensed under Creative Common (CC) BY 3.0. Available from: http://cnx.org/contents/FPtK1zmh@6.27:4CMef3D9@4/The-Vertebral-Column. With permission.)

site of failure and are, therefore, considered to be the weakest part of the disc (Oliver and Middleditch 1991; Bogduk and Mercer 2000; Cramer and Darby 2013).

11.2.5 CERVICAL MUSCLES

The cervical muscles typically have a short tendon and are seemingly attached directly to the membranes of the skull or vertebrae. In general, the larger muscles are located superficially while the smaller muscles are closer to the vertebral column, connecting the skull with the spine and individual vertebrae (Figures 11.8 and 11.9).

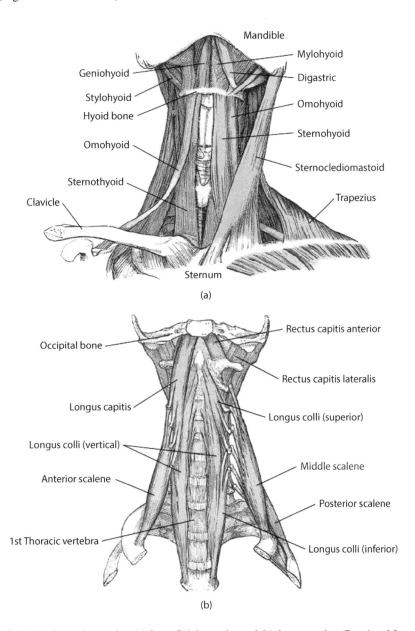

(a)

(b)

FIGURE 11.8 Anterior neck muscles. (a) Superficial muscles and (b) deep muscles. (Reprinted from Gray, H., *Anatomy of the human body*, Lea & Febiger, Philadelphia, PA, 1918, Copyright © 2015 by Bartleby.com. With permission.)

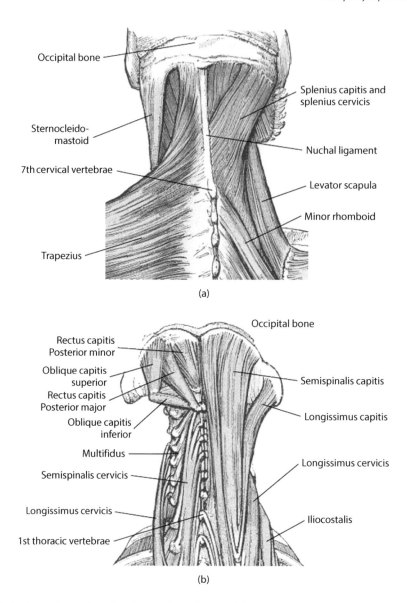

FIGURE 11.9 Posterior or dorsal neck muscles. (a) Superficial muscles and (b) deep muscles. (Reprinted from Gray, H., *Anatomy of the human body*, Lea & Febiger, Philadelphia, PA, 1918, Copyright © 2015 by Bartleby.com. With permission.)

Figure 11.8 shows all the major anterior cervical muscles. The most superficial muscles are the sternocleidomastoideus and the trapezius. Due to their long lever arms and their volumes, these two muscles generate large flexion and extension forces but are not expected to be involved during relaxed stabilisation. Both of these muscles consist of subpartitions that are most likely activated by different motor units to produce different motions. The trapezius acts as an extensor while the sternocleidomastoideus can be both flexor and extensor, depending on the initial head position. The anterior muscles are located anteriorly to the vertebral column and work as head flexors. Their combined physiological cross-sectional areas (PCSA) are smaller than the posterior or dorsal muscles, and the human neck is consequently weaker in flexion than in extension. The longus colli and longus capitis are located close to the vertebral column and produce only small motions of the head. The hyoid muscles run between the jaw and the hyoid bone and continue down to the sternum. They are considered to be mainly involved in the motions of swallowing and talking. However, it is assumed that they work as synergists together with other ventral muscles to produce head flexion.

The posterior or dorsal muscles, which are located posteriorly to the vertebral column, work against the gravitational forces and keep the head upright at normal positions. All the posterior muscles are shown in Figure 11.9. The splenius, semispinalis and longissimus are divided into two parts, namely, the capitis and cervicis, where the capitis is inserted in the skull base while the cervicis part is inserted into the upper vertebral column. The origins of the coupled muscles coincide, and the two parts are joined into one muscle. The semispinalis capitis is the largest muscle mass of the back of the neck and runs straight along the vertebral column. The splenius and longissimus muscles are slightly angled, running laterally from the origin on the vertebrae to the more proximal insertion on the skull.

The suboccipital muscles attach between the occipital bone and the two first cervical vertebrae. They have short lever arms and small PCSAs, which makes them suitable for stabilising the head rather than to exert large motions. The scalenes and levator scapula are the lateral muscles located on the side of the neck and are used in lateral motions and to lift the shoulders (Figures 11.8 and 11.9).

11.3 TYPES OF NECK INJURIES AND INJURY MECHANISMS

In general, cervical spine injury can be categorised into three main types based on the injured or damaged anatomical components. The most common type is fracture of the cervical vertebrae and rupturing of the cervical vertebral ligaments or muscles due to the excessive and complex combination of bending, shearing and axial loading generated by an indirect impact to the neck.

Secondly, damage to the spinal cord due to these combined loadings between cervical vertebrae, physical stretching of the spinal cord or direct impingement by the fractured cervical vertebrae, particularly in the atlas (C1) and axis (C2), can lead to permanent disability or be fatal. The last type of neck injury is trauma or damage to the soft tissues containing the trachea, larynx, oesophagus, as well as major arteries and veins to the head and brain, and can occur from either blunt or sharp objects, causing severe lacerations and penetrating injuries to the carotid artery and jugular vein (Payne and Patel 2001). Different types of head motion which the neck can be subjected to are shown in Figure 11.10.

Indirect loads can also result from relative displacement between the head and torso in either rear-end or side-impact automotive crashes, resulting in whiplash. One of these is a tension–extension mechanism commonly encountered in motor vehicle crashes: in the case of a rear impact to a vehicle, both the vehicle seat and the torso are accelerated rapidly forward and the head is accelerated via the neck, thus producing hyperextension if the rear seats have no head restraint before rebounding back to neutral position or hyperflexion.

A tension–flexion type of mechanism often occurs in frontal collisions when the torso is restrained and the neck stops the head from flexing. Commonly seen injuries due to a tension–flexion mechanism include hyperflexion sprain as well as bilateral facet dislocation. It should be noted that bilateral facet dislocation can occur in both tension–flexion and compression–flexion events, suggesting that flexion is the key contributor for the injury. Another possible mechanism in an automotive crash event is shearing effect of the cervical spine (Chen et al. 2009). In a rear impact on vehicle, the cervical spine is pushed forward by the seat, causing the 'S-shaped' spine to be straightened. At this instant, the weight of the head compresses onto the cervical spine. As the torso pulls the head forward, the compressed cervical vertebrae slide relative to one another, generating shear forces at the intervertebral joints of the cervical spine. This can cause the facet capsular ligaments to be stretched and torn (Deng et al. 2000). It can also cause the TL to rupture or even fracturing of the den (Braakman and Penning 1971). These injuries are often considered more severe due to the likelihood of lethal spinal cord impingement and the difficulties associated with stabilising the injuries through surgery (McElhaney et al. 2002). Other mechanisms such as lateral

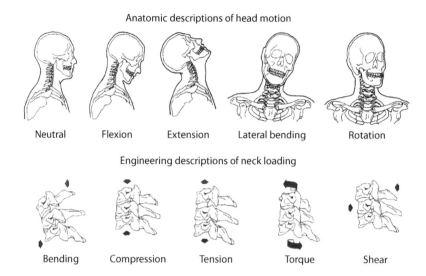

FIGURE 11.10   Anatomic descriptions of head motion and engineering descriptions of neck injury mechanisms. (Modified with kind permission from Springer Science+Business Media: *Accidental injury: Biomechanics and prevention*, Biomechanical aspects of cervical trauma, 2002, pp. 324–73, McElhaney, J. H., et al., Copyright © 2002.)

bending and lateral shearing are generally considered less common than those above mentioned mechanisms occurring in the sagittal plane due to the relatively lower incidence of side impacts than frontal crashes, and the neck's lesser compliance to flex in lateral bending. However, these lateral mechanisms can still produce severe injuries such as nerve-root avulsion and transverse process fractures.

Cervical spinal injury mechanisms commonly encountered in military scenarios such as fighter pilot ejection from a cockpit in an emergency and UB blast are attributed mainly to compression, compression–extension and compression–flexion, where extremely high vertical acceleration loads the spinal column over a very short span of time (Figure 11.11). Compression–extension can produce fractures of the spinal processes and causes spine dislocations, while compression–flexion can produce burst fractures of the vertebral wedge type (Payne and Patel 2001).

Finally, excessive bending, fractures and dislocations of the cervical vertebrae can impinge directly onto the spinal cord. This can cause reversible or irreversible damage to individual nerves leading to long-term discomfort or permanent disability. In its most severe form, cervical spine injury is usually life threatening or fatal, especially when the cervical vertebral fractures and dislocations occur at the upper cervical vertebrae (C1 and C2).

11.4   CURRENT METHODOLOGIES IN STUDYING NECK INJURY MECHANISMS

11.4.1   Experimental Approach

In the past few decades, tremendous effort has been spent on understanding head and neck injuries, generally in the area of materials characterisation as well as injury simulation studies using volunteers, cadavers, animal subjects and ATDs. The following paragraphs summarise the experimental studies conducted by two different approaches: (1) determining the material properties of bone, ligaments and muscles in the neck and (2) reproducing injuries, or the loading conditions which lead to neck injuries, by using volunteers, cadavers, animal subjects and ATDs.

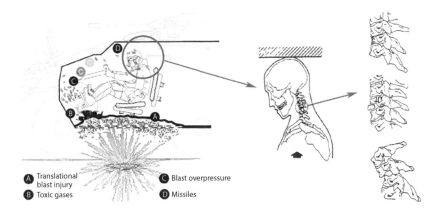

FIGURE 11.11   An underbelly blast event demonstrating how vertical acceleration can result in compressive spinal injuries to the neck due to a head impact with the interior vehicle roof. (Modified from *European Journal of Radiology*, 63(2), Vogel, H., and Dootz, B., Wounds and weapons, 151–166., Copyright 2007, with permission from Elsevier; modified with kind permission from Springer Science+Business Media: *Accidental injury: Biomechanics and prevention, Biomechanical aspects of cervical trauma*, 2002, pp. 324–73, McElhaney, J. H., et al., Copyright 2002.)

11.4.1.1   Material Characterisation Tests

Extensive research efforts have been made in determining constitutive material properties of the human body, and ligamentous and muscular tissues. Such quantitative knowledge is vital to understand the failure mechanisms for the different tissues in the spine.

11.4.1.1.1   *Compressive Mechanical Properties of Cancellous Bone and Constitutive Modelling*

The mechanical properties of cancellous bone can be attributed to differences in bone density (Keller 1994; McCalden et al. 1997; Kopperdahl et al. 1999). Bone density can be defined as 'fresh bone density', which corresponds to the mass of a fresh specimen averaged over its entire volume, or as 'apparent bone density', corresponding to the mass of defatted bone divided by the volume. It has been reported that mechanical properties can be described by a power-law function of density (Kaplan et al. 1985; Giesen et al. 2001). Investigations using Hopkinson bar techniques have focused on eliciting the mechanical properties of cortical bone under high strain rates (10^1-10^3/s) (Tennyson et al. 1972; Lewis and Goldsmith 1975; Katsamanis and Raftopoulos 1990). To understand the vulnerability of cancellous bone subjected to impact or shock loading, determination of its mechanical properties and characterisation of its load-bearing response are necessary.

In the recent work by Liu et al. (2013), cancellous bone from the vertebral bodies of C2–C7 was extracted from the cervical spines of eight male cadavers aged 40–79 years (Figure 11.12). To minimise damage, a precision diamond saw was used to cut raw samples at a low rotational speed (150–250 rpm) and with saline water irrigation, into cuboid specimens (5 × 5 × 8 mm). The fresh bone specimens were defatted by an ultrasonic bath of trichloroethylene, which was followed by centrifuging to remove liquid from the pores. The apparent bone density was then obtained and found to range from 0.3–0.9 g/cm^3, which corresponds to a fresh bone density of 0.4–1.3 g/cm^3. As the chemical defatting process may affect the properties of bone, Shim and colleagues also conducted experimental tests using fresh bone specimens, with the marrow and fat present (Shim et al. 2005, 2006; Liu et al. 2013). Loading was applied along the longitudinal axis of the spine, and the mechanical response under quasi-static compression was found to be essentially linearly elastic prior to failure.

(a)   (b)   (c)

(d)   (e)

FIGURE 11.12 Internal structural characteristics of cancellous bone: (a) sagittal cross section of the C2 vertebral body and (b) sagittal cross section of the C4–C5 vertebral bodies; (c) fresh and (d) defatted specimens and (e) magnification of its microstructure.

The compressive strength (in MPa), defined by a value 5% below the ultimate stress, can be described by a power-law function of the fresh bone density, according to $\sigma_c = 5.09\,\rho^{2.35}$, where the density ($\rho$) is in units of g/cm^3. The compressive Young's modulus E (in MPa) also increases with bone density and follows a power-law relationship, $E = 242\rho^{1.95}$. Sustained loading beyond the ultimate strength was accompanied by development of significant damage and was not of particular interest.

As part of the experimental work, Shim et al. (2005) also performed dynamic compressive tests at strain rates of 10^2–10^3/s using a split Hopkinson pressure bar device. Since the mechanical impedance of cancellous bone is much lower than that of steel or aluminium, magnesium input and output bars were utilised. From approximate equality of the signals generated by piezoelectric PVDF (polyvinylidene fluoride) sheets placed at the two sides of the specimen being tested, approximate stress equilibrium in a specimen could be attained. The dynamic test results indicate that the mechanical behaviour of cancellous bone is strain-rate dependent and non-linear. The dynamic strength can be expressed as a function of both the fresh bone density and strain-rate ($\sigma_c^d = 5.09\rho^{2.35} + 0.545\dot{\varepsilon}^{0.45}$). Assuming that deformation prior to attainment of bone strength is recoverable; the rate-dependent mechanical response can be described by viscoelasticity. Due to the small strain limit (up to 5%) imposed on the cancellous bone, movement of marrow and fat, through the porous cells was negligible and hence its effect on the viscoelastic response can be ignored. Thus, the following viscoelastic constitutive model was proposed:

$$\sigma = E_0\rho^\beta\varepsilon + \int\limits_0^t c_1\dot{\varepsilon}(\tau)\exp\left(-\frac{t-\tau}{\theta}\right)d\tau + \eta\dot{\varepsilon}^{1/2} \qquad (11.1)$$

The first term ($E_0\rho^\beta\varepsilon$) corresponds to density dependent linear elasticity, while the second ($\int_0^t c_1\dot{\varepsilon}(\tau)\exp\left(-\frac{t-\tau}{\theta}\right)d\tau$) and third ($\eta\dot{\varepsilon}^{1/2}$) terms define rate-dependent non-linear viscoelasticity via a Maxwell element in parallel with a non-linear Newtonian dashpot; a diagrammatic representation

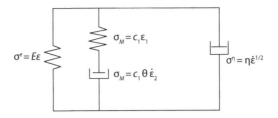

FIGURE 11.13   Schematic representation of proposed viscoelastic constitutive model for cancellous bone.

of this is shown in Figure 11.13. Stress–strain curves obtained from experiments were used to evaluate the parameters ($E_0 = 242$ MPa, $\beta = 1.95$, $c_1 = 354$ MPa, $\theta = 0.055$ ms and $\eta = 0.122$ MPas$^{1/2}$). A comparison between the fitted curves and test data is shown in Figure 11.14, demonstrating that the proposed model is able to describe the mechanical behaviour of cancellous bone under high loading rates (Shim et al. 2005).

In the same study, while the authors recognised that cancellous bone from the human cervical spine is anisotropic, an isotropic model was proposed as a first step to facilitate numerical modelling of the response to rapid loading and trauma (Shim et al. 2005). By replacing the one-dimensional strain and strain-rate variables by a function of the strain-rate tensor $\dot{\varepsilon}$, the one-dimensional relationship can be expanded to a three-dimensional constitutive equation, where the stress tensor σ is defined by Shim et al. (2005):

$$
\sigma = \frac{E_0 \rho^\beta}{1+\nu}\left(\frac{\nu}{1-2\nu}\mathrm{tr}(\varepsilon)\mathbf{I} + \varepsilon\right) + \frac{\eta}{(1+\nu)^{1/2}}\left(\frac{\nu}{1-2\nu}\mathrm{tr}(\dot{\varepsilon}(\tau))\mathbf{I} + (\dot{\varepsilon}(\tau))\right)^{1/2}
$$

$$
+ \int_0^t \frac{c_1}{1+\nu}\left(\frac{\nu}{1-2\nu}\mathrm{tr}(\dot{\varepsilon}(\tau))\mathbf{I} + \dot{\varepsilon}(\tau)\right)\exp\left(-\frac{t-\tau}{\theta}\right)d\tau \tag{11.2}
$$

where ν is Poisson's ratio, which is assumed to have a value of 0.3.

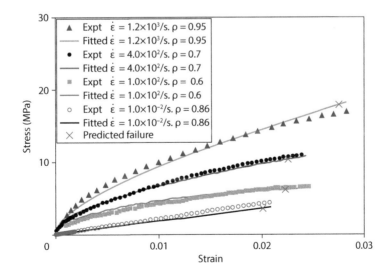

FIGURE 11.14   Comparison between fitted stress–strain curves and experimental data for different bone densities and strain rates. (Reprinted from *International Journal of Impact Engineering*, 32(1–4), Shim, V. P. W., et al., Characterisation of the dynamic compressive mechanical properties of cancellous bone from the human cervical spine, 525–540, Copyright 2005, with permission from Elsevier.)

This expression was subsequently revised and cast into the following form by Bekker et al. (2014):

$$\sigma(t) = k_1\varepsilon(t) + \eta_1(\dot{\varepsilon}_{eq})^{P-1}\dot{\varepsilon}(t) + \int\limits_0^t k_2\dot{\varepsilon}(\tau)e^{-\frac{t-r}{\theta_2}}d\tau \qquad (11.3)$$

where $\dot{\varepsilon}_{eq} = \sqrt{\frac{\dot{\varepsilon}:\dot{\varepsilon}}{1+2v^2}}$, P is an exponent, and k_1, k_2, η_1 and θ_2 are constants.

11.4.1.1.2 Tensile Testing of Cervical Ligaments

There have been several studies on the quasi-static mechanical properties of spinal ligaments (Chazal et al. 1985; Myklebust et al. 1988; Yoganandan et al. 2000). Since these soft tissues exhibit rate-sensitive behaviour, attempts to investigate their dynamic responses were motivated by concerns related to trauma generated from sudden rapid motion. Dynamic tests by piston-driven devices at extension rates up to 2.5 m/s have showed significant rate dependence in the mechanical response of ligaments (LF, ALL, TL and alar ligaments) in the human cervical spine (Yoganandan et al. 1989; Panjabi et al. 1998).

Shim et al. (2006) employed a split Hopkinson bar device to elicit the dynamic tensile properties of seven of the major ligaments, whereby a tubular striker was used to impact an anvil cap attached to the end of the input bar to generate a tensile loading pulse; quasi-static loading was applied using an Instron micro-tester. Clamping of the ends of a specimen for tensile testing is a major challenge in experimental tests on soft biological samples. Bone-ligament-bone units were therefore utilised as specimens for the short ligaments in the human cervical spine (Figure 11.15). The bone ends were inserted into the cavity of aluminium clamps, and dental cement was used to fill the remaining space to bind them together. Specimens were prepared in a way that the applied tension aligns with the physiological loading direction.

Ligaments are extremely soft and flexible and possess very low mechanical/wave impedance; consequently, tests on these materials pose significant difficulties. Therefore, in Shim et al. (2006)'s work, special techniques were developed to ensure the Hopkinson bar system was accurate and reliable for testing these materials. In order to obtain reliable data from the extremely small wave signals transmitted to the output bar, a lighter metal (magnesium) was employed for the input and output bars. Furthermore, highly sensitive semiconductor strain gauges, which improve the signal-to-noise ratio of the transmitted wave captured by more than fifty times compared to foil gauges, were used on the output bar.

A typical load-deformation curve for a soft tissue has an initial exponential-like toe regime, where the stiffness is quite low and a very small force induces significant deformation. This toe regime is associated with the physiological range and is an essential segment in the complete stress–strain curve for a ligament. To avoid initial pre-loading of the ligament specimen in a Hopkinson bar test, a procedure which incorporates the introduction of initial slack to a specimen was established to achieve a zero pre-load condition and to facilitate capture of the entire toe regime. The stretching of a specimen only commences after the initial slack in the specimen is removed by the moving input bar. Also, to stretch the compliant ligament to failure to obtain a complete stress–strain curve, a heavy striker was used to strike the thin input bar in order to generate a sufficiently long loading pulse (Figure 11.16).

Test results showed that the stress–strain curves of ligaments exhibit a characteristic non-linear initial toe region followed by an almost linear elastic response, which then culminates into rupture. Significant strain-rate sensitivity was observed by comparing the quasi-static and dynamic responses (Figure 11.17). As with most biological materials, there is noticeable experimental scatter in the results; however, the average response shows the overall trends for ligaments, and a power-law relationship was found to be suitable for describing both the static and dynamic tensile stress–strain curves (Table 11.1) (Shim et al. 2006). It is envisioned that the information obtained will be useful for studies related to neck injuries occurring from suddenly applied loads.

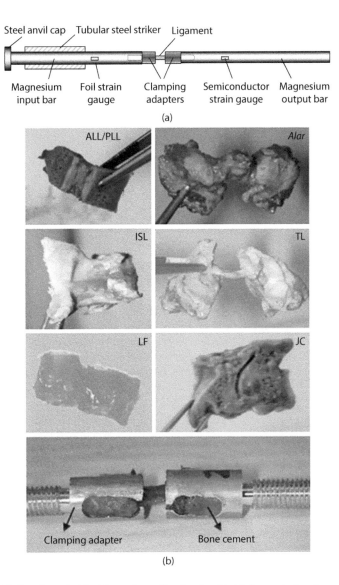

(a)

(b)

FIGURE 11.15 (a) Tensile split Hopkinson bar. (Reprinted with kind permission from Springer Science+ Business Media: *Experimental Mechanics*, A technique for dynamic tensile testing of human cervical spine ligaments, 46(1), 2006, 77–89, Shim, V. P. W., et al., Copyright 2006.) (b) Tensile test specimen of the seven primary ligaments from the human cervical spine, and the clamping method used. *Note*: ALL = anterior longitudinal ligament, alar = alar ligament, PLL = posterior longitudinal ligament, TL = transverse ligament, LF = ligamentum flavum, JC = facet joint capsule joint, ISL = interspinous ligament. (Reprinted from with kind permission from Springer Science+Business Media: *Experimental Mechanics*, A technique for dynamic tensile testing of human cervical spine ligaments, 46(1), 2006, 77–89, Shim, V. P. W., et al., Copyright 2006.)

11.4.1.2 Neck Injury Tests

Experimental injury studies have been performed using live human volunteers, cadavers, animal subjects, and ATDs in order to better understand the factors contributing to cervical spine injuries.

11.4.1.2.1 Experiments Simulating Dynamic Vertical Compressive Loading

Experiments simulating cervical spine injury due to vertical compressive loading can be traced back to around the 1980s. In 1983, Maiman et al. (1983) conducted vertical compressive loading tests

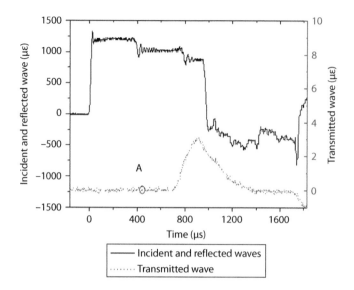

FIGURE 11.16 Illustration of a sufficiently long input and output loading wave generated in a tensile Hopkinson bar test on an interspinous ligament (Point A on the transmitted wave denotes possible initiation of stretching if the specimen has no initial slack). (Modified with kind permission from Springer Science+Business Media: *Experimental Mechanics*, A technique for dynamic tensile testing of human cervical spine ligaments, 46(1), 2006, 77–89, Shim, V. P. W., et al., Copyright 2006.)

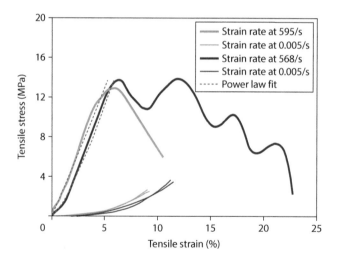

FIGURE 11.17 Static and dynamic tensile stress–strain curves for transverse ligament and power-law fits to the point of failure. (Modified with kind permission from Springer Science+Business Media: *Experimental Mechanics*, A technique for dynamic tensile testing of human cervical spine ligaments, 46(1), 2006, 77–89, Shim, V. P. W., et al., Copyright 2006.)

on three intact cadavers and ten isolated cadaveric cervical spinal columns from the skull base to T3. For the intact cadaveric whole-body tests, the specimens were subjected to vertical compressive loads by a mechanical testing machine, with a dynamic loading rate from 1.12 to 1.42 m/s, while the isolated cervical spinal column specimens were tested in three different pre-alignments (pre-flexion, pre-axial and pre-extension) before being compressed axially at a quasi-static rate of 0.23 m/s to dynamic rate of 1.52 m/s (Figure 11.18).

TABLE 11.1

Average Quasi-Static and Dynamic Mechanical Properties of Ligaments Tested

Ligament Type	Overall Dynamics Mechanical Properties			Overall Quasi-Static Mechanical Properties		
	Strain Rate (1/s)	Fitted Equation	Ultimate Tensile Stress (MPa) Average (SD)	Strain Rate (1/s)	Fitted Equation	Ultimate Tensile Stress (MPa) Average (SD)
Alar	0.005	$\sigma = 0.00065\varepsilon^{2.0675}$	4.30 (1.26)	1733	$\sigma = 0.0663\varepsilon^{1.5177}$	10.91 (1.26)
TL	0.005	$\sigma = 0.0092\varepsilon^{2.1309}$	7.65 (1.08)	570	$\sigma = 1.6324\varepsilon^{1.2344}$	15.15 (3.39)
ALL	0.005	$\sigma = 0.004883\varepsilon^{2.1366}$	7.92 (3.49)	3055	$\sigma = 0.7809\varepsilon^{1.2482}$	22.47 (5.72)
PLL	0.005	$\sigma = 0.004511\varepsilon^{2.1686}$	6.79 (6.75)	3603	$\sigma = 0.5759\varepsilon^{1.199}$	22.97 (12.95)
ISL	0.005	$\sigma = 0.0115\varepsilon^{1.3952}$	0.95 (1.18)	1321	$\sigma = 0.01846\varepsilon^{1.4638}$	1.92 (1.39)
LF	0.005	$\sigma = 0.003971\varepsilon \\ -0.00024\varepsilon^2 \\ +0.0000219\varepsilon^3$	3.05 (1.39)	2554	$\sigma = 0.1053\varepsilon \\ -0.00506\varepsilon^2$	8.14 (2.53)
JC	0.005	$\sigma = 0.0087\varepsilon^{1.47}$	3.44 (1.20)	1920	$\sigma = 0.04427\varepsilon^{1.409}$	7.22 (3.29)

Source: Modified with kind permission from Springer Science+Business Media: *Experimental Mechanics*, A technique for dynamic tensile testing of human cervical spine ligaments, 46(1), 2006, 77–89, Shim, V. P. W., et al., Copyright 2006.

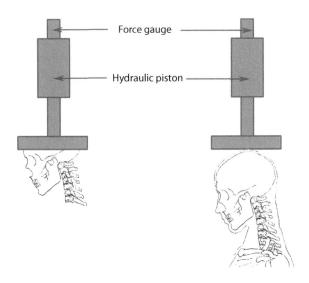

Force gauge

Hydraulic piston

FIGURE 11.18 Schematic representation of Maiman et al. (1983)'s isolated cervical spinal column tests and intact cadaveric whole-body tests. (Modified from *Journal of Biomechanics*, 22(5), Viano, D. C., et al., Injury biomechanics research: An essential element in the prevention of trauma, 403–417., Copyright 1989, with permission from Elsevier.)

Peak forces, vertical compressive displacements, failure modes, and locations were recorded in the study of Maiman et al. (1983). For the two intact specimens loaded under pure axial compression, injuries were observed in the C1–C2 segments, while the other intact specimen, which was pre-extended before compressed axially, failed at the C4–C6 segment. The peak recorded force for the intact specimen tests ranged from 1,512 to 2,936 N. The isolated cervical spinal column specimens under pure compression failed at 1,509–7,439 N and sustained injuries such as a burst fracture at C5, compressive fractures at the C2–C3 segments, a facet fracture ligament disruptions of the lower cervical spine. The injuries observed in the tests for the pre-flexed isolated specimen included C1–C2 dislocations and C4–C5 compression fractures associated with posterior column disruption (with the failure load of 645–3,000 N), while the pre-extended specimen sustained C4 fracture and anterior longitudinal ligament disruption at C5–C6 (with failure load of 667 N).

Yoganandan et al. (1991) established a method to evaluate the biodynamic strength and to determine the local kinematics of the cervical spine under a vertical compressive load. Six fresh human cadaveric head–neck complexes (the head to T2) were impacted axially at the head by the piston of an electrohydraulic testing device at varying speeds ranging from 5.28 m/s to 8.50 m/s. The test sequence was captured by high-speed camera, while the localised kinematic data were obtained from retroreflective targets which were placed on the bony landmarks (vertebral body, facet column and the tip of spinous process) at every spinal level. The input forces, accelerations, compressive displacement and force histories were recorded by a digital data acquistion system. The peak impact forces recorded ranged from 3,300 to 5,600 N. Post-test radiography, computed tomography and cryomicrotomy were performed for all the specimens for detailed injury pathology. The injuries sustained included wedge, burst and compression fractures of the mid-lower cervical vertebrae. Yoganandan et al. (1991) quantified the resulting localised kinematics of the cervical spine and suggested that the temporal deformation characteristics as well as a plausible sequence of cervical spinal injuries led to cervical spinal injuries. Moreover, they were the first to observe that the peak output compressive forces are of lower magnitude and longer impulsive duration at the distal end of the cadaveric complexes (T2) than the experimental input or impact forces on the top of the head, suggesting a possible decoupling of the head and neck.

Recently, Yoganandan et al. (2013) conducted a statistical analyses based on vertical loading tests conducted using the above method, with similar alignment and boundary conditions to the 1991 study, to quantify the effects of loading speed, sex and age on the peak compressive force obtained in experiments. The effect of loading rate on the peak impact force for specimens for different ages is shown in Figure 11.19a. It was found that the effect of loading rate was more pronounced in younger specimens than in older specimens. Force-based injury risk curves are shown in Figure 11.19b, indicating that the injury tolerance is lower for older than younger populations.

However, caution should be exercised when applying this data to military scenarios as these statistical curves were primarily derived for automotive environments and might be too conservative for situations encountered in military operations. Nevertheless, these injury data may serve as a first step for military applications.

11.4.1.2.2 Experiments Simulating Pilot Ejection

The incidence of acute neck injuries is a major concern in pilot ejection, and rocket sled tests have been used to mimic the ejection process (Arnaiz 1986; Burton 1999; Vasishta et al. 2003; David et al. 2011). However, rocket sled tests are highly expensive and there are limited facilities worldwide. There are also a limited number of reports on experiments that simulate vertical ejection on cadaveric head–neck complexes because of the practical difficulties in conducting these tests (Gaynor et al. 1962; Stemper et al. 2009). Hence, most pilot ejection sled tests have been performed using ATDs. To evaluate neck injury risk due to two helmet supported masses, Perry (1994) performed an experimental ejection on an Advanced Dynamic Anthropomorphic Manikin (ADAM), equipped with Concept VI and ANVIS 49/49 helmets and seated on an Advanced Concept Ejection Seat (ACES) II and B-52 seats using a vertical deceleration tower. It was found that both helmets did not induce any neck injury, as the neck loads are less than the maximum (1800 N) required for neck fracture. It was also found that a helmet supported mass with a weight of 2.7 kg or less and having a centre of gravity from −0.51 to 2.79 cm and 1.02 to 3.56 cm on the anatomical x and y axes, respectively, did not result in neck injury. Another series of Gz impact tests using an ATD was conducted by Shender et al. (2000), where the aim of the tests was to determine the effect of varying the helmet's weight and its centre of gravity position during pilot ejection. It was found that for a helmet weight of 1.4 kg to 2.5 kg, the +12 Gz manoeuvre would give rise to a neck load ranging from 1010 N to 1112 N and a bending moment ranging from 78 Nm to 112 Nm.

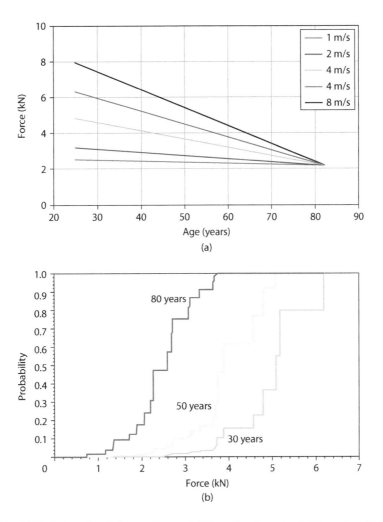

FIGURE 11.19   (a) Variation of neck force with age at different loading rate; (b) Force-based injury risk curve for vertical loading with age. (Reprinted from *Clinical Biomechanics* (*Bristol, Avon*), 28(6), Yoganandan, N., et al., Cervical spine injury biomechanics: Applications for under body blast loadings in military environments, 602–609., Copyright 2013, with permission from Elsevier.)

As mentioned earlier, whole human body cadaveric sled tests are sparse due to ethical issues and practical difficulties. In order to assess potential injury generated during pilot ejection and to validate numerical simulations, cadaveric head–neck structure tests and numerical simulations have been performed. For example, Shim et al. (2011) established experimental techniques to replicate the acceleration-time history corresponding to pilot ejection. A typical acceleration-time profile during an ejection event (Figure 11.20) involves an initial rapid increase in acceleration from 0 to around 12 G, after which a velocity about 12 m/s is attained, which corresponds to a vertical displacement of 0.6–0.7 m over an approximately 0.15 s duration (Ho et al. 2004).

To reproduce this on a head–neck specimen under laboratory conditions, a means of imposing such upward acceleration was devised by Shim et al. (2011), using a vertical ejection simulation tower which was designed and constructed to facilitate rapid upward acceleration of a test load (Figure 11.21). A steel wire rope was passed through a pulley system and connected to a specimen mounting fixture via an elastic cord. A mass released from a drop tower was used to generate the desired motion. The initial slack in the wire rope was incorporated to induce sudden traction, which was moderated by the elastic cord to achieve the desired force/acceleration profile. The force, acceleration, velocity and

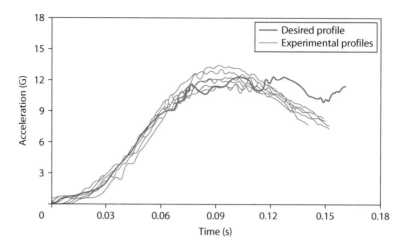

FIGURE 11.20   Typical acceleration-time profile and the profiles from experiments simulating vertical ejection. (Modified from *International Journal of Impact Engineering*, 38(8–9), Shim, V. P. W., et al., A technique to prescribe a vertical acceleration-time load on the human head–neck complex, 707–714., Copyright 2011, with permission from Elsevier.)

displacement history of the specimen were calculated in terms of the masses, stiffness, and initial velocity of the system components, e.g. the equation for the acceleration of specimen is:

$$\ddot{y}_2 = \frac{1}{M_2}(C_1 \cos \omega t + C_2 \sin \omega t + C_0) - g \tag{11.4}$$

where the constants C_0, C_1, C_2 and ω are defined by:

$$C_0 = M_2 g + ng\left(\frac{M_1 - nM_2}{M_1}\right)\Big/\left(\frac{n^2}{M_1} + \frac{1}{M_2}\right) \tag{11.5}$$

$$C_1 = -ng\left(\frac{M_1 - nM_2}{M_1}\right)\Big/\left(\frac{n^2}{M_1} + \frac{1}{M_2}\right) \tag{11.6}$$

$$C_2 = nv_0\Big/\sqrt{\left(\frac{n^2}{M_1} + \frac{1}{M_2}\right)\left(\frac{1}{k_1} + \frac{1}{k_2}\right)} \tag{11.7}$$

$$\omega = \sqrt{\left(\frac{n^2}{M_1} + \frac{1}{M_2}\right)\Big/\left(\frac{1}{k_1} + \frac{1}{k_2}\right)} \tag{11.8}$$

and n is twice the number of movable pulleys; k_1 and k_2 denote the stiffness of the steel rope and elastic cord, and v_0 is the velocity of the falling mass at the instant the specimen just begins to move.

Seven cadaveric head–neck complexes were tested using the vertical ejection simulation tower to simulate pilot ejection. The base of the specimen was encased by dental cement in a potting cup, which could be detached from the specimen after a prescribed ejection distance. The specimen was then decelerated and prevented from dropping after the ejection process. A load cell and an accelerometer can be attached below and on the side of the potting cup, respectively, in order to measure the force and acceleration experienced. The experimental acceleration profiles of the seven specimens demonstrated good agreement with that in a typical rocket sled test, and the desired G amplitude was also attained (Figure 11.20, shown earlier). A high-speed camera was used to capture visual images of

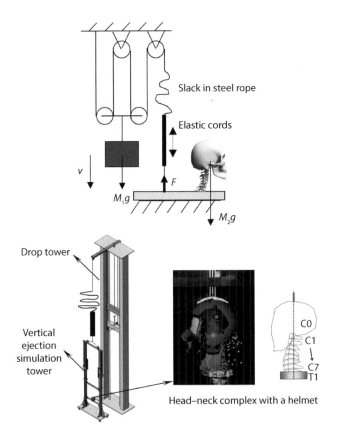

FIGURE 11.21   Schematic diagram of the pulley system and vertical ejection simulation tower to facilitate vertical ejection tests. (Modified from *International Journal of Impact Engineering*, 38(8–9), Shim, V. P. W., et al., A technique to prescribe a vertical acceleration-time load on the human head–neck complex, 707–714., Copyright 2011, with permission from Elsevier.)

the dynamic response of the cervical spine at a rate of 2,000 frames per second (Shim et al. 2011). Marker pins were inserted into each vertebra and used to track their relative angular movements, since the cervical spine was covered by muscle and skin, making direct observation impossible. The results showed that the angular displacements between adjacent vertebrae were not large during this short period of high G ejection and appear to be within the physiological range of head–neck flexion (Table 11.2). Thus, the amount of flexion induced during the tests is not likely to cause injuries. The injuries expected in a cervical spine subjected to ejection forces are spinal compression and fracture; this coincides with the descriptions of Vasishta et al. (2003). It was concluded that the test protocol established can be employed to investigate the dynamic response of specimens subjected to acceleration-time histories that correspond to pilot ejection.

11.4.1.2.3 Experiments Simulating Underbelly Blast

In recent years, there have been expanding military interest on survivability of armoured vehicles' occupants undergoing UB blast; many UB blast tests have been performed on armoured vehicles using ATDs.

From 1994 to 1996, Alem and Strawn (1996) from the US Army Aeromedical Research Laboratory (USAARL) performed landmine explosion tests on 5-tonne military trucks to investigate the effectiveness of an energy-absorbing seat in mitigating blast to the seated crew through the cabin floor and seat. In these tests, two ATDs were placed on the unmodified driver seat and the new energy-absorbing passenger seat. Accelerations at the vehicle structure, the ATDs' head,

TABLE 11.2

Limits of Relative Angular Displacement between Adjacent Vertebrae

Specimen No.	Range of Relative Angular Displacements between Adjacent Vertebrae during the Ejection Tests (< 0.15s)							
	C0–C1	C1–C2	C2–C3	C3–C4	C4–C5	C5–C6	C6–C7	C7–T1
1	−1.35–1.12	0.63–9.29	−12.13–−0.07	−4.37–1.40	−9.62–0.22	−0.73–2.68	−0.61–11.39	−0.35–13.33
2	−0.35–6.29	−2.94–10.78	−2.13–1.85	−1.90–−0.03	−4.45–1.02	−6.46–0.50	−1.41–0.66	–
3	−0.99–5.60	−0.59–13.18	−5.53–0.57	−4.16–0.24	−0.46–1.34	−2.34–1.16	−4.07–0	−6.54–0.40
4	−9.93–3.76	−2.98–6.66	−11.57–1.16	−3.34–8.62	−29.58–−0.21	−0.42–8.76	−5.26–3.60	−0.19–8.20
5	−0.96–3.98	−7.05–0.58	−6.98–4.39	−1.88–5.11	0–2.80	−0.29–2.28	−2.19–3.04	0–11.29
6	0–2.67	−5.06–8.04	−1.49–0.89	−0.77–3.20	−5.05–0	−0.08–2.35	−8.0–1.80	−0.96–17.46
7	0–7.26	−0.04–8.36	0–3.70	−2.25–3.71	−0.84–1.02	−3.27–0	−2.25–0	−4.51–1.29

Source: Reprinted from *International Journal of Impact Engineering*, 38(8–9), Shim, V. P. W., et al., A technique to prescribe a vertical acceleration-time load on the human head–neck complex, 707–714, Copyright 2011, with permission from Elsevier.

Note: Positive values denote forward flexion, while negative values denote rearward movement.

chest and pelvis, as well as forces at cervical and lumbar spine, were recorded. For the neck force signals, the initial spikes (Peak 1) occurred at about 3 ms after impact and had very short duration of less than 2 ms, while the primary pulses, whose peaks were designated as Peak 2, occurred at 25–30 ms after impact for a period of about 17 ms (Figure 11.22). As it was generally accepted that neck injuries occur when the neck experiences compressive forces greater than the 1112 N (250 lb) limit for long duration (about 30 ms), the initial peaks (Peak 1) were neglected and not believed to result in neck injury. Neck injury in both ATDs was considered unlikely to occur as the primary pulse duration was approximately half of the duration to trigger serious neck injury at this load. The highest peaks for the primary pulses were 1481 N (or 333 lb) and 1704 N (or 383 lb) for the driver and the passenger, respectively.

All the above mentioned experimental studies have been performed to better understand the factors underlying flight-induced or vertical acceleration-induced neck injury and to improve seat design and flight or vehicular devices. Despite the bulk of invaluable information and insights provided, these experimental studies are limited in usefulness as human volunteers have low tolerance limits, cadavers lack the presence of active muscular control. Also, the vast physiological and structural differences between human and animal subjects or ATDs limit the extrapolation of animal or ATD test data to humans. There are some controversial issues regarding the appropriateness of using ATDs and injury criteria that are primarily designed and validated for lower loading rates in civilian motor

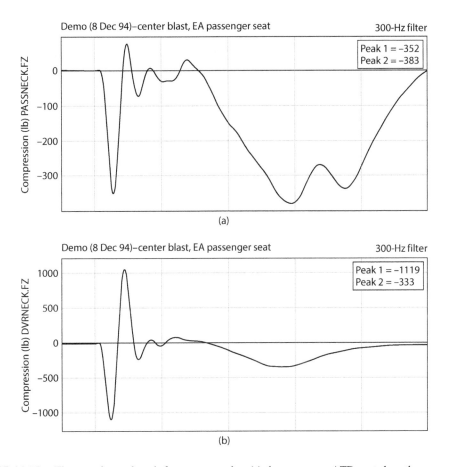

FIGURE 11.22   The experimental neck force measured at (a) the passenger ATD seated on the new energy-absorbing seat and (b) the driver ATD seated on the unmodified seat. (Reprinted from Alem, N. M., and Strawn, G. D., *Evaluation of an Energy Absorbing Truck Seat for Increased Protection from Landmine Blasts*, USAARL Report No. 96–06, United States Army Aeromedical Research Laboratory, 1996., Copyright 1996 by USAARL.)

vehicle accidents in UB blast injury prediction. Therefore, a gap may exist between these experimental studies and the actual situation in the event of pilot ejection or UB blast on armoured vehicles. Thus, numerical modelling can help to close the gap of information by enabling the biomechanical response of the human cervical spine under dynamic conditions to be simulated. Complemented by relevant experimental studies, computational modelling is a reliable tool for the analytical prediction of the dynamic response of the neck and head.

11.4.2 NUMERICAL MODELLING

Numerical modelling, which was considered to be a cost-effective alternative to experimental methods, has long been a tool in investigating the head–neck dynamics and biomechanical responses of the head–neck complex. It can be divided roughly into three different groups, namely the lumped-mass-spring-damper modelling, multibody (MB) modelling and finite element (FE) modelling.

11.4.2.1 Lumped Mass-Spring-Damper Modelling

While numerical modelling of the human head commenced as early as 1960s, attempts to model human neck dynamics did not begin until the 1980s. These earlier mathematical or analytical neck models were represented by simplified geometries using a lumped mass-spring-damper technique or multi-rigid body technique. Bowman et al. (1984) attempted to model the neck in the early 80 s using a 3D lumped mass-spring-damper head–neck model to simulate the behaviour of human head–neck dynamic responses. Other head–neck models using the lumped mass-spring-damper concept were subsequently developed by Bosio and Bowman (1986) and Wismans et al. (1986). These models simply consisted of two rigid bodies, each representing the masses of the spine, and the head connected by a spherical pinned joint and a rigid linkage at the cranial-cervical and cervical-thoracic junctions, respectively. These basic lumped mass-spring-damper models were validated against Naval Biodynamics Laboratory (NBDL)'s experimental volunteer sled test data (Ewing et al. 1976, 1978).

11.4.2.2 Multibody Modelling

Around the same time, Williams and Belytschko (1983) developed a mathematic head–neck model using rigid body elements for the bony head and vertebrae, while the intervertebral discs, ligaments, facet joints, and muscles were represented by finite elements: beam, spring, pentahedral continuum, and muscle elements, respectively (Figure 11.23). The model was validated for frontal and lateral impact accelerations against Ewing et al.'s vehicular experimental data from volunteers (Ewing and Thomas 1972; Ewing et al. 1978). The effects of the stretch reflex response on the dynamics of the head and neck under moderate acceleration were shown, and it was found that the stretch reflex response increases the peak compressive axial force by 20%–50% compared to the passive muscle model. Since this time, similar head–neck multi-rigid body models have been developed (Deng and Goldsmith 1987; Tien and Huston 1987). These head–neck models using multi-rigid bodies to represent head and vertebrae were a closer representation of the human head–neck complex than earlier models. Similar to the lumped mass-spring-damper models, these multi-rigid bodies were connected by joints, spring and damper elements with the equivalent elastic and viscoelastic properties of the intervertebral discs, facet joints, ligaments and muscles.

De Jager et al. (1996) developed two multibody models of human head and neck using the Mathematic Dynamic Model (MADYMO) software. They developed a global model and a detailed model, the latter of which included the essential material properties for active muscle behaviour (Figure 11.24). The global model consisted of a rigid head and vertebrae connected through non-linear viscoelastic intervertebral joints representing the lumped behaviour of the intervertebral disc, ligaments, facet joints, and muscles, while the detailed model comprised the rigid head and vertebrae connected through linear viscoelastic discs, non-linear viscoelastic ligaments, frictionless

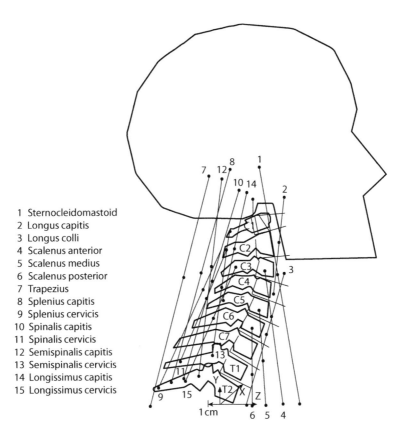

1 Sternocleidomastoid
2 Longus capitis
3 Longus colli
4 Scalenus anterior
5 Scalenus medius
6 Scalenus posterior
7 Trapezius
8 Splenius capitis
9 Splenius cervicis
10 Spinalis capitis
11 Spinalis cervicis
12 Semispinalis capitis
13 Semispinalis cervicis
14 Longissimus capitis
15 Longissimus cervicis

FIGURE 11.23 Deng and Goldsmith (Deng and Goldsmith 1987)'s multibody head–neck model. The cervical model connected to a simplified thoracolumbar model, with its major muscle and ligament elements. (Reprinted from *Journal of Biomechanics*, 20(5), Deng, Y. C., and Goldsmith, W., Response of a human head/neck/upper-torso replica to dynamic loading-II. Analytical/numerical model, 487–497., Copyright 1987, with permission from Elsevier.)

facet joints, and contractile muscles. The responses of both the models agreed reasonably well with those of the volunteer experiments.

De Jager et al. (1996)'s model was improved by van der Horst et al. (1997) with regards to the level of detail for the neck muscles. The model included all the muscles and ligaments surrounding the vertebrae, which were represented as connecting spring-dampers. The properties of these connecting spring-damper elements were either derived from experimental force-deflection curves or by validation of the entire model against experimental results. The model responses with maximum muscle activation to high severity frontal and lateral impacts agreed well with the experimental responses from volunteer sled tests, whereas a sub-maximum activation level or larger reflex delay provided better results for the low severity impacts. It was also found from the numerical simulations that the muscle contraction has significant influence on the head–neck dynamic responses.

Multibody or MB models have also been used intensively in the military applications. An example to illustrate this is an in-house full MB model of an ACES II seat and an aircraft pilot fitted with a HGU55/P aircrew helmet developed by Ho et al. (2004) to simulate the initial ejection process of the aircraft pilot (Figure 11.25). The anthropometry of the biofidelic model of the human pilot was benchmarked with established human databases such as GeBod or PeopleSize, which were derived from anthropometrical studies of Caucasian and Asian populations, respectively. Material properties of the seat, helmet and the vertebral joints were obtained either in-house or from proprietary sources, predominantly acquired through experiments involving cadavers. The head–neck pivot force during

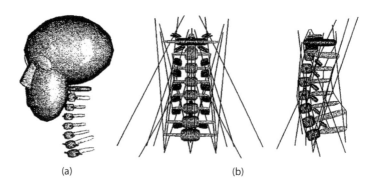

FIGURE 11.24 De Jager et al. (1996)'s multibody models of head and neck. (a) The global model and (b) the detailed model. (Reprinted from de Jager, M., et al., A global and a detailed mathematical model for head-neck dynamics, *Paper read at 40th Stapp Car Crash Conference, Albuquerque, NM*, 1996, Copyright 1996 by SAE International. With permission.)

the initial ejection process of the aircraft pilot was obtained from MB simulations and was validated against with experimental results of the 50th percentile Hybrid III ATD seated in a vertical accelerator apparatus (Ho et al. 2004). In addition, the MB model was used to evaluate the relative likelihood of cervical spinal injury when the pilot was equipped with three different helmet linings.

11.4.2.3 Finite Element Modelling

Numerical approaches such as the FE method make it possible to construct the anatomical geometry of human parts more realistically and to implement inhomogeneous human tissues, which can be derived from experimental specimen testing, more suitable for modelling the mechanical structures and responses of human anatomy. In the last few decades, advancements in technology (cheaper computing tools and faster codes) has led to the rapid development of more complex and anatomically realistic FE models of the cervical spinal column in the study of neck injury.

11.4.2.3.1 Early FE Neck Models

Similar to the head modelling described in the preceding chapter, early FE models of the human neck were generally simplified and considered to be anatomically unrealistic. One of the most comprehensive FE models of human neck developed in the 1980s was the 3D homomorphic model developed by Hosey and Liu (1981), which was attached to an FE model of the human head. While now considered geometrically simplified, this model consisted of the basic anatomic and material features of the major structural components such as the skull, brain, cerebrospinal fluid (CSF), the cervical spinal vertebrae and the intervertebral discs (Figure 11.26). Hosey and Liu (1981)'s model did not only include the basic anatomy of head, brain and neck but also factored the inertial and material properties of the head and neck. Despite being simplified and considered to be anatomically unrealistic, this earlier attempt in modelling the human head and neck did represent a first step in the numerical analysis of head and neck injury biomechanics.

Since the 1990s, when computing capability had sufficiently advanced, the development of more realistic and comprehensive 3D FE models of human components based on the specific geometry of the human was possible. In early 1990s, Kleinberger (1993) built a 3D FE model of human ligamentous cervical spine to study the mechanics of whiplash injuries in automotive crashes. The neck model was constructed using more than 13,000 solid hexahedral elements representing the cervical vertebrae, intervertebral discs, and ligaments, while the head was modelled as a rigid body to provide proper application of non-contact inertial loading. A series of simulations with various loading conditions such as axial compression and frontal flexion was performed using the model. Meanwhile, another FE neck model was developed by Dauvilliers et al. (1994) to study the dynamic responses of the human head and neck in frontal and lateral impacts. It comprised a rigid head and rigid vertebrae,

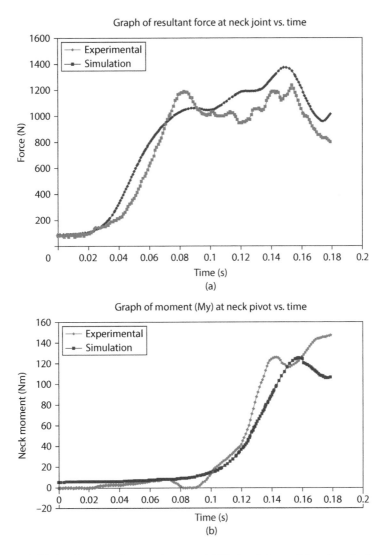

FIGURE 11.25 Validation between the experiment and MB simulations of the pilot ejection process using Nanyang Technological University (NTU)'s in-house MB model. (a) Neck resultant pivot force and (b) neck pivot moment in the flexural direction. (Reprinted from Ho, W. M., et al., Human modelling for injury assessment during ejection. *Paper read at 42nd SAFE Symposium*, Salt Lake City, UT, 2004, Copyright 2016 by SAFE Association. With permission.)

with intervertebral discs and spinal ligaments modelled as brick and spring elements. The model has been validated using the volunteer sled test data given by NBDL (Ewing et al. 1976, 1978). It should be noted that there was a restriction on the number of elements in the model as it was intended that the model be integrated to a whole human body FE model of an occupant for car crash simulations: each vertebra was modelled by 12 solid elements and each intervertebral disc was modelled by 8 solid elements. Ligaments, membranes and muscles were represented by spring-damper elements to keep calculation time low. This neck model was later used in a seated 50th percentile whole human body FE model presented by Lizee et al. (1998). This entire seated whole-body model with approximately 10,000 elements was then used in a vehicle environment to replicate more than 30 cadaver and volunteer experimental impactor, sled and belt compression tests with different impact energy levels and in different directions. The experimental results were then used in validating the FE model.

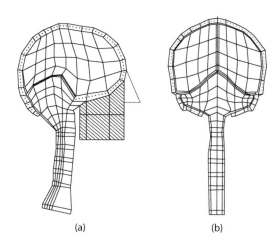

(a)   (b)

FIGURE 11.26 Finite element head model developed by Hosey and Liu in 1981 (Hosey and Liu 1981). (a) Sagittal cross section and (b) coronal cross section. (Reprinted with kind permission from Springer Science+Business Media: *Medical and Biological Engineering and Computing*, Finite-element models of the human head, 34(5), 1996, 375–381, Voo, L., et al., Copyright 2016.)

11.4.2.3.2   Wayne State University (WSU) FE Neck Model (1998)

In the late 1990s, after the development of the well-known Wayne State University Head Injury Model (WSUHIM), King and his team from Wayne State University (WSU) continued their efforts in extending the WSUHIM with an anatomically detailed FE model of the cervical column. The geometrical information for the neck was obtained from magnetic resonance imaging (MRI) scans of a 50th percentile male volunteer. The WSU FE neck model consists of the vertebrae from C1 through T1, including the intervertebral discs and the cervical ligaments. The vertebrae and intervertebral discs were modelled with 8-noded brick elements while all the cervical ligaments, namely the anterior and posterior longitudinal ligaments, facet JC ligaments, alar ligaments, transverse ligaments and anterior and posterior atlanto-occipital membranes were represented by non-linear bar elements or as tension-only membrane elements. The vertebrae and the intervertebral discs were assigned with linear elastic-plastic and linear viscoelastic material properties, respectively (Figure 11.27a).

The WSU FE neck model was then incorporated with the WSUHIM to study the mechanics of cervical spine when subjected to impact. The model has been validated with the experimental data from head-drop tests performed at Duke University and cadaver rear-end impact sled tests

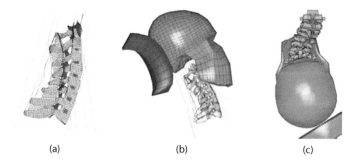

(a)   (b)   (c)

FIGURE 11.27 (a) The WSU FE neck model and its applications in simulating neck injuries seen in (b) a whiplash scenario in a frontal crash and (c) a rollover event. (Reprinted from *Advanced human modeling laboratory: Advanced virtual human for safety improvement*, Copyright 2016 by Albert King (Wayne State University, Available from: http://engineering.wayne.edu/pdf/humanmodelinglab.pdf. With permission.)

(Yang et al. 1998). Furthermore, the integrated WSU FE neck/WSUHIM has been amalgamated with the WSU torso model to predict head–neck kinematics and airbag pressure in a whiplash impact with a pre-deployed airbag, with the prediction comparing favourably to experimental data (Yang et al. 1998). The integrated WSU FE neck/WSUHIM model has been used extensively in automotive applications to study mechanisms and prevention of whiplash in frontal crashes and neck injury from rollover events. Different designs of the seat, seat belt and the roof interior were then evaluated and optimised during different crash scenarios (Figure 11.27b & c).

11.4.2.3.3   University of Cape Town (UCT) FE Neck Model (1999)

Jost and Nurick (2000) from University of Cape Town (UCT) reconstructed an FE model of human neck using the Patran-based pre-processor of Abaqus (Dassault Systemes SIMULIA, RI, USA), Abapre (a Patran-based pre-processor of Abaqus), to obtain the geometry of the cervical spine from physical and scanned human anatomical data. Each of the cervical vertebrae was meshed with 280 shell elements, while each of the intervertebral discs was modelled with 52 solid elements and 6 non-linear damper elements. All the relevant ligaments were modelled with spring and membrane elements, while only membrane elements were used to model the essential head–neck muscles (Figure 11.28). The UCT FE neck model had been validated against both the low G-level NBDL volunteer tests (Ewing et al. 1976, 1978) and high G-level LPB-APR cadaver tests (Bendjellal et al. 1987). It should be noted that the number of elements were limited as the UCT head–neck model was intended to be used as part of a whole-body FE model to simulate automotive crashes.

11.4.2.3.4   Nanyang Technological University (NTU) FE Neck Model (2004)

In 2004, another representative FE model of human head–neck was developed by Teo and his co-workers from Nanyang Technological University (NTU) of Singapore (Teo et al. 2004, 2006, 2007). Geometrical information of a 68-year-old male cadaveric head and a cervical spine specimen provided by Singapore General Hospital (SGH) were extracted using a FaroArm (Faro Technologies, Inc., Florida, USA), a 6 degree-of-freedom digitiser probe. The extracted surface information was then imported into Ansys (Ansys, Inc., Canonsburg, PA, USA) for 3D mesh generation. The intervertebral discs were modelled using the basic geometries derived from average values reported in literature (Gilad and Nissan 1986). The vertebral and intervertebral discs were meshed with 8-noded brick elements while the skull was meshed with 4-noded shell elements. The cervical ligamentous and muscular tissues were modelled with linkage or connector elements. The entire NTU FE head–neck model consisted of 27,712 elements and 31,749 nodes (Figure 11.29a). All the bony components

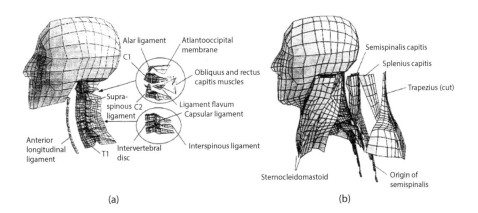

(a)   (b)

FIGURE 11.28   The UCT FE neck model. (a) The vertebrae, intervertebral discs, and ligaments; (b) the neck muscle. (Reprinted from Jost, R., and Nurick, G. N., *Int. J. Crashworth.*, 5(3), 259–270, 2000., Copyright 2000 by Taylor & Francis. With permission.)

+15° Inclination

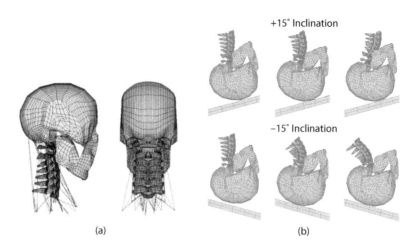

−15° Inclination

(a)   (b)

FIGURE 11.29   (a) The NTU FE head–neck model developed by Teo et al. 2004. (b) Neck responses during anterior and posterior impacts. (Reprinted from Teo, E. C., et al., *J. Musculoskel. Res.*, 8 (4), 155–165., Copyright 2004 by World Scientific Publishing Company. With permission.)

were modelled with linear elastic, homogeneous material properties, while the cervical ligamentous and muscular tissues were prescribed with non-linear constitutive relations derived from literature (Yoganandan et al. 2000). All the material properties were summarised and can be found in Teo et al. (2007).

Teo and his team (Teo et al. 2004) validated the NTU FE model of human head and neck against Nightingale et al. (1997)'s experimental data of free-drop tests of 18 cadaveric head and neck specimens at three inclination angles of +15° (anterior impact), 0° (vertex impact) and −15° (posterior impact) (Figure 11.29b). It should be noted that the muscular connector elements were excluded in the validation simulation to better replicate the experimental setting. The NTU FE head–neck model was also used to simulate pilot ejection (Teo et al. 2004). The predicted flexion angles of the cervical spine with and without muscles were also compared with the physiological limits provided by Panjabi et al. (2001), and the comparison was found to match reasonably well. The S-formed curvature of the neck during the acceleration onset stage caused the peak stress to be concentrated in the anterior portion of the intervertebral disc between C2 and C3 (upper cervical spine). After 80 ms, all the neck segments went into flexion, causing the peak stress to be at the anterior portion of the lower cervical spine (C6–C7) due to the restriction on C7 by the torso body mass. It was found that the activation of muscles effectively reduced the peak stress. In addition, the NTU FE head and neck model was also used to determine the kinematic responses of head–neck in automotive crash scenarios (Teo et al. 2006, 2007).

11.4.2.3.5   Simulated Injury Monitor (SIMon) FE Neck Model (2004)

The National Highway Traffic Safety Administration (NHTSA) developed the Simulated Injury Monitor (SIMon) FE neck model in 2004 with the intention of assessing the potential location and severity of neck injuries (Lee et al. 2004). The SIMon neck model was constructed based on 50th percentile male and consists of cervical vertebrae, intervertebral discs, cervical ligaments and muscles (Figure 11.30). The cervical vertebrae were modelled as rigid bodies with experimental mass and inertia properties and the intervertebral discs were modelled by 6-degrees-of-freedom joint elements with experimentally derived stiffnesses. The cervical ligaments and muscles were represented by 1D bar elements and discrete bar elements, respectively, with the active and passive muscle response modelled by a Hill-type algorithm (ESI 2002). Both the local segments and the entire column of the SIMon neck model have been validated against the tensile and bending responses of unembalmed

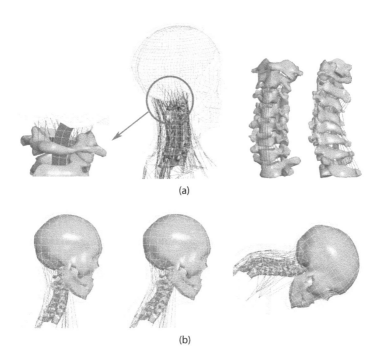

(a)

(b)

FIGURE 11.30 (a) The Simulated Injury Monitor (SIMon) FE neck model developed by the National Highway Traffic Safety Administration (NHTSA) and (b) its validation with the NBDL frontal sled test. (Reprinted from Lee, I.-H., et al., *Int. J. Automot. Technol.*, 5(1), 33–46, 2004, Copyright 2004 by KSAE. With permission.)

cadaveric cervical spine specimens using the tests performed by Duke University (Chancey et al. 2000; Nightingale et al. 2002). Frontal experimental sled tests by NBDL (Wismans et al. 1986; Thunnissen et al. 1995) as well as direct head pulling tests by Japan Automobile Research Institute (JARI) were then simulated using the SIMon neck model for validation of whiplash injury seen in motor vehicle accidents.

11.4.2.3.6   *University of Louis Pasteur (ULP) FE Neck Model (2004)*

Around the same time, Meyer et al. (2004a, 2004b) from the University of Louis Pasteur (ULP) developed a head–neck model consisting of a rigid body head and an FE neck (Figure 11.31a). Modal analysis had been performed using the ULP head–neck model to determine the natural frequencies and modal shapes (extension–flexion mode, lateral flexion mode axial rotation mode of the head, S-shaped anterior–posterior retraction mode, lateral retraction mode and vertical translation mode) of the head–neck system. In 2012, Willinger and Meyer (Meyer et al. 2012) coupled the FE neck model with the ULP FE head model developed in 1997 by Willinger and Kang (Kang et al. 1997) (Refer to Chapter 10). The ULP FE model of human head and neck has all the cervical muscles, and the face was improved using continuum brick elements, resulting in 25,596 elements in total. All the material properties of the head, the cervical vertebrae, discs and ligaments remained unchanged from the previous model, except for the cervical muscles, which were prescribed with viscoelastic material properties, while the active properties were represented by Hill spring elements. The complete ULP model of the human head and neck was then used to simulate NBDL's volunteer sled tests (frontal, lateral, oblique impact) (Ewing et al. 1976, 1978) and Ono et al. (1997)'s volunteer sled tests (rear impact) (Figure 11.31b). Additional validation in the frequency domain was performed and validation against experimental modal analysis reported by Bourdet and Willinger (2005) as well as Gunzel et al. (2009). Lastly, real-world accident reconstruction of a rear impact (Case Corolla 98 No. 29737) was performed using the ULP model to investigate the potential of predicting whiplash

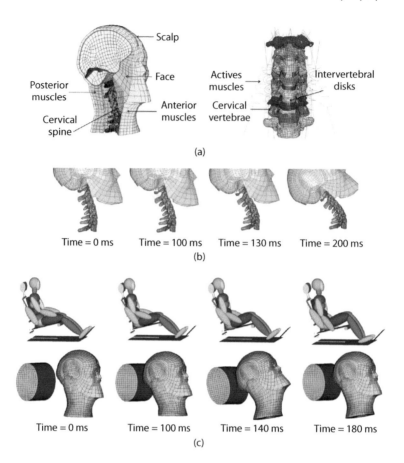

FIGURE 11.31 (a) The University of Louis Pasteur (ULP) FE neck model; (b) its validation with Ono et al. (1997)'s volunteer sled tests (rear impact) and (c) its application to real-world accident reconstruction of a rear impact (Case Corolla 98 No. 29737). (Modified from Meyer, F., et al., *Int. J. Crashworth.*, 18(1), 40–63, 2012, Copyright 2012 by Taylor & Francis. With permission.)

injuries (Figure 11.31c). It should be noted that the kinematics at T1 and headrest predicted by the MADYMO multibody model were used as boundary conditions for the FE accident reconstruction. Their simulation did not capture the 'S-shaped' movement of the cervical spine commonly seen in experimental cadaveric or volunteer tests before the cervical spine flexes and extends. Thus, it can be concluded that the shearing mechanism at different levels of the cervical spine may cause the injuries, which can be seen as the sum of the displacements of the cervical bodies along a horizontal direction (Meyer et al. 2012).

11.4.2.3.7 Kungliga Tekniska Högskolan (KTH) FE Neck Model (2007)

Another representative and comprehensive model of the human neck is the Kungliga Tekniska Högskolan (KTH) FE neck model, which was developed by Hedenstierna (2008). Unlike the discrete muscle models (discrete spring and damper elements) developed by previous researchers, all the cervical muscles as well as the thoracic muscles of the KTH neck model were reconstructed from digitised MRI images of a 50th percentile male subject by continuum 3D elements (linear brick and wedge elements) (Figure 11.32a–c). Each of the muscles was connected to a rigid endplate which was attached to the human skeleton. The KTH FE neck model had been coupled with the KTH FE head model developed by Kleiven in 2002 (Kleiven and Hardy 2002) (See Chapter 10). The material properties previously used in Kleiven and Hardy (2002) were adopted

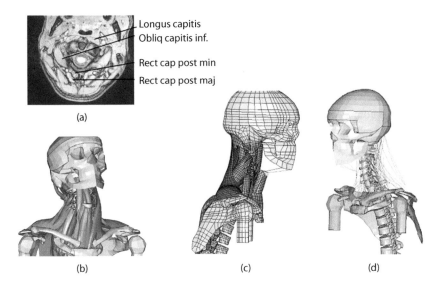

(a)

(b)   (c)   (d)

FIGURE 11.32   The Kungliga Tekniska Högskolan (KTH) FE head and neck model. (a) Segmentation from MRI; (b–c) the continuum muscle model and (d) the discrete muscle model. (Modified from Hedenstierna, S., *3D Finite Element Modeling of Cervical Musculature and Its Effect on Neck Injury Prevention*, Division of Neuronic Engineering, Royal Institute of Technology, Huddinge, Sweden, 2008, Copyright 2008 by Sofia Hedenstierna. With permission.)

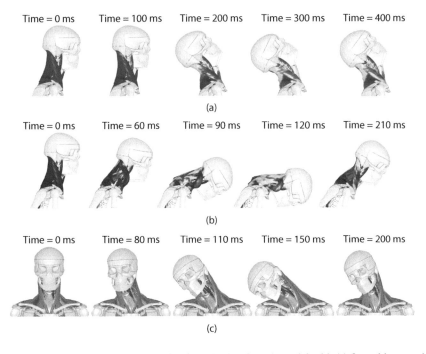

FIGURE 11.33   The validation simulations for the KTH head–neck model with (a) frontal impacts by Ewing et al. (1976); (b) rear impacts by Davidsson et al. (1999); and (c) lateral impacts by Ewing et al. (1976). (Modified from Hedenstierna, S., *3D Finite Element Modeling of Cervical Musculature and Its Effect on Neck Injury Prevention*, Division of Neuronic Engineering, Royal Institute of Technology, Huddinge, Sweden, 2008, Copyright 2008 by Sofia Hedenstierna.)

for all the of the head components, while the continuum muscles were modelled as a hyperelastic Odgen material, which is commonly used to described the non-linear stress–strain relationship of biological tissues.

The KTH FE model of human head and neck, which was fully meshed with continuum elements, was used to simulate the experimental frontal and rear impact sled tests conducted by Ewing et al. (1976) and Davidsson et al. (1999) (Figure 11.33a and b). The KTH model with all the continuum muscles was then compared with the same KTH model with muscles replaced by discrete Hill-type discrete elements (Hedenstierna et al. 2006). It was found that the differences in kinematic head responses between the discrete muscle model and continuum muscle model as well as between the passive model and active model was negligible in frontal impacts but significant in rear impacts, even though both were within reasonable range of the experimental corridors. Additional validation for the KTH continuum head–neck model was performed against Ewing et al. (1976)'s lateral impact data (Figure 11.33c).

11.5 CONCLUDING REMARKS AND FUTURE DIRECTIONS

In the past 40 years, particular emphasis has been placed on reducing the number and severity of cervical spine injuries sustained by automobile crash occupants and military aviators. However, more recent threats such as UB blast present new challenges for researchers. Despite the occupants experiencing a similar acceleration direction and load profile in both pilot ejection and UB blast events, their patterns of cervical spinal injury are different, suggesting that the two may not be mechanistically similar (Spurrier et al. 2015).

In this chapter, we highlighted the various research methods which have been used to understand cervical spine injury mechanisms, including spinal tissue material characterisation, cadaveric experiments and computational modelling. It is evident that advancing our understanding of cervical spine injury would require further work in all three interdependent areas. For example, in material characterisation, further research efforts should be made towards the advancement of more realistic material and failure models for cervical vertebrae, intervertebral discs, cervical ligaments and muscles under higher dynamic loading rate, primarily due to the high rate of loading commonly experienced in military scenarios, Similarly, post-mortem human specimen or cervical spine cadaveric tests need to be conducted using realistic loading and boundary conditions. Finally, all these experiments, including spine tissue characterisation, are critical to ensure that the computational models which are developed will be able to predict the cervical spine responses accurately. Future efforts should be focused on generating a seamless and efficient framework combining all three methods, enabling researchers to confidently predict the risk of injury for high-rate military events.

REFERENCES

Advanced human modelling laboratory: Advanced virtual human for safety improvement. Wayne State University Engineering, 2015 [cited 21st Jan 2016]. Available from: http://engineering.wayne.edu/pdf/human modelinglab.pdf.

Alem, N. M., and G. D. Strawn. 1996. *Evaluation of an energy absorbing truck seat for increased protection from landmine blasts.* United States Army Aeromedical Research Laboratory. USAARL Report No. 96–06.

Anatomy and function. 2003. University of Maryland Medical Center, 2003 [cited 5th Feb 2016]. Available from: http://umm.edu/programs/spine/health/guides/anatomy-and-function.

Anatomy & physiology: The vertical column. OpenStax CNX, 2014 [cited 15th April 2016]. Available from: http://cnx.org/contents/FPtK1zmh@6.27:4CMef3D9@4/The-Vertebral-Column.

Arnaiz, J. B. 1986. Helmet assembly aerodynamic interference effects on ACES II ejection seat operation. *SAFE Journal* 16:26–43.

Bekker, A., S. Kok, T. J. Cloete, and G. N. Nurick. 2014. Introducing objective power law rate dependence into a visco-elastic material model of bovine cortical bone. *International Journal of Impact Engineering* 66:28–36. doi: 10.1016/j.ijimpeng.2013.12.003.

Bendjellal, F., C. Tarriere, D. Gillet, P. Mack, and F. Guillon. 1987. Head and neck responses under high G-level lateral deceleration. In *Paper read at 37th Stapp Car Crash Conference*, San Antonio, TX.

Bogduk, N., and S. Mercer. 2000. Biomechanics of the cervical spine. I: Normal kinematics. *Clinical Biomechanics (Bristol, Avon)* 15 (9):633–48.

Bosio, A., and B. Bowman. 1986. Simulation of head-neck dynamic response in −Gx and +Gy. In *Paper read at 30th Stapp Car Crash Conference*, Warrendale, PA.

Bourdet, N., and R. Willinger. 2005. Human neck characterization under thoracic vibration—Inter-individual and gender influence. In *Paper read at International Conference on the Biomechanics of Impacts (IRCOBI)*, Prague, the Czech Republic.

Bowman, B., L. Schneider, L. Lustick, and W. Anderson. 1984. Simulation analysis of head and neck dynamic response. In *Paper read at 28th Stapp Car Crash Conference*, Chicago, IL.

Braakman, R., and L. Penning. 1971. Injuries of the cervical spine. Amsterdam, *Excerpta Medica*, 267 pages.

Bridwell, K. 2016. *Vertical column 2016* [cited 1st Apr 2016]. Available from: http://www.spineuniverse.com/anatomy/vertebral-column.

Burton, R. 1999. *Cervical spinal injury from repeated exposures to sustained acceleration.* NATO RTO Technical Report, Research and Technology Organization, RTO-TR-4. France: NORTH ATLANTIC TREATY ORGANIZATION Research and Technology Organization.

Cervical spine injury. University of Connecticut, 2016 [cited 2nd Feb 2016]. Available from: http://ksi.uconn.edu/emergency-conditions/cervical-spine-injury/.

Chancey, V. C., C. A. Van Ee, R. W. Nightingale, D. L. Camacho, and B. S. Myers. 2000. Understanding and minimizing error in cervical spine tensile testing. In *28th International Workshop on Injury Biomechanics Research*. San Antonio, TX.

Chazal, J., A. Tanguy, M. Bourges, G. Gaurel, G. Escande, M. Guillot, and G. Vanneuville. 1985. Biomechanical properties of spinal ligaments and a histological study of the supraspinal ligament in traction. *Journal of Biomechanics* 18 (3):167–76. doi: http://dx.doi.org/10.1016/0021-9290(85)90202-7.

Chen, Y. C., D. H. Smith, and D. F. Meaney. 2009. In-vitro approaches for studying blast-induced traumatic brain injury. *Journal of Neurotrauma* 26 (6):861–76. doi: 10.1089/neu.2008.0645.

Cramer, G.D., and S.A. Darby. 2013. *Clinical anatomy of the spine, spinal cord, and ANS.* Elsevier Health Sciences (Third Edition). Saint Louis (Location): Mosby (Publisher), 688 pages.

Dauvilliers, F., F. Bendjellal, M. Weiss, F. Lavaste, and C. Tarriere. 1994. Development of a finite element model of the neck. In *Paper read at 38th Stapp Car Crash Conference*, Fort Lauderdale, FL.

David, N. V., A. S. Sulian, and A. P. P. A. Majeed. 2011. Preliminary investigation of the impact resistance properties of a PASGT-type ballistic helmet. In *2011 International Symposium on Paper read at Humanities, Science & Engineering Research (SHUSER)*, 6–7 June 2011.

Davidsson, J., P. Lövsund, K. Ono, M. Y. Svensson, and S. Inami. 1999. A comparison between volunteer, BioRID P3 and Hybrid III performance in rear impacts. In *International IRCOBI Conference*, Sitges, Spain.

de Jager, M., A. Sauren, J. Thunnissen, and J. Wismans. 1996. A global and a detailed mathematical model for head-neck dynamics. In *Paper read at 40th Stapp Car Crash Conference*, Albuquerque, NM.

Deng, B., P. C. Begeman, K. H. Yang, S. Tashman, and A. I. King. 2000. Kinematics of human cadaver cervical spine during low speed rear-end impacts. *Stapp Car Crash J* 44:171–88.

Deng, Y. C., and W. Goldsmith. 1987. Response of a human head/neck/upper-torso replica to dynamic loading-II. Analytical/numerical model. *Journal of Biomechanics* 20 (5):487–97. doi: 10.1016/0021-9290(87)90249-1.

ESI. 2002. PAM-CRASH SAFE V2002 Solver Notes Manual. ESI Software.

Ewing, C. L., and D. J. Thomas. 1972. *Human head and neck reponse to impact acceleration.* NAMRL Monograph 21. Pensacola, FL: Naval Aerospace Medical Research Laboratory.

Ewing, C. L., D. J. Thomas, L. Lustick, W. H. Muzzy, G. Willems, and P. L. Majewski. 1976. The effect of duration, rate of onset and peak sled acceleration on the dynamic response of the human head and neck. In *Paper read at 20th Stapp Car Crash Conference*, San Diego, CA.

Ewing, C. L., D. J. Thomas, L. Lustik, G. C. Willems, III, W.H. Muzzy, E. B. Becker, and M. E. Jessop. 1978. *Dynamic response of human and primate head and neck to +Gy impact acceleration.* DOT HS-803-058. Pensacola, FL: Naval Aerospace Medical Research Laboratory.

Gaynor, F. E., H. R. Lissner, and L. M. Patrick. 1962. Acceleration induced strains in the intact vertebral column. *Journal of Applied Physiology* 17:405–9.

Giesen, E. B. W., M. Ding, M. Dalstra, and T. M. G. J. van Eijden. 2001. Mechanical properties of cancellous bone in the human mandibular condyle are anisotropic. *Journal of Biomechanics* 34 (6):799–803. doi: 10.1016/S0021-9290(01)00030-6.

Gilad, I., and M. Nissan. 1986. A study of vertebra and disc geometric relations of the human cervical and lumbar spine. *Spine (Phila Pa 1976)* 11 (2):154–7.

Gray, H. 1918. *Anatomy of the human body.* Bartleby.com, 2000. Philadelphia, PA: Lea & Febiger.

Gunzel, K., F. Meyer, N. Bourdet, and R. Willinger. 2009. Multi-directional modal analysis of the head–neck system and model evaluation. In *Paper read at International Conference on the Biomechanics of Impacts (IRCOBI)*, York, UK.

Hedenstierna, S. 2008. *3D Finite element modeling of cervical musculature and its effect on neck injury prevention.* Division of Neuronic Engineering, Royal Institute of Technology, Huddinge, Sweden.

Hedenstierna, S., P. Halldin, and K. Brolin. 2006. Development and evaluation of a continuum neck muscle model. In *6th European LS-DYNA Users' Conference.*

Ho, W. M., T. H. Ng, K. W. Tan, V. S. Lee, G. Leng, and E. C. Teo. 2004. Human modelling for injury assessment during ejection. In *Paper read at 42nd SAFE Symposium*, Salt Lake City, UT.

Hosey, R. R., and Y. K. Liu. 1981. A homeomorphic finite element model of the human head and neck. In *Finite elements in biomechanics*, edited by B. R. Simon, R. H. Gallagher, P. C. Johnson, and J. F. Gross, pp. 379–401. Wiley & Son.

Jost, R., and G. N. Nurick. 2000. Development of a finite element model of the human neck subjected to high g-level lateral deceleration. *International Journal of Crashworthiness* 5 (3):259–70. doi: 10.1533/cras.2000.0140.

Kang, H. S., R. Willinger, B. Diaw, and B. Chinn. 1997. Validation of a 3D anatomic human head model and replication of head impact in motorcycle accident by finite element modeling. In *Paper read at 41st Stapp Car Crash Conference*, Lake Buena Vista, FL.

Kaplan, S. J., W. C. Hayes, J. L. Stone, and G. S. Beaupre. 1985. Tensile strength of bovine trabecular bone. *Journal of Biomechanics* 18 (9):723–7.

Katsamanis, F., and D. D. Raftopoulos. 1990. Determination of mechanical properties of human femoral cortical bone by the Hopkinson bar stress technique. *Journal of Biomechanics* 23 (11):1173–84. doi: 10.1016/0021-9290(90)90010-Z.

Keller, T. S. 1994. Predicting the compressive mechanical behavior of bone. *Journal of Biomechanics* 27 (9):1159–68.

Kleinberger, M. 1993. Application of finite element techniques to the study of cervical spine mechanics. In *Paper read at 37th Stapp Car Crash Conference*, San Antonio, TX.

Kleiven, S., and W. N. Hardy. 2002. Correlation of an FE model of the human head with local brain motion–consequences for injury prediction. In *Paper read at 46th Stapp Car Crash Conference*, November, Ponte Vedra, FL.

Kopperdahl, D. L., A. D. Roberts, and T. M. Keaveny. 1999. Localized damage in vertebral bone is most detrimental in regions of high strain energy density. *Journal of Biomechanical Engineering* 121 (6):622–8.

Lee, I.-H., H.-Y. Choi, J.-H. Lee, and D.-C. Han. 2004. Development of finite element human neck model for vehicle safety simulation. *International Journal of Automotive Technology* 5 (1):33–46.

Lewis, J. L., and W. Goldsmith. 1975. The dynamic fracture and prefracture response of compact bone by split Hopkinson bar methods. *Journal of Biomechanics* 8 (1):27–40. doi: 10.1016/0021-9290(75)90040-8.

Literature review on biomechanics of injury causation. ForensisGroup, 09th May 2012 [cited 29th Jan 2016]. Available from: http://www.forensisgroup.com/literature-review-on-biomechanics-of-injury-causation-2/#sthash.DgYsXDfa.dpuf.

Liu, J. F., V. P. W. Shim, and P. V. S. Lee. 2013. Quasi-static compressive and tensile tests on cancellous bone in human cervical spine. In *Mechanics of Biological Systems and Materials, Volume 5: Proceedings of the 2012 Annual Conference on Experimental and Applied Mechanics*, edited by C. Barton Prorok, F. Barthelat, S. Chad Korach, J. K. Grande-Allen, E. Lipke, G. Lykofatitits, and P. Zavattieri, pp. 109–18. New York: Springer.

Lizee, E., S. Robin, E. Song, N. Bertholon, J. Y. L. Coz, B. Besnault, and F. Lavaste. 1998. Development of a 3D finite element model of the human body. In *Paper read at 42nd Stapp Car Crash Conference*, Tempe, AZ.

Mahoney, P. F., C. Buckenmaier, M. D. C. Buckenmaier, and Borden Institute. 2016. *Combat anesthesia: The first 24 hours*. Department of the Army.

Maiman, D. J., A. Sances, Jr., J. B. Myklebust, S. J. Larson, C. Houterman, M. Chilbert, and A. Z. El-Ghatit. 1983. Compression injuries of the cervical spine: A biomechanical analysis. *Neurosurgery* 13 (3): 254–60.

McCalden, R. W., J. A. McGeough, and C. M. Court-Brown. 1997. Age-related changes in the compressive strength of cancellous bone. The relative importance of changes in density and trabecular architecture. *Journal of Bone Joint Surgery America* 79 (3):421–7.

McElhaney, J. H., R. W. Nightingale, B. A. Winkelstein, V. C. Chancey, and B. S. Myers. 2002. Biomechanical aspects of cervical trauma. In *Accidental injury: Biomechanics and prevention*, edited by A. M. Nahum and J. W. Melvin, pp. 324–73. New York: Springer.

Meyer, F., N. Bourdet, K. Gunzel, and R. Willinger. 2012. Development and validation of a coupled head-neck FEM—Application to whiplash injury criteria investigation. *International Journal of Crashworthiness* 18 (1):40–63. doi: 10.1080/13588265.2012.732293.

Meyer, F., N. Bourdet, R. Willinger, F. Legall, and C. Deck. 2004a. Finite element modelling of the human head-neck: Modal analysis and validation in the frequency domain. *International Journal of Crashworthiness* 9 (5):535–45. doi: 10.1533/ijcr.2004.0309.

Meyer, F., R. Willinger, and F. Legall. 2004b. The importance of modal validation for biomechanical models, demonstrated by application to the cervical spine. *Finite Elements in Analysis and Design* 40 (13–14): 1835–55. doi: 10.1016/j.finel.2003.11.005.

Middleditch, A., and J. Oliver. 2005. *Functional anatomy of the spine*. Elsevier Butterworth-Heinemann, 359 pages.

Myklebust, J. B., F. Pintar, N. Yoganandan, J. F. Cusick, D. Maiman, T. J. Myers, and A. J. R. Sances. 1988. Tensile strength of spinal ligaments. *Spine* 13 (5):528–31.

Netter, F. H. 1997. *Atlas of human anatomy*. edited by A. F. Dalley, II. 2nd ed. East Hanover, NJ: Novartis.

Nightingale, R., V. Chancey, J. Luck, L. Tran, D. Ottaviano, and B. Myers. 2002. Accounting for frame and fixation compliance in cervical spine tensile testing. In *30th International Workshop on Injury Biomechanics Research*, Jacksonville, FL.

Nightingale, R. W., W. J. Richardson, and B. S. Myers. 1997. The effects of padded surfaces on the risk for cervical spine injury. *Spine (Phila Pa 1976)* 22 (20):2380–7.

Oliver, J., and A. Middleditch. 1991. *Functional anatomy of the spine*. Butterworth-Heinemann, 328 pages.

Ono, K., K. Kaneoka, A. Wittek, and J. Kajzer. 1997. Cervical injury mechanism based on the analysis of human cervical vertebral motion and head–neck–torso kinematics during low speed rear impacts. In *Paper read at 41st Stapp Car Crash Conference*, Lake Buena Vista, FL.

Pal, G. P., R. V. Routal, and S. K. Saggu. 2001. The orientation of the articular facets of the zygapophyseal joints at the cervical and upper thoracic region. *Journal of Anatomy* 198 (Pt 4):431–41. doi: 10.1046/j.1469-7580.2001.19840431.x.

Panjabi, M. M., J. J. Crisco, 3rd, C. Lydon, and J. Dvorak. 1998. The mechanical properties of human alar and transverse ligaments at slow and fast extension rates. *Clinical Biomechanics (Bristol, Avon)* 13 (2):112–20.

Panjabi, M. M., J. J. Crisco, A. Vasavada, T. Oda, J. Cholewicki, K. Nibu, and E. Shin. 2001. Mechanical properties of the human cervical spine as shown by three-dimensional load-displacement curves. *Spine (Phila Pa 1976)* 26 (24):2692–700.

Payne, A. R., and S. Patel. 2001. *Neck injury mechanisms and injury criteria*. Eurailsafe [cited on 1st Feb 2016]. Available from: http://www.eurailsafe.net/subsites/operas/HTML/Section3/Page3.3.2.htm.

Perry, C. E. 1994. *Vertical impact testing of two helmet-mounted night vision systems (Interim Report, Jun 1993–Dec 1993)*. United States ARMSTRONG Laboratory. Report No. AD-A293055.

Rhee, P., E. J. Kuncir, L. Johnson, C. Brown, G. Velmahos, M. Martin, D. Wang, et al. 2006. Cervical spine injury is highly dependent on the mechanism of injury following blunt and penetrating assault. *The Journal of Trauma* 61 (5):1166–70. doi: 10.1097/01.ta.0000188163.52226.97.

Schoenfeld, A. J., B. Sielski, K. P. Rivera, J. O. Bader, and M. B. Harris. 2012. Epidemiology of cervical spine fractures in the US military. *The Spine Journal* 12 (9):777–83. doi: 10.1016/j.spinee.2011.01.029.

Shender, B., G. Paskoff, G. Askew, R. Coughlan, and W. Isdahl. 2000. *Head and neck loads and moments developed during tactical and rotary wing +Gz stress.* Report No. AD-A385951. Naval Airi Warfare Center Aircraft Division.

Shim, V. P. W., P. V. S. Lee, J. F. Liu, and C. H. Cheong. 2011. A technique to prescribe a vertical acceleration-time load on the human head–neck complex. *International Journal of Impact Engineering* 38 (8–9):707–14. doi: 10.1016/j.ijimpeng.2011.04.001.

Shim, V. P. W., J. F. Liu, and V. S. Lee. 2006. A technique for dynamic tensile testing of human cervical spine ligaments. *Experimental Mechanics* 46 (1):77–89. doi: 10.1007/s11340-006-5865-2.

Shim, V. P. W., L. M. Yang, J. F. Liu, and V. S. Lee. 2005. Characterisation of the dynamic compressive mechanical properties of cancellous bone from the human cervical spine. *International Journal of Impact Engineering* 32 (1–4):525–40. doi: 10.1016/j.ijimpeng.2005.03.006.

Spurrier, E., J. A. G. Singleton, S. Masouros, I. Gibb, and J. Clasper. 2015. Blast injury in the spine: Dynamic response index is not an appropriate model for predicting injury. *Clinical Orthopaedics and Related Research®* 473 (9):2929–35. doi: 10.1007/s11999-015-4281-2.

Stemper, B. D., N. Yoganandan, F. A. Pintar, B. S. Shender, and G. R. Paskoff. 2009. Physical effects of ejection on the head-neck complex: Demonstration of a cadaver model. *Aviation, Space, and Environmental Medicine* 80:489–94.

Surange, P. N. 2010. *Anatomy of spine.* Slideshare, [cited on 28th Jan 2016]. Available from: http://www.slideshare.net/pankajnsurange/anatomy-of-spine.

Tennyson, R. C., R. Ewert, and V. Niranjan. 1972. Dynamic viscoelastic response of bone. *Experimental Mechanics* 12:502–7.

Teo, E. C., Q. H. Zhang, and R. C. Huang. 2007. Finite element analysis of head-neck kinematics during motor vehicle accidents: Analysis in multiple planes. *Medical Engineering Physics* 29 (1):54–60. doi: 10.1016/j.medengphy.2006.01.007.

Teo, E. C., Q. H. Zhang, and T. X. Qiu. 2006. Finite element analysis of head-neck kinematics under rear-end impact conditions. In *Paper read at Biomedical and Pharmaceutical Engineering, 2006. ICBPE 2006. International Conference on*, 11–14 Dec 2006.

Teo, E. C., Q. H. Zhang, K. W. Tan, and V. S. Lee. 2004. Effect of muscles activation on head-neck complex under simulated ejection. *Journal of Musculoskeletal Research* 8 (4):155–65.

The Back. Pocket Dentistry, 2015 [cited 29th Feb 2016]. Available from: http://pocketdentistry.com/2-the-back/.

Thunnissen, J., J. Wismans, C. L. Ewing, and D. J. Thomas. 1995. *Human volunteer head-neck response in frontal felxion: A new analysis.* In Paper read at 39th Stapp Car Crash Conference, San Diego, CA.

Tien, C. S., and R. L. Huston. 1987. Numerical advances in gross-motion simulations of head/neck dynamics. *Journal of Biomechanical Engineering* 109 (2):163–8. doi: 10.1115/1.3138660.

Torretti, J. A., and D. K. Sengupta. 2007. Cervical spine trauma. *Indian Journal of Orthopaedics* 41 (4):255–67. doi: 10.4103/0019-5413.36985.

van der Horst, M., J. Thunnissen, R. Happee, R. van Haaster, and J. Wismans. 1997. The influence of muscle activity on head-neck response during impact. In *Paper read at 41st Stapp Car Crash Conference*, Lake Buena Vista, FL.

Vasishta, V.G., L.J. Pinto, and S. Gambhir. 2003. A pilot study of bone scintigraphy in the evaluation of spinal injuries in aircraft accidents. *Indian Journal of Aerospace Medicine* 47:45–50.

Viano, D. C., A. I. King, J. W. Melvin, and K. Weber. 1989. Injury biomechanics research: An essential element in the prevention of trauma. *Journal of Biomechanics* 22 (5):403–17. doi: 10.1016/0021-9290(89)90201-7.

Vogel, H., and B. Dootz. 2007. Wounds and weapons. *European Journal of Radiology* 63 (2):151–66. doi: 10.1016/j.ejrad.2007.04.026.

Voo, L., S. Kumaresan, F. Pintar, N. Yoganandan, and A. Sances. 1996. Finite-element models of the human head. *Medical and Biological Engineering and Computing* 34 (5):375–81. doi: 10.1007/bf02520009.

WHO. 2013. *International perspectives on spinal cord injury.* [cited on from 2nd Feb 2016]. Available from: http://apps.who.int/iris/bitstream/10665/94190/1/9789241564663_eng.pdf.

Williams, J. L., and T. B. Belytschko. 1983. A three-dimensional model of the human cervical spine for impact simulation. *Journal of Biomechanical Engineering* 105 (4):321–31. doi: 10.1115/1.3138428.

Wismans, J., H. van Oorschot, and H. Woltring. 1986. Omni-directional human head-neck response. In *Paper read at 30th Stapp Car Crash Conference*, Warrendale, PA.

Yadla, S., J.K. Ratliff, and J. S. Harrop. 2008. Whiplash: Diagnosis, treatment, and associated injuries. *Current Reviews in Musculoskeletal Medicine* 1 (1):65–8. doi: 10.1007/s12178-007-9008-x.

Yang, K., F. Zhu, F. Luan, and L. Zhao. 1998. Development of a finite element model of the human neck. In *Paper read at 42nd Stapp Car Crash Conference*, Tempe, AZ.

Yoganandan, N., F.A. Pintar, A. Sances, Jr., J. Reinartz, and S.J. Larson. 1991. Strength and kinematic response of dynamic cervical spine injuries. *Spine (Phila Pa 1976)* 16 (10 Suppl):S511–7.

Yoganandan, N., F. Pintar, J. Butler, J. Reinartz, A. Sances, Jr., and S.J. Larson. 1989. Dynamic response of human cervical spine ligaments. *Spine (Phila Pa 1976)* 14 (10):1102–10.

Yoganandan, N., S. Kumaresan, and F.A. Pintar. 2000. Geometric and mechanical properties of human cervical spine ligaments. *Journal of Biomechanical Engineering* 122 (6):623–9. doi: 10.1115/1.1322034.

Yoganandan, N., B. D. Stemper, F. A. Pintar, D. J. Maiman, B. J. McEntire, and V. C. Chancey. 2013. Cervical spine injury biomechanics: Applications for under body blast loadings in military environments. *Clinical Biomechanics (Bristol, Avon)* 28 (6):602–9. doi: 10.1016/j.clinbiomech.2013.05.007.

12 The Thoracolumbar Spine and Pelvis

Injury Criteria for Traumatic Spinal and Pelvic Injury and Their Current Applications

Melanie Franklyn and Brian D. Stemper

CONTENTS

12.1   INTRODUCTION

Descriptive and epidemiological investigations are used in trauma-oriented clinical literature to understand the incidence and risk of injuries, delineate the influence of external environments for trauma causation, and examine the role of biomechanics on trauma mechanisms. In the civilian domain, computerised hospital-based databases such the US National Trauma Database (NTDB) and the Crash Outcomes and Data Evaluation System (CODES) are used (Hanrahan et al. 2009; Richards et al. 2006; Schinkel et al. 2006; Wang et al. 2009). The US National Automotive Sampling System (NASS) has been employed for over three decades to analyse traumatic injuries from motor vehicle crashes (Cobb et al. 2005; Evans 1994; Sivak and Schoettle 2010; Storvik et al. 2009). Data are gathered in a randomised manner, and national estimates are obtained using statistical methods to improve occupant safety. The Fatality Analysis Reporting System (FARS) is another database from the United States focused solely on fatalities (Briggs et al. 2005; Clark and Ahmad 2005; Tsai et al. 2008). From a regulatory perspective, NASS and FARS are commonly used to advance automotive crashworthiness standards. A newer database, Crash Injury Research and Engineering Network (CIREN), which was initiated in 1996, focuses on automotive trauma, with data selected from a group of consented patients from Level One Trauma Centres (Brown et al. 2006; Nirula and Pintar 2008; Smith et al. 2005; Pintar et al. 2008; Yoganandan et al. 2004). Because it has considerably more medical information than NASS and FARS, clinicians are increasingly using the CIREN database in association with NTDB and CODES to better understand injuries and treatments resulting from automotive and other trauma. While these databases have been invaluable in providing information so different injury mechanisms can be delineated in a civilian context, these mechanisms are not always directly applicable for military populations.

Military populations are relatively younger and healthier than civilian populations, wear protective clothing and equipment, and are exposed to more extreme environmental stresses. These environmental stresses may result in higher injury rates and more severe injuries, including but not limited to the musculoskeletal system. Because excessive mechanical loading is one of the primary causes for acute trauma and ensuing sequelae, any change in the pattern of insult alters load transmission to the musculoskeletal structures. Altered load paths can change the spectrum of trauma, and this can be defined using descriptive and epidemiological studies specific to military environments.

A number of trauma databases exist for military injuries sustained in combat. For example, the US Department of Defense Trauma Registry (DoDTR), formally the Joint Theater Trauma Registry (JTTR), was established in 2004 to record demographic, mechanistic, physiologic, diagnostic, therapeutic and outcome data from all injured patients presenting to military hospitals in Iraq and Afghanistan, including injured combatants from coalition partners such as the UK, Canada, Australia and Singapore. As of the end of 2014, the DoDTR comprised over 130,000 trauma records. The UK collects data as part of their own JTTR registry, which is comprised of data of all injured personnel, their injuries and the circumstances in which the injuries occurred (Spurrier et al. 2015). Nevertheless, these databases are less well established than civilian automotive trauma databases, and there is still a paucity of information in certain areas of military trauma.

In recent conflicts of war, injuries to the thoracolumbar spine and pelvis have become a major problem for combatants in military vehicles due to the increasing threat from underbelly (UB) blast events. Spinal injuries are most commonly observed at the thoracolumbar junction and in the lumbar spine, with injury types including Chance fractures, compression and burst fractures (Freedman et al. 2014; Lehman et al. 2012; Possley et al. 2012; Ragel et al. 2009). Pelvic fracture incidence and causation, however, has been much less well quantified, although it is known that complex pelvic fractures occur in UB blast events from vertical loading (Alvarez 2011; Bailey et al. 2013, 2015a, 2015b; Yoganandan et al. 2014), and these fractures are similar to those seen in civilian populations due to individuals falling or jumping from a height.

In the current chapter, spinal and pelvic anatomy is discussed in context with the clinical, biomechanical and military literature on spinal and pelvic injuries sustained in aircraft ejection, helicopter crash and UB blast events, with a focus on UB blast. These three mechanisms represent common modes in which vertebral column trauma have been reported, albeit not an exhaustive list. Lastly, injury criteria for UB blast are presented in addition to the current standards used by different nations and some methods of assessment.

12.2 SPINAL AND PELVIC ANATOMY AND INJURY TYPES IN UNDERBELLY BLAST EVENTS

12.2.1 ANATOMY OF THE SPINE

The human vertebral column is comprised of five distinct regions: the cervical, thoracic, lumbar sacrum and coccyx spines, where there are 24 articulating vertebrae and nine fused vertebrae to form a total of 33 vertebrae. Each of these regions is contoured to form either a lordotic (convex anteriorly) or a kyphotic (convex posteriorly) curve (Gray 2000; Moore et al. 2015), as shown in Figure 12.1. These curves are important in order to allow spine to support the weight of the body, as the

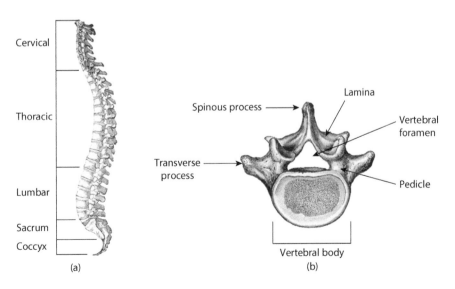

FIGURE 12.1 (a) Lateral view of the lumbar spine and (b) superior view of the fifth lumbar vertebra demonstrating the main anatomical features. (Modified from Gray, H., *Gray's Anatomy. Anatomy of the Human Body*, Lea and Febiger, Philadelphia, PA, 1918, 2000.)

curves increase the resistance to axial compression; thus, a greater weight can be supported than if the spine was straight.

The articulating vertebrae of the spine is comprised of seven cervical (C1–C7), twelve thoracic (T1–T12) and five lumbar (L1–L5) segments, while the fused bones include five sacral (S1–S5) and four coccygeal vertebrae (C1–C4). The thoracic and lumbar components form the thoracolumbar region of the spine, while the thoracolumbar junction (T10–L1) is the region where the spinal curvature changes from kyphotic to lordodic (Gray 2000; Moore et al. 2015).

The thoracolumbar junction is particularly susceptible to fracture as it is under significant biomechanical stress due to the articulation of the relatively rigid thoracic segment, through its connections to the ribcage and sternum, with the more mobile lumbar region (Rajasekaran et al. 2015). The vertebral body is the main weight-bearing region of the vertebrae, and its flat superior and inferior surfaces are designed to support vertical loading (Bogduk 2012).

12.2.2 SPINAL INJURIES AND FRACTURE CLASSIFICATION

Thoracolumbar spinal fractures can be classified according to a number of different systems, with four of the most common categorisations being:

1. Denis classification (also known as the three-column concept)
2. Magerl classifications
3. McAfee classification
4. Thoracolumbar injury classification and severity score (TLICS)

Perhaps the most widely used, Francis Denis (Denis 1983) developed an anatomical classification system of the spine by dividing the thoracolumbar vertebrae into three distinct regions when the vertebrae are viewed laterally: the anterior, middle and posterior regions (Figure 12.2). Called the three-column concept of the spine, it has now become widely used in clinical radiology to describe the severity of a fracture and its level of stability.

In the Denis system, the anterior column comprises the anterior half of the vertebral body and the anterior longitudinal ligament; the middle column consists of the posterior half of the vertebral body, the posterior longitudinal ligament and the posterior annulus fibrosis; and the posterior column includes facet joints and posterior spinal ligaments (Denis 1983; Smith et al. 2010).

Denis classified fractures to the posterior column, such as transverse process (TP) fractures, as minor, while fractures to the anterior and middle columns, such as compression, burst and Chance

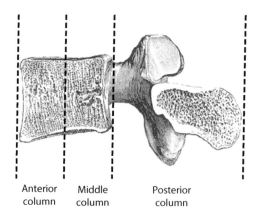

Anterior   Middle   Posterior
column   column   column

FIGURE 12.2   The Denis three-column concept of the spine as described by Denis 1983 (Modified from Gray, H., *Gray's Anatomy. Anatomy of the Human Body*, Lea and Febiger, Philadelphia, PA, 1918, 2000.)

fractures, as major. These fractures are discussed in further detail later. In the Denis system, minor fractures do not lead to acute instability, while major fractures are unstable, or have the potential to lead to instability, i.e. the potential to lead to neurological injury.

The Magerl classification was developed by Magerl and colleagues in order to classify spinal fractures based on healing and recovery (Magerl et al. 1994) and are categorised according to the mechanism of injury. Type A fractures are those due to axial loading (e.g. compression and burst fractures), Type B are distraction fractures and Type C are fractures due to rotation. Each of these three main fracture types also has various subgroup classifications. The main limitations of the classification system are its complexity (due to the number of subgroup classifications) and that the neurological status of the patient not being factored in the injury classification (Joaquim and Patel 2014).

The McAfee classification, developed in 1983 like the Denis system, also uses a three-column concept of the spine. McAfee proposed spinal injuries be categorised into six different types: wedge compression, stable burst, unstable burst, flexion–distraction, Chance and translational. One of the main differences between the Denis and McAfee classifications is that McAfee proposed that burst fractures can be either stable or unstable, with a stable burst fracture involving the anterior and middle columns but with no loss of integrity of the posterior column. This differs to Denis, who considered a burst fracture to be always unstable. (Gomleksiz et al. 2015; Joaquim and Patel 2014).

In the thoracolumbar injury classification and severity score (TLICS), the severity score assigned is based on three different categories: injury morphology (fracture type), the integrity of the posterior ligamentous complex, and neurological involvement, with higher scores given for greater injury severity. The total score is the sum of the three individual grades from each category, and the score can be used to guide surgeons and physicians on the management and treatment of the injury. A score < 4 means the spinal injury can be usually treated non-operatively, a score = 4 may or may not need operative management, while a score > 4 usually requires operative management.

In the following sections, the common fracture types relevant to combatants in a UB blast event are discussed. These fracture types include TP, Chance, compression and burst fractures (Blair et al. 2012; Ragel et al. 2009; Schonfield et al. 2012; Spurrier et al. 2015, 2016). Although the documented incidence of these different fracture types vary in the literature and are often limited by the small number of patients analysed, larger datasets such as the US and UK JTTR databases indicate that wedge compression and burst fractures are the most common fracture types sustained (e.g. Spurrier et al. 2015). The incidence of different fracture types is discussed further in Section 12.3.

12.2.2.1 Transverse Process Fractures

This fracture is uncommon, resulting from rotation or extreme sideways (lateral) bending, and usually not affecting stability. Injuries due to lateral bending are, for instance, observed after automotive side impact. While they are only AIS 2 fractures, they often occur in conjunction with more severe injuries and thus are an indication of other potential major trauma. Lumbar TP fractures often occur in conjunction with abdominal injuries (Patten et al. 2000) and, for this reason, are an indicator of potential underlying abdominal trauma (Miller et al. 2000).

12.2.2.2 Chance Fractures

Chance fractures were first described by the British radiologist George Quentin Chance in 1948 in relation to the spinous process fracture caused by load on lumbar spine when landing upright from a fall (Chance 1948). A similar type of fracture was later identified from automotive crashes, where it was described as a 'splitting apart' fracture of the vertebral body due to the lap belt acting as a fulcrum, splitting the vertebral body into two parts (Howland et al. 1965). These flexion–distraction-type fractures were soon known as '"seat-belt fractures," "seat-belt injuries" or "seat-belt syndrome"' as the two-point belt became the most common cause of this injury after these restraints were introduced in passenger vehicles: occupants in frontal crashes would flex forward, causing the posterior elements of the vertebrae to separate, particularly if the belt was not correctly worn over the pelvis.

Chance fractures were initially treated as a separate type of fracture; however, they later became synonymous with the flexion–distraction fractures, although they are still sometimes treated as separate entities, for example, in the McAfee classification of spinal injuries. A Chance fracture can be currently considered to be a flexion–distraction fracture of the spine due to rotation about a fulcrum located within the vertebral body, so the anterior component of the vertebral body fails under compression and the posterior elements of the vertebrae and the posterior vertebral body fail under tension. It is classified as 'multiple fractures of the same vertebrae' in AIS and has an injury severity of AIS 2 (AIS 2008).

The introduction of the sash component of seatbelts and three-point belts (i.e. a lap-sash belt as one unit) resulted in Chance fractures now only occurring infrequently in automotive crashes. In civilian settings, most of these fractures now result from falls or crush-type injuries where the thorax is acutely hyperflexed. However, Chance fractures are now being sustained by combatants in military vehicles subjected to improvised explosive device (IED) events (Ragel et al. 2009).

Ragel et al. (2009) proposed three different mechanisms of Chance fracture in UB blast events. These scenarios include the body hyperflexing around the lumbar spine while being subjected to a vertical load, either in forward flexion or due to the lower body submarining under the belt, with the combinations including with or without body armour being worn. Although the mechanisms of Chance fracture in frontal automotive crashes and UB blast events are clearly different, the exact mechanisms of Chance fractures in UB blast events are still unknown.

12.2.2.3 Compression Fractures

Vertebral compression fractures can be classified into numerous categories based on morphology as they can occur in situations other than trauma, e.g. osteoporosis. In the trauma scenario, the most common type of compression fracture is the wedge fracture (Figure 12.3), which results from high vertical or axial loads and may or may not involve a degree of flexion. A wedge fracture is characterised by a loss a vertebral height in the anterior column and does not involve the posterior vertebral body (Joaquim and Patel 2014). It is generally considered to be a stable fracture (Denis 1983).

Compression forces are mainly absorbed by the vertebral body as the nucleus pulposus of the intervertebral disc is essentially incompressible, being a liquid, while the fibrous annulus does not bulge significantly (Roaf 1960). Thus, these forces cause the vertebral body to collapse, with the severity of the injury determined by the amount of compression. A compression fracture with $\leq 20\%$ loss of vertical height is considered to be a minor compression fracture and is classified as an AIS 2 injury,

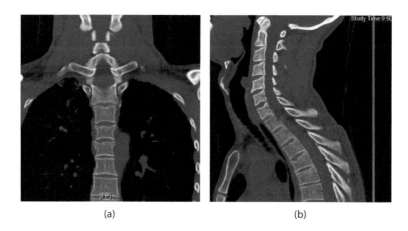

(a)   (b)

FIGURE 12.3 A 34-year-old male who fell from a motorbike at low speed, sustaining T4 and T5 superior endplate fractures/wedge compression (a) anterior and (b) lateral views. (Courtesy of Royal North Shore Trauma Service, NSW, Australia.)

whereas a compression fracture with > 20% loss of vertical height is a major compression fracture and coded as an AIS 3 injury (AIS 2008).

12.2.2.4 Burst Fractures

A burst fracture is a type of compression fracture where the vertebral body is severely compressed from a high magnitude axial load caused by the intervertebral disk being forced into the inferior aspect of the vertebral body above. This results in a comminuted fracture of the vertebral body where fragments of the fracture may be retropulsed into the spinal canal, potentially compromising the spinal cord (Figure 12.4).

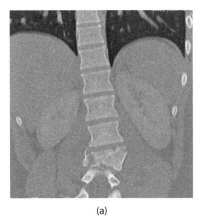 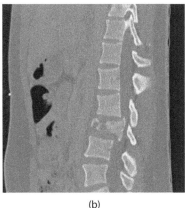

(a)   (b)

FIGURE 12.4 Burst fracture. An 18-year-old male fell 15 m off a cliff sustaining an L3 Burst fracture with neurological deficit (a) anterior and (b) lateral views. (Courtesy of Royal North Shore Trauma Service, NSW, Australia.)

Generally considered to be unstable (according to the Denis classification system), burst fractures can involve associated injury to the posterior ligamentous structures, especially if there is a combination of axial loading and flexion at the time of injury. Indications of posterior ligamentous disruption include a >50% loss of anterior vertebral body height, interspinous process widening and >20% kyphosis (Alpantaki et al. 2010; Gomleksiz et al. 2015).

12.2.3 ANATOMY OF THE PELVIS

The pelvis is comprised of four bones which together form the pelvic ring: two pelvic bones (the pelvic girdle), the sacrum and the coccyx (Figure 12.5). The pelvic bones form part of the appendicular skeleton, which connects the spine to the lower limbs, while the sacrum and coccyx are part of the axial skeleton, which are the bones of the head and trunk.

The main functions of the pelvis are to support and transfer weight from the axial to the appendicular components of the skeleton, to provide attachment points for muscles and ligaments used while walking, and to protect the abdominal and pelvic contents (Moore et al. 2015). Some of the salient features of the female pelvis, when compared to that of the male, are its greater width, with more flared iliac wings and a larger more oval-shaped aperture as opposed to the more circular or heart-shaped inlet in the male. These features of the female pelvis enable the weight of the foetus to be supported during pregnancy and facilitate childbirth.

The sacrum supports the lumbar spine vertically, while laterally, it transmits forces from the vertebral column transversely to the lower limbs and vice versa. The sacroiliac (SI) joint, which lies between the articular surfaces of the sacrum and ilium bones, is covered by two different types of cartilage (hyaline cartilage and fibrocartilage) in addition to a series of strong ligaments which help maintain its stability (Bogduk 2012).

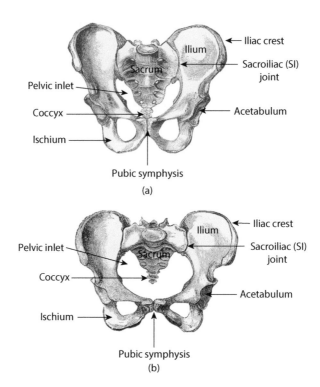

FIGURE 12.5 Anatomy of the (a) male and (b) female pelvis. (Modified from Gray, H., *Gray's Anatomy. Anatomy of the Human Body*, Lea and Febiger, Philadelphia, PA, 1918, 2000.)

The function of the SI joint includes enabling the pelvis some limited rotation when walking, and more importantly, as a stress-relieving joint at the point of maximum torsional stress when the pelvis is rotating about its transverse axis while walking (Bogduk 2012). Thus, the SI joint needs to be strong enough to withstand and transmit forces, but be compliant enough to allow some limited motion and therefore stop any potential fractures developing in the bone.

While sitting, the lumbosacral angle alters, so the superior aspect of the pelvis tilts posteriorly, thereby decreasing the lordodic curve of the lumbar spine. In military vehicles, while combatants should maintain an upright posture to reduce injury potential (discussed further later); this is not always possible due to factors such as soldier fatigue and the body armour weight.

12.2.4 Pelvic Injuries and Fracture Classification

Pelvic fractures are usually described using one of two classification systems: the Tile Classification System or the Young-Burgess System, where the former is based on the integrity of the pelvis while the latter is based on the fracture pattern and mechanism of injury. With each of these systems, fractures are classified according to three main categories; where there are also some further subcategories for some of these fracture types:

- Tile System: Stable, partially or rotationally unstable (vertically stable), rotationally and vertically unstable (totally unstable)
- Young-Burgess System: Anterior–posterior compression, lateral compression, vertical shear

Stable fractures do not disrupt the ligamentous structures of the pelvic ring, for example, an isolated fracture of the iliac wing (Tile et al. 2003). In a partially or vertically stable fracture, the ligamentous structures are intact or the ligaments and muscles of the pelvic floor are intact, the most common type of which is the open book type of fracture (Tile et al. 2003). Open book fractures cause disruption to the pubic symphysis and often occur in frontal automotive crashes due to anterior–posterior compression forces.

Fractures of the acetabulum are classified according to Letournel, who developed the concept of anterior and posterior columns in order to classify acetabular fractures morphologically (Letournel 1980).

Using the AIS classifications for pelvic injuries, which are based on the US Orthopaedic Trauma Association categorisations from the work of Pennal, Letournel, Tile and Judet, and some of the Young-Burgess classifications (Marsh et al. 2007), the pelvis can be considered as two structures: the pelvic ring and the acetabulum. The severity of a pelvic ring fracture is based on its stability, where it can be classified into three different categories: stable, partially or vertically stable, or totally unstable (AIS 2008). The SI joint is fundamental in the severity of pelvic injuries as the stability of the pelvis depends on an intact SI complex (Tile et al. 2003).

Vertical shear pelvic fractures, which are of particular concern in UB blast events, are totally unstable, or vertically unstable, and are characterised by multiple fractures and vertical disruption to the pelvic ring, complete disruption of the posterior arch and pelvic floor, and SI disruption. They can be a major threat to life due to haemorrhage and as such, they are classified as an AIS 4 (severe) injury, unless accompanied by extensive blood loss, in which case they are classified as an AIS 5 (critical) injury (AIS 2008). Pelvic shear fractures require surgery to stabilise the pelvis and can result in long-term impairment such as altered gait or difficulty in walking.

12.3 PATTERNS OF SPINAL AND PELVIC INJURIES IN MILITARY SCENARIOS

12.3.1 EARLY STUDIES ON PILOT EJECTION

Due to historical frequency, aviator ejection from military aircraft is one of the most comprehensively studied military injury scenarios. As early as 1910, people contemplated 'assisted escape' from aircraft. During WWII, the need for ejection seats became imperative, and a number of technologies were investigated. Prior to that time, an aviator's only option of escape from an incapacitated aircraft was to physically launch themselves out of the cockpit. When aircraft were developed that could fly higher and faster, this means of egress became more difficult. From early testing, it was found that a type of high-energy boost mechanism was required to remove the aviator safely from the cockpit. This led to the development of ballistic catapults, which provided a large amount of energy over a very short time period. Fortunately, ejection seat technology advanced in parallel with the biomechanical understanding of human tolerance. Therefore, whereas early ejection seats were designed simply to remove the aviator from the aircraft as quickly as possible, contemporary seat designs have focused on a higher level of safety for the aviator. The increased focus on safety is evident in recent fatality rates reported in the clinically-orientated ejection literature.

Earlier ejection reports, compiling data prior to the 1980s, commonly indicated fatality rates near or above 20% (Harrison 1979; Hearon et al. 1982; Smelsey 1970; Smiley 1964). However, ejections incorporating more recent seat designs have reported a far superior level of safety. A considerable decrease in mortality was evident from the mid-1970s (Table 12.1), about the time of the introduction of the MK-10 ejection seat. Accordingly, studies conducted around the mid-to-late 1970s and into the 1980s reported mortality rates less than 15% (see Sandstedt, Sturgeon, Moreno-Vasquez papers), although as low as 10% has been documented (see Rowe paper). More recent iterations have demonstrated even lower rates. In a study published in 1996, out of 199 United States Navy ejections reported, only four (2%) resulted in fatalities (Edwards 1996). Other studies reported similarly low mortality rates, albeit with smaller samples (see Werner, Williams, Osborne, Deroche papers). Although the ejection protocol has become more survivable for the aviator, an increased focus is being paid toward minimising severe injuries sustained during these events as 'yesterday's fatalities are today's injured survivors' (Werner 1999).

In terms of factors influencing injury outcomes during ejections, altitude at the time of ejection was shown to have a major effect on the likelihood of survival (Milanov 1996). Higher fatality rates were

TABLE 12.1

Rates of Ejection Fatalities and Vertebral Fractures in Published Literature

Primary Author	Published	Country	Study Years	No. of Ejections	% Fatalities	% Vertebral Fracture
Smiley	1964	Canada	1952–1961	281	24	70
Ewing	1966	United States	1959–1963	616	NR	10
Chubb	1965	United States	1960–1964	729	NR	4
Smelsey	1970	United States	1949–1968	3,528	17	NR
Shannon	1969	United States	1967–1968	101	1	NR
Delahaye	1975	France	1960–1974	258	18	NR
Harrison	1979	United States	1971–1977	583	19	14
Hearon	1982	United States	1967–1980	100	20	23
Rowe	1984	Canada	1972–1982	77	7	0
McCarthy	1988	United States	1973–1985	830	18	NR
Sandstedt	1989	Sweden	1967–1987	92	10	23
Sturgeon	1988	Canada	1975–1987	78	14	18
Visuri	1992	Finland	1958–1991	17	6	18
Newman	1995	Australia	1951–1992	84	8	35
Williams	1993	United States	1991–1992	4	0	0
Milanov	1996	Bulgaria	1953–1993	60	17	13
Edwards	1996	United States	1989–1993	199	2	4
Moreno-Vazquez	1999	Spain	1979–1995	48	14	NR
Collins	1997	United States	1981–1995	691	12	8
Osborne	1997	United States & Italy	1991–1995	18	0	33
Deroche	1999	France	1987–1996	63	6	27
Werner	1999	Germany	1981–1997	86	2	19
Lewis	2006	Great Britain	1973–2002	232	11	29
Nakamura	2007	Japan	1956–2004	140	23	6
Manen	2014	France	2000–2008	36	0	42
Pavlović	2014	Serbia	1990–2010	52	0	10

Note: 'NR' indicates data not reported in the manuscript.

associated with low altitude at ejection, as the ejection seat is not able to function correctly. Other studies have reported that aviator position at the time of ejection had an effect on injury rates (Chubb et al. 1965; Moreno et al. 1999; Sandstedt 1989; Visuri and Aho 1992). Chubb et al. reported that 19 of 133 occupants (14%) who ejected in a non-erect position sustained compression fractures (Chubb et al., 1965). Conversely, only 6 of 531 occupants (1%) who ejected in an erect posture sustained vertebral fracture. Prior training using an ejection tower was credited with assisting the occupants to attain an erect posture prior to ejection. Aviators are instructed to maintain an upright position with their shoulders against the seatback and head against the head restraint prior to ejection. Biomechanical studies have demonstrated that position of the spine prior to impact greatly affects spinal motions, loads and injury tolerance (Curry et al. 2015; Liu and Dai 1989; Maiman et al. 2002; Stemper et al. 2011, 2009) (N.B. posture and injury risk also discussed further under the section *DRI and Spinal Alignment* in Section 12.4.5.3). However, occupants may not always have time or be able to obtain the desired pre-ejection position, which may result in greater injury risk. Another factor identified to modulate injury risk was aircraft speed at the time of ejection (Milanov 1996). In general, higher speed ejections were more likely to result in major injuries. Milanov reported that the risk of unfavourable outcome following ejection increased for ejections that occurred with flight speeds of 600 to 800 km/h (Milanov 1996). Some studies have indicated that aviator anthropometry may be an influencing factor for increased injury risk, with heavier and taller aviators more likely to sustain spinal fracture (Edwards 1996). Other factors including aircraft control may also influence injury risk during ejection.

Aircrew ejections tend to be violent events, designed to vertically displace the occupant at a very high rate to avoid contacting the posterior structures of the aircraft. In earlier studies, it was found that restraint by only a lap belt was insufficient and injuries at acceleration levels as low as 3.5 to 4 times the acceleration due to gravity (g) were observed (Anonymous 1956; Marin 1955). Optimal body support was achieved through a combination of lap and shoulder restraints. Additionally, the proper use of armrests was beneficial in reducing the overall load borne by the lumbar spine. Variations in onset rate were later investigated as well as peak acceleration. An initial onset rate of 180 g/s resulted in an easily tolerable total velocity change of 17.8 m/s over 1.0 m of travel. However, increasing the rate to 500 g/s, even with a lower peak g, resulted in a jolt that was highly uncomfortable to the occupant (Watts et al. 1947). Modern ejection seat catapults operate between 150–220 g/s, although this number is also affected by the aviator's weight as well as the design of the system. For example, the United States Naval Aircrew Common Ejection Seat (NACES) SJU-17, manufactured by Martin-Baker Aircraft Co. Ltd. (Denham, Buckinghamshire, UK), comprises the bulk of the United States Navy ejection-capable aircraft. Qualified to an aviator weight range of 62–97 kg, the NACES provides a means of safe escape within an airspeed range of 0–600 knots, including zero speed/zero altitude. These factors indicate that loading rate, restraints and body posture are the likely factors that require biomechanical consideration for an improved understanding of ejection-related injuries.

Vertebral fractures are the most common severe injury sustained during ejection (Newman 1995; Osborne and Cook 1997; Werner 1999). However, the reported incidence of these injuries due to ejection has varied widely from study to study. For example, research focused on ejections occurring in the 1950s and 1960s reported vertebral fractures in 4% of 729 ejections in one paper (Chubb et al. 1965) and 70% of 281 ejections in another paper (Smiley 1964). Likewise, two studies focusing on injuries sustained during ejections in Operation Desert Storm reported vertebral fractures in 33% of 18 ejections (Osborne and Cook 1997) and no fractures in four ejections (Williams 1993). This difference in the rate of vertebral fractures is likely due, at least in part, to differences in inclusion criteria. For example, vertebral fractures and spinal column injuries can occur during the initial catapult phase, during parachute opening, or as the occupant strikes the ground during landing. In at least one study, there was differentiation between injuries sustained during the different phases (Chubb et al. 1965). This difference may be important as characteristics of the catapult phase acceleration can be altered to decrease injury risk, whereas parachute landing presents approximately similar injury risks for all ejection seats.

12.3.2 Injury Causation and Mechanisms Due to Vertical Acceleration

The rate of vertebral fracture due to military aircraft ejection has widely varied; however, the distribution of affected spinal levels has remained fairly consistent (Figure 12.6). Grouping data across eight studies accounting for 381 vertebral fractures during ejection, 2% of fractures affected the cervical spine, 78% of fractures affected the thoracic spine and 19% of fractures affected the lumbosacral spine (Milanov 1996; Nakamura 2007; Newman 1995; Osborne and Cook 1997; Sandstedt 1989; Sturgeon 1988; Werner 1999). This distribution is generally in agreement with similar research reporting level-by-level fracture data (Delahaye et al. 1975). From the data presented in Figure 12.6, it is evident that vertebral fractures primarily affect the thoracolumbar junction (T11–L1) and the mid-thoracic spine (T6–T9). Very few fractures occurred at the cranial and caudal levels. Specifically for ejections, 28% of vertebral fractures occurred at the T12 and L1 levels and 38% of fractures occurred between T6 and T9. In a review of clinical ejection data, this distribution was also identified (James 1991). The bimodal distribution of fractures may be the result of differing ejection systems, with some systems jettisoning the canopy prior to accelerating the aviator and other systems launching the aviator through the canopy (Yacavone et al. 1992). Vertical acceleration-deceleration of the seated occupant without head/shoulder contact has traditionally resulted in spinal fractures at the thoracolumbar junction (Hsu et al. 2003; Inamasu and Guiot 2007; Richter et al. 1996). However,

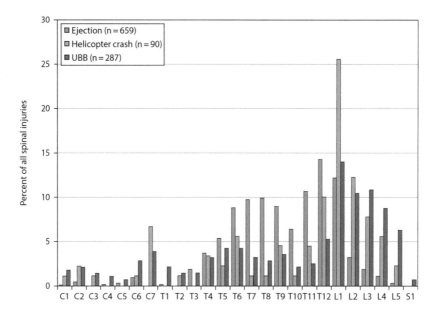

FIGURE 12.6 Distribution of spinal injuries from studies in the literature on military loading environments.

compressive loads applied through the head/shoulder are more likely to result in mid-thoracic spine injuries (Maiman 1986). Although specific injury descriptions were not provided in many studies, in general, compression and anterior wedge fractures were the most common lumbar spine injuries, with transverse/spinous process and dens fractures also occurring at lower rates (Chubb et al. 1965; Edwards 1996; Lewis 2006; Sandstedt 1989; Visuri and Aho 1992; Werner 1999). In the thoracic spine, retro-hyperflexion and hyperflexion fractures were most common (Kazarian et al. 1979). These mechanisms tend to support the theory on canopy-mediated injuries, wherein seat-bottom thrust resulting in pure axial loading would occur when the canopy was jettisoned. However, contact with the canopy would lead to bending-type injuries, such as the retro-hyperflexion and hyperflexion injuries which were reported to be most common by Kazarian et al. (1979).

During vertical accelerative loading, the pelvis has a tendency to rotate since the ischial tuberosities are eccentric to the axis of pelvis rotation. This pelvis rotation decreases the natural lumbar lordosis and places the thoracolumbar column in a pre-flexed state. The upper and middle thoracic spine has a tendency to resist flexion and transfer its inertial load compressively to the thoracolumbar junction. At this site, the stability gained by the rib cage is lost, and the spine becomes less resistant to flexion due to the increasingly sagittal-oriented facet joints. While the vertebrae of the thoracolumbar junction have characteristics relatively similar to the lumbar spine, they are structurally smaller and weaker than L3–L5 in axial loading. Theoretically, this setting will leave T11–T12 predisposed to anterior wedge fractures. Clinical literature supports this assertion by identifying that (anterior wedge) compression fractures are the most common injury to occur at this location along the spinal column. The effect of the loading pulse, however, may also predispose certain levels of the spine to injury, as shown in Figure 12.6 where different types of military events appear to change the fracture location.

12.3.3 HELICOPTER CRASHES

In addition to ejection, military personnel are at risk of sustaining lumbar spine injuries in other scenarios, including helicopter crashes. Although less frequent than ejection seat incidents, clinical studies have reported spinal injury patterns resulting from military helicopter crashes (Bruckart et al. 1993; Chaudhri et al. 1993; Italiano 1966; Scullion et al. 1987; Shanahan and Shanahan 1989a; Shand 1978). In a case report of a single helicopter crash, 11 of 18 occupants sustained injury

(Scullion et al. 1987). Of the 11 occupants sustaining injury, six occupants sustained a thoracolumbar vertebral fracture (33% of all occupants), with two occupants sustaining multiple vertebral fractures. Other studies indicate smaller vertebral fracture incidence rates of approximately 10% or lower (Italiano 1966; Shand 1978). Shanahan and Shanahan (1989a) reported 82 vertebral fractures in 611 injured occupants between 1975 and 1989. Figure 12.6 demonstrates the distribution of injured spinal levels for 90 vertebral fractures resulting from helicopter crashes published in the literature. In terms of regional distribution, 12% of helicopter-related fractures affected the cervical spine, 34% of fractures affected the thoracic spine, and 53% of fractures affected the lumbosacral spine. This distribution indicates a considerable caudal shift in affected levels from the ejection-related injury distribution reported above. Additionally, compared to aviator ejections, the distribution of vertebral fractures resulting from helicopter crashes is concentrated at a fewer number of spinal levels. Specifically, over 61% of all spinal fractures occurred between T12 and L4, and over 25% occurred at L1. This change in distribution is likely the effect of increased loading rate. In a study of 298 mishaps involving 303 aircraft, it was reported that the 95th percentile survivable vertical change in velocity for a helicopter crash was 11.2 m/s and that the 95th percentile survivable horizontal change in velocity was 25.5 m/s (Shanahan and Shanahan 1989b). According to vector mechanics, this would result in a total change in velocity of 27.9 m/s, which is over 50% more severe than the typical ejection pulse as described above. It is also important to note that just as ejection seat types perform differently, rotary wing aircraft and the pilot and troop seats they are installed in have different characteristics of intrinsic crashworthiness. Additionally, rotary wing aircraft crashes can occur in all three axes, although the primary acceleration is directed vertically as in aircraft ejection.

12.3.4 UNDERBELLY BLAST EVENTS

Use of IEDs during the contemporary conflicts in Iraq and Afghanistan and elsewhere has resulted in different spinal injury patterns than have traditionally been encountered in mounted military personnel, primarily due to the rise in vehicle UB blast events (Helgeson et al. 2011; Poopitaya and Kanchanaroek 2009; Ragel et al. 2009; Ramasamy et al. 2011). One example of this unique injury pattern was reported by Helgeson et al. (2011). In a retrospective review of orthopaedic surgery logs from Operation Iraqi Freedom (OIF) and Operation Enduring Freedom (OEF), the authors noted 23 lumbosacral dissociation (LSD) injuries. These injuries are rare in the civilian environment and involve traumatic high-grade spondylolisthesis, disruption of iliolumbar and SI ligaments, and multiple lumbar TP fractures that contribute to mechanical instability between the pelvis and spine. In the civilian environment, case reports indicate that these injuries have resulted from traumatic events, such as a woman falling off a motorcycle and being run over by an 18-wheel tractor-trailer, or a man falling 30.5 m from a radio tower onto a roof (Vresilovic et al. 2005).

Lumbar spine fractures have also been documented following IED events. Ragel et al. (2009) performed a retrospective review of soldiers admitted to military hospitals with orthopaedic spine fractures subsequent to IED attack. In their study, 12 soldiers with 16 thoracolumbar fractures that occurred over a 4.5-month period were identified. Compression fractures were most common (44%), followed by flexion–distraction fractures (37.5%) and burst fractures (19%). In a recent study by Spurrier et al. (2015), where spinal injury patterns in UB blast from the UK JTTR database were compared with historical data from ejection seat occupants, T7 to T12 was found to be the primary fracture region in ejection seat incidents, whereas the thoracolumbar (T12–L5; 53%) and mid-thoracic (T4–T9; 22%) spine were the main regions that sustained fractures in UB blast events. In the UB blast events, the most common types of fractures in the thoracolumbar region were wedge compression and burst fractures. In another recent study by the same research group and using the same database, the majority of fractures from UB blast events were identified in the thoracolumbar region, followed by the cervical spine (Spurrier et al. 2016). The thoracolumbar fractures were predominately wedge compression, whereas the cervical spine fractures were mainly compression–extension injuries.

As with fractures sustained from helicopter crashes compared to ejections, the distribution of lumbar spine fractures from UB blast events, accounting for 266 fractures from three studies (Poopitaya and Kanchanaroek 2009; Ragel et al. 2009; Spurrier et al. 2015) is more focused and occurs caudally when compared to helicopter crashes and ejections (Figure 12.7). In terms of regional distribution, 14% of fractures affected the cervical spine, 35% of fractures affected the thoracic spine and 51% of fractures affected the lumbosacral spine.

While there are other papers in the literature documenting spinal injuries sustained by military combatants, for some of these studies, it is difficult to quantify the incidence of these different fracture types in UB blast events, as often combined data for mounted and dismounted soldiers is reported. For example, while Lehman et al. (2012) found that the lower lumbar levels were the most common site of burst fractures between T12 to L5, the 32 soldiers whose medical files were reviewed in the study comprised both mounted and dismounted combatants.

Both similarities and differences exist in spinal injury patterns between ejection and UB blast events, which may be due to a number of reasons. Firstly, as mentioned earlier, it is known that the thoracolumbar junction of the spine is particularly susceptible to injury due to the articulation of the more rigid thoracic regions with the mobile lumbar region. Thus, this may account for the high number of fractures at T11/T12/L1 regions for ejection, helicopter crash and UB blast events. Secondly, a number of authors, for example, Lehman et al. (2012), have postulated that one of the reasons for the differences in spinal fracture location for ejection seats and UB blast events can be attributed to the rigid body armour worn by military vehicle occupants, as the body armour may transfer the region most susceptible to fractures from the thoracolumbar junction to the lower lumbar regions. Lastly, Ragel et al. (2009) proposed several mechanisms of Chance fractures in UB blast events, where the spine essentially acts as a fulcrum, either due to body armour as discussed above or from the restraint system.

While there are also differences in loading rate between pilot ejection and UB blast events, this has not been studied extensively enough to know if this is a contributing factor to the differences in spinal injury location from these two scenarios. While floor velocities of a medium-sized armoured vehicle can exceed 100 g or 12 m/s (Wang et al. 2001) in a UB blast event, floor accelerations exceeding 300 g over a 2.5 ms rise time have been recorded (Wang et al. 2001). Although measured at the floor and not on the seat base, accelerations recorded during their study far exceed the peak acceleration and rate of onset for ejections and helicopter crashes. The severe types of lumbosacral injuries sustained during extremely high loading rate events clearly require more biomechanical information with regards to the response of the human body and spine and the decreasing injury tolerance.

The effect of loading rate on burst fracture mechanisms, however, has been studied in a number of experimental studies using vertical loading on spinal components. For example, Tran et al. (1995) found that for higher loading rates (mean time to peak load = 20 msec), fractures with significant canal encroachment result, whereas at lower loading rates (mean time to peak load = 400 msec), there is minimal canal encroachment.

More recently, Finite Element (FE) models have been used to better understand the mechanisms of compression and burst fractures due to high-rate vertical loading (e.g. Langrana et al. 2002; Lee et al. 2000; Wilcox et al. 2004). The results of these studies have provided insights into how these fractures occur; however, further computational studies are needed in order to better understand the mechanisms of these fractures under different loading rates and magnitudes, particularly at higher rate loads representing UB blast events.

Pelvic injuries in UB blast events have not yet been studied extensively, despite the fact that they have been identified as a significant problem in these scenarios. For example, analysis of data from the Joint Trauma Analysis and Prevention of Injury in Combat (JTAPIC) database demonstrates that pelvic injuries are a significant problem, particularly in soldiers who are killed in action (Alvarez 2011). Consequently, there has been a number of recent cadaveric studies performed in order to better understand the injury mechanisms of pelvic fractures due to vertical loading (Bailey et al. 2013, 2015a, 2015b; Yoganandan et al. 2014). From this work, it is known that pelvic shear (Figure 12.7) and other types of complex pelvic fractures can result

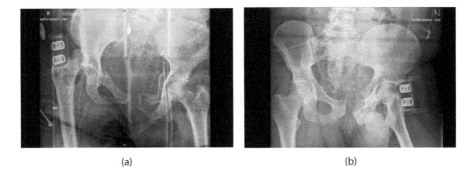

(a)   (b)

FIGURE 12.7 Vertical shear pelvic fracture. (Courtesy of Royal North Shore Trauma Service, NSW, Australia).

from vertical loads, similar to the types of pelvic injuries sustained in individuals falling or jumping from a height in civilian populations.

As in the case of spinal fractures, data for pelvic injuries sustained by military populations have also been reported as combined data for mounted and dismounted soldiers. For example, in the analysis by Alvarez (2011) mentioned above, 5% of soldiers wounded in action (WIA) and 46% of soldiers killed in action (KIA) sustained pelvic injuries; however, the data was not further divided into mounted and dismounted combatants. Thus, using this type of data, it is difficult to determine the exact injury mechanisms in UB blast events.

Prevention and clinical treatment of spinal and pelvic injuries depend upon understanding the types of injuries sustained, affected locations, occupant-related factors, injury mechanisms and biomechanical factors. Additional clinical data being collected and analysed during the current conflicts will provide a better understanding of injury types/distributions and assist in the development of appropriate safety technology. However, in the interim, it is challenging to design specific injury countermeasures for these injuries, as the exact causal mechanisms are largely unknown. In addition, other design considerations need to be factored, such as the need to limit the overall height of the vehicle for combat reasons, and a low centre of gravity requirement to prevent rollover. A better understanding of spinal and pelvic injury mechanisms in UB blast events, including the effects of factors such as body armour, restraints and seat design, will lead to improved injury criteria for the assessment of these injuries.

12.4 INJURY CRITERIA AND THRESHOLDS FOR THE LUMBAR SPINE

12.4.1 INTRODUCTION

Early injury tolerance data for the spine originated during WWII when the Germans recognised there was a need to better understand spinal tolerance in jet aircraft and aeroplane crashes. These initial experiments were followed by further research from the 1950s to the 1970s, when work was also conducted by organisations in other countries, such as the National Aeronautics and Space Administration (NASA) in the United States, that led to a number of injury models, injury risk curves, tolerance levels and injury assessment reference values which are still in use today. However, while this early pioneering work was important in establishing levels of tolerance for spinal and other injuries, since the 1960s and early 1970s, there has been little research on either improving the existing injury criteria, or developing new injury criteria, for the spine.

From the early 1970s, when General Motors developed the Hybrid II and then later the Hybrid III anthropomorphic test device (ATD) for automotive crash research, injury criteria have been used to relate recorded mechanical responses to human injury tolerance. Early ATD development focused on compliance testing vehicles fitted with passive restraints and establishing injury criteria for frontal collisions, while side and rear impact ATDs and corresponding injury criteria were established later. As vertical loading was not an important loading mechanism in automotive crashes, there was little need to use or mandate any injury criteria or injury thresholds for vertical loading. However, the more recent need to assess the emerging threats from UB blast events has resulted in aircraft vertical loading injury thresholds and criteria being applied to situations they were not designed for. This is discussed in more detail below.

12.4.2 EARLY TOLERANCE DATA ON THE SPINE

Early data on the tolerance of the spine to loading was collected by testing individual vertebrae and vertebral segments. These data provided an early understanding of the tolerance of the spine to different loading conditions and are mentioned briefly here as they are not directly incorporated into any current military standards.

Ruff summarised some of the early tests performed by Geertz during WWII, which involved finding the compressive strength of either individual vertebra or vertebral complexes of cadaveric specimens (Ruff 1950). Yorra later developed load-deflection curves for the spine in his research on lower back pain (Yorra 1956). The Ruff and Yorra data were then later used by Stech and Payne in the development of the DRI (Stech and Payne 1969).

Some of the other early testing was performed by Sonoda (1962) and Hutton (1978). Sonoda tested the strength of isolated cadaveric vertebral bodies from T8 to L5 for subjects between 22 and 79 years of age, while Hutton tested cadaveric vertebral bodies from L1 to L5 for subjects between 17 and 65 years of age. While this early work provided initial data on the tolerance of the spine to load, the limited number of tests performed and the lack of details on some of the experimental methods makes it difficult to use the data for current applications.

While several other injury models are available, they are not widely used and thus are only mentioned briefly here. For example, Hirsch (1964) developed a tolerance curve of seated, unrestrained males in an upright position subjected to shock motion of short duration. Designed for the protection of individuals in ships, Hirsch concluded that for a single degree of freedom system, the tolerable load was defined by peak acceleration for a pulse which was long compared to the natural period of the system, but defined by peak velocity change if the pulse was short compared to the natural frequency of the system. Hirsch derived acceleration thresholds of 15 to 30 g and peak velocity changes from 4 to 8 m/s (HFM-90). Another model was developed by Tremblay and colleagues in 1998, who proposed that the tolerance of the lumbar spine for pulse durations from 0 to 30 ms decreased linearly from 6,673 to 3,800 N, and for pulse durations >30 ms, the tolerance was 3,800 N (HFM-90).

12.4.3 THE RUFF TOLERANCE CURVE

The Ruff tolerance curve, developed in 1950 (Ruff 1950), was an initial attempt to factor in the time-dependent effect of spinal injuries (Figure 12.8). Including data for static tolerance from WWII experiments conducted by Geertz on cadaveric spinal segments between 19 and 46 years of age, dynamic tolerance data from catapult experiments, and German data on cardiovascular effects, Ruff developed a curve to factor in the rate dependency of human tolerance to injury in the seat-to-head direction, i.e. vertically upwards in the +Z direction (Ruff 1950). As there was no capacity to measure local effects using an instrumented device such as an ATD, the global acceleration from the seat or seatpan was used; however, correlations between the global acceleration and local effects were not analysed at the time.

Ruff concluded that for exposure periods of less than 5 ms, the structural tolerance of the spine was determined by the dynamic strength of the most susceptible vertebrae, while for exposure periods of 5 milliseconds to 1 second, the structural tolerance was determined by static compressive strength. For time durations greater than 1s, the tolerance limit was defined by circulatory disturbances (Henzel 1967; Ruff 1950), which can cause a rapid drop in blood pressure and heart rate, potentially leading to effects such as cardiac irregularities and loss of consciousness (Gauer 1950).

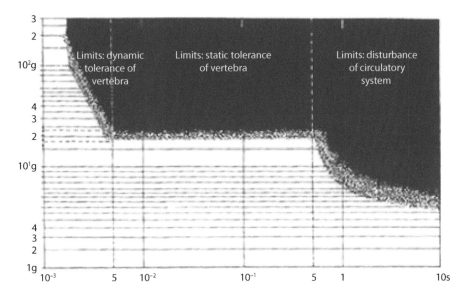

Limits: dynamic tolerance of vertebra

Limits: static tolerance of vertebra

Limits: disturbance of circulatory system

FIGURE 12.8   The Ruff tolerance curve. (Modified from Ruff, S., *German Aviation Medicine*, World War II, Vol. I, Chapter VI-C, 584–598, Department of the Air Force, 1950.)

12.4.4   THE EIBAND CURVE

The Eiband curve was developed by Martin Eiband of NASA in 1959 (Eiband 1959), who analysed the data available in the literature at the time on human and animal tolerance to rapidly applied acceleration. As in the case of the Ruff tolerance curve, much of this data was derived from experiments conducted in WWII and resulted in a series of whole-body tolerance curves for different directions when a trapezoidal-shaped pulse is applied from a seat or platform.

Eiband developed tolerance curves for acceleration in different directions: the Eiband curve for acceleration vertically upwards (Figure 12.9) is divided into three regions, where a tolerable acceleration was defined to be where the occupant is uninjured, a moderate injury was not life-threatening, and a severe injury was dangerous to life, where the injury descriptions were based on the Cornell Crash Injury Scale, a precursor to the AIS. From these data, Eiband identified some of the characteristics which affected human tolerance to acceleration, i.e. the type of restraint system, the direction and magnitude of the acceleration, the duration of the loading, and the rate at which the acceleration is applied.

In the Eiband curve, the boundary between the uninjured region and the area of moderate injury was obtained from catapult (ejection) seat experiments on human volunteers wearing restraints, where it was found that humans could endure 16 g up to 4 seconds. Data from Geetz and Ruff was used to establish the 20 g safe design limit for ejection seats (Figure 12.9b), where it was concluded that 20 g could be sustained between 0.005 and 0.5 seconds, or higher magnitudes could be endured if the duration was less than 0.005 seconds. The line dividing the moderate and severe injury regions was defined using experimental data from domestic pigs and chimpanzees, where pigs could sustain 110 g for 0.002 seconds without permanent injuries. Rocket-propelled sledtesting on chimpanzees lying supine resulted in 42 g for 0.048 second and 28 g for 0.14 second without permanent injuries.

One of the main limitations of the Eiband curve is that the tolerance to severe injuries has been defined by animal rather than human data, while the curve itself relates to whole-body tolerance rather than local effects or tolerance to specific injuries such as spinal fractures. Lastly, it is unknown if the curve is applicable for input pulses which are not trapezoidal.

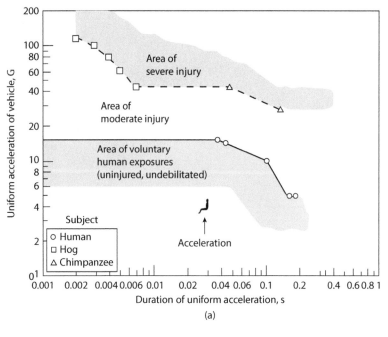

(a)

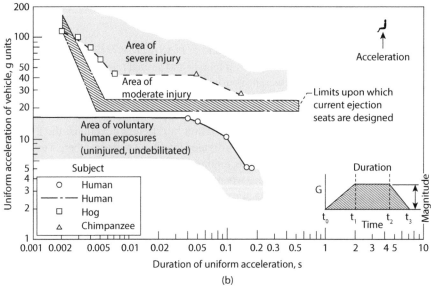

(b)

FIGURE 12.9 (a) Eiband curve and (b) Eiband curve demonstrating the limits for which ejection seats are designed. (Reproduced from Lawrence, C., et al., *The Use of a Vehicle Acceleration Exposure Limit Model and a Finite Element Crash Test Dummy Model to Evaluate the Risk of Injuries during Orion Crew Module Landings*, NASA TM 215198, National Aeronautics and Space Administration, OH, 2008.)

12.4.5 THE DYNAMIC RESPONSE INDEX

12.4.5.1 Development of the DRI

While the problem of assisted escape was identified as early as 1910, as described earlier, the requirement became critical during WWII, when pilots found it difficult to affect an emergency escape from the exposed cockpit or exit of high-speed aircraft due to issues such as wind pressure and centrifugal forces. The need for ejection seats, which apply a load so the pilot could clear the aircraft but not be subjected to high accelerations, was paramount. In order to examine the issue of human tolerance to short duration vertical accelerations, Latham of the Royal Air Force Institute of Aviation Medicine (active 1945–1994) first introduced the concept of a mechanical spring-mass model under vertical loading (Latham 1957) to represent the pilot and the seat.

This idea was later developed further by Payne (1962) and then Stech and Payne (1969), the latter authors of whom introduced the term 'dynamic response index', or DRI, to express the non-dimensional parameter of peak force in pounds by the body weight in pounds. Stech and Payne derived values for spring and damping constants for the spine using earlier cadaveric data by Ruff (1950)—discussed earlier in Sections 12.3.2 and 12.3.3.

Brinkley and Shaffer (1971) presented the analytical single degree of freedom mass-spring-damper model (Figure 12.10) where the equation of motion could be expressed as a second-order differential equation, which could then be solved to find the DRI, where the equation of motion is:

$$\frac{d^2z}{dt^2} = \frac{d^2\delta}{dt^2} + 2\zeta\omega_n\frac{d\delta}{dt} + \omega_n{}^2\delta$$

which can be also be represented as:

$$\ddot{z}(t) = \ddot{\delta} + 2\zeta\omega_n\dot{\delta} + \omega_n{}^2\delta$$

where $\ddot{z}(t)$ is the input acceleration, δ is the maximum deflection or displacement of the model (maximum spinal displacement, where $\delta > 0$ means compression), ω_n is the natural frequency and ζ is the damping ratio, where $\omega_n = \sqrt{(k/m)}$ and $\zeta = c/2m\omega_n$ and k is the spring constant (96,601 N/m), c the damping coefficient (818.1 ns/m) and m is the mass of the upper torso (34.52 kg). Solving the equation for δ results in the DRI:

$$DRI = \frac{\omega_n{}^2\delta_{max}}{g}$$

Thus the DRI is the ratio of the total compression in the spine during a dynamic event to the static compression in the spine due to gravity (3.5 mm):

$$DRI = \frac{\delta_{dyn}}{\delta_{static}} = \frac{\delta_{dyn}}{3.5}$$

where the DRI is a dimensionless parameter describing the ratio between the dynamic and static compression of the spine.

The properties of the model were derived by Stech and Payne (1969), where $\zeta = 0.224$ and $\omega_n = 52.9$ rad/s were calculated for pilots with a mean age of 27.9 years, which represented the age of the US Air Force flying population at that time, at an acceleration level of ~20g. For this age, they calculated a 50% probability of injury at a DRI value of 21.3.

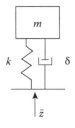

FIGURE 12.10   The Brinkley and Shaffer DRI model.

Assuming that the DRI values are normally distributed and the relationship between vertebral breaking strength and body weight is random. Brinkley and Shaffer (1971) calculated the relationship between the DRI and the probability of injury, which resulted in the laboratory graph (shown in Figure 12.11).

The DRI model was not validated using operational data until the late 1960s and early 1970s when Baumann et al. (1968) and Brinkley and Shaffer (1971) obtained data from pilot ejections. Using a pilot of 50th percentile weight with their personnel equipment, compression fractures (AIS 2 and AIS 3 injury severity) which could be attributed to ejection acceleration were analysed in catapult test programmes. This resulted in the 'operational' graph (also shown in Figure 12.11), which is the linear laboratory (cadaveric) graph translated to fit the ejection data, where the data was based on 361 non-fatal ejections from six different types of aircraft. It should be noted that while the pelvis is assumed to be the point of initiation for the laboratory data, the seat is the point of initiation for the operational graph. Consequently, the acceleration should be measured from the seat when using the operational graph and measured from the pelvis when using the laboratory graph.

Thus, the DRI relates to the deflection of the model or compression of the spine, and in essence, was a way to describe the probability of injury on an ejectee in terms of spinal compression due to an input acceleration. The DRI is also denoted as DRI_Z when referring to loading in the vertical direction (for which it was developed) in order to distinguish it from the DRI calculated in other directions, which were derived later.

The DRI operational graph was developed with a number of important assumptions: firstly, it is assumed that the pilot is upright (vertically aligned) on a rigid seat; secondly, the input acceleration is a trapezoidal pulse from the seatpan; thirdly, the ejected mass is the mass of the pilot and their personal equipment, where the torso mass is 34.52 kg and the equipment includes a helmet weighing

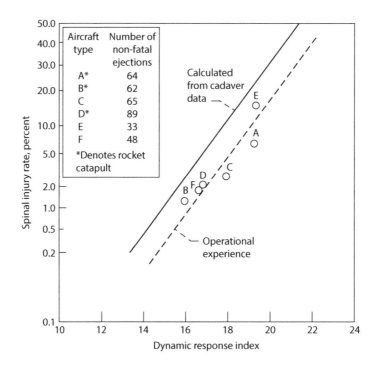

FIGURE 12.11 The Brinkley operational data with the DRI laboratory graph. (Reproduced from Lawrence, C., et al., *The Use of a Vehicle Acceleration Exposure Limit Model and a Finite Element Crash Test Dummy Model to Evaluate the Risk of Injuries during Orion Crew Module Landings*, NASA TM 215198, National Aeronautics and Space Administration, OH, 2008.)

about 2.3 kg, a survival vest and personal effects (discussed further later). Lastly, for the operational graph, it is assumed that the pilot is restrained, thereby ensuring the gap between the seat and the test subjects is a minimum so the occupant and seat moves as a lumped mass (Caldwell et al. 2012; Lawrence et al. 2008; Somers et al. 2013). Increasing the gap can allow higher contact forces, leading to higher local vertebral compression forces; thus, the risk of injury becomes greater.

Despite these limitations, the DRI model could reliably predict spinal injury potential for the situations it was designed for, and in a time period before the widespread use of computation methods, it was an analytical tool which successfully resulted in the reduction of spinal injuries in ejection scenarios. Consequently, after the DRI was verified with operational data, the model was included in the US Air Force specification used to develop escape systems and to quantify ejection catapults. In the mid-1970s, the model was also incorporated into air standards in Australia, New Zealand, Canada, the United Kingdom and the US Air Force and Navy (Lawrence et al. 2008).

Since this time, the majority of the work to develop the DRI further has involved exploring issues such as the effect of body armour mass on the DRI thresholds (discussed later). One recent analysis by Spurrier et al. (2015), which has been mentioned earlier (see Introduction), involved comparing injury patterns in historical data from ejection seats to UB blast data from the UK JTTR database. They concluded that the DRI was not applicable to blast scenarios as the injury patterns from blast and ejection seats were different, therefore, different prediction models were required for each scenario. However, as the authors correctly pointed out, their study was not designed to prove the applicability of the DRI model to blast events, but rather, to demonstrate the injury patterns in UB blast differed to those in ejection seat scenarios.

12.4.5.2 The DRI and the Risk of Spinal Injury

As calculated by Stech and Payne (1969) and later reported by Brinkley and Shaffer (1971), a 50% probability of spinal injury for a pilot of age 27.9 years had a corresponding DRI of 21.3. In the 1980s, DRI threshold values for low, medium and high risk of spinal injuries were established by Brinkley (Brinkley 1984; Lawrence et al. 2008) and then later adopted in the Advisory Group for Aerospace Research and Development (AGARD) in 1996 (Table 12.2).

These DRI values represent a deconditioned astronaut, i.e. one who has been in space for a period of time (Lawrence et al. 2008) based on injury probabilities of 0.5%, 5% and 50% from the operational graph (Brinkley et al. 1984). The low risk (0.5%) corresponds to probability level without injury in volunteer tests conducted at the US Air Force Aerospace Medical Research Laboratory, the medium injury risk (5%) corresponds to the current ejection seat designs, while the high risk (50%) was chosen as it represented the highest spinal injury rate which had been observed in US Air Force ejection seats, but with no associated spinal cord damage, and it also allowed for multiple exposures (Brinkley 1984).

TABLE 12.2

DRI Values and Risk of Spinal Injury Included in AGARD (1996)

DRI	Risk of Spinal Injury	Percent Risk
15.2	Low	0.5
18.0	Medium	5
22.8	High	50

12.4.5.3 The DRI and Spinal Alignment

In the DRI model, it is assumed that the spine is in alignment with the acceleration vector: if the spine is more than 5 degrees out of alignment relative to the load vector, the risk of injury increases significantly. This was first noticed for the F-4 ejection seat, where a DRI of 19 was predicted to result in a spinal injury rate of ~9% according to the operational graph (Brinkley and Shaffer 1971). However, the actual spinal injury rate was ~41% of the non-fatal ejections, where ~34% of the spinal injuries could be attributed to the ejection force (Figure 12.12). Brinkley and Shaffer (1971) attributed the difference in the predicted and actual rates of spinal injury to the seat alignment, as the pilot's vertebral column was not aligned with the catapult acceleration vector in the F-4 ejection seat. Thus, increasing the angle between the seat (spine) and the acceleration vector increases the probability of spinal injuries for the same DRI value.

Differences in alignment between the seat and acceleration vector were analysed in a number of studies performed around the same time and also in some later work. Mohr et al. (1969) investigated the seated postures of different-sized pilots in several types of egress seats, including the F-4 ejection seat, to examine the variation between spinal/seat alignment and the acceleration vector. Using radiography to image the spines, they found that the vertebral column was essentially parallel (within 5 degrees) of the catapult thrust axis for a number of different seat types; however, in the case of the F-4 seat, the difference in alignment was as high as 14 degrees forward of the catapult axis.

Bosee and Payne (1961) examined the association between load alignment and spinal injuries, finding that the optimal spinal alignment occurs when a normal seated posture is adopted, when the vertebral bodies are located on top of each other and there is an equal distribution of the compressive force over each vertebra. However, according to Bosee and Payne, any deviation from this posture increases the load concentration, which increases the potential for compression fracture. They postulated that this may occur due to several different factors, including differences in alignment between the seat and the acceleration vector, or forward flexion of the spine due to alterations in posture. For example, in the latter case, flexion can put increased pressure on the anterior part of the

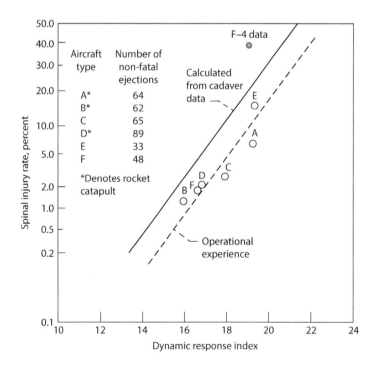

FIGURE 12.12   The F-4 ejection seat data on the Brinkley graph. (Modified from Lawrence, C., et al., *The Use of a Vehicle Acceleration Exposure Limit Model and a Finite Element Crash Test Dummy Model to Evaluate the Risk of Injuries during Orion Crew Module Landings*, NASA TM 215198, National Aeronautics and Space Administration, OH, 2008.)

vertebral body, making the pilot more susceptible to a wedge fracture. Spinal flexion may also be induced by submarining of the pelvis, which was demonstrated in the radiographic study by Mohr et al. (1969), highlighting the need for the pilot to maintain an optimal posture.

The relationship between seat angle and spinal loading was investigated by Perry et al. (1991), who conducted 82 experiments on 18 subjects, 16 males and 2 females, subjected to up to 10g in a drop-test machine. With an aim of establishing if differences in spinal injury potential were due to spinal alignment or seatback angle (as seatpan acceleration may alter for differently inclined seats), the authors measured various loads and accelerations on the subjects, comparing different seatback angles: +5 degrees (aft of vertical), −5 degrees and −10 degrees (forward of vertical). The results demonstrated that as the seatback angle was altered, the resultant seat loads measured were not statistically significantly different (although there was a trend for the seat load to increase as the seat was inclined forward). However, head acceleration data in conjunction with high-speed photographic motion analysis showed that as the seat angle became more negative, cervical spine flexion increased. Although the results of this work indicate that spinal position, rather than seatback angle, was the determining factor in spinal injury causation, the authors tested only a small number seatback angles, and it is not known if these results apply for seatback angles larger than 10 degrees.

In summary, the ability of the spinal column to withstand vertical acceleration can be significantly increased with optimal spinal column alignment, including the use of a restraint, and ensuring that the acceleration vector is aligned with the spinal column. The DRI can be used to predict spinal injuries only when the spine is aligned along the acceleration vector (or within 5 degrees of the acceleration vector), otherwise, the DRI value underestimates the true injury rate.

12.4.5.4 The DRI, Torso and Body Armour Mass, and Posture

The original laboratory DRI model incorporated an upper torso mass of 34.52 kg, and thus does not factor in body armour. The operational graph, however, was based on a pilot of 50th percentile weight with their personnel equipment: the pilots were wearing similar clothing to non-combat crews, which includes a survival vest, their own personal effects and a helmet weighing approximately 2.3 kg. The weight of the survival vest, which may also contain items such as a radio, firearm and rations, could weigh up to about 10 kg (J. Brinkley, pers. Comm., 2015).

Thus, the DRI model does not factor the mass of the body armour (laboratory graph), or the entire mass of the body armour (operational graph), worn by military vehicle combatants, which can often be 20 to 30 kg. However, live-fire UB blast testing as well as drop testing (described later) is often performed with the ATD wearing similar body armour to that of a mounted combatant. The addition of a torso mass will result in greater compression of the spine (Lammers and Dosquet 2006), consequently increasing spinal injury potential, which risks exceeding the critical spinal compression of 62 mm.

However, FE simulations, drop test and UB blast test data has shown that the addition of body armour *decreases* the DRI, although not necessarily by a large amount, whereas the additional body armour mass increases the lumbar compression (and therefore increases the injury potential), as would be expected (Dosquet 2010; Iluk 2013; Thyagarajan et al. 2014). For example, DST Group drop test data presented in Table 12.3 shows that similar values of delta-V (Tests 1 and 2, Tests 3 and 4) result in a decrease in DRI when body armour is used. This seemingly contradictory result is because the DRI equations do not factor in the effect of body armour. In live-fire UB blast tests (or drop tests) where the ATD is subjected to a vertically-applied input acceleration, the additional mass of the body armour increases the inertia of the ATD; thus, the ATD moves more slowly and achieves a lower peak acceleration. Consequently, the DRI, which is measured from the pelvic accelerometer, is lower. This may therefore lead to the erroneous conclusion that an additional torso mass results in a lower probability of spinal injuries when, in fact, the DRI model has not been modified to factor in the body armour. However, the model can be adapted to account for the greater torso mass by altering the spring and damping constants to factor in the new upper torso mass. The mass of the upper torso in the DRI model, m, is 34.52 kg; thus if a nominal body armour mass of 20 kg is

TABLE 12.3

The Effect of Body Armour on the DRI in Drop Testing

Test No.	Delta-V (m/s)	Body Armour	DRI
1	5.69	N (baseline)	22.50
2	5.71	Y	22.01
3	5.80	N (baseline)	21.97
4	5.83	Y	20.94

Source: Reproduced with permission from DST Group, Defence Australia.

added, the spring and damping coefficients remain constant while the new torso natural frequency will be:

$$\omega_n = \sqrt{\frac{k}{m}} = \sqrt{\frac{96601}{34.52 + 20}} = 42.1 \text{ rad/s}$$

and the damping ratio is:

$$\xi = \frac{c}{2\sqrt{km}} = \frac{818.1}{\sqrt{96,601 + (34.52 + 20)}} = 0.1816 \text{ Ns/m}$$

By altering the DRI equations as above, a modified (higher) DRI can be calculated. Cheng et al. (2010) conducted a mathematical analysis where they computed a modified DRI for different masses of body armour and pulse durations (the latter of which will be affected by the seat stroke; a large seat stroke will increase the pulse duration). They found that for all pulse durations, the DRI will increase by approximately 0.33 per kilogram of body armour weight added to the ATD. Thus, if 10 kg of equipment was added to the combatant, for example, the DRI would be predicted to increase by 3.3, which is consistent with the spinal injury potential increasing due to additional body armour mass. Figure 12.13 demonstrates the torso mass versus predicted DRI for a square pulse of 4.90 m/s and a pulse duration of 40 ms. Thus, using a baseline value for an ATD with no body armour in an UB blast event where the DRI is calculated to be ~16 (pass), the addition of 25 kg of body armour is predicted result in a true DRI of ~25 (fail).

Although the above study was based on a theoretical model, the model has shown to have validity in practise. For example, DST Group live-fire blast tests, where both the DRI and the modified DRI (i.e. modified the DRI model to account for the additional torso mass) were calculated, demonstrated a significant increase in DRI due to body amour mass. In these tests, ATD 1 had a DRI of 10.6 and a modified DRI of 13.5, while ATD 2 had a DRI of 13.1 and a modified DRI of 16.5. While the DRI was under the 17.7 threshold in these specific examples, it demonstrates the higher more realistic values of DRI with the addition of body armour, which in many cases will result in a DRI under the threshold value increasing to above the threshold value.

If this analysis is taken a step further, the DRI can be calculated and compared for cases where the body armour is, or is not, factored in the DRI equation. For example, if the DRI equals the 17.7 threshold value when no body armour is worn, the addition of 25 kg of body armour is predicted to result in a DRI of ~25.5; however, if DRI the equation is not adjusted for body armour, the DRI will be just over 16 (Figure 12.14). While this is a theoretical model and needs further validation, it highlights the influence of body armour on the DRI and the need to consider this effect when evaluating spinal injury potential.

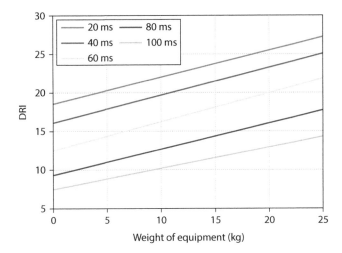

FIGURE 12.13 Graph showing the calculated effect of torso mass on the DRI for a square pulse of 4.9 m/s. The dotted line is the 17.7 DRI threshold. (Data obtained from Cheng, M., et al., Use of the dynamic response index as a criterion for spinal injury in practical applications, *Paper Presented at the Personal Armour System Symposium (PASS)*, Quebec City, September 13–17, 2010.)

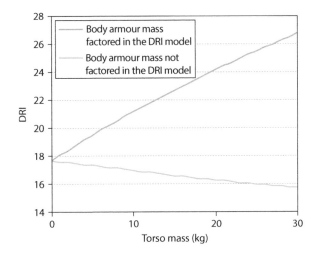

FIGURE 12.14 Effect of DRI when the body armour mass is, and is not, factored into the DRI model. (Analysis and data provided by Colin Nelson, Defence Materiel Organisation, Department of Defence, Australia, 2015 [personal communication]).

A similar effect can be seen if the torso mass is higher due to a large-sized occupant, although in this case, the situation is more complex, as the natural frequency and damping ratio of the system for the larger occupant is essentially unknown. In addition, a larger individual would also be expected to have bigger vertebrae and thus increased resistance to load. While the spinal injury potential in larger occupants has yet to be studied in great detail, modifying the DRI equations to account for the greater torso mass of a larger occupant is unlikely to accurately reflect the true spinal injury potential.

Further issues which compound the DRI calculation include the stiffness of the ATD spine; thus, differences between the DRI values measured from cadaveric tests (laboratory data), the DRI from pilots with their personal equipment (operational data), the FAA HIII (Federal Aviation

Administration Hybrid III) and the automotive HIII would be expected. In addition, variation in human and ATD anthropometry will have an effect on DRI and lumbar compression criteria as well as the spinal injury potential. A 50th percentile ATD is currently used to measure pelvic acceleration (for DRI) or lumbar compression; however, this size does not represent the size of the current 50th percentile solider; who is considerably larger. An alternative option is using a 95th automotive HIII; however, the effect of the additional mass on the DRI equations has to be considered, while the lumbar compression criterion was designed to be used with a straight-spine ATD, and the 95th automotive ATD has a curved spine. A second alternative is to use a 95th percentile aerospace ATD, which differs from the FAA HIII as the spine was derived from the pedestrian ATD; thus, there are differences in geometry, and the spine of the aerospace ATD is made from natural rubber instead of butyl rubber like the HII (Pellettiere et al. 2011). However, the aerospace series was recently deleted by Humanetics as it is seldom used in any present-day testing.

The DRI was developed for a pilot who is seated upright on a rigid seat and appropriately restrained. While the DRI measured from an ATD may not differ significantly between a straight and curved spine, as it is calculated from the pelvic acceleration, the spinal injury potential will differ. For a combatant seated in a military vehicle, the weight of the body armour can alter the posture, for example, causing him or her to lean forward, particularly if a restraint is not worn. This can alter the pattern of spinal injuries sustained and can also increase the likelihood of spinal injuries. As seen earlier for the F-4 ejection seat, when the acceleration vector is not aligned with the spine, the DRI underestimates the probability of spinal fracture. Thus, the DRI cannot be used to assess issues such as the effect of different postures on spinal injury outcome.

12.4.5.5 The DRI in Other Directions

The injury thresholds in the Z direction were expanded by Brinkley in the 1980s (Brinkley 1984; Brinkley et al. 1990; Lawrence et al. 2008) to give the dynamic response (DR) values for injury in different directions (Table 12.4 and Figure 12.15). These values were developed for the Orion spacecraft in order to define tolerance limits for astronauts subjected to rapid acceleration (Lawrence et al. 2008).

The values for the $-Z$ direction (vertically downwards) were based on laboratory experiments on downward ejection seats with volunteers and from downward ejections in operational aircraft, while the limits for the $+X$, $-X$ and $\pm Y$ directions are essentially whole-body injuries and were predominately

TABLE 12.4
The Limiting Values for the Dynamic Response in Different Directions

Axis	Direction	DR Limiting Value (g)		
		Low	Medium	High
+X	Forward	35	40	46
−X	Reverse	28	35	46
+−Y (with side restraints)	Lateral	15	20	30
+−Y (no side restraints)	Lateral	14	17	22
+Z	Vertically up	15.2	18.0	22.8
−Z	Vertically down	13.4	16.5	20.4

Source: Data from Lawrence, C., et al. *The Use of a Vehicle Acceleration Exposure Limit Model and a Finite Element Crash Test Dummy Model to Evaluate the Risk of Injuries during Orion Crew Module Landings,* NASA TM 215198, National Aeronautics and Space Administration, OH.

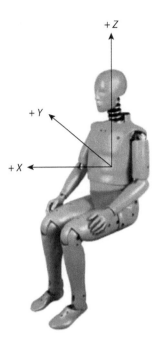

FIGURE 12.15 Coordinate axes for the dynamic response (DR) in different directions. (ATD image courtesy of Humanetics Innovative Solutions, Plymouth, MI.)

based on cardiovascular shock (Brinkley, personal communication, 2015; Somers et al. 2013). Specific injuries for the $-X$ axis for the high risk level include cardiovascular shock, retinal haemorrhage, and a minor fracture in the subject's neck hyoid bone (Somers et al. 2013). For the $\pm Y$ axis, injuries include adverse cardiovascular responses, a knee injury, bradycardia and syncope.

Thus, injurious effects in the other directions (i.e. the non- $+Z$ direction) focus heavily on cardiovascular responses, and as such, do not translate to the UB blast environment. In a military vehicle, typical injuries in the fore-aft (X) and lateral (Y) directions are similar to those in an automotive crash, i.e. impact injuries from hitting the interior components of the vehicle, or potential injuries from flying objects. In a military vehicle, while the accelerations in the X and Y directions are generally lower than the $+Z$ axis acceleration, high X and Y accelerations are unlikely to produce the cardiovascular responses observed in astronauts subjected to rapid acceleration in spacecraft. Thus, the use of the X and Y thresholds for UB blast testing is questionable.

The values of the DR for the other directions (X and Y) have been incorporated into the US Army Occupant Protection Handbook in the recommended criteria for landmine testing; however, the HFM panel does not recommend its use. This is discussed further in Section 12.4. The individual DR values are evaluated against the DR limits in Table 12.4 in order to determine the individual injury risk for each axis. The method was also expanded to examine the probability of injury due to combined x, y and z accelerations and angular velocities and thus find the general risk of injury β (ARGARD 1996). Known as the acceleration exposure limit method, it is calculated using the DR values for each axis and the DR limiting values from Table 12.4 from the equation:

$$\beta = \sqrt{\left(\frac{DR_x}{DR_{x(\lim)}}\right) + \left(\frac{DR_y}{DR_{y(\lim)}}\right) + \left(\frac{DR_z}{DR_{z(\lim)}}\right)}$$

where the value of β must be <1 to pass the criteria for a very low or low risk of injury. However, the DR values in this equation (other than the Z direction) were primarily developed for cardiovascular responses, as previously described. In a military vehicle, injuries resulting from lateral and frontal loading directions, when they occur, are impact injuries from hitting the interior components of the vehicle. Thus, while the acceleration exposure limit may have good validity for tolerances to acceleration in spacecraft, its applicability to UB blast events is debatable.

12.4.5.6 DRI Summary

For nearly 50 years, the DRI has been shown to have practical utility and in providing a standard for the assessment of and prevention of spinal injuries. Like the Head Injury Criterion (HIC), it has been heavily criticised; however, it was developed before the widespread use of computational models. Furthermore, it was designed for specific loading scenarios, and most of its limitations are not with the model itself, but with its application to situations it was not designed for. Significantly, validation of the DRI has involved hundreds of tests on live volunteers and ejectees, and it is unlikely that such a large number of tests on live subjects will be performed again in future. While research is currently being conducted under the US WIAMan programme to develop a new spinal injury criterion for UB blast events, it will take some years to establish and validate a new criterion. In the meantime, the DRI (and the lumbar compression) will continue to be used in military standards and live-fire testing.

12.4.6 The Lumbar Compression Load

The lumbar compression load originated from the General Aviation Safety Panel (GASP), established by the FAA in 1982 in order to improve civil aviation safety for small aeroplanes (Chandler 1985). CAMI (Civil Aeromedical Institute) found the use of the DRI model in civil aircraft to be problematic, as the flexible lightweight seat structure in small aeroplanes, as opposed to the rigid military aircraft seat, made it difficult to consistently measure a single seat acceleration (Chandler 1993). Furthermore, there were other issues such as the restraint system in civil aircraft differing from military restraints. Therefore, a method was required to determine spinal injury potential using an input acceleration from the ATD rather than the seatpan.

Consequently, CAMI conducted tests in order to correlate the DRI to the lumbar compression using a lumbar load cell at the base of the lumbar spine in the Hybrid II ATD (Chandler 1985, 1993). This resulted in a correlation graph (Figure 12.16) between the DRI and the lumbar load, which was based on 12 tests. A DRI of 19 was selected for the standard, as the ejection seat data, i.e. the operational graph from the DRI model, indicated that a DRI of 19 would correspond to ~9% chance of a 'detectable' injury (and therefore close to the nominal 10% risk of injury). Using the correlation graph, a DRI of 19 corresponded to a lumbar compression load of 1500 lbs (6.67 kN), which was then chosen for the standard.

The GASP panel submitted its recommendations to the FAA in 1984 and they were formally proposed by the FAA in 1986. This included the lumbar compression load limit, and the FAA later adopted these recommendations in 1988 (Chandler 1993; Gowdy et al. 1999). At the time, the Hybrid II, which has a straight spine, was used for regulation testing. However, development of the new more biofidelic Hybrid III ATD for automotive testing and regulation and the subsequent deletion of the Hybrid II from the NHTSA regulations in 1997 meant that an alternative ATD was required for the FAA, as the Hybrid III had a curved spine.

In order to address this issue, a lumbar spine modification to the Hybrid III was developed through a collaboration between the FAA, CAMI, the Applied Safety Technologies Corporation, and Robert A. Denton Inc. The aim of the spine modification was for the Hybrid III to produce similar responses to the Hybrid II, and consequently make the Chandler criterion still relevant. This resulted in the FAA Hybrid III, which has both HII and automotive HIII components (Gowdy et al. 1999), and has a more upright posture than the automotive HIII (Figure 12.17). The FAA HIII was subsequently

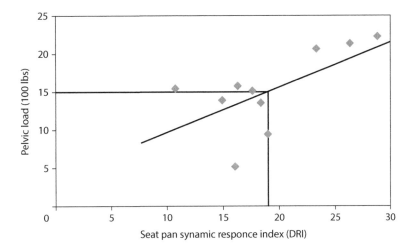

FIGURE 12.16 Correlation between the DRI and lumbar compression based on 12 tests. (Data obtained from Chandler, R.F., Data for the development of criteria for general aviation seat and restraint system performance, SAE Technical Paper 850851, *Presented at the General Aviation Aircraft Meeting and Exposition*, Wichita, KS, 1985.)

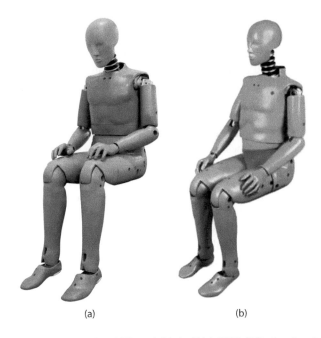

(a)   (b)

FIGURE 12.17 (a) The HIII Automotive ATD and (b) the FAA HIII ATD showing the differences in spinal alignment. (Courtesy of Humanetics Innovative Solutions, Plymouth, MI.)

adopted in the FAA standard, which is still currently used. This standard, FAA 25.562: *Emergency Landing Dynamic Conditions for the Certification of Airline Passenger Seats*, states that the maximum compression between the pelvis and lumbar spine should be <1500 lbs (6.67 kN) with an ATD weighing 77 kg, seated in the normal upright position (Code of Federal Regulations FAA 25.562, 2016).

The differences between the FAA and automotive HIIIs are not limited to the overall posture: there are also some other dissimilarities such as the design of the spine (including material properties), the spinal load cell orientation, the overall mass, the mass of different body components (e.g. the torso) and the seated height (i.e. the distance from the inferior aspect of the buttocks to the crown of the head). Additionally, the transfer of load from the seat to the ATD depends on which HIII is used (straight or curved spine) and the type of military seat (straight back or curved back), as this will affect the posture of the ATD during the test. These differences may have important implications for blast or drop testing, although this is not yet fully understood.

In a series drop tests, Polanco and Littell (2011) found that the lumbar compression was lower in the automotive HIII than the HII or the FAA HIII for several different input pulses. The authors hypothesised this may be due to different load paths from the structure of the ATD spine itself, or alternatively, the difference in load cell orientation between the ATDs: the lumbar spine load cell of the automotive ATD is inclined at an angle of 22 degrees from the vertical, whereas the HII and FAA HIII both have a vertically orientated load cell. The load cell orientation would suggest that the loads measured in the automotive HIII would be lower than those measured in the FAA HIII for the same input accelerations.

Differences in component weights may also affect the loads being measured by each ATD, although this has yet to be studied in any detail. For example, the upper torso mass of the FAA HIII is 24.9 kg, while for the automotive HIII it is 17.19 kg, which may affect the load being measured in the spinal load cell, and there are also slight differences in overall weight (74.4 kg for the FAA HIII and 77.7 kg for the automotive HIII). In a series of sled tests on 5th, 50th and 95th percentile ATDs which were modified to have a straight spine, Rapaport et al. (1997) found that the lumbar load correlated to upper torso weight rather than the weight of the ATD: the 95th percentile ATD, which had a higher overall weight but lower torso weight than the 50th percentile ATD (due to different abdominal inserts used), had a lower lumbar load than the 50th percentile ATD. However, the effect of the different ATD weights needs to be investigated in future studies, and needs to be taken in consideration with other differences such as load cell orientation.

Rapaport et al. (1997) also tested straight spine 5th, 50th and 95th percentile HIII's using a rigid seat (as opposed to Chandler, who used a stroking seat) with the aim of translating the lumbar compression threshold to a crashworthiness scenario for crash resistant seats, and to develop a relationship between occupant weight versus lumbar load. However, these values are not widely used and there appears to be no data in the literature correlating the threshold values for different-sized occupants to injury potential, although the values are presented as injury thresholds in the US Army Occupant Protection Handbook.

As discussed earlier, the effect of adding body armour is to lower the DRI as the DRI equations do not taken into account this additional mass; however, the lumbar compression increases with the addition of body armour (Thyagarajan et al. 2014), which is consistent with physical reality. In fact, due to the increased risk of spinal injuries due to body armour, Lammers and Dosquet (2006) recommend that, in general, body armour should not be used during transfers or missions in IED-protected vehicles, but instead should be worn in unprotected vehicles or light armoured vehicles (LAVs), although this recommendation needs to be considered in conjunction with the specific tactical situation.

One significant issue with the lumbar compression criterion is that it was not empirically derived, but was calibrated against the DRI. Additionally, it is evident from the graph that the correlation between the lumbar compression and the DRI is quite low (~0.4), which means that the injury predictions from the lumbar load do not have a high degree of confidence, except perhaps near the injury threshold of 1500 Ibs (6.67 kN), for which is was calibrated against the DRI. Nevertheless, as a result of this work, the lumbar compression was introduced into a number of military standards. This can lead to a new problem when one criterion passes while the other fails (discussed further below).

12.4.7 The Relationship Between DRI and Lumbar Compression and the Use of Multiple Injury Criteria

In an analysis investigating the relationship between DRI and lumbar compression, Thyagarajan et al. (2014) recently plotted the DRI against the lumbar compression for ~1,200 UB blast tests of different vehicle and seating positions where the occupants were restrained and most of whom were wearing body armour. They found there was no overall governing relationship between the lumbar compression and DRI, a finding which is not surprising based on the low correlation found by Chandler (~0.40) between the two parameters. MADYMO and LS-DYNA simulations performed by the same authors revealed that when the DRI and lumbar compression are compared for specific seat types, the correlation was much higher, suggesting that the seat properties have a significant influence on the relationship between the two variables.

Drop testing conducted by DST Group has also demonstrated that the seat type influences the relationship between lumbar compression and DRI. For example, in a series of drop tests for four different seat heights (Table 12.5), a table acceleration of 300g resulted in a DRI of 21.6 for both seats but different values of lumbar compression, highlighting the fact that the relationship between the criteria depends on seat type or seat stroke.

Comparison of all four tests highlights a further problem with using both the DRI and the lumbar compression for spinal injury prediction. While the use of multiple injury criteria can increase the level of confidence in spinal injury assessment to account for the limitations with each, it can also introduce a new problem of one criterion passing while the other one fails, as shown in this data.

TABLE 12.5

Seat Test Results from Drop Tower Testing Using Both the Lumbar Compression and DRI Injury Thresholds (DST Group, Department of Defence, Australia)

Table accl	Delta-V	Seat type	Lumbar Compression 6.67 kN	Result	DRI Limit 17.7	Result
140g	4.3	Rigid Aluminium	10	Fail	15.2	Pass
300g	7.2	Collapsible	5.6	Pass	21.6	Fail
300g	7.2	Spring damped	5.91	Pass	21.6	Fail
170g	5.3	Flexible Base	4.68	Pass	18	Fail

Source: DST Group, Department of Defence, University of Melbourne, Australia.

12.4.8 The Spinal Injury Criterion

Dosquet et al. (2010) recently proposed a new spinal injury force-based criterion for UB vehicular blast events, which was denoted as the spinal injury criterion (SIC). Using a similar analysis to Chandler in that they correlated a number of different parameters, including the lumbar compression, to the DRI, Dosquet et al. (2010) analysed datasets of German (WTD 91), Dutch (TNO), Canadian (DRDC Valcartier), US (ARL/ATC) and South African (CSIR) full-scale blast tests from military vehicles, some with and some without body armour, to identify representative loading regimes for a variety of different UB blast scenarios. Using an automotive HIII, Dosquet and colleagues subsequently replicated the loading regimes using a drop-test tower with at least six tests for each impact condition, resulting in over 200 tests.

The aim in developing the SIC was to overcome some of the limitations with the DRI and the lumbar compression by establishing a force-based criterion (like lumbar compression) which had a high correlation to the DRI, but is applicable to a wide range of blast-test loading regimes. In essence, the intention was to apply Chandler's approach but using a more sophisticated statistical method in order to obtain higher correlations between the parameter of interest and the DRI. For the Dosquet (2010) study, the DRI was correlated with:

- The minimum lumbar spine compression force (F_z)
- The maximum vertical pelvis acceleration (a_z)
- The impulse ($I = \int F_z dt$)
- The effective impulse $I_{eff} = \int F_z(\Delta t).d(\Delta t)$, i.e. the area under the duration analysis graph of the lumbar spine compression force

The data was analysed to find the best correlation coefficient for the function:

$$F(t_2 - t_1, x) = \sup_{t_1, t_2} \left\{ \left(\frac{1}{t_2 - t_1} \int_{t_1}^{t_2} F.dt \right)^x (t_2 - t_1) \right\}$$

for different time intervals up to 40 ms and exponents between 1 and 3. The highest correlation (0.97) was found when t = 40 ms and the exponent was 1.2. This yielded the following equation, which was denoted as the SIC:

$$SIC = \sup_{t_1, t_2} \left\{ \left(\frac{1}{t_2 - t_1} \int_{t_1}^{t_2} F.dt \right)^{1.2} (t_2 - t_1) \right\}$$

where the maximum time interval is 40 ms.

In 2012, Dosquet et al. (2012) presented an optimised SIC by factoring the differences in alignment of the lumbar spine load cell and the automotive Hybrid III ATD. They also considered an extension of the possible time interval for calculating SIC, which was up to 100 ms. Due to the differences in alignment of the lumbar spine load cell and the ATD coordinate system (22 degrees), they used the measured Fz^* as well as the transferred compression force Fz^*, where Fz^* is the force obtained when the lumbar spine load cell coordinate system is rotated 22 degrees about the y-axis (see Figure 12.15 for axes). The transferred compression force was then used to analyse the correlation between the abovementioned parameters and the DRI. The highest correlation (0.99) for the SIC was obtained using a time interval of 55 ms and an exponent of 1.1. The time interval and exponent were simplified to 50 ms and 1.0 respectively; this still resulted in a correlation of 0.99 between the SIC and the DRI. This yielded the equation:

$$SIC = \max_{t_2 - t_1 \leq 50ms} \int_{t_1}^{t_2} F_z^*(t)dtz$$

Thus, the SIC is a measure of the impulse of the axial force through the thoracolumbar spine. The SIC criterion is relatively new and as a result, has not yet been widely used, nor has it been validated against cadaveric data. The main issue is that it was not empirically derived; thus, like the lumbar compression, it is still based on the DRI. However, its advantages are that it is applicable to a wider range of loading regimes than the lumbar compression, and it also has a significantly higher correlation to the DRI than the lumbar compression criterion (the lumbar compression ~0.4).

Like the DRI and the lumbar compression, the mass of the body armour has not been factored in the SIC (Dosquet et al. 2010), although in the case of the DRI, it is possible to modify the relevant equations to factor in the body amour mass, as discussed earlier in this chapter. The SIC is still under development; however, the intention is to validate it with cadaveric tests in the near future (F. Dosquet, pers. Comm., 2016). Nevertheless, it has recently been recommended by NATO as an additional assessment criterion with the DRI.

12.4.9 Tolerance of the Pelvis to Vertical Acceleration

Very little work has been completed on the tolerance of the pelvis to vertical loading. Early studies were completed by Evans and Lissner (1955) and Fasola et al. (1955). Evans and Lissner analysed light aeroplane crash injury data from the Cornell Crash Injury Group database, and also investigated the effects of impact loading on the spine and pelvis by tests on cadaveric pelves. From the real-world Cornell data, they found that pelvic injuries were relatively rare: 3.2% of the 800 survivors sustained pelvic fractures, and in seven individuals they attributed the fractures to vertical loading. Cadaveric tests involved finding regions of tensile strain using both static and dynamic tests to the pelvis, most specimens of which were embalmed.

Fasola et al. (1955) tested specimens which were comprised of the pelvis and lower lumbar spine, also using a combination of embalmed and unembalmed specimens. By dropping a weight on to the lumbar spine, where the lower portion of the pelvis was potted in cement, they found that fractures of the lateral wall of the pelvis or acetabulum do not produce fractures or disruption of the SI joint; alternatively, primary fractures of the pubic ramus result in fractures or disjunction of the SI joint by disrupting the anterior/posterior wall stability. The main limitations of these earlier studies include the use of both embalmed and unembalmed cadaveric specimens, and the lack of instrumentation available at the time to provide more detailed data. Isolated pelvic impacts, while being a controlled repeatable type of test, have the disadvantage of limiting the movement of the bone; hence, they may not be the same dynamic movements which occur *in vivo*.

Since these earlier studies, the tolerance of the pelvis to dynamic loading has focused on reconstructing automotive impacts; hence, research has concentrated on loading in the lateral and frontal directions. In the military context, spinal injuries have been the primary concern and thus the focus of studies on aircraft and ejection seats. Consequently, while vertical loading on the occupant in a UB blast event can result in high loads to the spine and pelvis, the association between spinal and pelvic injury potential is not well understood. For example, it might be expected that the high vertical loading causing spinal fractures would usually also result in pelvic injuries; however, this does not occur, and spinal injuries are often sustained without concomitant pelvic fractures (Patzkowski et al. 2012).

Pilot ejection studies have shown that while compression and burst fractures of the spine have been a significant issue in ejectees, this is not the case for pelvic fractures, suggesting there are fundamental differences in the applied loading between pilot ejection and UB blast events. The loading to the combatant in a military vehicle, while affected by factors such as vehicle type, blast magnitude and the position of the occupant relative to the blast, is highly dependent on the seat design, because the behaviour of the seat can significantly alter the acceleration pulse to the occupant. Thus, the mechanism and rate of seat stroking is critical to controlling factors such as the magnitude, pulse type and rate of onset of the acceleration to the combatant.

Recent preliminary work by Pintar (2012) involved investigating of the effect of different pulse types on spinal and pelvic fractures. While initial conclusions indicated that softer seats producing sigmoid pulses may protect the pelvis but injure the spine, ejection seats, which are hard or rigid, produce spinal rather than pelvic injuries, suggesting that there are other characteristics with the impact acceleration which may affect pelvic injury potential. In subsequent work, which is discussed in more detail in Chapter 13, the authors suggested that fast loading rates produce pelvic injuries while slower loading rates result in spinal fractures (Pintar et al. 2012). Indeed, the rate of change of acceleration (the jerk) is known to be considerably higher in blast tests than in pilot ejection (Stemper et al. 2012), and given the very short duration of an UB blast event, it is probable that that characteristics such as loading rate are more important than other pulse characteristics such as the pulse shape. However, it is clear that more research is needed to elucidate the specific relationships between impact acceleration pulse characteristics, pelvic fractures and spinal injuries.

12.5 CURRENT STANDARDS

12.5.1 THE NORTH ATLANTIC TREATY ORGANIZATION

The North Atlantic Treaty Organization (NATO) publishes the AEP-55 (Allied Engineering Publication-55), Volumes 2 and 3, which consist of the specifications for live-fire blast testing of military vehicles (NATO 2006, 2014). The standards specify that a 50th percentile Hybrid III ATD be used, where the clothing and footwear represents that of a real crew member, including personal protective equipment (PPE) if relevant. The type of Hybrid III is not explicitly specified, but references are made to the US Code of Federal Regulations and the ECE (Economic Commission for Europe), suggesting that an automotive HIII should be used.

For the spine and pelvis regions, the only mandatory injury criterion, the DRI, is calculated from the pelvis linear acceleration in the vertical direction (A_z). For a DRI of 17.7 (STANAG AEP-55 value), which equates to ~10% probability of spinal injury using the laboratory data:

$$17.7 = \frac{\delta_{dyn}}{3.5}$$

thus:

$$\delta_{dyn} = 62 \text{ mm}$$

While additional instrumentation is highly recommended to allow further analysis of the ATD response, it is not mandatory. Additional recommended injury criteria for the spine are the lumbar axial compression (F_z), the shear force (F_x) and the bending moment (M_y), as recommended by HFM-90 (discussed later). More recently, the SIC criterion has been advocated.

12.5.2 THE NATO HUMAN FACTORS AND MEDICINE PANEL

The NATO Human Factors and Medicine Panel (HFM) HFM-090 was created to advise NATO of appropriate injury assessment criteria for the assessment of landmine attack on vehicles. They provided suitable recommendations for an injury assessment methodology for the qualification of light armoured and logistic vehicle protection systems against blast loadings. HFM-148 later verified the approach taken by HFM-90 (HFM-148, 2012).

For the spine and pelvis regions, the NATO Human Factors and Medicine (HFM) Panel (HFM-90) has recommended that the DRI be the mandatory injury criterion for the spine, as it takes the duration of the load into account. The panel has advised not using other criteria (such as the lumbar compression) for the assessment of spinal injuries, as the other criteria are either not time-dependent, or are not validated. Moreover, based on the relatively low risk of spinal injury in the x and y directions, the panel recommended that only spinal injuries in the vertical direction be assessed. (HFM-90). The decision to assess only the z direction has good validity: as discussed earlier, the DRI in the other directions is related primarily to cardiovascular responses, which are not applicable to UB blast scenarios.

HFM-90 decided to use the automotive Hybrid III ATD for blasts which are under, at the front or at the rear of the vehicle for three reasons: (1) because the curved spine gives a better representation of seating postures in military vehicles, (2) as the Hybrid III is more commonly used worldwide, and (3) because the spinal configuration has no, or very limited influence on the DRI when the pelvis acceleration is used as the input. Also, by using an automotive HIII, the injury criteria for the other body regions are still applicable. Using a 50th percentile automotive HIII, the vertical acceleration of the pelvis is used as input into the model and the 17.7 threshold is employed (i.e. based on the laboratory data), which equates to a 10% risk of an AIS 2+ injury based on the Brinkley laboratory data. This more conservative laboratory graph was recommended as it takes into account seats which might have significant alignment differences from the vertical, i.e. angled differently to the acceleration vector (HFM-90).

HFM-90 also recognised that pelvic injuries may also occur due to the high vertical loadings, but assumed thoracolumbar spine injuries to be the more predominant injury type. In addition, they consider spinal injuries to be an indicator of pelvic injury (Dosquet 2003; HFM-90 2007).

Measurements from the seat cushion are sometimes implemented in the calculation of the DRI; hence, Dosquet examined the correlation between acceleration values measured from the seat cushion and accelerations measured from the ATD pelvis in 63 tests performed in different test facilities (Dosquet 2004). The tests were all live-fire blasts conducted by different test centres in five nations, where different vehicles and seat types were used, and the seats were either energy absorbing or energy dissipating rather than rigid (Dosquet 2016, pers. Comm., 2016). While the DRI values measured from the pelvis were slightly lower than the DRI values from the seat cushion, the differences were not significant (Dosquet 2004). There was a high correlation (almost 90%) between the acceleration values measured from the seat, seat cushion and the pelvis, and as a result, HFM-90 TG-25 (Task Group 25) decided to use the pelvis measurements in the DRI calculation (Dosquet 2004; HFM-90 2007).

HFM-148, a later HFM panel, extended the approach of HFM-90 by considering a wider range of UB blast scenarios, but recommended using the DRI only. HFM-148, however, advised using the SIC in addition to the DRI for injury assessment for configurations where restraints and/or PPE are used (Frank Dosquet, personal communication, 2016).

The use of a EuroSID-2re ATD was recommended for blasts which are located laterally with respect to the ATD.

12.5.3 US Standards

The United States produces the Department of Army Occupant Protection Handbook (2000), which is not an official manual to US Army specifications, but a testing guide to support an informed selection of requirements for vehicle-specific applications. In addition to providing information on the standard automotive tests such as the requirements for frontal, lateral and rear impacts, the manual provides recommended injury criteria for landmine blast testing of high-mobility wheeled vehicles. In the Occupant Protection Handbook, injury tolerance levels established by the US Army's Aberdeen Test Centre are provided for the 50th percentile male, 5th percentile female and the 95th percentile male. The thresholds and criteria relevant to the spine and pelvis of a 50th percentile ATD are shown in Table 12.6.

TABLE 12.6

Recommended Injury Thresholds Criteria for the Spine and Pelvis from the US Army Occupant Protection Handbook (2000)

Injury Criterion	Symbol	SAE Filter	IARVs
Pelvis forward acceleration	Ax (g)	CFC 180	40 g @ 7 ms
Pelvis lateral acceleration	Ay (g)	(300 HZ)	23 g @ 7 ms
Pelvis vertical acceleration	Az (g)		23 g @ 7 ms
Seat (Pelvis) forward DRI	DRI – x (g)	CFC 180	35, 40, 46 gx (low, med, high risks)
Seat (Pelvis) lateral DRI	DRI – y (g)	(300 HZ)	14, 17, 22 gy (low, med, high risks)
Seat (Pelvis) vertical DRI	DRI – z (g)		15, 18, 23 gz (low, med, high risks)
Lumbar spine shear force	Fx or Fy (N)	CFC 1000	3800 N (45 ms), 5200 N (25–35 ms),
		(1650 HZ)	10700 N (0 ms)
			3800 N (30 ms), 6673 N (0 ms),
Lumbar spine axial compression force	−Fz (N)		3800 N (45 ms), 10200 N (35 ms)
Lumbar spine axial tension force	+Fz (N)		12700 N (0 ms)

The approach used is that the DRI is more robust for injury prediction if applied in conjunction with the Eiband curve and lumbar compression, as using multiple injury criteria can increase the degree of confidence in the final assessment of the vehicle, i.e. they employ a multicriteria approach to predicting injuries. This is similar to an earlier recommendation made by Rapaport et al. (1997), who concluded that the lumbar force can support other injury indicators such as the Eiband curve and the DRI.

The DRI values in the x and y directions were derived from the work of Brinkley and implemented in the ARGARD standards, as discussed earlier. However, the thresholds relate mainly to cardiovascular effects sustained by astronauts subjected to rapid acceleration, and thus do not relate to spinal injuries, or to any specific injuries; thus, their use for prediction of spinal injuries in live-fire testing should be considered with caution.

12.5.4 AUSTRALIAN STANDARDS

Australian live-fire testing involves the measurement of a large number of injury thresholds and criteria to inform the survivability assessment of the vehicle. In relation to the spine and pelvis, while acceleration, force and moment channels are considered in the assessment, the potential for spinal injuries is evaluated using both the DRI and the lumbar compression in conjunction, where a DRI of 17.7 and a lumbar compression threshold of 6.67 kN are used. While this can be implemented for most cases of live-fire (or drop tower) testing, it can result in the issue, discussed earlier, when one criterion passes and the other fails. Both criteria are required to pass for a successful test result. At present, the modified DRI is not routinely used. More research needs to be conducted before a single spinal injury assessment criterion is accepted and modifications are made to account for body armour.

For the pelvis, the pelvis forward and lateral accelerations are measured ($A_x = 40$ g @ 7 ms and $A_y = 23$ g @ 7 ms), and the two vertical acceleration pelvic thresholds derived from the Eiband curve ($A_z = 23$ g @ 7 ms and 100 g @ 2 ms) are used for pelvic injury assessment (i.e. similar to the US but with an additional value for A_z).

12.5.5 SUMMARY

NATO, the United States and Australia have each used a slightly different approach to evaluate spine and pelvic injuries in live-fire UB blast testing. The justification for using the DRI only is that it is a time-dependent criterion, and in addition to indications that there is a degree of time-dependency in spinal injury causation, the time-dependence also factors in the wide range of loading regimes characterised by UB blast events. However, the DRI was designed for ejection seat scenarios where the seat is rigid, and not for UB blast events. In addition, the body armour is usually not factored in the model, even though body armour is often worn by ATDs in live-fire tests and drop testing; thus, the true DRI is generally underestimated.

The lumbar compression has the advantage of being designed to assess different types of seats, particularly flexible seats, and is a more direct measurement of spinal injury potential. It was designed for civilian aircraft crashes, with its primary limitation being that it was correlated to the DRI rather than being empirically derived, and the correlation is not high. Thus, the lumbar compression in its current form is not well validated; however, being a more direct measurement of spinal compression injuries, is worth considering for future injury criterion development.

Due to the limitations of the DRI and the lumbar compression, both criteria can be used, potentially also in conjunction with the Eiband curve, to give more confidence in the results. The main disadvantage of this approach is that one criterion can pass while the other fails; thus making the results difficult to interpret. More recently, the SIC criterion has been developed, but it is still relatively new and needs further validation. The Eiband curve can be used for the assessment of pelvic injuries, although it based on whole-body acceleration; however, there are currently no other alternatives for assessing pelvic injuries due to vertical acceleration.

Although there are arguments for either using the DRI alone, or the DRI in conjunction with the lumbar compression and/or the Eiband curve, it is difficult to justify the use of the DRI in other directions to assess spinal injuries in UB blast events. The DRI in the x and y directions were designed for spacecraft where the types of injuries focus heavily on cardiovascular responses, and while valid for that environment, they do not translate to the UB blast event, where the injuries are impact based.

12.6 METHODS OF ASSESSMENT

12.6.1 LIVE-FIRE BLAST TESTING

The dynamics of the blast event has previously been described in Chapter 3. In relation to spinal injuries, two phases of the blast event are important: firstly, the hull response and internal vehicle dynamics, where deformation of the hull and floorpan results in spinal injuries, and secondly, the vehicle slamdown (or set-down), where additional loading may result in spinal injuries. However, the initial response is the predominant loading mode in most cases and thus usually the cause of the injury, or the most severe injuries, from vertical loading.

12.6.2 DROP-TOWER TESTING

A drop-test tower (Figure 12.18) is a method which can be used to evaluate spinal injury potential for different military seat and cushion designs by providing an idealised loading scenario representing a UB blast test acceleration pulse. It has the advantages of being able to apply a consistent repeatable acceleration pulse to the platform in addition to being relatively cheap and easy to perform; hence, a large number of tests can be conducted in a short period of time.

The drop test involves raising the platform, including the seat and ATD, to a specific vertical height; then releasing the system so it falls under gravity until it rapidly decelerates as it impacts the base. The maximum acceleration or delta-V can be altered by using different drop heights, while the pulse profile and the time to peak acceleration depend on the platform payload and the seat and cushion characteristics.

(a)   (b)

FIGURE 12.18 (a) Drop-test rig and (b) drop tower platform showing the restrained ATD. (Courtesy of DST Group, Department of Defence, University of Melbourne, Australia.)

While a drop tower does not exactly replicate a UB blast test, as there are slight differences in inertia (a vertical acceleration pulse applied to ATD moving with gravity compared to a vertical acceleration pulse applied to a stationary ATD in a UB blast test), it provides a tool to rapidly assess new seat and cushion prototypes. Spinal injury potential can be evaluated using the DRI or the lumbar compression; however, if body armour is worn by the ATD in the drop test, the same issues arise as with a live-fire blast test, i.e. if the mass of the body armour is not considered. Conversely, if the aim is to evaluate a series of military seats or seat cushions, the relative injury criteria values can be compared.

12.7 CONCLUSIONS

The characteristics of thoracolumbar spinal and pelvic injuries sustained in military environments have been outlined. Descriptive studies have demonstrated that axial loading due to ejection, helicopter crash and UB blast events can lead to markedly different injury patterns.

For spinal injuries, ejection studies have demonstrated a bimodal distribution of injuries concentrated in the mid-thoracic spine and thoracolumbar junction; higher rate loading scenarios have focused the distribution of these injuries to caudal locations. Additionally, clinical studies and pilot ejection experiments have shown that a number of occupant-related factors influence spinal injury type and severity, including subject anthropometry and occupant posture at the time of loading. Less is known about pelvic injuries in UB blast events, but pelvic shear and other types of complex pelvic fractures are known to occur from vertical loading.

Injury criteria and injury thresholds currently available for assessing spinal and pelvic injuries in UB blast events have been developed for other scenarios, such as pilot ejection. As such, although they have been shown to have good validity in those scenarios, their applicability to UB blast events is less reliable.

Recent research has focused on developing alternative criteria such as the SIC, performing whole-body cadaveric tests for spine and pelvic injury, spinal component testing and finite element modelling. Much of this work has been part of the WIAMan project, aimed at developing new injury criteria specifically for UB blast events. However, the current injury assessment thresholds will continue to be used until new criteria are developed specifically for UB blast events.

ACKNOWLEDGEMENTS

The authors would like to gratefully acknowledge Mr James Brinkley, Aerospace Consultant and formally with NASA, for his expert advice relating to the development of the DRI, Mr Colin Nelson of Defence Materiel Organisation, Department of Defence, Australia, for permission to present his graph and discussions on his personal research relating to the modified DRI. We would also like to thank Dr Stephen Cimpoeru of DST Group, Department of Defence, Australia for his comments and advice relating to the current NATO standards, Dr Peter Shoubridge of DST Group, Department of Defence, Australia for his approval to use DST Group drop tower photos and data, and Mr Frank Dosquet of the Bundeswehr Technical Center for Weapons and Ammunition, Meppen, Germany for reviewing the sections on injury criteria and providing invaluable comments on some of the early spinal testing and development of the DRI. Lastly, we appreciate the help of Mr Jeff Copeland and Ms Sheridan Laing for checking and formatting the references and proofreading.

REFERENCES

AGARD (Advisory Group for Aerospace Research and Development). 1996. *Anthropomorphic dummies for crash and escape system testing.* AGARD-AR-330. Neuilly-Sur-Siene, France: AGARD.
AIS (Abbreviated Injury Scale). 2008. Association for the Advancement of Automotive Medicine. (AIS 2005 Update 2008). Eds. T. A.. Gennarelli and E.. Wodzin. Barrington, IL.

Alpantaki, K., A. Bano, D. Pasku, A. F. Mavrogenis, P. J. Papagelopoulos, G. S. Sapkas, D. S. Korres, and F. Katonis. 2010. Thoracolumbar burst fractures: A systematic review of management. *Orthopedics*. 33(6): 422–429. DOI: 10.3928/01477447-20100429-24.

Alvarez, J. 2011. Epidemiology of blast injuries in current operations. In RTO-MP-HFM-207—Survey of Blast Injury across the Full Landscape of Military Science. *Proceedings of RTO Human Factors and Medicine Panel (HFM) Symposium*, Halifax, Nova Scotia, October 3–5.

Anonymous, D. 1956. Developing the Martin-Baker ejection seat. *The Aeroplane*. 90(2324):141–143.

Bailey, A. M., J. J. Christopher, F. Brozoski, and R. S. Salzar. 2015a. Post mortem human surrogate injury response of the pelvis and lower extremities to simulated underbody blast. *Annals of Biomedical Engineering*. 43(8):1907–1917. DOI: 10.1007/s10439-014-1211-5.

Bailey A. M., J. J. Christopher, K. Henderson, F. Brozoski, and R. Salzar. 2013. Comparison of Hybrid-III and PMHS response to simulated underbody blast loading conditions. *Presented at the International Research Council on the Biomechanics of Injury (IRCOBI) Conference*, Gothenburg, Sweden, September 11–13.

Bailey, A. M., J. J. Christopher, R. S. Salzar, and F. Brozoski. 2015b. Comparison of Hybrid-III and postmortem human surrogate response to simulated underbody blast loading. *Journal of Biomechanical Engineering*. 137(5):051009. DOI: 10.1115/1.4029981.

Baumann, R. J., J. W. Brinkley, and A. Brandau. 1968. Back injuries experienced during ejection seat testing. *Presented at the 39th Annual Aerospace Medical Association*. AMRL-TR-68-82. Aerospace Medical Research Laboratory, Wright-Patterson Air Force Base, OH.

Blair, J. A., D. R. Possley, J. L. Petfield, A. J. Schoenfeld, R. A. Lehman, and J. R. Hsu. Skeletal Trauma Research Consortium. 2012. Military penetrating spine injuries compared with blunt. *The Spine Journal*. 12(9):762–768.

Bogduk, N. 2012. *Clinical and radiological anatomy of the lumbar spine*. 5th ed. Edinburgh: Elsevier Churchill Livingston.

Bosee, R. A., and C. F. Payne, Jr. 1961. Theory on the mechanism of vertebral injuries sustained on ejections from aircraft. AD0256378. *Presented at the meeting of the Aerospace Medical Panel of the Advisory Group for Aeronautical Research and Development*, Aerospace Crew Equipment Lab., Naval Air Engineering Center, Philadelphia, PA.

Briggs, N. C., R. S. Levine, W. P. Haliburton, D. G. Schlundt, I. Goldzweig, and R. R. Warren. 2005. The Fatality Analysis Reporting System as a tool for investigating racial and ethnic determinants of motor vehicle crash fatalities. *Accident; Analysis and Prevention*. 37(4):641–649. DOI: 10.1016/j.aap.2005. 03.006.

Brinkley, J. W. 1984. Personal protection concepts for advanced escape system design. In *Human factors considerations in high performance aircraft, AGARD Conference Proceedings*, no. 371.

Brinkley, J. W., and J. T. Schaffer. 1971. *Dynamic simulation techniques for the design of escape systems: Current applications and future Air Force requirements*. AMRL-TR-71-291971. OH: Aerospace Medical Research Laboratory, Wright-Patterson Air Force Base.

Brinkley, J. W., L. J. Specter, and S. E. Mosher. 1990. Development of acceleration exposure limits for advanced escape systems. In *Implications of Advanced Technologies for Air and Spacecraft Escape, AGARD Conference Proceedings*, no. 472.

Brown, J. K., Y. Jing, S. Wang, and P. F. Ehrlich. 2006. Patterns of severe injury in pediatric car crash victims: Crash Injury Research Engineering Network database. *Journal of Pediatric Surgery*. 41(2):362–367. DOI: 10.1016/j.jpedsurg.2005.11.014.

Bruckart, J. E., B. J. McEntire, J. R. Licina, D. K. Brantley, and D. F. Shanahan. 1993. Fatal mishap report: First SPH-4B flight helmet recovered from a U.S. Army helicopter mishap. *Aviation, Space and Environmental Medicine*. 64(8):755–759.

Caldwell, E., M. Gernhardt, J. T. Somers, D. Younker, and N. Newby. 2012. *NASA evidence report: Risk of injury due to dynamic loads*. Houston, TX: National Aeronautics and Space Administration, Lyndon B. Johnson Space Center.

Chance, G. Q. 1948. Note on a type of flexion fracture of the spine. *British Journal of Radiology*. 21(249): 452–453. DOI: 10.1259/0007-1285-21-249-452.

Chandler, R. F. 1985. Data for the development of criteria for general aviation seat and restraint system performance. SAE Technical Paper 850851. *Presented at the General Aviation Aircraft Meeting and Exposition*, Wichita, Kansas, April 16–19. DOI: 10.4271/850851.

Chandler, R. F. 1993. Development of crash injury protection in civil aviation. In *Accidental injury: Biomechanics and prevention*, eds. A. M. Nahum and J. W. Melvin, pp. 151–185. Springer-Verlag, New York.

Chaudhri, K., G. R. Cybulski, D. T. Pitkethly, B. Al Hemsi, and M. C. Johnson. 1993. Multiple cervical spine fractures without neurologic deficit after a helicopter crash. Case report. *Spine*. 18(14):2135–2137.

Cheng, M., J. P. Dionne, and A. Makris. 2010. Use of the dynamic response index as a criterion for spinal injury in practical applications. *Paper presented at the Personal Armour System Symposium (PASS)*, Quebec City, QC, September 13–17.

Chubb, R. M., W. R. Detrick, and R. H. Shannon. 1965. Compression fractures of the spine during USAF ejections. *Aerospace Medicine*. 36(10):968–972.

Clark, D. E. and S. Ahmad. 2005. Is the number of traffic fatalities in American hospitals decreasing? *Accident; Analysis and Prevention*. 37(4):755–760. DOI: 10.1016/j.aap.2005.03.015.

Cobb, J. D., P. A. MacLennan, G. McGwin, Jr., S. G. Metzger, and L. W. Rue. 2005. Motor vehicle mismatch-related spinal injury. *The Journal of Spinal Cord Medicine*. 28(4):314–319.

Code of Federal Regulations. 2016. *FAA 25.562: Emergency Landing Dynamic Conditions*. US Government Publishing Office, Washington, DC.

Collins R., G. W. McCarthy, I. Kaleps, and F. S. Knox. 1997. Review of major injuries and fatalities in USAF ejections, 1981–1995. *Biomedical Sciences Instrumentation*. 33:350–353.

Colin Nelson, Defence Materiel Organization, Department of Defence, Australia, 2015 (personal communication).

Curry, W. H., F. A. Pintar, N. B. Doan, H. S. Nguyen, G. Eckardt, J. L. Baisden, D. J. Maiman, G. R. Paskoff, B. S. Shender, and B. D. Stemper. 2016. Lumbar spine endplate fractures: Biomechanical evaluation and clinical considerations through experimental induction of injury. *Journal of Orthopeadic Research*. 34(6): 1084–1091. DOI: 10.1002/jor.23112.

Delahaye, R. P., B. Vettes, and R. Auffret. 1975. Lesions observees apres ejections a grade vitesse dans l'armee de l'air francalse. *Med Armees*. 3:706–712.

Denis, F. 1983. The 3 column spine and its significance in the classification of acute thoracolumbar spinal injuries. *Spine*. 8(8):817–831. DOI: 10.1097/00007632-198311000-00003.

Department of Army. 2000. *Occupant crash protection handbook for tactical ground vehicles*. Warren, MI.

Deroche J., J. P. Taillemite, J. Y. Grau, and P. Devaulx. 1999. Ejections des pilotes de l'armee de l'air Francaise de 1987 a 1996. *Medecine et Armees*. 1999:487–494.

Dosquet, F. 2003. Thoracolumbar spine and pelvis criteria for vehicular mine protection. *NATO-RTOHFM-090, 3rd Meeting*, TARDEC, Warren, MI.

Dosquet, F. 2004. *Analysis of thoraco-lumbar spine and pelvis criteria for vehicular mine protection. NATO-RTO-HFM-090, 4th Meeting*, WTD 91, Meppen, Germany.

Dosquet F., C. Lammers, Y. Schneider, D. Soyka, M. Hennemann, and Joel McLain. 2012. Impact responses of Hybrid III ATD in under-belly vehicular protection applications to analyse thoraco-lumbar spine injury risks. *Paper presented at the Personal Armour System Symposium (PASS)*, Nuremberg, Germany, September 17–21.

Dosquet, F., C. Lammers, D. Soyka, and M. Henneman. 2010. Hybrid III thoraco-lumbar spine impact response in vehicular protection applications for vertical impacts. *Paper presented at the Personal Armour System Symposium (PASS)*, Quebec City, QC, September 13–17.

Edwards, M. 1996. Anthropometric measurements and ejection injuries. *Aviation, Space and Environmental Medicine*. 67(12):1144–1147.

Eiband, A. M. 1959. *Human tolerance to rapidly applied accelerations: A summary of the literature*. NASA-MEMO-5-19-59E. Cleveland, OH: National Aeronautics and Space Administration.

Evans, F. G., and H. R. Lissner. 1955. Studies on pelvic deformations and fractures. *Anatomical Record*. 121(2): 141–165. DOI: 10.1002/ar.1091210203.

Evans, L. 1994. Driver injury and fatality risk in two-car crashes versus mass ratio inferred using Newtonian mechanics. *Accident; Analysis and Prevention*. 26(5):609–616. DOI: 10.1016/0001-4575(94)90022-1.

Ewing, C. L. 1966. Vertebral fracture in jet aircraft accidents: A statistical analysis for the period 1959 through 1963, U. S. Navy. *Aerospace Medicine*. 37(5):505–508.

Fasola, Alfred F., Rollo C. Baker, and Fred A. Hitchcock. 1955. *Anatomical and physiological effects of rapid deceleration*. Ohio State Univ Research Foundation, Columbus, OH.

Freedman, B. A., J. A. Serrano, P. J. Belmont, Jr., K. L. Jackson, B. Cameron, C. J. Neal, R. Wells, C. Yeoman, and A. J. Schoenfeld. 2014. The combat burst fracture study—Results of a cohort analysis of the most prevalent combat specific mechanism of major thoracolumbar spinal injury. *Archives of Orthopaedic and Trauma Surgery.* 134(10):1353–1359. DOI: 10.1007/s00402-014-2066-9.

Gauer, O. 1950. The physiological effects of prolonged acceleration. In *German Aviation Medicine, World War II*, Vol. I, Chapter VI-B, 554–583. Department of the Air Force, WA.

Gomleksiz, C., E. Egemen, S. Senturk, O. Yaman, A. L. Aydin, T. Oktenoglu, M. Sasani, T. Suzer, and A. F. Ozer. 2015. Thoracolumbar fractures: A review of classifications and surgical methods. *Journal of Spine.* 4(4). DOI: 10.4172/2165-7939.1000250. https://scholar.google.com.au/scholar?hl=en&q=Thoracolumbar+fractures%3A+A+review+of+classifications+and+surgical+methods&btnG=&as_sdt=1%2C5&as_sdtp=

Gowdy, V., R. DeWeese, M. Beebe, B. Wade, J. Duncan, R. Kelly, and J. L. Blaker. 1999. A lumbar spine modification to the Hybrid III ATD for aircraft seat tests. SAE Technical Paper 1999-01-1609. *General, Corporate and Regional Aviation Meeting and Exposition*, Wichita, KS, April 20–22.

Gray, H. 2000. *Gray's anatomy. Anatomy of the human body.* Lea and Febiger, Philadelphia, PA, 1918; Bartleby.com.

Hanrahan, R. B., P. M. Layde, S. Zhu, G. E. Guse, and S. W. Hargarten. 2009. The association of driver age with traffic injury severity in Wisconsin. *Traffic Injury Prevention.* 10(4):361–367. DOI: 10.1080/15389580902973635.

Harrison, W. U. 1979. Aircrew experiences in USAF ejections, 1971–1977. In *Proceedings of the 16th Survival and Flight Equipment Association Annual Symposium*, San Diego, CA, vol. 8(12).

Hearon, B. F., H. A. Thomas, and J. H. Raddin, Jr. 1982. Mechanism of vertebral fracture in the F/FB-111 ejection experience. *Aviation, Space and Environmental Medicine.* 53(5):440–488.

Helgeson, M. D., R. A. Lehman, Jr., P. Cooper, M. Frisch, R. C. Anderson, and C. Bellabarba. 2011. Retrospective review of lumbosacral dissociations in blast injuries. *Spine.* 36(7):E469–E475. DOI: 10.1097/BRS. 0b013e3182077fd7.

Henzel, J. H. 1967. *The human spinal column and upward ejection acceleration: An appraisal of biodynamic implications.* AMRL-TR-66-233. Aerospace Medical Research Laboratory, Wright-Patterson Air Force Base, OH.

HFM-90, TG-25. 2007. *Test methodology for the protection of vehicle occupants against anti-vehicular landmine effects.* North Atlantic Treaty Organisation (NATO). Brussels: Belgium.

HFM-148. 2012. *Test methodology for the protection of vehicle occupants against anti-vehicular landmine effects.* North Atlantic Treaty Organisation (NATO). Brussels: Belgium.

Hirsch, A. E. 1964. *Man's response to shock motions.* Report 1797, Washington, DC: David Taylor Model Basin, Structural Mechanics Laboratory.

Howland, W. J., J. L. Curry, and C. B. Buffington. 1965. Fulcrum fractures of the lumbar spine: Transverse fracture induced by an improperly placed seat belt. *Journal of the American Medical Association.* 193(3):240–241. DOI: 10.1001/jama.1965.03090030062025.

Hsu, J. M., T. Joseph, and A. M. Ellis. 2003. Thoracolumbar fracture in blunt trauma patients: Guidelines for diagnosis and imaging. *Injury.* 34(6):426–433. DOI: 10.1016/S0020-1383(02)00368-6.

Hutton, W. C., B. M. Cyron, and J. R. Stott. 1979. The compressive strength of lumbar vertebrae. *Journal of anatomy.* 129(Pt 4):753.

Iluk, A. 2013. Influence of the additional inertial load of the torso on the mine blast injury. *Presented at the International Research Council on the Biomechanics of Injury (IRCOBI) Conference*, Gothenburg, Sweden, September 11–13.

Inamasu, J. and B. H. Guiot. 2007. Thoracolumbar junction injuries after motor vehicle collision: Are there differences in restrained and nonrestrained front seat occupants? *Journal of Neurosurgery. Spine.* 7(3): 311–314.

Italiano, P. 1966. [Vertebral fractures of pilots in helicopter accidents]. *Rivasta di Medicina Aeronautica e Spaziale.* 29(4):577–602.

James, M. R. 1991. Spinal fractures associated with ejection from jet aircraft: Two case reports and a review. *Archives of Emergency Medicine.* 8(4):240–244.

Joaquim, A. F., and A. A. Patel. 2014. Thoracolumbar fractures. In *Skeletal trauma: Basic science, management and reconstruction*, eds. B. D. Browner, J. B. Jupiter, C. Krettek, and P. A. Anderson, pp. 911–919. 5th ed. Philadelphia, PA: Elsevier Saunders.

Kazarian, L. E., K. Beers, and J. Hernandez. 1979. Spinal injuries in the F/FB-111 crew escape system. *Aviation, Space and Environmental Medicine.* 50(9):948–957.

Lammers, C., and F. Dosquet. 2006. Effects of personal protective equipment on occupant safety. In *Proceedings of the 3rd European Survivability Workshop*, Toulouse, France, May 16–18.

Langrana, N. A., R. D. Harten, Jr., D. C. Lin, M. F. Reiter, and C. K. Lee. 2002. Acute thoracolumbar burst fractures: A new view of loading mechanisms. *Spine*. 27(5):498–508.

Latham, F. 1957. A study in body ballistics: Seat ejection. *Proceedings of the Royal Society of London. Society B: Biological Sciences*. 147(926):121–139.

Lawrence, C., E. L. Fasanella, A. Tabiei, J. W. Brinkley, and D. M. Shemwell. 2008. *The use of a vehicle acceleration exposure limit model and a finite element crash test dummy model to evaluate the risk of injuries during Orion crew module landings*. NASA TM 215198. Cleveland, OH: National Aeronautics and Space Administration.

Lee, C.-K., Y. Eun Kim, C.-S. Lee, Y.-M. Hong, J.-M. Jung, and V. K. Goel. 2000. Impact response of the intervertebral disc in a finite element model. *Spine*. 25(19):2431–2439.

Lehman, R. A., Jr., H. Paik, T. T. Eckel, M. D. Helgeson, P. B. Cooper, and C. Bellabarba. 2012. Low lumbar burst fractures: A unique fracture mechanism sustained in our current overseas conflicts. *Spine Journal*. 12 (9):784–790. DOI: 10.1016/j.spinee.2011.09.005.

Letournel, E. 1980. Acetabulum fractures: Classification and management. *Clinical Orthopaedics and Related Research*. 151:81–106.

Lewis, M. E. 2006. Survivability and injuries from use of rocket-assisted ejection seats: Analysis of 232 cases. *Aviation, Space and Environmental Medicine*. 77(9):936–943.

Liu, Y. K and Q. C. Dai. 1989. The second stiffest axis of a beam-column: Implications for cervical spine trauma. *Journal of Biomechanical Engineering*. 111(2):122–127.

Magerl, F., M. Aebi, S. D. Gertzbein, J. Harms, and S. Nazarian. 1994. A Comprehensive classification of thoracic and lumbar injuries. *European Spine Journal*. 3(4):184–201. DOI: 10.1007/BF02221591.

Maiman, D. J. 1986. Experimental trauma of the human thoracolumbar spine. In *Mechanisms of Head and Spine Trauma*. eds. A. Sances Jr., D.J. Thomas, C.L. Ewing, and S.L. Larson, Aloray, New York. pp. 489–504.

Maiman, D. J., N. Yoganandan, and F. A. Pintar. 2002. Preinjury cervical alignment affecting spinal trauma. *Journal of Neurosurgery*. 97(1):57–62.

Manen, O., J. Clement, S. Bisconte, and E. Perrier. 2014. Spine injuries related to high-performance aircraft ejections: A 9-year retrospective study. *Aviation, Space and Environmental Medicine*. 85(1):66–70.

Marin, J. 1955. *Ejection from High Speed Aircraft*. Denham, UK: Martin-Baker Aircraft.

Marsh, J. L., T. F. Slongo, J. Agel, S. Broderick, W. Creevey, T. A. DeCoster, and L. Prokushi, et al. 2007. Fracture and dislocation classification compendium, Orthopaedic Trauma Association classification, database and outcomes committee. *Journal of Orthopaedic Trauma*. 21(10):1–164.

McCarthy, G. W. 1988. USAF take-off and landing ejections, 1973–85. *Aviation, Space and Environmental Medicine*. 59(4):359–362.

Milanov, L. 1996. Aircrew ejections in the Republic of Bulgaria, 1953–93. *Aviation, Space and Environmental Medicine*. 67(4):364–368.

Miller, C. D., P. Blyth and I. D. S. Civil. 2000. Lumbar transverse process fractures—A sentinel marker of abdominal organ injuries. *Injury-International Journal of the Care of the Injured*. 31(10):773–776. DOI: 10.1016/S0020-1383(00)00111-X.

Mohr, G., J. W. Brinkley, L. Kazarian, and W. Millard. 1969. *Variations in spinal alignment in egress systems and their effect*. AMRL-TR-67-232. Aerospace Medical Research Laboratory, Wright-Patterson Air Force Base, OH.

Moreno Vázquez, J. M., M. R. Durán Tejeda, and J. L. García Alcón. 1999. Report of ejections in the Spanish Air Force, 1979–1995: An epidemiological and comparative study. *Aviation Space and Environmental Medicine*. 70(7):686–691.

Moore, K. L., A. M. R. Agur, and A. F. Dalley. 2015. *Essential Clinical Anatomy*, 5th ed. Baltimore, MD: Wolters Kluwer-Lippincott Williams and Wilkins.

Nakamura, A. 2007. Ejection experience 1956–2004 in Japan: An epidemiological study. *Aviation, Space and Environmental Medicine*. 78(1):54–58.

NATO. 2006. *Allied Engineering Publication AEP-55: Procedures for Evaluating the Protection Level for Logistics and Light Armoured Vehicles*, vol. 2 for Mine Threat, ed. 1. Brussels: Belgium.

NATO. 2014. *Allied Engineering Publication AEP-55: Procedures for Evaluating the Protection Level of Armoured Vehicles—IED Threat*, vol. 3, part 1, ed. C, ver. 1. Brussels: Belgium

Newman, D. G. 1995. The ejection experience of the Royal Australian Air Force: 1951–92. *Aviation, Space and Environmental Medicine*. 66(1):45–49.

Nirula, R., and F. A. Pintar. 2008. Identification of vehicle components associated with severe thoracic injury in motor vehicle crashes: A CIREN and NASS analysis. *Accident; Analysis and Prevention*. 40(1):137–141. DOI: 10.1016/j.aap.2007.04.013.

Osborne, R. G., and A. A. Cook. 1997. Vertebral fracture after aircraft ejection during Operation Desert Storm. *Aviation, Space and Environmental Medicine*. 68(4):337–341.

Patten, R. M., S. R. Gunberg, and D. K. Brandenburger. 2000. Frequency and importance of transverse process fractures in the lumbar vertebrae at helical abdominal CT in patients with trauma. *Radiology*. 215(3): 831–834.

Patzkowski, J. C., J. A. Blair, A. J. Schoenfeld, R. A. Lehman, and J. R. Hsu. 2012. Multiple associated injuries are common with spine fractures during war. *The Spine Journal*. 12(9):791–797. DOI: 10.1016/j.spinee. 2011.10.001.

Pavlović, M., J. Pejović, J. Mladenović, R. Cekanac, D. Jovanović, R. Karkalić, D. Randjelović, and S. Djurdjević. 2014. Ejection experience in Serbian Air Force, 1990–2010. *Vojnosanitetski Pregled*. 71(6):531–533.

Payne, P. R. 1962. Dynamics of human restraint systems. In *Impact Acceleration Stress: A Symposium*, Washington, DC, November 27–29.

Pellettiere, J. A., D. Moorcroft, and G. Olivares. 2011. Anthropomorphic test dummy lumbar load variation. *Paper presented at the 22nd International Technical Conference on the Enhanced Safety of Vehicles (ESV)*, Washington, DC, June 13–16.

Perry, C. E., D. M. Bonetti, and J. W. Brinkley. 1991. *The effect of variable seat back angles on human response to +Gz impact accelerations*. AR-TR-1991-0110. OH: Aerospace Medical Research Laboratory, Wright-Patterson Airforce Base.

Pintar, F. A. 2012. *Biomedical analyses, tolerance, and mitigation of acute and chronic trauma*. ADA590473. Report prepared for U.S. Army Medical Research and Materiel Command, Fort Detrick, MD.

Pintar, F. A., N. Yoganandan, and D. J. Maiman. 2008. Injury mechanisms and severity in narrow offset frontal impacts. *Annals of Advances in Automotive Medicine*. 52:185–189.

Pintar, F. A., N. Yogananda, and J. Moore. 2012. Spine and Pelvis Fractures in Axial Z-Acceleration. *Injury Biomechanics Research: Proceedings of the Fortieth International Workshop*. NHTSA, Washington.

Polanco, M. and J. D. Littell. 2011. Vertical Drop Testing and Simulation of Anthropomorphic Test Devices. *Presented at the American Helicopter Society 67th Annual Forum*, Virginia Beach, VA, May 3–5.

Poopitaya, S., and K. Kanchanaroek. 2009. Injuries of the thoracolumbar spine from tertiary blast injury in Thai military personnel during conflict in southern Thailand. *Journal of the Medical Association in Thailand*. 92(Suppl 1):S129–S134.

Possley, D. R., J. A. Blair, B. A. Freedman, A. J. Schoenfeld, R. A. Lehman, and J. R. Hsu. 2012. The effect of vehicle protection on spine injuries in military conflict. *Spine Journal*. 12(9):843–848. DOI: 10.1016/j.spinee.2011.10.007.

Ragel, B. T., C. D. Allred, S. Brevard, R. T. Davis and E. H. Frank. 2009. Fractures of the thoracolumbar spine sustained by soldiers in vehicles attacked by improvised explosive devices. *Spine*. 34(22):2400–2405. DOI: 10.1097/BRS.0b013e3181b7e585.

Rajasekaran, S., R. M. Kanna, and A. P. Shetty. 2015. Management of thoracolumbar spine trauma: An overview. *Indian Journal of Orthopaedics*. 49(1):72–82. DOI: 10.4103/0019-5413.143914.

Ramasamy, A., A. M. Hill, R. Phillip, I. Gibb, A. Bull, and J. C. Clasper. 2011. The modern "deck-slap" injury-calcaneal blast fractures from vehicle explosions. *Journal of Trauma*. 71(6):1694–1698. DOI: 10.1097/TA.0b013e318227a999.

Rapaport M., E. Forster, and A. Schoenbeck. 1997. *Establishing a Spinal Injury Criterion for Military Seats*. Patuxent River, MD: Naval Air Warfare Center.

Richards, D., M. Carhart, C. Raasch, J. Pierce, D. Steffey, and A. Ostarello. 2006. Incidence of thoracic and lumbar spine injuries for restrained occupants in frontal collisions. *Annual Proceedings: Association for the Advancement of Automotive Medicine*. 50:125–139.

Richter, D., M. P. Hahn, P. A. Ostermann, A. Ekkernkamp, and G. Muhr. 1996. Vertical deceleration injuries: A comparative study of the injury patterns of 101 patients after accidental and intentional high falls. *Injury*. 27(9):655–659. DOI: 10.1016/S0020-1383(96)00083-6.

Roaf, R. 1960. A study of the mechanics of spinal injuries. *Journal of Bone and Joint Surgery*. 42–B(4):810–823.

Rowe, K. W., and C. J. Brooks. 1984. Head and neck injuries in Canadian Forces ejections. *Aviation, Space and Environmental Medicine*. 55(4):313–315.

Ruff, S. 1950. Brief acceleration: Less than one second. In: *German Aviation Medicine, World War II*, Vol. I, Chapter VI-C, 584–598. Department of the Air Force, WA.

Sandstedt, P. 1989. Experiences of rocket seat ejections in the Swedish Air Force: 1967–1987. *Aviation, Space and Environmental Medicine*. 60(4):367–373.

Schinkel, C, T. M. Frangen, A. Kmetic, H. J. Andress, and G. Muhr. 2006. Timing of thoracic spine stabilization in trauma patients: Impact on clinical course and outcome. *Journal of Trauma*. 61(1):156–160. DOI: 10.1097/01.ta.0000222669.09582.

Schoenfeld, Andrew J., Gens P. Goodman, and Philip J. Belmont. 2012. Characterization of combat-related spinal injuries sustained by a US Army Brigade Combat Team during Operation Iraqi Freedom. *The Spine Journal*. 12(9):771–776.

Scullion, J. E., S. D. Heys, and G. Page. 1987. Pattern of injuries in survivors of a helicopter crash. *Injury*. 18(1): 13–14. DOI: 10.1016/0020-1383(87)90376-7.

Shanahan, D. F., and M. O. Shanahan. 1989a. Injury in U.S. Army helicopter crashes October 1979–September 1985. *Journal of Trauma*. 29(4):415–423.

Shanahan, D. F., and M. O. Shanahan. 1989b. Kinematics of U.S. Army helicopter crashes: 1979–85. *Aviation, Space and Environmental Medicine*. 60(2):112–121.

Shand, L. D. 1978. Comparative injury patterns in U.S. army helicopters. In *Operational Helicopter Aviation Medicine*. ed. Knapp, S.C, 54.1–54.7. London, UK: Technical Editing and Reproduction Ltd.

Shannon, R. H and V. Ferrari. 1969. *USAF ejection experience in the combat environment, 1 Jan 1967–30 June 1969*. Directorate of Aerospace Safety, Deputy Inspector General for Inspection and Safety, USAF. Norton Air Force Base, CA.

Sivak, M., and B. Schoettle. 2010. Toward understanding the recent large reductions in U.S. road fatalities. *Traffic Injury Prevention*. 11(6):561–566. DOI: 10.1080/15389588.2010.520140.

Smelsey, S. O. 1970. Study of pilots who have made multiple ejections. *Aerospace Medicine*. 41(5):563–566.

Smiley, J. R. 1964. Rcaf ejection experience: Decade 1952–1961. *Aerospace Medicine*. 35:125–129.

Smith, H. E., D. G. Anderson, A. R. Vaccaro, T. J. Albert, A. S. Hilibrand, J. S. Harrop, and J. K. Ratliff. 2010. Anatomy, biomechanics, and classification of thoracolumbar injuries. *Seminars in Spine Surgery*. 22(1):2–7. DOI: 10.1053/j.semss.2009.10.001.

Smith, J. A., J. H. Siegel, and S. Q. Siddiqi. 2005. Spine and spinal cord injury in motor vehicle crashes: A function of change in velocity and energy dissipation on impact with respect to the direction of crash. *Journal of Trauma and Acute Care Surgery*. 59(1):117–131.

Somers, J. T., D. Gohmert, and J. W. Brinkley. 2013. *Application of the Brinkley dynamic response criterion to spacecraft transient dynamic events*. NASA/TM-2013-217380. Houston, TX: National Aeronautics and Space Administration.

Sonoda, T. 1962. Studies on the strength for compression, tension and torsion of the human vertebral column. *Journal of Kyoto Prefectural University of Medicine*. 71:659–702.

Spurrier, E., I. Gibb, S. Masouros, and J. Clasper. 2016. Identifying spinal injury patterns in underbody blast to develop mechanistic hypotheses. *Spine*. 41(5):E268–E275. DOI: 10.1097/BRS.0000000000001213.

Spurrier, E., J. A. Singleton, S. Masouros, I. Gibb, and J. Clasper. 2015. Blast injury in the spine: Dynamic response index is not an appropriate model for predicting injury. *Clinical Orthopaedics and Related Research*. 473(9):2929–2935. DOI: 10.1007/s11999-015-4281-2.

Stech, E. L., and P. R. Payne. 1969. *Dynamic models of the human body*. AMRL-TR-66-157. OH: Aerospace Medical Research Laboratory, Wright-Patterson Air Force Base.

Stemper, B. D., J. L. Baisden, N. Yoganandan, F. A. Pintar, J. DeRosia, P. Whitley, G. R. Paskoff, and B. S. Shender. 2012. Effect of Loading Rate on Injury Patterns During High Rate Vertical Acceleration. *Presented the International Research Council on the Biomechanics of Injury (IRCOBI) Conference*, Dublin, Ireland, September 12–14.

Stemper, B. D., S. G. Storvik, N. Yoganandan, J. L. Baisden, R. J. Fijalkowski, F. A. Pintar, B. S. Shender, and G. R. Paskoff. 2011. A new PMHS model for lumbar spine injuries during vertical acceleration. *Journal of Biomechanical Engineering*. 133(8):081002.

Stemper, B. D., N. Yoganandan, F. A. Pintar, B. S. Shender, and G. R. Paskoff. 2009. Physical effects of ejection on the head-neck complex: Demonstration of a cadaver model. *Aviation Space and Environmental Medicine*. 80(5):489–494.

Storvik, S. G., B. D. Stemper, N. Yoganandan, and F. A. Pintar. 2009. Population-based estimates of whiplash injury using nass cds data—Biomed 2009. *Biomedical Sciences Instrumentation*. 45:244–249.

Sturgeon, W. 1988. *Canadian forces aircrew ejections, descent, and landing injuries, 1 January 1975–31 December 1987*. DCIEM 88-RR-56. Downsview, ON, Canada: Defense and Civil Institute of Environmental Medicine.

Thyagarajan, R., J. Ramalingam, and K. B. Kulkarni. 2014. Comparing the Use of Dynamic Response Index (DRI) and Lumbar Load as Relevant Spinal Injury Metrics. ADA591409. *Presented at the Workshop on Numerical Analysis of Human and Surrogate Response to Accelerative Loading*, Aberdeen, MD, January 7–9.

Tile, M., T. Hearn, and M. Vrahas. 2003. Biomechanics of the Pelvic Ring. In: *Fractures of the Pelvis and Acetabulum*. eds. M. Tile, D.L. Helfet, and J.F. Kellam, 32–45. 3rd ed. Philadelphia, PA: Lippincott, Williams and Wilkins.

Tran, N. T., N. A. Watson, A. F. Tencer, R. P. Ching, and P. A. Anderson. 1995. Mechanism of the burst fracture in the thoracolumbar spine—The effect of loading rate. *Spine*. 20(18):1984–1988.

Tsai, V. W., C. L. Anderson, and F. E. Vaca. 2008. Young female drivers in fatal crashes: Recent trends, 1995–2004. *Traffic Injury Prevention*. 9(1):65–69. DOI: 10.1080/15389580701729881.

Visuri, T., and J. Aho. 1992. Injuries associated with the use of ejection seats in Finnish pilots. *Aviation, Space and Environmental Medicine*. 63(8):727–730.

Vresilovic, E. J., S. Mehta, R. Placide, and A. Milam. 2005. Traumatic spondylopelvic dissociation. A report of two cases. *The Journal of Bone and Joint Surgery. American Volume*. 87(5):1098–1103. DOI: 10.2106/JBJS.D.01925.

Wang, J., R. Bird, B. Swinton, and A. Krstic. 2001. Protection of lower limbs against floor impact in army vehicles experiencing landmine explosion. *Journal of Battlefield Technology*. 4(3):8–12.

Wang, M. C., F. Pintar, N. Yoganandan, and D. J. Maiman. 2009. The continued burden of spine fractures after motor vehicle crashes. *Journal of Neurosurgery: Spine*. 10:86–92. DOI: 10.3171/2008.10.SPI08279.

Watts, D. T., E. S. Mendelson, and A. T. Kornfield. 1947. Human tolerance to accelerations applied from seat to head during ejection seat tests. TED No. NAM 25605, Report No. 1. Philadelphia, PA: Naval Air Material Center.

Werner, U. 1999. Ejection associated injuries within the German Air Force from 1981–1997. *Aviation, Space and Environmental Medicine*. 70(12):1230–1234.

Wilcox, R. K., D. J. Allen, R. M. Hall, D. Limb, D. C. Barton, and R. A. Dickson. 2004. A dynamic investigation of the burst fracture process using a combined experimental and finite element approach. *European Spine Journal*. 13(6):481–488. DOI: 10.1007/s00586-003-0625-9.

Williams, C. S. 1993. F-16 pilot experience with combat ejections during the Persian Gulf War. *Aviation, Space and Environmental Medicine*. 64(9):845–847.

Yacavone, D. W., R. Bason, and M. S. Borowsky. 1992. Through the canopy glass: A comparison of injuries in Naval Aviation ejections through the canopy and after canopy jettison, 1977 to 1990. *Aviation, Space and Environmental Medicine*. 63(4):262–266.

Yoganandan, N., J. Moore, M. W. Arun, and F. Pintar. 2014. Dynamic responses of intact post mortem human surrogates from inferior-to-superior loading at the pelvis. *Stapp Car Crash Journal*. 58:123–143.

Yoganandan, N., F. Pintar, J. Baisden, T. Gennarelli, and D. Maiman. 2004. Injury biomechanics of C2 dens fractures. *Annual Proceedings/Association for the Advancement of Automotive Medicine. Association for the Advancement of Automotive Medicine*. 48:323–337.

Yorra, A. J. 1956. The Investigation of the Structural Behaviour of the Intervertebral Discs. Master's dissertations, MIT.

13 The Thoracolumbar Spine
Cadaveric Testing under Vertical Loading

Brian D. Stemper, Narayan Yoganandan and Frank A. Pintar

CONTENTS

13.1 BIOMECHANICAL FACTORS AND EXPERIMENTAL STUDIES OF THORACOLUMBAR SPINE TRAUMA

It is well known that injuries, mechanisms, tolerance and risk functions are modulated by various factors related to the occupant and loading scenario. From an anatomical perspective, for example, injury types can change from burst to compression–flexion mechanism for the same loading vector depending on the orientation or application of the insult relative to the spine. A vector oriented vertically along the 'neutral axis' of the spinal column predisposes the spine to axial loading, often resulting in burst fracture, while an anterior-eccentric vertical vector introduces bending in addition to the vertical load. This changes the quantification of injury: peak compressive forces may define burst fractures while compressive force along with flexion moment may characterise wedge fractures with posterior element distractions. Likewise, characteristics of the external insult can also alter injuries. Loading rate or rate of acceleration onset (jerk), pulse duration, change in velocity and the amplitude of maximum acceleration also influence the onset and type of injury. From an occupant protection perspective, ejection seats, aircraft designs and the ejection process try to account for these variables. The rostral migration of injuries seen in some descriptive ejection studies compared to the thoracolumbar junction concentration in civilian populations (discussed further in Chapter 12) may be due to differences in these anatomical and environmental parameters. Civilian trauma to the junction occurs at 'lower' (not quasi-static) rates and usually from caudal to cephalad loading vector from impact to the buttocks (as with a fall). Studies delineating the specific roles of these biomechanical variables are beginning to evolve. Numerous biomechanical investigations have been performed to characterise spinal fractures due to dynamic vertical loading. Most often, the specific aims were to clarify the mechanism, observe the fracture pattern, measure the spinal canal occlusion, compare surgical or conservative management outcomes, or examine the general mechanical behaviour of the injury.

To achieve these aims, experimental models were designed (Cain et al. 1993; Cotterill et al. 1987; Curry et al. 2015; Duma et al. 2006; Fredrickson et al. 1992; Hongo et al. 1999; Hoshikawa et al. 2002; Kazarian and Graves 1972; Langrana et al. 2002; Ochia et al. 2003; Panjabi et al. 1995; Shirado et al. 1992; Tran et al. 1995; Wilcox et al. 2003; Willen et al. 1984). These models involve fixing the caudal end of the specimen and applying a vertical load to the cranial end. Although this artificial end condition does not accurately replicate the inertially driven compression of the lumbar spine experienced during high-rate vertical acceleration, important injury tolerance metrics have been derived from these studies, including preliminary tolerance levels and factors affecting injury tolerance. These fixed-end condition studies have quantified the effect of orientation on lumbar spine fracture tolerance (Duma et al. 2006), quantified the contribution of the posterior elements (Langrana et al. 2002), and studied the effect of loading rate (Ochia and Ching 2002), in addition to a variety of other clinically and biomechanically related studies.

More recently, a new experimental model has been developed with the intent of replicating the inertial contribution of the body mass during vertical acceleration/deceleration scenarios such as avia-tor ejection, helicopter crashes and underbelly blast (UB blast) (Stemper et al. 2011). The model is unique in its capacity to control lumbar spine orientation and specific characteristics of the accelera-tion versus time pulse to induce clinically-relevant injuries. Specifically, the model induced multiple non-contiguous spinal fractures (i.e. L1 burst fracture and L4 wedge fracture) with an acceleration pulse of 21 G and 488 G/s (Stemper et al. 2011). The model was also used to quantify differing thor-acic and lumbar spine injury patterns based on acceleration inputs (Stemper et al. 2014; Yoganandan et al. 2013, 2015). Importantly, this model can also be compared to recent studies with whole-body post-mortem human subjects (PMHS). Descriptions of those models and relevant experimental find-ings are provided below due to the novelty of the models, their applications and their findings. A more thorough summary of fixed-end condition lumbar spine models and their associated biomechanical findings can be found in other reviews (Stemper et al. 2015).

13.1.1 Component-Level Spine Studies

Component-level spine studies were conducted to identify the injuries, injury mechanisms and injury metrics which can be used to understand the dynamic biomechanics of the human dorsal spine and define tolerance and injury criteria from tests on upper and lower thoracic and lumbar spinal columns. The protocol for those studies incorporated the use of spine segments obtained from unembalmed PMHS. An essential part of post-test injury determination was to obtain a complete set of pre-test ima-ging scans that were compared to imaging scans obtained following dynamic testing to identify any bony or catastrophic soft tissue injuries. Computed tomography (CT) scans were obtained prior to the initiation of any dynamic testing and anterior–posterior and lateral radiographs were obtained prior to each dynamic test. Component-level studies discussed in this section were grouped into upper (T2–T6) and lower (T7–T11) thoracic spines and thoracolumbar (T12–L5) columns to assess injury risk in the thoracic or lumbar spinal regions. Experimental preparation used in these tests consisted of mounting superior and inferior ends of the individual columns in polymethylmethacrylate (PMMA) fixative. The mid-plane of the intervertebral disc at the mid-height of the columns was maintained parallel to the ground. The prepared specimen was then connected to the custom testing device (Figure 13.1).

The experimental setup and methodology were designed to simulate vertical impact loading (along the z-axis). A mass simulating the inertial load of the torso on the lumbar spine during vertical accel-erative loading was attached to the cranial platform and the base of the specimen was attached to a decoupled caudal platform. The load was applied to the superior aspect of the specimen through a metal cylinder, which could be moved anterior-posteriorly to produce different eccentricities of the loading vector to the specimen. Varying masses were added to the superior end of the spine. The model incorporated a six-axis load cell located at the inferior end of the specimen to measure forces and moments, along with redundant accelerometers attached to the cranial and caudal platforms.

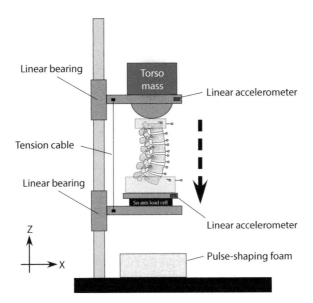

FIGURE 13.1 Experimental test setup for component-level spine testing. The thoracolumbar segment specimen is shown.

The preparation was held in place by a cable connecting the upper and lower metal platforms. The two platforms were connected to the vertical track of the custom drop tower using independent carriages with low-friction bearings. The specimen along with the platforms was dropped as one unit to sustain the impact loading from the caudal end of the spinal column, allowing the torso mass to apply an inertial load to the cranial aspect of the specimen.

The testing sequence typically involved two low-energy tests along with a final high-energy test. Anterior–posterior and lateral radiographs were conducted prior to the initiation of testing and following each dynamic test, with CT scans obtained prior to and following the completion of testing. Medical imaging scans along with specimen palpation are used to identify bony fracture (imaging) or soft tissue failure (palpation) and characterise fracture type. Low-energy (Group-A) tests were done by attaching a mass of 15 kg to the superior end of the preparation and dropping the specimen at different heights: 0.40 and 0.60 m for the upper and lower thoracic columns and 0.46 and 0.82 m for the lumbar columns. High-energy (Group-B) tests were conducted by dropping the thoracic and lumbar columns from a height of 1.5 m or higher and modulating the foam interface material at the base of the drop tower to control rate of acceleration onset. The resulting injuries were classified using pre- and post-test images based on the three-column concept and instabilities were assessed by a neurosurgeon. The biomechanical data were analysed as follows. The peak magnitudes of accelerations and forces along the z-axis were determined along with the times of occurrences. In addition, bending moments in the sagittal plane were obtained at the time of the development of peak axial forces. Data were evaluated for both sub-failure and the failure tests. Logistic regression techniques were used for the present data set using upper and lower thoracic columns and combining the thoracic and lumbar spines as two different data sets, and results were compared with literature. Upper and lower thoracic and lumbar spines and tested levels are used synonymously in the text and illustrations. Whereas biomechanical testing often incorporates older PMHS specimens, demographics of the PMHS used in these studies were selected to be more in-line with a military population in terms of age, stature and body weight.

13.1.1.1 Dynamic Biomechanics of Thoracic and Lumbar Columns

A recent study incorporating this model was focused on determining injuries, injury mechanisms and injury metrics applicable for the study of dynamic biomechanics of the thoracic and lumbar spines during vertical accelerative environments (Yoganandan et al. 2013). The experimental matrix resulted in a total of 24 tests: nine for the lumbar and upper thoracic columns and six for the lower thoracic spinal columns. Group-A consisted of six tests for the upper thoracic and lumbar spines and four tests for the lower thoracic spines, and Group-B consisted of three tests for the upper thoracic and lumbar spines and two tests for the lower thoracic columns. The mean peak axial force data for the Group-A tests for the upper thoracic spine were 1600 N for the 40 cm drop and 1700 N for the 60 cm drop. The lower thoracic spine data were 1300 N for the 40 cm drop and 1800 N for the 60 cm drop. In both cases forces increased with increasing drop height. The mean peak force data for the Group-A tests for the lumbar spine were 1400 N for the 46 cm drop and 2900 N for the 82 cm drop. As in the case of the thoracic spines, forces increased with increasing drop height. The mean peak axial forces for Group-B tests for the upper thoracic, lower thoracic and lumbar spines were 4,000, 5,500 and 6,200 N, respectively. The peak forces increased from the upper thoracic spine to the lumbar spine. The mean bending moment data for the Group-B tests for the upper thoracic, lower thoracic and lumbar spine tests were 90, 20 and 35 Nm, respectively.

All specimens sustained injuries during the final test. The upper thoracic spinal columns sustained comminuted fractures of the T4 vertebral body with bipedicular fractures and dislocation, and additional injuries in one specimen consisted of comminuted fracture of the T3 body with disruption of posterior elements and a minor fracture of the anterior region of the T5 body with intact posterior elements. The lower thoracic spine columns demonstrated comminuted fractures of the T8 vertebral body with bipedicular fracture in one case, and in the other case, the T8 body fracture was not associated with the bipedicular fracture although another fracture was identified to the T9 vertebral body and right pedicle. The lumbar spinal columns sustained comminuted fractures of the L1 vertebral body with bipedicular fractures in all three cases, and in one case a lamina fracture was also identified. These injuries were associated with dislocation. The injury mechanism in all cases was determined to be compression-related, and this was based on an examination of images and according to the clinical literature. The anterior and posterior columns of the spine were involved in these fractures, resulting in a clinically and biomechanically unstable column. Figure 13.2 shows a post-test CT axial view of a specimen demonstrating fractures to the vertebral body and its posterior elements.

Recent literature has highlighted an increased risk of thoracic and lumbar spine injury in military environments, given the high number of land-based vehicle UB blast events that have occurred during the current military conflicts in Iraq and Afghanistan (Stemper et al. 2011, 2015, 2016). Protecting military service members relies on accurate biomechanical information that can be used to design and implement more effective vehicle safety enhancements. There is a relative paucity of data available in literature regarding dynamic lumbar spine biomechanics in vertical accelerative environments, both injurious and non-injurious. The study discussed above quantified the physiologic and traumatic biomechanics of the lumbar spine in this type of environment.

13.1.1.2 Effect of Vertical Acceleration Characteristics

Another study was conducted using this model and incorporating five lumbar spine specimens to characterise differences in injury patterns resulting from variation in accelerative loading characteristics (Stemper et al. 2012; Yoganandan et al. 2015). Specimens 1 and 2 sustained peak accelerations of 182.1 and 218.1 m/sec^2 and were classified as belonging to high-speed aircraft ejection-type loadings. Specimens 3 to 5 sustained 347.9 to 549.2 m/sec^2 and

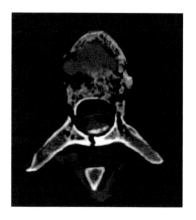

FIGURE 13.2 Axial CT image of post-test scan of L1 vertebra. Note comminuted fractures with bipedicular and lamina involvement.

were classified as belonging to helicopter crash-type loadings. Peak axial forces for high-speed air-craft ejection-type loadings ranged from 4.8 to 5.6 kN. Peak forces for helicopter crash-type loadings ranged from 5.3 to 7.2 kN. The first two specimens subjected to ejection-type loading responded with single-level injuries to the L1 vertebra; one injury was stable and the other was unstable. The stable injury was confined to the anterior column, whereas the unstable injury involved the anterior, middle and posterior columns. These loadings were associated with lower deceleration rates and lower forces. In contrast, the other three specimens subjected to helicopter crash-type loadings responded with differing injury patterns. The third specimen sustained bilateral facet dislocation with disc involvement at the L4–L5 level with associated widening of the interspinous ligament. The disc involvement was defined as rupture of the anterior and posterior longitudinal ligaments associated with instability. The other two specimens responded with compression-related vertebral fractures. Injuries at multiple caudal levels occurred at L1, L3 and L4 for specimen 4 and L1, L2 and L3 for specimen 5. All three specimens were considered clinically unstable at least at one spinal level. Figure 13.3 shows the schematic injuries sustained on a specimen-by-specimen basis based on the three-column (anterior, middle and posterior) approach.

These studies were based on recent field data wherein injuries have been identified to the dorsal spinal column in recent epidemiological and descriptive databases (Stemper et al. 2011), and the lack of understanding of the biomechanics of this region of the human vertebral column as applied to motor vehicle incidents (Pintar et al. 2012). The objectives were accomplished by impacting unembalmed PMHS spinal columns using a custom drop tower apparatus, recording the forces and sagittal bending moments, analysing these data for their peak magnitudes, and identifying the resulting injuries using pre- and post-test radiographs and CT scans by the neurosurgical faculty. The drop tower model simulates the loading from the seat of the vehicle to the pelvis of the occupant and to the lumbar spine in the vertical direction. This type of dynamic load transfer from the vehicle structure to the seated occupant has been hypothesised to be a mechanism of injury to the lumbar spine (Aslan et al. 2005; Bowrey et al. 1996; Pintar et al. 2012; Zaidel et al. 1992). Consequently, this experimental model was chosen for these studies.

It is known that degeneration of any spinal element affects the biomechanical response including the segmental mobility, injuries, injury mechanisms and tolerances/criteria (Board et al. 2006; Fujiwara et al. 2000; Stemper et al. 2010; Yoganandan et al. 2007). The influence of disc degeneration, an age-related process in the human spinal column, on the mechanism of compression fractures of the human thoracolumbar spine has been examined in one study (Shirado et al. 1992). The degeneration of disc components such as dehydration of the nucleus, degradation of the annulus fibers, decrease in disc height and formation of outward bony growths on the anterior regions of the spine segment(s)

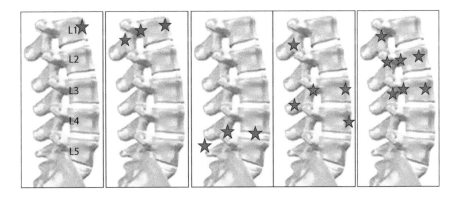

FIGURE 13.3 Injuries sustained in five lumbar spine specimens subjected to differing accelerative loading environments (Specimens 1 to 5 from left to right). Blue and red stars indicate injury locations resulting from acceleration tests designed to simulate aviator ejection and helicopter crash pulses, respectively.

contribute to decreased mobility and load-carrying capacity (Kumaresan et al. 2001). Because the studies described above used normal spine specimens from relatively younger age groups, age-related effects are considered negligible, and hence, the current results are relevant to military populations with emphasis on vertical dynamic load applications.

Injuries sustained by all preparations included fractures of the anterior and posterior regions of respective spinal vertebrae. The common presence of bipedicular fractures along with body fractures indicated that the load was shared by the entire vertebra. The presence of lamina injuries in some specimens further confirmed this analysis. Thus, injuries were attributed to the compressive mechanism and considered clinically unstable due to the involvement of anterior and posterior columns of the spine (Maiman and Pintar 1992; Yoganandan et al. 2012). However, disruptions of the posterior elements secondary to ligamentous complex involvement occurred from the flexion bending moment imparted to the specimen during the loading process, and from this perspective, flexion can be associated as the secondary mechanism in addition to the primary compressive force transmission. Thus, these studies have produced compression-related injuries to the spines from impact loading along the vertical (z-axis) direction, and as discussed earlier, offer support to previous postulates of motor vehicle-related spinal trauma.

13.1.1.3 Thoracolumbar Spine Injury Risk Curves

Because non-fracture and fracture data were available from the experimental protocol incorporated in the studies discussed above, it is possible to examine the possibility of deriving injury risk curves. Traditionally, from these types of data, functions such as Weibull and logistic regressions were used to derive probability-based curves (Cavanaugh et al. 1993; Pintar et al. 1997; Yoganandan et al. 1993, 1996a, 1996b). Logistic regression analysis of data from the upper and lower thoracic spines indicated that a peak force of 3.4 kN was associated with 50% risk for fracture, and when data were combined with the lumbar spine tests, the force increased to 3.7 kN at the same level of probability. An analysis of literature data for the thoracic spine vertebral bodies indicated a peak force of 2.7 kN (Kazarian and Graves 1977; Maiman et al. 1986; Sonoda 1962; Yoganandan et al. 1988), although other studies of lumbar spine vertebral body fracture have demonstrated considerably higher tolerance. For example, Stemper and colleagues reported mean fracture force between 8.1 and 10.8 kN for lower and higher rate dynamic compressions (Stemper et al. 2014). That study identified significant rate and gender dependence, with male specimens and higher compressive rates producing higher tolerance, up to 10.8 ± 3.1 kN for male vertebral bodies tested between 1.4 and 4.0 m/s. The injury risk curves for all cases are shown in Figure 13.4.

The likely reasons for the increased fracture force in the present study incorporating partial thoracic column specimens compared to the thoracic vertebral body literature data include the loading rate and the type of experimental model. Previous thoracic vertebral body tests were limited to loading rates of approximately 1 m/s. In contrast, present failure tests were conducted at a velocity of 5.4 m/s (measured at the specimen base), with associated vertebral body compression rates as high as 4.0 m/s (Stemper et al. 2014). Increasing loading rate increases the load-carrying capacity of bones, and this is true for the human vertebral column (Kazarian and Graves 1977; Ochia et al. 2003). Furthermore, the forces obtained from vertebral body tests do not include contributions from the posterior elements of the vertebra. The difference between the current fracture force of 3.4 kN and the literature force of 2.7 kN represents the share of the posterior components and includes rate effects, and this result is in-line with other biomechanical spine studies (Kumaresan et al. 1998).

However, eliminating the rate effect and comparing the lumbar spine fracture tolerance risk curve (Figure 13.4b) to rate-matched vertebral body tests reveals that fracture tolerance from column tests was actually lower than isolated vertebral body tests: 50% probably of column fracture was approximately 4.5 kN compared to 8.9 to 10.8 kN for higher rate tests on vertebral bodies. This finding highlights the complex fracture mechanics of the lumbar column compared to the isolated vertebral body. While vertebral body testing produced a homogeneous axial load around the cortex of the body, resulting eventually in cortical failure, column testing produced burst fractures through a mechanism

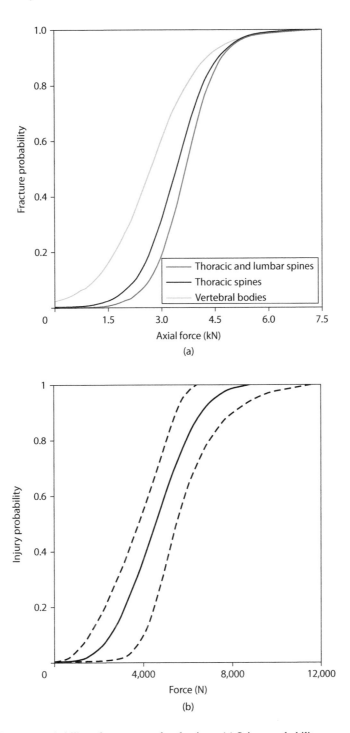

FIGURE 13.4 Fracture probability of component-level spines: (a) Injury probability curves for thoracic and lumbar spines, and vertebral bodies and (b) only lumbar spines.

involving fracture of the endplate due to compressive loads from the nucleus and to a lesser extent the annulus. This resulted in increased intra-body pressures that fractured the vertebral body outward from within along a more perpendicular vector, compared to axial loads applying compressive forces in-line with the cortical shell in the vertebral body tests. The endplate fracture tolerance was shown to be approximately 4 kN (Curry et al. 2015), which is more in-line with whole lumbar column tests and supports this theory.

In addition to the effects of loading rate and fracture mechanism, discussed above, the increase in the peak axial force level with the inclusion of the lumbar spine data indicates that the load-carrying capacity of the lumbar spine is greater than its rostral counterpart, a finding supported by the anatomical and geometrical differences between the two spines. However, because the sample size was small in the present study, if only lumbar spine data were to be included with additional specimens, as shown in Figure 13.4b, the level of fracture increases to approximately 4.5 kN at 50% probability. From this viewpoint, additional studies are needed to reinforce these observations and obtain robust lumbar spine-only injury criteria for developing and validating anthropomorphic test devices to predict injuries and advance crashworthiness studies.

13.1.1.4 Fracture Classification

Although other fractures occurred, some of the fracture types common to multiple specimens are discussed in more detail here. Fracture types can be defined according to the three-column concept, as described by Denis (Denis 1984), wherein the anterior column consists of the anterior vertebral body, anterior intervertebral disc and anterior longitudinal ligament, the middle column consists of the posterior vertebral body, the posterior intervertebral disc and the posterior longitudinal ligament, and the posterior column consists of the posterior ligamentous complex including the pedicles, facet joints, laminae and spinous processes (discussed further in Chapter 12). Due to the involvement of anterior and posterior columns or the middle column, these injuries can be attributed to axial compression. Vertical fractures affecting the anterior or posterior cortex were sustained at L2 and L3 by specimens subjected to the most severe acceleration pulses, as shown in Figure 13.5. Those fractures occurred in the caudal-cranial direction at approximately 1/3 the anterior–posterior body depth from the anterior or posterior cortex and involved the anterior/posterior cortex and cranial endplate, although integrity of the caudal endplate was not affected. One of those specimens also sustained a longitudinal fracture through the lamina and proximal aspect of the spinous process, and also affecting the pars interarticularis (Figure 13.5, right).

Burst fractures affecting anterior and posterior cortices as well as cranial and caudal endplates were not confined to one loading type (i.e., ejection catapult or helicopter crash) and occurred at different spinal levels. Specimens subjected to the ejection catapult acceleration sustained a burst fracture at L1 that included retropulsion of bony fragments into the spinal canal. Likewise, specimens subjected to the helicopter crash accelerations sustained a massive burst fractures in the middle lumbar spine (e.g. L3) that also included retropulsion of fragments into the canal (Figure 13.6). Burst fractures occur under pure compression mechanisms due to uniform load applied to anterior and posterior

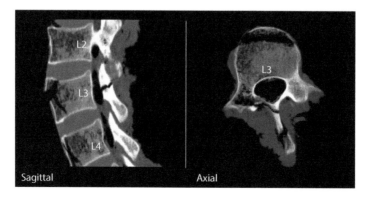

FIGURE 13.5 Fractures sustained at the L3 spinal level during vertical deceleration simulating helicopter crash.

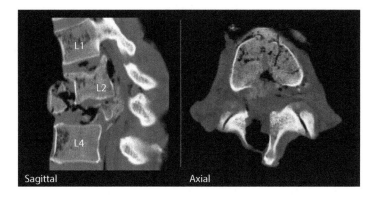

FIGURE 13.6 Fractures sustained at L2 (left & right) and L3 (left) spinal levels during the Helicopter Crash Test.

cortices and resulting in fracture of both regions. Some evidence of compression-bending was also evident from post-test CTs. Specimens subjected to the helicopter crash accelerations also sustained an anterior body compression fracture (i.e., wedge fracture) that included greater compression of the anterior than the posterior cortex. The mechanism for this type of fracture is compression–flexion, resulting in higher loading concentration on the anterior cortex.

13.1.1.5 Summary of Component-Level Studies

This section presented biomechanical findings from a novel acceleration-based thoracic and lumbar spine injury model used to experimentally replicate the vertical acceleration environment experienced in a variety of military injury scenarios including aviator ejections, helicopter crashes and UB blasts. Fracture types produced by this experimental model, as highlighted above, are consistent with the types of lumbar spine fractures identified following axial loading sustained in military environments (Spurrier et al. 2016). These studies have quantified physiologic and traumatic biomechanics of thoracic and lumbar columns, quantified injury tolerance and identified factors that can affect injury tolerance including loading rate, gender and spinal region. Another important finding from these studies is that injury mechanisms and tolerance are different between column and isolated vertebrae models, indicating a need for continued column-based experimentation to outline the dynamic biomechanics of the thoracic and lumbar regions.

13.1.2 Whole-Body PMHS Studies

In events such as UB blasts due to improvised explosive devices (IED), the accelerative impact vector is oriented more along the inferior-to-superior direction than other axes. These events are more commonly reported in recent conflicts and are a subject of biomechanics research (Yoganandan et al. 2015). While not epidemiological, reports indicate the importance of the inferior-to-superior load vector in the production of injuries to the pelvis-sacrum-lumbar spine complex. Inferior-to-superior loading to the pelvis-sacrum-lumbar spine complex initiates from the seat to a belted military vehicle occupant in events. The harness tends to restrain the occupant such that the impact loading emanating from the seat traverses along the longitudinal axis of the human vertebral column via the pelvic ring and ischial tuberosities. Consequently, injuries can occur to the pelvis and/or spine structures. The identification of effective parameters characterising the applied external insult is critical for developing countermeasures and safety devices including anthropomorphic test device (ATD) design and injury criteria. Any external insult can be characterised in the form of variables such as amplitude, duration and shape. These variables might play a role in the load transmission and injury production to the pelvis-sacral-lumbar spine complex. This concept needs an examination for inferior-to-superior loading of the pelvis because of a lack of published information on injuries to these body regions. Intact/whole-body PMHS (cadaver) experimental models subjected to inferior-to-superior accelerative loading to the pelvis structures can be used to examine potential variables that might describe injuries to the pelvis and spine, individually or in combination. Whereas the component-level studies highlighted above are ideal for defining injury tolerance and factors associated with increased injury risk due to well-controlled loading and boundary conditions, intact PMHS are the most biofidelic models for studying injury production and development of safety enhancements.

Intact PMHS testing commonly involves the application of inferior-to-superiorly directed accelerations using an experimental sled apparatus with the specimen seated in a supine position (Salzar et al. 2009; Yoganandan et al. 2011), as depicted in Figure 13.7. As with component tests, external examination and pre-test radiographs were obtained to ensure absence of musculoskeletal trauma. Subjects were dressed in tight-fitting leotards and a mask covered the head–face complex and positioned on a custom buck, fixed to the platform of an acceleration sled (Seattle Safety Systems, Seattle, WA). An adjustable sled buck was designed to position each subject. The vertical and horizontal spacing of the seatpan and foot plate were also adjustable to accommodate varying femur and tibia-fibula lengths. The seatpan and footpan were instrumented with accelerometers and load cells.

Subjects were positioned supine with the torso and leg horizontal, and the thigh vertical and orthogonal to the torso and leg (Figure 13.7). This 90-90-90 posture representing the orientation of the leg, thigh and torso is similar to the posture of an occupant seated in a military vehicle. It should be noted that the seat structure of military vehicles is different from the automobile bucket-type seats and the experimental setup reflected those differences. A six-point belt configuration was used: two belts crisscrossed the chest and a separate lap belt was used. An experimental test series was conducted to determine the effect of lumbar spine orientation on occupant kinematics and injuries during inferior-to-superior accelerative loading of intact PMHS specimens. For the first two specimens, the body was placed with the pelvis, back and head resting on the horizontal back plate. This defined the neutral posture for the spine and pelvis. For the remaining tests, after placing the subject on the plate, bolsters were introduced underneath the scapulae and sacrum to reduce lumbar lordosis, meant to represent the posture of a seated solider with personal protective equipment. The sternum and pelvic angles were measured with respect to the horizontal. The angulation difference was approximately 40 degrees between the two postures. In all except the first specimen, the specimen-seat interface was padded with 50-mm thick energy-absorbing material (honeycomb, compressive strength 150 kPa). Specimens were instrumented with accelerometers and optical marker mounts at specific anatomical landmarks.

FIGURE 13.7   Schematic of the whole-body test setup.

All tests were conducted using the acceleration sled. Triangular and sigmoidal accelerative input pulses varied in shape, duration and maximum acceleration (Table 13.1). Multiple tests were conducted on each specimen. Palpation, clinical assessments and radiographs were obtained in between tests to confirm the integrity of the musculoskeletal structures. Following the final test when injuries were identified, radiographs, CT imaging and detailed autopsies were done. In automotive frontal impacts, loading rates are about 300 N/ms with knee bolster forces reaching their peaks 20–60 ms (Rupp et al. 2002). Loading rates in this study ranged from 226 to 3,080 N/ms.

All temporal data were normalised to the midsize male anthropometry using the impulse-momentum approach (Mertz 1984). Details of normalisation including equations are summarised elsewhere (Yoganandan et al. 2014a, 2014b). Tests were grouped based on nominal velocities (Table 13.1). Seatpan forces were used to compute the impulse variable. Peak normalised seat force, impulse computed as the integral of a force-time curve, power computed as the product of force and velocity in the time domain, and the loading rate were obtained.

Specimens 1–3 were subjected to triangle-shaped acceleration pulses (except Test 3 for Specimen 1, wherein the unloading portion resembled the saw-tooth shape), and Specimens 4 and 5 were tested by applying the sigmoid pulse to examine the role of the shape. Each specimen was tested using the same shaped pulse with different durations (except Test 3). Testing was stopped in all but one specimen after injury was detected. Testing was concluded for Specimen 3 upon reaching the limits of the loading system. The change in velocities ranged from 2.8 to 12.5 m/s for all tests. Figure 13.8 shows normalised forces versus time responses from the pelvis, and resultant sacrum and spine accelerations on a temporal basis for each specimen. Normalised peak magnitudes of resultant sacrum and spine accelerations along with peak forces and other variables are shown on a test-by-test basis (Table 13.2). The peak forces ranged from 4,305 to 20,541 N for normalised data. Peak normalised resultant sacrum and spine accelerations ranged from 10.4 to 204.2 g and 12.1 to 84.6 g, respectively.

Times to peak force were shorter for Specimens 1 and 2 with pelvis-only injury (11.1 and 13.1 ms) than in Specimen 4 wherein the injury was associated with the spine (27.0 ms) and in Specimen 5

TABLE 13.1
Test Matrix Describing Pulse Characteristics

PMHS No	Test ID	Input Pulse Characteristics			Velocity (m/s)
		Shape	Peak (g)	Time to Peak (ms)	
1	1	Triangle	28	21	6.09
	2	Triangle	44	10	4.93
	3	Saw-tooth	60	6	6.42
2	4	Triangle	28	21	5.98
	5	Triangle	43	10	4.95
3	6	Triangle	10	30	2.84
	7	Triangle	9	40	3.69
	8	Triangle	19	31	6.02
	9	Triangle	19	40	7.73
	10	Triangle	29	31	8.97
4	11	Sigmoid	10	34	2.96
	12	Sigmoid	21	32	6
	13	Sigmoid	32	30	8.97
5	14	Sigmoid	10	42	4.13
	15	Sigmoid	19	37	8.05
	16	Sigmoid	29	46	12.35
	17	Sigmoid	39	38	12.46

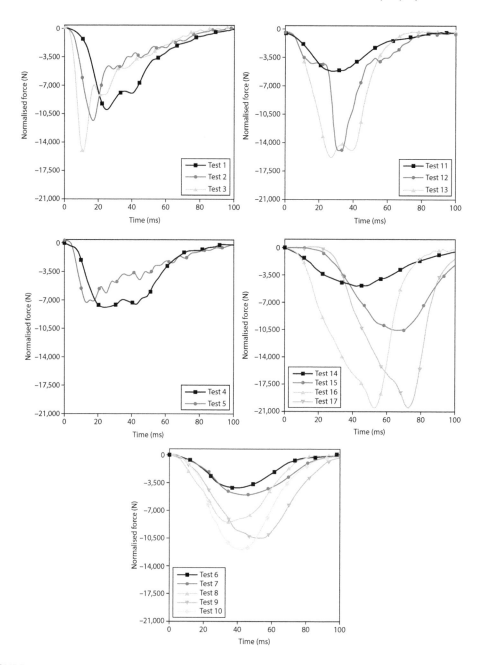

FIGURE 13.8   Normalised forces shown in sequence from upper left to lower right: PMHS 1 through PMHS 5.

wherein the spine was the only injured region (43.0 ms) (Figure 13.9). A similar trend was also found for the impulse metric. The time to peak accelerations of the sacrum and spine were such that the sacrum attained its peaks earlier in Specimens 1, 2 and 4 (11.8, 16.5 and 35.4 ms for the sacrum and 18.7, 28.5 and 41.7 ms for the spine), with pelvis injury and later in Specimen 5 with spine (without pelvis) injury (43.0 and 36.8 ms for sacrum and spine).

Specimen 3 did not sustain pelvis or spine injuries, while all other specimens sustained injuries. Specimen 5 sustained lumbar spine-only injury, Specimen 4 sustained lumbar spine and pelvis injuries, and Specimens 1 and 2 sustained only pelvis injuries. Injuries were coded using the latest AIS

TABLE 13.2
Normalised Data for Each Test

ID	Force (N)	Power (kW)	Imp. (N-s)	Time to Peak Sacrum-g (ms)	Sacrum-g	Time to Peak Spine-g (ms)	Spine-g	Force Rate (N/ms)	Eff. Mass (kg)	SF-Force	SF-Time	SF-g
1	9981.8	51.2	333.7	19.7	48.5	30.7	72.3	949.0	51.3	1.051	1.117	0.882
2	11336.0	58.4	272.0	13.5	81.0	24.4	68.1	1336.5	52.7	1.059	1.094	0.927
3	15254.3	86.5	323.6	11.8	204.2	18.7	84.6	3079.6	56.0	0.954	1.069	0.874
4	7982.7	47.3	337.7	14.8	28.5	41.1	30.8	753.4	62.9	0.957	1.020	0.984
5	7367.2	35.7	266.5	16.5	34.3	28.5	34.9	1134.7	59.8	0.980	1.037	0.974
6	4305.5	11.7	176.5	33.0	11.5	40.1	19.0	243.6	44.9	1.223	1.225	0.862
7	5177.4	16.3	247.0	33.1	11.3	38.4	18.0	232.3	51.0	1.185	1.157	0.937
8	8585.9	44.1	352.9	43.2	19.0	29.7	24.9	379.7	47.2	1.103	1.171	0.851
9	10598.0	73.9	515.6	59.3	20.4	34.3	32.1	372.7	51.2	1.142	1.168	0.886
10	12023.4	99.7	476.4	51.5	77.7	–	–	773.7	45.5	1.089	1.149	0.873
11	4893.1	16.3	178.2	34.3	15.7	30.4	15.5	285.0	55.0	1.120	1.048	0.967
12	14875.1	101.7	346.0	38.8	55.4	36.2	30.1	2776.2	46.9	1.172	1.112	0.899
13	15731.9	140.1	489.1	35.4	41.2	41.7	49.9	1066.2	43.2	1.181	1.116	0.899
14	4892.1	20.0	253.4	48.4	10.4	24.5	12.1	225.5	50.0	1.105	1.132	0.855
15	12367.6	91.1	498.3	46.1	64.8	52.4	49.6	402.4	52.1	1.123	1.122	0.886
16	20540.9	240.4	760.5	37.8	85.3	45.1	58.5	751.0	49.8	1.111	1.135	0.856
17	19627.0	241.5	734.0	47.8	52.4	36.8	46.8	1346.8	53.9	1.058	1.091	0.883

SF = scale factor.

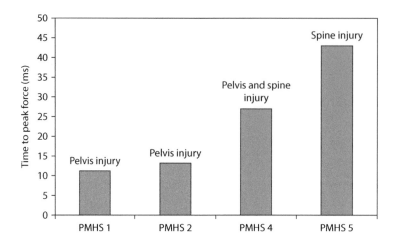

FIGURE 13.9 Time to peak force for injured specimens.

version (Abbreviated Injury Scale 2005). While the AIS 1990–1998 update was used in previous studies, it should be noted that not every fracture in the pelvis is assigned a separate AIS injury code in the latest version and this includes the sacroiliac region (Yoganandan et al. 2013).

In this study, the dynamic responses of the intact PMHS under inferior-to-superior loading applied to the pelvis were determined because data under this mode are sparse and injuries have been reported in certain military crash environments. The effect of pulse characteristics on the resulting injury and injury patterns to the pelvis and spine, alone or in combination, was delineated. This was accomplished using intact PMHS and applying impact loading using sled equipment. Normalised data were obtained in the form of sacrum/pelvis and spine accelerations, and pelvic forces were obtained using the seatpan load cell. Injuries were identified by palpation and radiographs between successive tests, and CT and detailed autopsy were done following the final test.

Researchers have used the repeated testing protocol to obtain more data and optimise specimen use. Recently, Pintar and colleagues conducted two to five sled tests using five PMHS for frontal impact studies (Pintar et al. 2010). However, early seat-to-head loading studies used greater number of repeated tests on each specimen. Evans and colleagues conducted 170 tests on 3, Ewing and colleagues reported 75 tests on 12, King and Vulcan reported 100 tests on 8, Prasad and colleagues reported 70 runs on 4, and Vulcan and colleagues reported 66 runs on 4 PMHS (Evans et al. 1962; Ewing et al. 1972; King and Vulcan 1971; Prasad and King 1974; Vulcan et al. 1970). Since repeated testing tends to weaken the specimen, limiting the number of tests for each specimen, as done in the present study, is more realistic.

Radiographs were obtained in between tests and after the final test. The intent was to stop further loading if the previous test resulted in detectable injury. Whole-body radiographs do not conclusively document/identify all hairline fractures, and in addition, some closed non-displaced fractures also elude radiographic diagnosis (Yoganandan et al. 1996). Thus, any abnormality due to their presence may influence response to subsequent loading. Repeated loading on a weakened specimen reduces tolerance. From this perspective, current loads are considered as lower bound/conservative estimates of human injury tolerance.

Intact specimens were used because the study of isolated components such as the pelvis alone does not incorporate coupling and injuries to adjacent structures. Mass recruitment effects are automatically included in the whole-body experimental model. However, the selection of this model does not allow the use of sensors such as acoustic sensors and/or strain gauges inside the pelvic ring without compromising its integrity, and hence, strains or fracture times could not be assessed, although methods exist to process signals from acoustic sensors (Arun et al. 2014). From this viewpoint, forces corresponding to the initiation of fracture were not obtained. However, impact loads transmitted to the pelvis and spine structures were recorded using the seatpan load cell, and accelerations of the sacrum and spine were recorded. These data serve as a first step in the characterisation of the dynamic responses of the intact PMHS from inferior-to-superior loading at the pelvis. Because the study used a horizontal acceleration sled as the load-delivering device, the specimens were positioned supine to induce loading along the desired axis. This is in contrast to the seated position of the occupant during the impacting event. A limitation of this approach is the absence of the axial pre-load inherently present on the spine joints due to gravity. However, effects of ignoring this load may be minimal as the restraint system applied some pre-load, likely in the range of the 1g, present on the erect vertebral column. Reducing the normal lumbar lordosis with additional flexural rotation also contributes to the pre-load.

The maximum forces in the present study were considerably greater than previous data (Bailey et al. 2013). This may be attributed to loading and rate differences because specimens were similar in terms of demographics in all studies. It should be noted that in the current study, only male specimens were used. Injury tolerance is known to be higher at higher loading rates in biological structures (McElhaney 1966; Pintar et al. 1998; Stemper et al. 2007, 2014; Yoganandan et al. 1989, 1990, 1991). Acknowledging that the study used the repeated loading protocol with a small sample size of five PMHS and 17 tests, the following discussions are advanced to provide basic insights into the role of the pulse characteristics on pelvis and lumbosacral spine injury biomechanics. In fact, this is the reason to examine many parameters using seatpan force-time and acceleration-time histories of the sacrum and spine. Specimens 1, 2 and 3 tolerated triangular sled pulses up to 30 and 1 g/ms. It should be noted that Specimens 1 and 2 had pelvis-only injuries, Specimen 4 had both pelvis and spine injuries and Specimen 5 had only spine injury. Specimen 3 remained intact following the entire test series. The time to peak for fracture force in Specimens 1, 2, 4 and 5 showed that with increasing time, these pulses reduce the risk of injury closer to the loading surface (i.e. pelvis) because of mass recruitment (Figure 13.9). This implies that a softer pulse tends to recruit more mass because of the extended time of wave travel and spares structures closer to the loading surface. With a shorter pulse, structures close to the loading surface (pelvis) tend to absorb the entire impact thus contributing to their injuries and sparing the spine from fracture. This implies that a softer and longer pulse spares structures closer to the loading surface and endangers structures further from it.

Rupp and colleagues gave another explanation based on approximately 250 knee-thigh-hip complex tests conducted with five male PMHS under various initial conditions for impacts (Rupp et al. 2008). The tests considered different types of padding and velocities and whole and segmented PMHS. The explanation was that high-rate short-duration loading to the knee by a rigid surface more likely induces injuries to the knee-distal femur region. Sparing of injury to the hip-proximal femur complex was attributed to the delay in reaching high enough forces further from the loading surface and to compliance of the entire complex. Pulses with longer times to peak produce higher forces further from the loading surface exceeding the lower tolerance of the hip-proximal femur region. This occurs before forces reach the higher tolerance of knee-distal femur region, closer to the loading surface. This is a compliance, timing and difference-in-injury-tolerance issue. Results from this study support these observations. Thus, all these factors should be considered when examining injury criteria.

Another variable is to compare the time of attainment of the sacrum and spine accelerations with the hypothesis that early peaking of the sacrum acceleration suggests predisposing the pelvis to impact loads and susceptibility to fracture, while delayed or parallel peaking of the sacrum and spine represents the coupled interaction between the two structures, mass recruitment effects and associated with pulse shape. This concept appears to be true in this limited specimen test series as sacrum accelerations reached their peaks in pelvis-only fracture tests earlier than spine (Figure 13.10). This parameter indicates the delay between the two components of the musculoskeletal system for inferior-to-superior impacts to the pelvis. Again, these observations are based on described limitations.

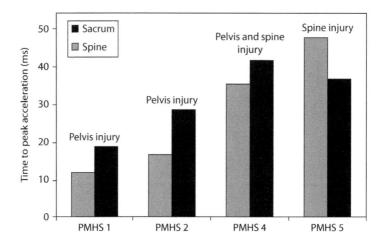

FIGURE 13.10   Time to peak for sacrum and spine resultant accelerations for injured specimens.

It is appropriate to examine the profiles of sacrum and spine accelerations for injured specimen tests. In Specimen 1, the rise time and peak sacrum acceleration were the shortest and greatest because of the use of the triangular pulse and no padding on the seat. It should be noted that padding is not always used in military vehicles, although soldiers wear personal protective equipment. In contrast, retaining the triangular pulse shape and adding an energy-absorbing material to the seat structure delayed and reduced the peak acceleration in Specimen 2.

Further, the sigmoid pulse inputted to Specimens 4 and 5 simulated the effect of a more extensively padded seat. However, in both cases, peak sacrum accelerations were greater and earlier than spine accelerations, and this may be related to the pelvis-only injury in these specimens. While Specimen 4 was padded like Specimen 2, the pulse shape was sigmoid, and even in this test, sacrum acceleration preceded spine, and spine acceleration reached above 45 g later, sustaining both pelvis and spine injuries. In contrast, under the same sigmoid shaped pulse, Specimen 5 sustained lumbar-spine-only injury, and characteristics were such that developments of the sacrum and spine kinematics were very similar, but the spine acceleration reached a greater magnitude than the sacrum. Mass recruitments effects may have played a role in the development of local kinematics. These initial observations appear to implicate that the inferior-to-superior loading mechanism and injury to the pelvis and spine are associated with pulse shape and not just the peak velocity. This is because, actual changes in velocities imparted to these specimens for fracture tests were 6 and 5 m/s for pelvis-only (Specimens 1 and 2), 9 m/s for pelvis and spine (Specimen 4) and 12 m/s for spine (Specimen 5) injuries (Table 13.1).

Regarding the choice of pulses, the study was initiated with the purpose of producing lumbar fractures in a rigid seat with a simple triangular pulse designated by velocity and time-to-peak metrics. After the first specimen sustained AIS 4 level pelvic injuries and no lumbar injury, the seat was padded with honeycomb and the outcome was still pelvic injury. For the third specimen, the triangular pulse was lengthened, with the hypothesis that padded seats in military vehicles lengthens the time of contact. The outcome was no injury. Although details of military vehicle seat characteristics were not available to the authors from published studies, the pulse was changed to a sigmoid shape in the fourth specimen. This choice was based on the bottoming out effect of (some) energy-absorbing seats. The outcome was pelvic and lumbar injuries. For the fifth specimen, the sigmoid pulse was further lengthened to induce lumbar injury without pelvic fracture. This result was produced in the final test. Precise pulse characteristics can be inputted when they are made available to further examine their roles on pelvic and or lumbar spine injuries, and the present experimental model can be used for this purpose.

A recent study used a similar supine positioning of five PMHS with a focus on lower leg injuries and the 90-90-90 posture allowed the researchers to load the pelvis along the inferior-to-superior direction (Bailey et al. 2013). Out of the five PMHS tested at higher magnitudes of accelerations, one did not sustain injuries, the other four sustained pelvis injuries without any lumbar fractures. In contrast, the present study included lumbar spine injuries in two PMHS. The onset time for the spine consistently lags the sacrum. Lag times seem to be influenced both by the specimen and pulse. For example, Specimen 3 had the longest delay in spine acceleration. While delays are expected because of mass recruitment effects from sacrum to spine, individual specimen compliance may also play a role, and it is not clear which is the dominant factor in this limited data set.

While the present analysis provides an initial assessment of mass recruitment and specimen compliance issues, to better understand the kinematics, it may be necessary to instrument the most inferior region of the pelvis and compare data with perhaps multiple levels of the spine. This will be a future study. These preliminary analyses indicate that the pelvis–spine complex response to inferior-to-superior accelerative loads may follow classic analyses done in impact biomechanical studies for over 60 years, and additional tests delineating the role of some variables discussed in this study are necessary for developing injury criteria and ATD design for this load vector.

13.2 SUMMARY

The clinical, epidemiological and biomechanical studies highlighted in this chapter have described the human response to vertical loading experienced in military environments. The specific environments highlighted here include aviator ejection catapult, helicopter crash and UB blast, organised in order of increasing severity. Injury mechanisms and tolerance in the thoracolumbar spine was shown to be dependent on a number of variables including acceleration versus time characteristics (i.e., rate of loading) and posture. Higher loading rates have, in general, produced injuries that occur at more cephalad spinal levels and may be more severe and unstable. Prevention of these injuries in military and civilian environments requires a comprehensive biomechanical understanding of the loading environment, mechanisms, and tolerance for injury, and specific factors that may influence the onset or severity of injury. The military environment, in particular, presents a unique scenario with extreme loading scenarios not commonly experienced in the civilian realm. These scenarios require complex solutions from clinical and safety engineering standpoints. Continued biomechanical research can be used to further understand these different injury scenarios and protect warfighters.

ACKNOWLEDGEMENTS

This work was supported by cooperative agreements through the US Army Aeromedical Research Laboratory, Fort Rucker, AL, USA, and by the Office of Naval Research through Naval Air Warfare Center Aircraft Division. This material is the result of work supported with resources and use of facilities at the Zablocki VA Medical Center, Milwaukee, Wisconsin. The authors are part-time employees of the Zablocki VA Medical Center, Milwaukee, Wisconsin. Any views expressed in this article are those of the authors and not necessarily representative of the funding organisations.

REFERENCES

Abbreviated Injury Scale. 2005. *Abbreviated Injury Scale*. Arlington Heights, IL: Association for the Advancement of Automotive Medicine. Update 2008.

Arun, M.W., Yoganandan, N., Stemper, B.D. and Pintar, F.A. 2014. A methodology to condition distorted acoustic emission signals to identify fracture timing from human cadaver spine impact tests. *J Mech Behav Biomed Mater*. 40:156–160.

Aslan, S., Karcioglu, O., Katirci, Y., Kandis, H., Ezirmik, N. and Bilir, O. 2005. Speed bump-induced spinal column injury. *Am J Emerg Med*. 23:563–564.

Bailey, A.M., Henderson, K., Brozoski, F. and Salzar, R.S. 2013.Comparison of Hybrid III and PMHS response to simulated underbody blast loading conditions. *International IRCOBI Conference on the Biomechanics of Impact*, 158–170. September 11–13.

Board, D., Stemper, B.D., Yoganandan, N., Pintar, F.A., Shender, B. and Paskoff, G. 2006. Biomechanics of the aging spine. *Biomed Sci Instrum*. 42:1–6.

Bowrey, D., Thomas, R., Evans, R. and Richmond, P. 1996. Road humps: Accident prevention or hazard? *J Accid Emerg Med*. 13:288–289.

Cain, J.E., Jr., DeJong, J.T., Dinenberg, A.S., Stefko, R.M., Platenburg, R.C. and Lauerman, W.C. 1993. Pathomechanical analysis of thoracolumbar burst fracture reduction. A calf spine model. *Spine (Phila Pa 1976)*. 18: 1647–1654.

Cavanaugh, J.M., Zhu, Y., Huang, Y. and King, A.I. 1993. Injury and response of the thorax in side impact cadaveric tests. *Stapp Car Crash Conference*. San Antonio, TX, 199–221.

Cotterill, P.C., Kostuik, J.P., Wilson, J.A., Fernie, G.R. and Maki, B.E. 1987. Production of a reproducible spinal burst fracture for use in biomechanical testing. *J Orthop Res*. 5:462–465.

Curry, W.H., Pintar, F.A., Doan, N.B., Nguyen, H.S., Eckardt, G., Baisden, J.L., Maiman, D.J., Paskoff, G.R., Shender, B.S. and Stemper, B.D. 2015. Lumbar spine endplate fractures: Biomechanical evaluation and clinical considerations through experimental induction of injury. *J Orthop Res*. 34(6):1084–1091.

Denis, F. 1984. Spinal instability as defined by the three-column spine concept in acute spinal trauma. *Clin Orthop Relat Res*. 189:65–76.

Duma, S.M., Kemper, A.R., McNeely, D.M., Brolinson, P.G. and Matsuoka, F. 2006. Biomechanical response of the lumbar spine in dynamic compression. *Biomed Sci Instrum*. 42:476–481.

Evans, F.G., Lissner, H.R. and Patrick, L.M. 1962. Acceleration-induced strains in the intact vertebral column. *J Appl Physiol*. 17:405–409.

Ewing, C.L., King, A.I. and Prasad, P. 1972. Structural considerations of human vertebral column under +Gz acceleration. *J Aircraft*. 9:84–90.

Fredrickson, B.E., Edwards, W.T., Rauschning, W., Bayley, J.C. and Yuan, H.A. 1992. Vertebral burst fractures: An experimental, morphologic, and radiographic study. *Spine (Phila Pa 1976)*. 17:1012–1021.

Fujiwara, A., Tamai, K., An, H.S., Kurihashi, T., Lim, T.H., Yoshida, H. and Saotome, K. 2000. The relationship between disc degeneration, facet joint osteoarthritis, and stability of the degenerative lumbar spine. *J Spinal Disord*. 13:444–450.

Hongo, M., Abe, E., Shimada, Y., Murai, H., Ishikawa, N. and Sato, K. 1999. Surface strain distribution on thoracic and lumbar vertebrae under axial compression. The role in burst fractures. *Spine (Phila Pa 1976)*. 24: 1197–1202.

Hoshikawa, T., Tanaka, Y., Kokubun, S., Lu, W.W., Luk, K.D. and Leong, J.C. 2002. Flexion-distraction injuries in the thoracolumbar spine: An in vitro study of the relation between flexion angle and the motion axis of fracture. *J Spinal Disord Tech*. 15:139–143.

Kazarian, L. and Graves, G.A. 1972. Dynamic response characteristics of the human vertebral column. An experimental study on human autopsy specimens. *Acta Orthop Scand Suppl*. 43S:1–186.

Kazarian, L. and Graves, G.A. 1977. Compressive strength characteristics of the human vertebral centrum. *Spine (Phila Pa 1976)*. 2:1–14.

King, A.I. and Vulcan, A.P. 1971. Elastic deformation characteristics of the spine. *J Biomech*. 4:413–429.

Kumaresan, S., Yogananan, N. and Pintar, F.A. 1998. Posterior complex contribution to the axial compressive and distraction behavior of the cervical spine. *J Musculoskeletal Res*. 2:257–265.

Kumaresan, S., Yoganandan, N., Pintar, F.A., Maiman, D.J. and Goel, V.K. 2001. Contribution of disc degeneration to osteophyte formation in the cervical spine: A biomechanical investigation. *J Orthop Res*. 19: 977–984.

Langrana, N.A., Harten, R.R., Lin, D.C., Reiter, M.F. and Lee, C.K. 2002. Acute thoracolumbar burst fractures: A new view of loading mechanisms. *Spine (Phila Pa 1976)*. 27:498–508.

Maiman, D.J. and Pintar FA. Anatomy and clinical biomechanics of the thoracic spine. *Clin Neurosurg* 1992;38: 296–324.

Maiman, D.J. and Sances, A., Jr., Myklebust, J.B., Chilbert, M., Yoganandan, N., Pintar, F.A. and Larson, S.J. 1986. Trauma of the human thoracolumbar spine. *Mechanisms of Head and Spine Trauma*. ed. A. Sances Jr., D.J. Thomas, C.L. Ewing, S.L. Larson and F. Unterharnscheidt. Aloray Publishers. New York. pp. 489–504.

McElhaney, J.H. 1966. Dynamic response of bone and muscle tissue. *J Appl Physiol*. 21:1231–1236.

Mertz, H.J. 1984. *A Procedure for Normalizing Impact Response Data*. Warrendale, PA: Society of Automotive Engineers.

Ochia, R.S. and Ching, R.P. 2002. Internal pressure measurements during burst fracture formation in human lumbar vertebrae. *Spine (Phila Pa 1976)*. 27:1160–1167.

Ochia, R.S., Tencer, A.F. and Ching, R.P. 2003. Effect of loading rate on endplate and vertebral body strength in human lumbar vertebrae. *J Biomech.*, 36:1875–1881.

Panjabi, M.M., Oxland, T.R., Kifune, M., Arand, M., Wen, L. and Chen, A. 1995. Validity of the three-column theory of thoracolumbar fractures. A biomechanic investigation. *Spine (Phila Pa 1976)*. 20:1122–1127.

Pintar, F.A., Yoganandan, N., Hines, M.H., Maltese, M.R., McFadden, J., Saul, R., Eppinger, R.H., Khaewpong, N. and Kleinberger, M. 1997. Chestb and analysis of human tolerance in side impact. *Stapp Car Crash Conference*. Lake Buena Vista, FL, 63–74.

Pintar, F.A., Yoganandan, N. and Maiman, D.J. 2010. Lower cervical spine loading in frontal sled tests using inverse dynamics: Potential applications for lower neck injury criteria. *Stapp Car Crash J*. 54:133–166.

Pintar, F.A., Yoganandan, N., Maiman, D.J., Scarboro, M. and Rudd, R.W. 2012. Thoracolumbar spine fractures in frontal impact crashes. *Ann Adv Automot Med*. 56:277–283.

Pintar, F.A., Yoganandan, N. and Voo, L. 1998. Effect of age and loading rate on human cervical spine injury threshold. *Spine (Phila Pa 1976)*. 23:1957–1962.

Prasad, P. and King, A.I. 1974. An experimentally validated dynamic model of the spine. *J Appl Mech*. 41: 547–550.

Rupp, J.D., Miller, C.S., Reed, M.P., Madura, N.H., Klinich, K.D. and Schneider, L.W. 2008. Characterization of knee-thigh-hip response in frontal impacts using biomechanical testing and computational simulations. *Stapp Car Crash J*. 52:421–474.

Rupp, J.D., Reed, M.P., Van Ee, C.A., Kuppa, S., Wang, S.C., Goulet, J.A. and Schneider, L.W. 2002. The tolerance of the human hip to dynamic knee loading. *Stapp Car Crash J*. 46:211–228.

Salzar, R.S., Bolton, J.R., Crandall, J.R., Paskoff, G.R. and Shender, B.S. 2009. Ejection injury to the spine in small aviators: Sled tests of manikins vs. post mortem specimens. *Aviat Space Environ Med*. 80: 621–628.

Shirado, O., Kaneda, K., Tadano, S., Ishikawa, H., McAfee, P.C. and Warden, K.E. 1992. Influence of disc degeneration on mechanism of thoracolumbar burst fractures. *Spine (Phila Pa 1976)*. 17:286–292.

Sonoda, T. 1962. Studies on the strength for compression, tension, and torsion of the human vertebral column. *J Kyoto Pref Med Univ*. 71:13.

Spurrier, E., Gibb, I., Masouros, S. and Clasper, J. 2016. Identifying spinal injury patterns in underbody blast to develop mechanistic hypotheses. *Spine (Phila Pa 1976)*. 41:E268–E275.

Spurrier, E., Singleton, J.A., Masouros, S., Gibb, I. and Clasper, J. 2015. Blast injury in the Spine: Dynamic response index is not an appropriate model for predicting injury. *Clin Orthop Relat Res*. 473:2929–2935.

Stemper, B.D., Baisden, J.L., Yoganandan, N., Pintar, F.A., Derosia, J., Whitley, P., Paskoff, G.R. and Shender, B.S. Effect of loading rate on injury patterns during high rate vertical acceleration. *Proceedings of the International IRCOBI Conference on the Biomechanics of Impact*. Dublin, Ireland, September 12–14, 2012.

Stemper, B.D., Board, D., Yoganandan, N. and Wolfla, C.E. 2010. Biomechanical properties of human thoracic spine disc segments. *J Craniovertebr Junction Spine*. 1:18–22.

Stemper, B.D., Pintar, F.A. and Baisden, J. 2015. Lumbar spine injury biomechanics. Accidental Injury: *Biomechanics and Prevention*. ed. N. Yoganandan, A.M. Nahum, J.W. Melvin. 3rd ed. Spring. New York.

Stemper, B.D., Storvik, S.G., Yoganandan, N., Baisden, J.L., Fijalkowski, R.J., Pintar, F.A., Shender, B.S. and Paskoff, G.R. 2011. A new PMHS model for lumbar spine injuries during vertical acceleration. *J Biomech Eng*. 133:081002.

Stemper, B.D., Yoganandan, N., Baisden, J.L., Umale, S., Shah, A.S., Shender, B.S. and Paskoff, G.R. 2014. Rate-dependent fracture characteristics of lumbar vertebral bodies. *J Mech Behav Biomed Mater*. 41:271–279.

Stemper, B.D., Yoganandan, N., Paskoff, G.R., Fijalkowski, R.J., Storvik, S.G., Baisden, J.L., Pintar, F. and Shender, B.S. 2011. Thoracolumbar spine trauma in military environments. *Minerva Ortopedica E Traumatologica*. 62:397–412.

Stemper, B.D., Yoganandan, N. and Pintar, F.A. 2007. Mechanics of arterial subfailure with increasing loading rate. *J Biomech*. 40:1806–1812.

Tran, N.T., Watson, N.A., Tencer, A.F., Ching, R.P. and Anderson, P.A. 1995. Mechanism of the burst fracture in the thoracolumbar spine. The effect of loading rate. *Spine (Phila Pa 1976)*. 20:1984–1988.

Vulcan, A.P., King, A.I. and Nakamura, G.S. 1970. Effects of bending on the vertebral column during +Gz acceleration. *Aerosp Med.* 41:294–300.

Wilcox, R.K., Boerger, T.O., Allen, D.J., Barton, D.C., Limb, D., Dickson, R.A. and Hall R.M. 2003. A dynamic study of thoracolumbar burst fractures. *J Bone Joint Surg Am.* 85-A:2184–2189.

Willen, J., Lindahl, S., Irstam, L., Aldman, B. and Nordwall, A. 1984. The thoracolumbar crush fracture. An experimental study on instant axial dynamic loading: The resulting fracture type and its stability. *Spine (Phila Pa 1976).* 9:624–631.

Yoganandan, N., Arun, M.W., Humm, J. and Pintar, F.A. 2014. Deflection corridors of abdomen and thorax in oblique side impacts using equal stress equal velocity approach: Comparison with other normalization methods. *J Biomech Eng.* 136:101012.

Yoganandan, N., Arun, M.W. and Pintar FA. 2014. Normalizing and scaling of data to derive human response corridors from impact tests. *J Biomech.* 47:1749–1756.

Yoganandan, N., Arun, M.W., Stemper, B.D., Pintar, F.A. and Maiman, D.J. 2013. Biomechanics of human thoracolumbar spinal column trauma from vertical impact loading. *Ann Adv Automot Med.* 57:155–166.

Yoganandan, N., Haffner, M., Pintar, F.A. 1996. Hampson, D: Facial injury: A review of biomechanical studies and test procedures for facial injury assessment. *J Biomech.* 29:985–986.

Yoganandan, N., Halliday, A., Dickman, C. and Benzel, E. 2012. Practical anatomy and fundamental biomechanics. *Spine Surgery: Techniques, Complication Avoidance, and Management.* ed. E. Benzel, 93–118, 3rd ed. Churchill Lingstone. New York.

Yoganandan, N., Morgan, R.M., Eppinger, R.H., Pintar, F.A., Sances, A., Jr. and Williams, A. 1996. Mechanisms of thoracic injury in frontal impact. *J Biomech Eng.* 118:595–597.

Yoganandan, N., Pintar, F.A., Boynton, M., Begeman, P., Prasad, P., Kuppa, S., Morgan, R.M. and Eppinger, R.H. 1996. Dynamic axial tolerance of the human foot-ankle complex. J *SAE Trans.* 105:1887–1989.

Yoganandan, N., Pintar, F., Butler, J., Reinartz, J., Sances, A., Jr. and Larson, S.J. 1989. Dynamic response of human cervical spine ligaments. *Spine (Phila Pa 1976).* 14:1102–1110.

Yoganandan, N., Pintar, F.A., Humm, J.R., Stadter, G.W., Curry, W.H. and Brasel, K.J. 2013. Comparison of AIS 1990 update 98 versus AIS 2005 for describing PMHS injuries in lateral and oblique sled tests. *Ann Adv Automot Med.* 57:197–208.

Yoganandan, N., Pintar, F.A., Reinartz, J. and Sances, A., Jr. 1993. Human facial tolerance to steering wheel impact: A biomechanical study. *J Safety Res.* 24:77–85.

Yoganandan, N., Pintar, F.A., Sances, A., Jr., Maiman, D., Myklebust, J.B. and Harris, G. 1988. Biomechanical investigations of the human thoracolumbar spine. *Society of Automotive Engineers.* SAE Paper Number: 881331.

Yoganandan, N., Pintar, F.A., Sances, A., Jr., Reinartz, J. and Larson, S.J. 1991. Strength and kinematic response of dynamic cervical spine injuries. *Spine (Phila Pa 1976).* 16:S511–S517.

Yoganandan, N., Pintar, F.A., Stemper, B.D., Wolfla, C.E., Shender, B.S. and Paskoff, G. 2007. Level-dependent coronal and axial moment-rotation corridors of degeneration-free cervical spines in lateral flexion. *J Bone Joint Surg Am.* 89:1066–1074.

Yoganandan, N., Sances, A., Jr., Pintar, F., Maiman, D.J., Reinartz, J., Cusick, J.F. and Larson, S.J. 1990. Injury biomechanics of the human cervical column. *Spine (Phila Pa 1976).* 15:1031–1039.

Yoganandan, N., Stemper, B.D., Baisden, J.L., Pintar, F.A., Paskoff, G.R. and Shender, B.S. 2015. Effects of acceleration level on lumbar spine injuries in military populations. *Spine J.* 15:1318–1324.

Yoganandan, N., Stemper, B.D., Pintar, F.A. and Maiman, D.J. 2011. Use of postmortem human subjects to describe injury responses and tolerances. *Clin Anat.* 24:282–293.

Zaidel, D., Hakkert, A.S. and Pistiner, A.H. 1992. The use of road humps for moderating speeds on urban streets. *Accid Anal Prev.* 24:45–56.

14 The Thorax and Abdomen
Behind Armour Blunt Trauma

Cameron R. 'Dale' Bass

CONTENTS

14.1   INTRODUCTION TO BEHIND ARMOUR BLUNT TRAUMA

The evolution of ballistic body armour includes an early series of inventions in 1861 when thin steel plates were enclosed in military jacket materials to protect against sabre attacks and bullets (Peterson 1950). Whether or not to use protective armour was a personal choice and depended on cost ($5–7), weight (2 kg) and appearance of 'unmanliness'. The use of vests ceased after the American Civil War and did not reappear in earnest until 100 years later, when the US government began to supply law enforcement and public officials with protection from handgun bullets.

There are two major types of personal body armour, soft and hard. Soft armour is designed to protect against shrapnel resulting from explosions and low-velocity, low-energy bullets (e.g. 9 mm or 38 calibre). Hard armour contains an additional hard plate made of polyethylene or ceramic material to defeat high-velocity threats such as the 7.62 mm (30 calibre) and 12.7 mm (50 calibre) rifle bullets. The original work in standardising body armour focused on soft armour, but currently the threats for both war fighters and law enforcement personnel are from shrapnel and projectiles of higher energy and higher velocity than anticipated 35 years ago.

Modern body armour can defeat incoming pistol and rifle rounds, trading energy and momentum deposition into the armour for deformation of the armour. This backface deformation (BFD) includes direct deformation of the body armour with soft body armour and deformation with fracture in hard body armour. This deformation, however, has the potential for creating injuries in the thorax behind the armour that may be generally characterised as blunt trauma. These injuries are often termed behind armour blunt trauma (BABT).

BFD of the thorax under these defeated rounds is sufficient to cause local and distant fractures, contusions and haemorrhage as has been demonstrated in numerous animal studies (e.g. Clare 1975a; Prather et al. 1977; Cooper et al. 1982; Suneson et al. 1987; Lidén et al. 1988; Knudsen and Gøtze 1997; Sarron et al. 2000; Mayorga et al. 2010, among others). These injuries are the result of the transfer of a stress wave through the thorax as well as the physical deformation of the back face of the armour. One key unknown about BABT is the extent to which there may be significant injury to organs and structures more distant from the point of impact, such as the brain, heart, spinal cord and intestines, due to significant pressure waves that occur when the body armour is impacted by the deforming armour backface.

The injury risk from BFD will generally depend on the type and configuration of the armour, the round and the delivered energy of the round that results in an impact displacement and profile. This impact displacement and profile also depend on physical characteristics of the person wearing the body armour. For both the body armour and the thorax, the impact location is important, and for the thorax, the effect of rate sensitivity of the impact may be large.

A bullet is a localised source of energy that can cause high local compression and shear forces, penetrating protective layers. The most effective bullets deposit energy/shear/momentum rapidly in the target. One general strategy for protection is to blunt the penetration of the incoming round, picking up as much mass as possible in the body armour while decreasing the round energy and increasing the contact area. Hence, the protective effect of any ballistic protective vest is provided by increasing the area of impact, thus transferring energy and momentum to the vest spreading the impact energy over a longer time. However, effective transfer of large amounts of energy and momentum from the incoming round into the body armour generally implies some deformation of the rear or backface of the body armour.

The BFD is generally larger under soft body armour for a given incoming round. An interesting comparison of energy and momentum scales may be seen by comparing characteristics of various rounds as shown in Table 14.1. Energy varies by a factor of over 30 between the relatively slow 9 mm handgun round and the 50-calibre rifle round.

A further elucidative comparison may be made between the impact energy and momentum scales of lower rate blunt trauma events such as automobile impacts and high-rate impact events such as BABT. The energy and momentum for various potential blunt trauma situations are shown in Table 14.2 and are plotted in Figure 14.1. It is apparent that the non-penetrating ballistic impact involves much lower total momentum transfer than typical low-rate blunt impact. However, energy transfer is comparable, depending on the round and impact velocity. This implies increased localisation of energy transfer and shorter interaction time, and likely increased localisation of injury.

At high rates, blast impacts may occur with very small impact momentum and energy over very short timescales. As the duration of such impacts becomes very short, an interesting comparison may be made with damage using ultrasonic energy. At high rate (~4 MHz), less than 20 cycles of acoustic energy delivered to lung tissue with a peak pressure of approximately 1 MPa will cause tissue damage (cf. Raeman et al. 1996). It is uncertain if these high-frequency effects occur with BABT.

Figure 14.2 shows a high-speed radiograph of deformation of hard body armour under rifle round impact. The chest deformation shown here may lead to trauma to ribs, lungs, heart, liver and other organs. Data are needed to determine the optimum vest design that provides protection to the body, potentially including organs remote from the site of impact *while minimising weight that the solider must carry.*

TABLE 14.1
Typical Muzzle Parameters – Various Rifle and Handgun Rounds

Round	Muzzle Velocity (m/s)	Mass (g)	Muzzle Energy (kJ)	Muzzle Momentum (kg m/s)
9 mm	358	8	0.5	2.86
5.56 × 45 M193 ball	991	3.6	1.7	3.57
7.62 × 51 NATO ball	838	9.6	3.4	8.13
0.50 M2	890	42	16.6	37.4

TABLE 14.2
Energy/Momentum – Various Typical Thoracic Trauma Situations

Action	Velocity (m/s)	Mass (g)	Energy (kJ)	Momentum (kg m/s)
US football block	5	100,000	2.5	500
Automobile thoracic dash impact	5	50,000	0.6	250
Automobile head impact	5	5,000	0.06	25
Blast	~300	–	0.0004[a]	0.0013
Ultrasound damage	~1500	–	0.00001[b]	6.7×10^6

[a] Based on assumed total lung volume of 3000 mL.

[b] Based on applied lung volume of 300 mL.

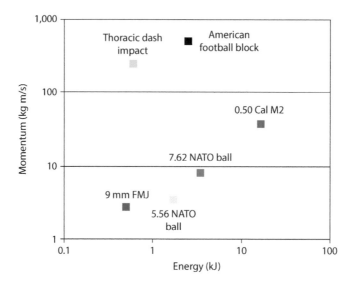

FIGURE 14.1 Initial energy and momentum – ballistics and other blunt impacts.

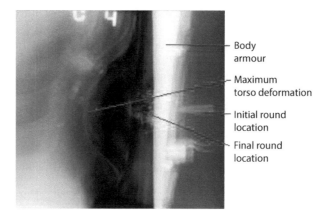

FIGURE 14.2 Superimposed high-speed radiographs of the initial shock wave and deformation of the thorax during a 7.62 mm projectile live-fire test in a pig protected by hard body armour. (From Mayorga, M.A., et al., *Thoracic Response to Undefeated Body Armor*, Research and Technology Organisation, North Atlantic Treaty Organization (NATO), BP 25, F-92201 Neuilly-sur-Seine Cedex, France, Task Group HFM-024 TG-001, 2010. Copyright clearance from Mayorga.)

14.1.1 ERGONOMICS OF BODY ARMOUR

Until the eighteenth century, combat infantry soldiers are estimated to have carried about 15–30 kg (33–66 lb) with the remaining equipment/supplies being transported in a baggage train (e.g. Knapik et al. 2004). Recently, however, this load carriage has substantially increased to 60% or more of the lean body mass (e.g. Birrell et al. 2007). For a US 50th percentile male of 80 kg (175 lb), this implies load carriage of over 48 kg (106 lb). The negative consequences of substantial load carriage include potential heat stress (Cadarette et al. 2001; Barwood et al. 2009), decrements in psychomotor performance (Bensel and Lockhart 1975) and ergonomic factors that may limit mobility (e.g. Harman 1992). Beekley et al. (2007) found significant increases in oxygen consumption, ventilation and heart rate for loads up to 70% of lean body mass in male US Army personnel.

The current body armour mass basic system weight (US 2014) is 7.1 kg (15.7 lb), providing 15% of typical maximum load carriage with a maximum weight of 15.1 kg (33.3 lb) or 31% of maximum load carriage. Clearly, reductions in body armour mass are a potential method for reducing the total load carriage and increasing the mobility on the battlefield.

14.1.2 Injury Criteria and Experimentation

There are a number of potential sources of BABT from BFD: these include the initial contact shock and the subsequent displacement of the thoracic wall. The initial shock may occur at high frequency with a relatively low resulting displacement. This shock pressure peak occurs because of a mismatch of impedance between the rear of the body armour and the thoracic wall. Subsequent bulk displacement may occur following significant local momentum transfer between the back of the body armour and the body. There has been extensive investigation into the relative effect of the initial shock and resulting displacements. Animal experiments at Oksboel, discussed below, were designed to investigate this (Sarron et al. 2000). This issue, however, is not completely resolved.

Pressure profiles have been measured in tissue simulants for impacts behind body armour. As shown in Figure 14.3, the pressure peaks may be large (>6000 kPa) with short duration (<1 ms). In addition, stress wave propagation and concentration of reflected and refracted waves may enhance injury (Liu 1996). This is discussed further in the Oksboel animal section below.

The experimental basis for high-rate impact is not well established. Variables that may significantly affect injury potential include delivered energy, thoracic wall displacement, contact area or contact profile, loading rate and location of impact. It is clear that the more localised the energy, the greater the potential deformation into and through the body armour. Large local displacements may cause destructive local shearing or compression of tissue. Area or profile, distributed thoracic loading has a different injury pattern than localised loading (Crandall et al. 1998). Indeed, relatively sharp profiles may be assumed by penetrations behind soft body armour (Lewis, E. 2001, Personal communication) or behind full penetrations of ceramic armour captured in the soft backing material (DeMaio 2001). The effect of the rate sensitivity of human tissue may be large, and the location of impact is important. Additional factors may include gender, age, body mass, stature and other

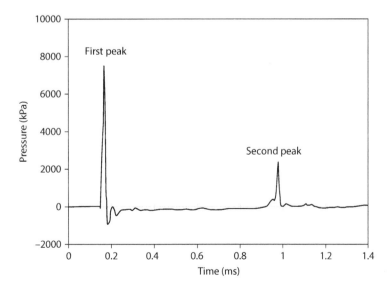

FIGURE 14.3 Pressure in gelatine simulant behind 6.4 mm ceramic (24 kg/m²) with aramid composite (10 kg/m²) and fragment protective vest. (From van Bree, J., and Gotts, P. *The 'Twin Peaks' of BABT*, Personal Armor Systems Symposium, Colchester, UK, 2000. Copyright clearance from van Bree.)

anthropometric parameters. Experimental programmes designed to develop standard injury test methodologies often focus first on a single relevant subject population. This may imply a focus on mid-sized males as appropriate for a military population. However, it is perhaps necessary to consider sex-related size differences for general applications.

Injury mechanisms and mechanical correlates with injury are discussed in the next sections. These sections are followed by a discussion of animal, cadaveric and epidemiological experimentation for assessment of BABT.

14.1.3   BABT Injury Mechanisms

The principal anatomical features of the thorax, such as the heart, major blood vessels, lungs and other major organ systems, represent areas of potential vulnerability to BABT from ballistic insults and BFD (Figure 14.4). Indeed, ballistic protective measures have been designed specifically to protect this region and portions of this region. The majority of organ systems necessary for life are located in the thorax. The mediastinal region is particularly important. Notable structures in the mediastinum include major blood vessels (aorta, pulmonary artery and pulmonary vein), branch points of the lungs (trachea), entry into the gastrointestinal system (oesophagus) and the heart. In addition, the thorax includes the lungs, offering a large impact surface area, and the part of the liver under the rib cage. In the posterior region of the thorax, the spine presents an additional impact location that is potentially debilitating or life-threatening.

There are a number of potential mechanisms for injury that have been seen in animal, cadaver or human epidemiological studies. These include physiological mechanisms such as cardiac and pulmonary disruptions. Examples of these mechanisms include commotio cordis (Link et al. 1999) and acute respiratory distress syndrome (ARDS, e.g. Miller et al. 2002). Commotio cordis, however, is thought to occur in a small portion of the cardiac cycle. Hence, the risk of such electrical disruptions may be small.

Several types of injuries may be attributed to displacement, including large displacements that may cause shear or crushing. One of these injury mechanisms is puncture caused by bony fracture. An example of this is a fractured rib penetrating into a lung causing disruption of the plural cavity that causes a pneumothorax or haemothorax. Fung et al. (1988) suggested that lung injuries may be related to compression of the alveoli under mechanical stress. However, at high rate, for ultrasonic forcing, the influence of local cavitation has been suggested. Indeed, negative pressure has been seen recently in ballistic animal experiments (Sarron et al. 2001).

Pulsation mechanisms similar to those seen in ultrasound tissue damage may be applicable to BABT tissue damage. Thermal mechanisms that are unlikely contributors to BABT are proposed for long duration continuous wave ultrasound forcing; however, for short duration ultrasound, stress wave mechanisms and cavitation mechanisms have been proposed that may be

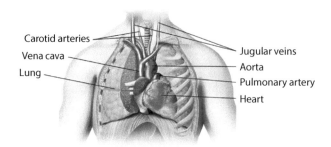

Carotid arteries
Vena cava
Lung
Jugular veins
Aorta
Pulmonary artery
Heart

FIGURE 14.4   Thoracic anatomy (http://www.alamy.com/stock-photo-man-anatomy-thorax-cutaway-with-heart-with-main-blood-veins-arterias-53055679.html).

BABT injury mechanisms. Ultrasound impulses involve low displacement with relatively high frequency (50–1000 kHz): this forcing is possible in shock wave interactions generated between deforming body armour and the body surface. The effect of exposure duration on the threshold injury pressure is important as shown by Carstensen et al. (1990) and Bass et al. (2006). Thresholds for damage in lung tissue in the murine model are significantly lower than those of other tissues. The threshold has been found to be frequency dependent (1,000–4,000 kHz) with the threshold:

$$P = 630 \, f^{0.54} \, (\text{kHz}).$$

Unfortunately, no 'single pulse' thresholds are currently known; current thresholds involve five cyclic pulses. Of the non-thermal mechanisms, cavitation from repeated cycling and spalling from an interfacial impedance mismatch, cyclic cavitation is not likely a BABT injury mechanism. However, spalling may be a BABT injury mechanism. Spalling may occur at alveolar surfaces resulting in local haemorrhage.

For high-rate BFD, there is a potential for ablation injuries caused by the friction of the impacting surface on the body. This has been seen in epidemiological studies by Mirzeabassov et al. (2000), and cadaveric experiments by DeMaio (2001) and Bass et al. (2002). These ablation injuries may be quite severe. DeMaio et al. noted large deep bilateral chest wounds under certain circumstances, and Bass et al. observed injury from ablation behind helmets with relatively low-velocity incoming projectiles.

A final class of BABT includes the potential for injuries remote from the direct backface contact. The earliest observations on the effects of penetrating injuries to the nervous system remote from the site of penetration were case studies from the Civil War of temporary and sometimes long-term motor and sensory paralysis (Mitchell et al. 1864; Akimov 1993). The occurrence of damage from transmission of kinetic energy from the point of impact on the torso to remote body organs in humans has been observed in a number of cases (Chamberlin 1966; Carroll and Soderstrom 1978; Sperry 1993; Cannon 2001; Krajsa 2009), and these corroborate Civil War case studies (Mitchell et al. 1864). A report on human trauma from BFD in law enforcement personnel emphasised the fact that protection from penetration does not protect from significant thoracic trauma (Wilhelm and Bir 2008).

14.1.4 Mechanical Correlates with Injury

To be useful, an injury criterion may be used with a mechanical correlate with injury and a surrogate test device to evaluate injury. The earliest criteria were based on global acceleration. These may not be appropriate for injuries with sufficient localisation. Beyond the correlates used for the blunt criterion discussed above (Sturdivan 1976), several mechanical correlates that have been investigated are discussed below.

14.1.5 Human Epidemiology for Battlefield BABT

There is a strong need for the investigation of epidemiological studies appropriate for behind armour effects. Indeed, the limited number of high-rate epidemiological studies limits evaluation of the protective capabilities of current and future ballistic protection. The value of epidemiological research is seen in many fields. Use of such studies in the field of anaesthesia substantially lowered the incidence of death from adverse events over the last 20 years (cf. Hawkins et al. 1997). Epidemiological and retrospective studies are used in aircraft accidents and automobile crashes worldwide. Such self-examination has been useful in designing countermeasures for injury.

14.1.5.1 Injury Scales

An organised and robust injury scale is necessary for the evaluation of injuries using a common basis for animal, cadaveric and epidemiological experiments. There are several extant injury scores, but none are completely appropriate for use in scoring BFD injury. One standard for the assessment of thoracic injury is the Abbreviated Injury Scale (AIS) of the Association for the Advancement of Automotive Medicine (AAAM), which is discussed further in Chapter 4. The numerical rating system ranges from 1 (minor injury) to 6 (maximum, virtually unsurvivable). According to the 2008 update of the 2005 revision (AAAM, 2008), ≥3 rib fractures are coded as an AIS 3. Lung lacerations are coded separately from rib fractures and range from AIS 3 to 5 depending on injury severity and extent. One of the most serious thoracic injuries is aortic laceration that is ranked from AIS 4 to 6 (e.g. Cavanaugh 1993). As discussed below, however, this criterion is not generally sufficiently discriminative for research BABT work.

The most widely used criteria for thoracic injury are the compression criterion and the viscous criterion for frontal impacts, and the Thoracic Trauma Index (TTI) for side impacts. The compression criterion relates the relative chest deformation C with respect to the chest depth to the level of injury. According to Kroell et al. (1971, 1974), 30% and 40% chest compressions cause injuries that may be assessed as AIS 2 injuries and AIS 4 injuries, respectively (using AIS 76). In the viscous criterion, the rate dependency of soft tissue injury is taken into account and VC_{max}, the maximum product of velocity of deformation (V) and relative compression (C), is proposed as an effective predictor of injury risk. In an analysis of 39 unembalmed cadaver sternal impacts, a VC_{max} of 1.3 m/s was associated with a 50% probability of AIS >3 injury (Viano and Lau 1985). Eppinger et al. (1984) analysed a large number of side impact test results and proposed the TTI injury criterion, a function that is proportional to age, mass and the average of the peak values of fourth struck-side rib and twelfth thoracic vertebra accelerations.

The Injury Severity Score (ISS) (Baker et al. 1974), discussed further in Chapter 4, may be used as an overall score for multiple injuries. Each body region is assigned an AIS score for the six ISS body regions (head, face, chest, abdomen, extremities and external). The highest AIS score for each ISS body region is selected, and the highest three of these are squared to produce the ISS score as:

$$ISS = AIS_{max1}^2 + AIS_{max2}^2 + AIS_{max3}^2.$$

The ISS score value ranges from 1 to 75. This score has been correlated with outcome for thoracic trauma, potential chest, abdomen and external injuries.

14.1.5.2 Mirzeabassov et al. (2000)

Mirzeabassov et al. (2000) have developed a scale for combat-specific BFD trauma. This scale is shown in Table 14.3, with an associated injury scale in Table 14.4, and is based on levels of damage sustained from BFD.

The Mirzeabassov study is the most significant epidemiology on BABT that is openly available. It includes data from 17 military personnel hit in the thoracic region wearing body armour with either 1.25 mm titanium (6B2) or 6.5 mm titanium (6B3TM) plates. The data were acquired during the Soviet experience in Afghanistan. Injury data include location of impact, injury and the long-term consequences of the impact and age of the patient. The ballistics data include the type of weapon fired (reportedly either 7.71 mm Enfield or 7.62 AKM in all cases), firing distance and impact kinetic energy.

Bullet kinetic energy plotted against injury severity was derived for both human epidemiological data and animal experiments, as shown in Figure 14.5. The lighter body armour (6B2) had a significantly lower threshold for the onset of severe injuries. The most serious injuries reported were haemopneumothoraces in two patients that progressed to abscesses. In addition, an impact in the left rib was reported to develop into a large ecchymosis that extended from the groin to the knee.

TABLE 14.3
Description Levels of Thoracic Trauma

Level of Trauma	Nature of Injuries
I (slight)	Scratches on the skin, ecchymosis and restricted subcutaneous haematomas. Isolated focal subpleural haemorrhages
II (medium gravity)	Contused cutaneous wounds. Focal intramuscular haemorrhages. Plural focal subpleural haemorrhages. Isolated focal haemorrhages into the intestinal mesentery
III (grievous)	Closed and open rib fractures. Lacerations of the pleura, haemorrhages into the pulmonary parenchyma. Subepcardial or subendocardial haemorrhages. Subcapsular haematomas of parenchymal organs of the abdominal cavity and retroperitoneal space. Subserous haemorrhages into the intestines, ruptures of the mesentery. Restricted haemopneumothorax and haemoperitoneum. Vertebral fractures without injury to the spinal cord
IV [extremely grievous (lethal)]	Ruptures and crushing of internals. Closed trauma of the vertebral column followed by an injury to the spinal cord

Source: Mirzeabassov, T., et al., Further investigation of modelling for bulletproof vests, *Personal Armor Safety Symposium*, Colchester, UK, 2000.

TABLE 14.4
Combat Effectiveness Versus Levels of Thoracic Trauma

Gravity Level of the Trauma	Characteristic of the Loss of Fighting Efficiency	Probability of Rehabilitation	Class of Losses
I (light)	Loss of fighting efficiency for 1–3 minutes. Limited fighting efficiency for 15 minutes. Complete restoration within 24 hours	99%	Left in action
II (medium)	Loss of fighting efficiency for 3–5 minutes. Limited fighting efficiency up to 10 days. Complete restoration within 15–20 days	85%	Combat sanitary (recoverable) losses
III (high)	Complete loss of fighting efficiency, limited fighting efficiency within 15–20 days, complete restoration within 30–60 days. Possible fatal outcome	25%	Combat sanitary (recoverable) losses
IV (extremely high)	Immediate death. Death caused by complications. Invalidism and complete loss of fighting efficiency in surviving persons	0%	Unrecoverable losses

Source: Mirzeabassov, T., et al., Further investigation of modelling for bulletproof vests, *Personal Armor Safety Symposium*, Colchester, UK, 2000.

Coverage of the plates is not reported; however, substantial injuries occurred to the back of the torso from relatively low-energy impacts, implying that the impacts may have occurred in an area without plate coverage. Mirzeabassov et al. developed a scale that relates the initial bullet kinetic energy with the severity of injury in both the human epidemiology and animal experiments, as shown in Figure 14.5.

To understand the Russian epidemiological data, it is important to consider details of the Soviet military medical service at the time. Extensive data analysis is available for mine trauma victims from the Soviet experience in Afghanistan (Nechaev et al. 1995). Over 90% of patients were

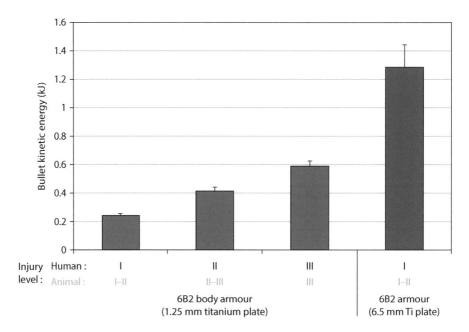

FIGURE 14.5   Kinetic energy versus injury severity based on classifications from Table 14.4. (Based on Mirzeabassov, T., et al., Further investigation of modelling for bulletproof vests, *Personal Armor Safety Symposium*, Colchester, UK, 2000.)

evacuated by air, but only 4% were delivered to the Central Military Hospital (CMH) within 6 hours of injury (Figure 14.6a). This was the initial treatment received for more than 80% of the wounded patients (Nechaev et al. 1995). Furthermore, the severity of the wounds was distributed as shown in Figure 14.6b. It is likely that the distribution of the severity of injuries may have changed with the time to treatment. Increased time to treatment would exacerbate the severity of the injuries and also tend to decrease the number of severe injuries since these personnel would have died before they reach the hospital. This may be seen in the types of injuries reported in the epidemiology study.

Indeed, no injuries were seen that would be considered grave if they occurred in an automobile collision. It may be speculated that there may have been serious injuries that became fatal with delayed treatment. However, the development of infection in the two pneumothoraces increased the threat-to-life substantially. Typical acute modern treatment, including antibiotics, substantially reduces the incidence of infection. One may further speculate that soldiers with more severe BABT would have an extremely large risk of death prior to reaching the CMH. This work is almost the singular instance of comprehensive epidemiology for BABT trauma available in public that is correlated with well-described body armour. There are, however, limitations as with all epidemiological studies. The ranges were estimated by the authors, and battlefield trauma care may have been significantly different between Soviet and western militaries, changing the distribution of injuries.

14.1.5.3 Current Epidemiology for Battlefield/Law Enforcement BABT

A recent US National Academies Report (National Academies 2014) emphasised that 'there have been no known US soldier deaths due to small arms and shrapnel that were attributable to a failure of the issued ceramic body armour for threats for which the armour was designed'. Based on this, the US military has fielded hard body armour with adequate survivability characteristics for soldiers in combat. Furthermore, since the National Institute of Justice undertook the responsibility to standardise personal body armour for law enforcement personnel in 1973, over 2,900 lives have been saved (e.g. http://www.dupont.com/kevlar/lifeprotection/survivors.html?NF=1). The tragic failure

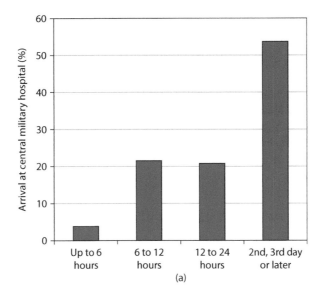

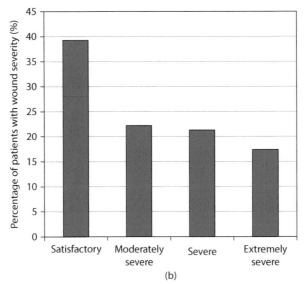

FIGURE 14.6 (a) Time of delivery of wounded to CMH (average 1983–1984) (Nechaev 1995). (b) State of severity of wounded to CMH (average 1983–1984). (Based on Nechaev, E., et al., *Mine Blast Trauma*, Russian Ministry of Public Health and Medical Industry, St. Petersburg, FL, 1995.)

of zylon body armours in a small number of well-publicised cases in law enforcement (e.g. Walsh et al. 2006) further emphasises the rarity of armour failures and the success of the testing programs based on clay.

14.2 LARGE ANIMAL EXPERIMENTS FOR BABT

Animal experiments may be used for the development of injury criteria for blunt trauma including BABT. However, animals used in such experimentation, typically livestock, have significant differences in anatomy from humans. Consequently, scaling data from animals to humans must be performed with techniques that may be uncertain or non-existent.

Although there are a number of experiments where thoracic penetration and BABT with soft body armour have been investigated (cf. Prather 1977; Carroll and Soderstrom 1978; Lidén et al. 1988) and hard body armour using animals, this section concentrates on those studies that are applicable to BABT with hard body armour. This includes several direct impact studies at high rate.

Large animal studies have been specifically designed to assess injury resulting from non-penetrating ballistic impacts on body armour. These were performed by American, Canadian, Swedish, Danish, Dutch and French teams, mostly in anaesthetised pigs and using various models of body armour and threats ranging from 0.38 calibre to 0.30 calibre (7.62 mm) and 50 calibre (12.7 mm). Many of these studies were reported by the NATO Task Group on Thoracic Response to Undefeated Body Armour (Mayorga et al. 2010).

14.2.1 Lovelace Foundation (Bowen et al. 1966)

Bowen et al. reported experiments on dogs with the lateral thorax impacted by non-penetrating missiles, as shown in Figure 14.7. Cup-shaped missiles were constructed of aluminium with variable masses. The impact occurred on a flat cylindrical end with a diameter of 7 cm. Impactor masses varied from 63 to 381 g, impact velocities varied from 18.9 to 91.4 m/s and the dog mass varied from 12.2 to 23.1 kg. The dogs were positioned so that impacts were produced at the right lateral chest wall near the mid-thorax. Ribs fractured include the fourth rib through eighth rib implying impact locations near the fifth or sixth rib. Animals that survived the immediate post impact time were sacrificed at 30–40 minutes from the impact time.

Scaling techniques were developed for transferring canine values to equivalent human values. These scaling relations, however, are uncertain as the impact area of the missile was not scaled in the experiments with the body mass of the dog (Bowen et al. 1966). The mass of the impacted lung was compared with the mass of the lung on the contralateral side. This ratio is generally correlated with the severity of injury; the threshold value of the ratio of the right mass to the left for fatalities in the period under observation was approximately 2.3. In addition, the scaled energy of impact

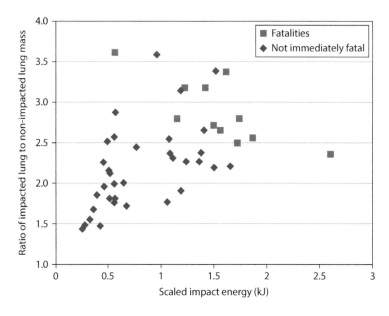

FIGURE 14.7 Impact energy (scaled to a 75-kg man) versus increased lung mass. (Based on Bowen, I.G., et al., *Ann. N. Y. Acad. Sci.*, 152(1), 122–146, 1966.)

(scaled to a 75-kg man) correlated well with fatalities. In Figure 14.7, the increased impact lung mass is plotted against the scaled impact energy. As an injury measure, increased lung mass measured post-mortem has limitations. The principal limitation is that injury may not be directly associated with bleeding since a surviving animal may bleed more than an animal that died earlier. So, this injury measure should be used for injury and not for fatality. However, over the limited time that the study followed the test animals, the correlation was relatively good.

The impactor mass of 63 g may represent behind armour impact in some regimes. For body armour with 24 kg/m² areal density, the effective impact diameter of approximately 37 mm reported by Mirzeabassov et al. (2000) implies an additional 84 g added mass owing to the induced motion of the body armour. In addition, the time scale is similar to BABT forcing. Hinsley et al. (2002) reported approximately 2 ms to full displacement of approximately 4 cm with a peak acceleration of approximately 28,000 g.

14.2.2 DANISH ARMY COMBAT SCHOOL (KNUDSEN AND GØTZE 1997)

Animal experiments were performed to assess the potential for thoracic injury from BABT behind undefeated body armour. Projectiles included 5.56 mm NATO ball at 921 m/s, 7.62 mm NATO ball at 848 m/s and 12.7 mm AP rounds with reduced load at 463–595 m/s. Tests were performed using twenty 60 kg swine, with one used as a control. The swine were placed in a standing posture, and lateral impacts were performed at the level of the xiphoid cartilage. Physiological monitoring of ECG, spirometer and pulse oxymetry was performed for the hour long observation period prior to sacrifice.

Two of the four animals at the largest kinetic energy were sacrificed for ethical reasons before the end of the 60-minute period. Injuries were assigned as shown in Table 14.5. Minimal injuries were seen at kinetic energies below 1.7 kJ while fatal injuries were expected above 8 kJ for the body armour selected for testing. Cardiac lesions were seen in many of the test animals and in the controls, suggesting this damage was an experimental artefact. There was no consistent correlation seen between skin damage and lung damage. No specific injury criterion was developed that was independent of the experimental setup.

TABLE 14.5
Bullet Specifications and Injury Outcome

Round	Impact Velocity (m/s)	Kinetic Energy (kJ)	Injuries
5.56 mm NATO ball	9920–9922	1.693–1.700	Minimal
7.62 mm NATO ball	838–861	3.248–3.429	Minimal to moderate
12.7 mm M2	463–595	4.839–7.992	Severe to fatal

Source: Knudsen, P.J., and Gøtze, H., *Behind Armour Blunt Trauma*, Trials report 02/97, Danish Army Combat School, Oksbøl, Denmark.

14.2.3 OKSBOEL TRIALS – NATO

As a more extensive follow-on to the Danish Army Combat School tests, trials were performed using extensive instrumentation in Oksboel in 1999. The trials used porcine specimens similar to those used in the earlier Danish Army Combat School tests. The protective equipment tested was provided by Denmark, France, the United Kingdom and the Unites States. The Danish and French developed body armour systems that defeat the 7.62 × 51 mm threat, while the UK and US systems defeat the more energetic 12.7 mm (50 calibre) sniper rifle threat. Three designs of body armour were used so that the trials were separated into three groups of pigs with eight animals in each group. There were three control animals. The impact site was selected in the right thorax in the middle of the eighth rib. Instrumentation included pressure and acceleration measurements in the thoracic wall and physiological measurements for half an hour before euthanasia, which in retrospect was too short a time period.

The data were obtained from pressure transducers, accelerometers, ECG, blood oxygen saturation, respiratory rate observations and post-experimental autopsy examinations. The gross and microscopic studies included lungs, liver and, in some studies, the heart and kidneys (Mayorga et al. 2010). Unfortunately, no intestine, brain, spleen, aorta or spinal cord studies were reported. Each of four separate experiments (i.e. Danish, United Kingdom and two French studies) followed similar protocols for animal anaesthesia and physiological observations, but the selection of protective armour varied widely across studies, with each comparing multiple types of armour.

The extensive test series was designed to discern the cause of wounding, separating the effects of the initial large, short-duration pressure peak and a secondary displacement pressure peak, presumably caused by the deformation of the body wall behind the body armour. The body armour was designed as shown in Figure 14.8. The first type is a typical body armour with a steep first pressure peak and a more extended later displacement peak (G1). The second would have a second peak only (G2), and the third would have only a first peak (G3).

The armour did not perform as expected. As shown in Figure 14.9, the 'first peak only' armour not only decreased the second peak but also reduced the first peak pressure. 'Second peak only' armour significantly reduced both the first peak and the second peak, but not so much that the first peak was less than the second peak. The 'first peak only' armour is a difficult engineering problem, implying infinite rigidity in the armour system, significantly limiting resulting backface displacement. This 'second peak only' is a very difficult problem, implying a match in impedance between the rear of the armour and the tissue.

All animals tested were injured. Post-mortem lung mass, as shown in Figure 14.10, appears to be a reliable injury measure when animals were sacrificed within 60 minutes. In general testing, however,

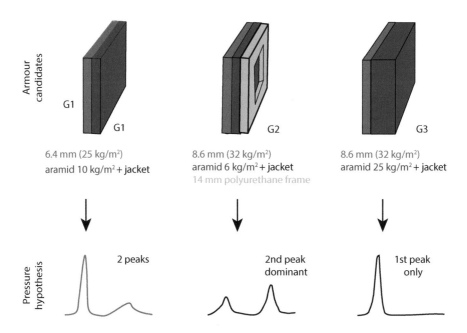

FIGURE 14.8 Body armour for Oksboel trials. (From Sarron, J.C., et al., Physiological results of NATO BABT experiments, *Personal Armour Systems Symposium (PASS 2000)*, Colchester, UK, 2000. Copyright clearance from Sarron, personal communication.)

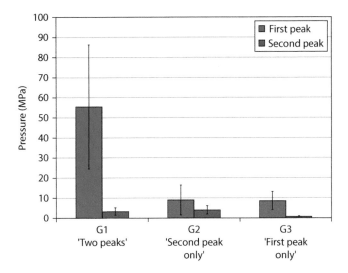

FIGURE 14.9   Average first and second peak pressure – Oksboel trials. (Based on Sarron 2000.)

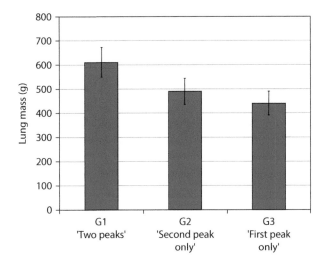

FIGURE 14.10   Average post-mortem lung mass – Oksboel trials. [Based on Sarron, J.C., et al., Physiological results of NATO BABT experiments, *Personal Armour Systems Symposium (PASS 2000)*, Colchester, UK, 2000.]

different times of expiration and the dynamic effects of haemorrhage may confound lung mass measurements.

The effect of the first peak only may be best studied by comparison with shock impingement. To get a qualitative idea of the effect of this pressure peak, the Oksboel pressure peak can be compared with the Bowen curves as shown in Figure 14.11. While the use of this internal pressure is not appropriate for the Bowen curves, it is likely conservative for the assessment of an injury threshold. The Oksboel experiments show pressure peaks in excess of 30 MPa with durations of 0.3–0.4 ms. The 100% blast lethality curve (Bass et al. 2008) is far below this value. This suggests that the local lung damage may occur as a result of a transmission of a high amplitude pressure wave. Indeed, lung mass injury from the Oksboel tests scales directly with first peak only.

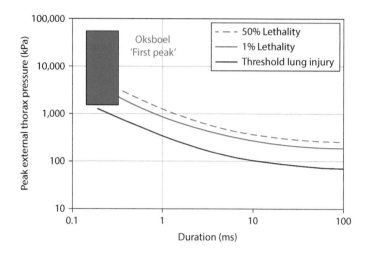

FIGURE 14.11 Oksboel first peak on Bowen curve. [Based on Sarron, J.C., et al., Physiological results of NATO BABT experiments, *Personal Armour Systems Symposium (PASS 2000)*, Colchester, UK, 2000.]

The results showed significant injury and a high mortality for most of the study groups, along with surprisingly significant thorax and lung injuries from hard armour having areal densities up to 24 kg/m^2 with foam backing. Pressure and accelerometer recordings were incomplete as saturation of the instruments with pressures exceeding 34 MPa was experienced in the majority of studies. Nevertheless, some very important observations and correlations were made for high-velocity ballistic impacts and the subsequent deformations of the back face of body armour. From the extensive studies in anaesthetised pigs by the French investigators and as reported in a NATO summary of international studies (Mayorga et al. 2010), a synopsis of the relationships between lung contusion areas, recorded pressures and deformation measurements can be made. These results are summarised in Figures 14.12 and 14.13.

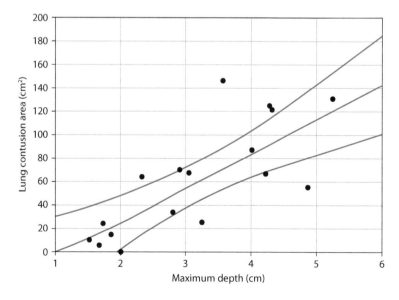

FIGURE 14.12 Relationship between area of lung surface contusion and maximum back face deformation of body armour. (Based on Sarron, J.C., et al., Displacements of thoracic wall after 7.62 non-penetrating impact, *Presentation to the NATO BABT Task Group*, Copenhagen, 2001.)

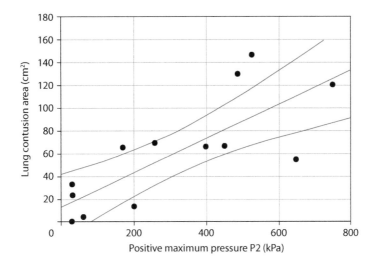

FIGURE 14.13   Relationship between area of lung surface contusion and pressure 6 cm from point of impact. (Based on Sarron, J.C., et al., Displacements of thoracic wall after 7.62 non-penetrating impact, *Presentation to the NATO BABT Task Group*, Copenhagen, 2001.)

Since Figure 14.2 presents the superimposition of a series of cine-radiographs, the timing of the chest deformation relative to the BFD cannot be related to the maximum deformation shown in Figure 14.12, or to the pressure measurements shown in Figure 14.13.

14.2.4   DGA – France (Sendowski et al. 1994; Sarron et al. 2001)

An extensive test series using porcine subjects was performed by Direction générale de l'armement (DGA) during the 1990s and in the 2000s. The model selected was female swine of mass 60 kg (±5 kg). Four impact areas selected were as follows:

1. *Pulmonary near the seventh right dorsal vertebra.* The animals were observed for 2 hours after trauma.
2. *Lateral cardiac area at the fifth left thoracic vertebra.* The animals were observed for 2 hours after trauma. This condition was chosen to investigate cardiac contusion.
3. *Mediastinum opposite the apex of the heart.* The animals were observed for 15 minutes prior to euthanasia. This condition was chosen to investigate induced ventricular fibrillation.
4. *Liver.* The animals were observed for 15 minutes.

As with most extant animal experiments, the DGA animals were not observed for a substantial time post-test. This prevented the investigation of injuries such as ARDS and others that require long-term physiological monitoring.

All shots occurred at the end of the inhale cycle, except for the lateral cardiac tests in which shots occur at the end of the exhale cycle. The body armour used included ultra-high molecular weight polyethylene (UHMWPE), while the test rounds involved a 7.62 mm round at 829 m/s and a 5.56 mm round at 989 m/s. Five tests were performed for each of the rounds for each test condition (i.e. 40 tests in total).

The instrumentation included an accelerometer on the rib and impact pressure measured using a balloon gauge inside the thoracic oesophagus. The accelerometer, however, did not function for most of the tests, and the systemic pulmonary pressures were very low (~1.8 mmHg). Physiological measurements include respiration and electrocardiogram. Vascular pressure measurements include pulmonary arterial pressure, occlusion pulmonary arterial pressure, abdominal aortic pressure and vena cava pressure. Blood gases and cellular enzymes 'creatine phosphokinase lactate dehydrogenase' were measured.

Post-test measurements included extensive grading of locations of skin damage. It was found that the cutaneous wound was statistically significantly greater for the 7.62 series than for the 5.56 series. The average diameter of the cutaneous wound was found to be about 14 cm.

Pulmonary wounds were assessed from the right thoracic shots. Phenomenologically, a bruise developed under the pleura surrounded by a region with inflammation. In some tests, emphysema developed at the centre of the impact. It was found that the percentage of injured lung for the 7.62 mm (13%) and for the 5.56 mm (7.4%) is statistically significantly different. Cardiac wounds were assessed using the lateral cardiac area shots and the mediastinal shots. The tests were highly variable, likely owing to the shape of the sternum.

Initial apnoea was seen in 30 tests, averaging 15 seconds in the 7.62 mm tests and 8 seconds in the 5.56 mm tests. This includes all shots in the sternal area. Secondary apnoeas occur in several of the tests. These are generally associated with advance shock. The number of deaths observed in these tests is shown in Figure 14.14.

Experiments at the French DGA (Sarron et al. 2001) using 7.62 mm test rounds into laminated UHMWPE body armour concentrated on investigating physical measurements of the thoracic wall using a flash radiography technique with lead markers as shown in Figure 14.15. This technique has been found to provide a good representation of the motion of the chest wall up to a maximum velocity of 30 m/s. In these studies, a large negative pressure (~3 MPa) was observed in preliminary data, suggesting that a cavitation injury mechanism may be possible. It was found that the shock wave arrives before significant displacement, and that the pressures are not well associated with the local displacements.

In order to find a correlation between intrathoracic pressure, BABT and high-velocity versus low-velocity bullets, 20 pigs, protected by a National Institute of Justice Level 3 or 4 bulletproof vest, were shot with 7.62 mm NATO bullets (2.4 and 3.2 kJ) and 10 unprotected pigs were shot by air gun with

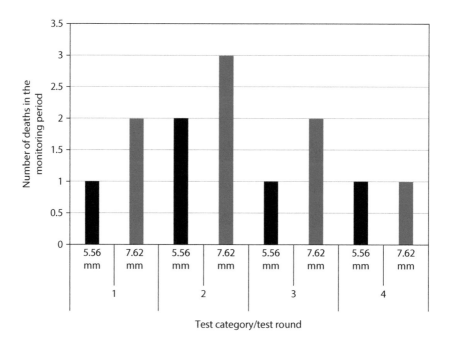

FIGURE 14.14 Animal fatalities during monitoring period. (Based on Sendowski, I., et al., *Rear Side Effects of Protective Vests in Non-Perforating Model. Tolerance Norms Research*, Institute of Tropical Medicine of the Armed Forces Health Service, DGA, France, 1994.)

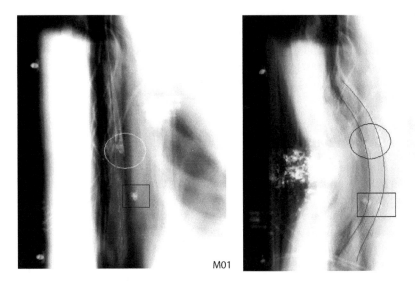

M01

FIGURE 14.15 Behind armour blunt trauma flash radiography. Left – initial time, right – 0, 1.5, 2 and 2.5 ms. (From Sarron, J.C., et al., Displacements of thoracic wall after 7.62 mm non-penetrating impact, *Presentation to the NATO BABT Task Group*, Copenhagen, 2001. Copyright clearance from Sarron for underlying CT.)

40 mm rubber projectiles (0.07–0.2 kJ) (Prat et al. 2010). Rib fractures occurred in 21 of the 30 animals with no correlation to the projectile kinetics, but intrathoracic peak pressures showed a good correlation with the volume of lung contusions. Other studies by the French are reviewed under NATO studies described below.

14.2.5 US Army Aeromedical Research Laboratory (Haley et al. 1997)

Tests were performed to evaluate the effect of PVC foam at standoffs of 14, 21 and 27 mm. Seventeen pigs (91 kg) were tested using ceramic body armour with a 12.7 mm test round. Three control pigs were used, and the animals were monitored for 3 hours after the tests. The subjects were instrumented with accelerometers on the top and bottom of the armour plate, and load cells were placed behind the pig to evaluate the global force. In initial testing, eight thin load cells were placed between the thorax and the armour. The presence of the load cells decreased the local standoff. Six load cells were used in the remaining tests. Heart rate and respiration rate were monitored. The standoff foam was found to be rate sensitive, transmitting large pressures to the thorax. The researchers concluded that a 25 mm standoff is necessary for effective protection of the thorax.

14.2.6 Anter Corporation (Mirzeabassov et al. 2000)

Anter Corporation reported an investigation of BABT on small dogs with various vest types; these included two thicknesses of titanium plates and one vest of ceramic body armour. Rounds included 7.62 mm with two impact energies, a 0.45 calibre M11911A1 and a 9 mm round. They reported impact kinetic energy ranging from approximately 0.3 to 3.2 kJ on 21 canine subjects. Tests were performed under high-speed flash radiography. Injuries sustained included superficial wounds, rib fractures, haemorrhages and deep lacerations. The canine tests were correlated to both cadaveric tests and a limited epidemiological test discussed below, though extensive information on instrumentation of the animal testing is not reported.

The researchers found that BFD with soft body armour has a maximum depth of penetration (H) that is comparable to the diameter (L) of the maximum area of contact (S). In contrast, the action of ceramic plates tends to lower the depth of penetration relative to the contact diameter (L). For these experiments, the depth and area of contact are measured using a high-speed flash radiography. The volume was inferred assuming an ellipsoid of revolution as $W = S^2/H$.

14.2.7 SWEDISH STUDIES (2000s)

Several Swedish studies were conducted that show, for live pigs, the 40 mm BFD criterion may not be protective to higher velocity projectiles (e.g. Gryth et al. 2007). Gryth and colleagues exposed 22 pigs that were protected by armour to impacts from 7.62 × 51 mm (800 m/s) rounds. After the backface impacts, acute physiological parameters for respiration, circulation and blood chemistry were monitored in the anaesthetised animals. Following sacrifice, histopathological studies were performed. Findings include the observation that the mortality rate for acute injury was 50% in the group in which the tested armour clay BFD values were <40 mm. In heavier armour with tested clay values <34 mm BFD, 25% of the animals died. It is unclear, however, what the clay type and temperature of the clay are from these studies.

In other Swedish studies, the experimental pigs were protected in soft armour (various Kevlar constructions with and without foam rear layer) and the torso was exposed to 9 mm, 44 magnum and 12-gauge shotgun slug projectiles.

In these studies, BFD into a soap simulant with the animal pathology was compared (Lidén et al. 1988). An interesting study with hard armour investigated electroencephalographic response in armour-protected pigs impacted with 7.62 × 51 mm (800 m/s) bullets. Though there were no penetrations, all pigs had lung injuries from the BFD and five of eight pigs showed transient changes in EEG response (Drobin et al. 2007).

In more recent Swedish studies, the effect of an impedance mismatched light layer to decrease the transmissions of pressure waves across hard body armour was determined (Sondén et al. 2009), now called trauma-attenuating backing (TAB). The effect of this backing on the performance of ceramic/aramid body armour was studied in 24 pigs, with TAB (n = 12) or without TAB (n = 12) following BFD from a 7.62 × 51 mm projectile. Based on this study, the TAB used decreased the area of lung contusion, significantly decreased haemoptysis and reduced impact peak pressures by 91%.

14.2.8 FRENCH STUDIES (2010)

In French research, Prat et al. (2010) attempted to develop a correlation between intrathoracic pressure, trauma and high-velocity versus low-velocity bullets. In this study, a National Institute of Justice Level 3 or 4 ballistic vest protected 20 pigs. The ballistic threats included 7.62 × 51 mm NATO rounds (2.4 and 3.2 kJ) (20 pigs) and 40 mm rubber projectiles shot with air guns (0.07–0.2 kJ) (10 pigs). Prat et al. found no correlation between rib fractures (occurred in 21 of the 30 animals) and projectile kinetics. However, measured intrathoracic peak pressures were correlated with lung contusion volume.

14.2.9 UNITED KINGDOM STUDIES ON CARDIAC TRAUMA THRESHOLD

The UK performed extensive live-animal tests of pathophysiological response behind ballistic protective body armour in the 1980s. Forty-eight anaesthetised pigs were instrumented with rib accelerometers and internal pressure transducers. Kinematic parameters and the deformation mechanics of the armour and chest of the animals were measured using high-speed photography and cineradiography. Projectiles representing armour BFD (0.14 kg, 0.38 kg) were fired into the anterior sternum and other locations with an air gun at velocities between 20 and 74 m/s (64–363 J impact energy) (Cooper et al. 1982). Impactor projectile diameters (3.7 cm, 10 cm) are intended to represent BFD diameters behind hard armour. They found that the severity of cardiac injury was correlated with the chest wall displacement normalised by anterior–posterior chest diameter. This was found to be proportional to the impact energy normalised by the product of the impact diameter and the body mass.

Cooper et al. also determined displacement criteria using imaging methods with cineradiography (Cooper et al. 1984). Small steel balls were surgically implanted in the heart and mediastinum to determine displacement response. These studies provide fundamental response parameters for cardiac damage based on impact energy, area of impact and BFD. The range of input energies ranged from 3 to 13 kJ and may be useful for design against thoracic injury from BFD.

14.3 CADAVERIC EXPERIMENTS FOR BABT

A limited number of cadaveric experiments with ballistic behind armour impact have been performed (e.g. Bass et al. 2006). For injury association using animal experiments, such cadaveric experimentation is likely crucial as there are significant differences between the anatomy of typical animal surrogates (e.g. pigs and other models) and human anatomy, especially in the mediastinal region.

14.3.1 ANTER CORPORATION (MIRZEABASSOV ET AL. 2000)

In addition to the animal experimentation reported in Mirzeabassov et al. (2000), 13 cadaver experiments were also included in their study. The tests used various vest types, which included two thicknesses of titanium plates and ceramic body armour. Rounds include 7.62 mm with two impact energies, a 0.45 calibre M11911A1 and a 9 mm round. The researchers reported impact kinetic energy ranging from approximately 0.3 to 3.2 kJ on 13 cadaveric subjects. Details regarding the preparation of the cadavers and conditions under which the tests were conducted are not reported. The authors state some differences in tolerance between animal and cadaver results for the internal organs; however, the differences are not quantified.

14.3.2 DEMAIO ET AL. (2001)

In a study of body armour, researchers instrumented 17 cadaveric specimens, including 6 females and 11 males, with a mean age of 73 years. Various body armours with different velocity regimes were reported by DeMaio (2001). The instrumentation included accelerometers at the sternum, T7, carina and ligamentum arteriosum, as well as pressure in the right and the left ventricles and the left chest. There was an attempt to measure the impact pressure between the armour and chest wall that was reportedly unreliable. Pressurisation of the lungs and cardiovascular system was performed. Post-test evaluation was performed for exterior wounds, sternal and rib fractures, cardiac bruising and other injuries including pulmonary injuries and spinal fractures.

Three body armour systems were tested, a soft vest with 9 mm test rounds, a light plate with a 7.62 mm round at two representative velocities and a heavy plate with a 7.62 mm round at one representative velocity. Three injury levels were defined. These are survivable (with minimal trauma), immediately survivable (non-lethal if treated within 1 h) and lethal (fatal even with treatment within 1 h). Injuries sustained ranged from light surface friction to deep extensive bilateral open chest wounds. The most severe injuries arose from complete plate penetration; an estimated three quarters of the cases where the round went fully through the plate were lethal.

Use of data from instrumentation for this study poses several difficulties. First, the accelerometers were attached to the sternum using suture material. However, accelerations were seen in excess of 1,000 g in some tests. The effective mass of a typical shock accelerometer (nominal mass ~1 g) at this acceleration level exceeds 1 kg. To ensure repeatable results, some rigid connection is exceedingly desirable. Second, the measurement of uniaxial accelerations on viscoelastic components within the thorax such as the carina and ligamentum arteriosum is questionable. Results will vary significantly based on local details of mounting and local viscoelastic behaviour of the compliant structures of the carina and ligamentum arteriosum. While it may be possible to use these measurements for qualitative estimation of the gross arrival of local tissue deformation owing to the significantly increased density of the accelerometer compared with the surrounding tissue, intraspecimen comparisons are questionable. Therefore, this work is most useful for quantitative injury performance of the articles tested. No development of an injury risk function has been performed in terms of measured dynamic variables.

14.3.3 BASS ET AL. (2006)

This study used a highly deforming hard body armour to estimate the mechanical correlates with BFD injury in nine cadavers and two ATDs (AUSMAN, clay). A range of velocities were tested including low-severity impacts, medium-severity impacts and high-severity impacts based upon risk of sternal fracture. Thoracic injuries ranged from minor skin abrasions (AIS 1) to severe sternal fractures (AIS 3+) and were well correlated with impact velocity and bone mineral density. Eight male cadavers were used in the injury risk criterion development. A 50% risk of AIS 3+ injury corresponded to a peak impact force of 24,900 ± 1,400 N. Spinal impacts behind body armour in a single test were also investigated in this study. Correlation of the injuries to the sternum and spine in the same specimen under the same threat round velocity suggests that sternal impact may not be the worst case for behind hard armour impact. Preliminary data from a single specimen with a matched sternal and spinal impact behind the body armour suggest that additional spinal impact research would be of significant value. However, impacts to the ribs or more general loading conditions with body armours of different characteristics were not assessed in this study.

14.4 BABT TEST METHODOLOGIES

A dummy or surrogate for human response is generally used to provide a reliable and inexpensive test methodology. This surrogate may allow repeatable characterisation of the performance of ballistic protective gear. The existing surrogate ballistic impact simulators may be divided into three classes as shown in Figure 14.16. The first class may be characterised as *bulk tissue simulants*; they are comprised of a single layer of material and are characterised by deformation response, for example, post-test residual clay penetration depth or dynamic gelatine penetration depth. The second class, or *instrumented response elements*, generally includes a simplified thoracic wall that simulates the motion of the surface and may incorporate multiple layers. The third class is *instrumented detailed anatomical surrogates*. These generally include some form of thoracic viscera and are designed to investigate a wide range of forcing. It is generally defined below as a research device if it includes internal thoracic organs.

To develop BABT test methodologies, all three types may be useful. There are trade-offs as devices run from simpler to more complex; this usually implies less expensive to more expensive, but the more complex devices generally have the potential to assess more detailed injury criteria where appropriate. For instance, it is difficult to evaluate complex BABT interactions, especially for very lightweight body armour systems, without detailed anatomical surrogates. However, it is advisable to use a relatively inexpensive bulk tissue simulant or instrumented response element for production testing to avoid the potential for penetrating events to destroy a test device worth thousands or tens of thousands of dollars. Each class of device may have an advantage for a given test condition. The three classes of devices are discussed below.

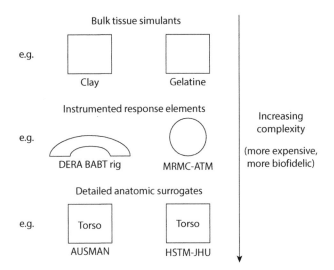

FIGURE 14.16   BABT assessment devices/methodologies including single-layer bulk tissue simulants, more complex instrumented response elements with more complex geometry and physical construction and detailed anatomical surrogates.

1. Bulk Tissue Simulants

Bulk tissue simulants are characterised by a single ideally isotropic and homogeneous layer. Though a number of bulk tissue simulants have been considered, including those investigated in the original Prather study (discussed below), only clay and ballistic gelatine are discussed below.

i. *Clay*

As exhaustively discussed above, the residual deformation of the body armour backface into Roma Plastilina No. 1 (oil-based modelling clay) is used as a standard for evaluating the performance of the body armour. The principal advantages of this material are the potential for rapid and inexpensive tests and the large historical database of body armour tests using the clay simulant. The principal disadvantages are the limitations of the clay as a human thorax simulant, the very limited validation against human injury response, especially against hard body armours, and the inability to consider more complex BFD scenarios. For instance, the methodology was validated using goat experiments with a single velocity regime; the test is not likely to be appropriate over wide ranges of deformation. Furthermore, the standard does not account for the effect of local penetrations (pencilling effect). Such a pencilling effect is frequently seen in soft body armour backing (Lewis, E. 2001, Personal communication).

ii. *Ballistic gelatine*

Hydrated gelatine (Kind & Knox, Sioux City, IA) is a commonly used test device for the evaluation of ballistic penetration in tissues. Introduced by Harvey et al. (1962), gelatine at 20% by mass at 24°C was used to investigate ballistic penetration. Fackler and Malinowski (1985) introduced a test methodology using 10% ballistic gelatine by mass at 4°C. For this formulation, the ballistic temporary cavity, permanent cavity and penetration were found to be similar to those seen in living tissue. Correlation of gelatine with goat response and injury is discussed above.

Both 10% and 20% gelatine are linear materials, with little viscoelastic response (Bass et al. 1998). However, human tissue has a strong viscoelastic character. If high shear rate response from penetrating impacts is similar to that seen in living tissues, the lower rate blunt impact properties of the ballistic gelatine are unlikely to be similar to human tissue.

As a practical test device, gelatine is difficult to handle and degrades over time, although it is still used regularly as a test device by the Army Research Lab (Kinsler, R. 2010, Personal communication). Furthermore, the properties may vary with production technique. Gelatine blocks are not generally reusable, and the local properties change with large deformation impacts. This may limit the usefulness of the gelatine for repeated testing.

However, gelatine may be useful for simulating viscera, although work is necessary on skin or membrane encapsulation. In tests with non-lethal munitions, Bir (2000a) found difficulties with round penetration. Further issues include the fact that local ballistic tissue failures seen for blunt tests are unlikely to be representative of human tissue if the penetrating properties are similar.

Ballistic gelatine at 20% by weight has been used in cooperative testing with Natick Soldier Center. The tests include body armour with a ceramic plate and a UHMWPE laminate. The test round was a 7.62 mm M80 ball round at 838 m/s nominal velocity. These tests were matched with those performed with the DERA BABT simulator to assist with validation of the physical models.

2. Instrumented Response Elements

The instrumented response element is designed to simulate the average behaviour of the thoracic wall. Below, the characteristics of a number of existing devices are reported, with the parameters measured and injury validation used. The trade-offs involved in constructing such a device include simplicity and ease of use versus biofidelity. Laboratory instrumentation is used to provide time histories or responses, including deformation, acceleration or other physical parameters that have been correlated with injury response.

3. Detailed Anatomical Surrogates

Detailed anatomical surrogates provided enhanced representations of anatomical response and biofidelity at the expense of complexity and cost. These devices should be seen as an evolution of instrumented response elements. For such devices, biofidelic skin, musculature, fat, bones and organ structures may be used with advanced local and global instrumentation to assess local injuries. Such models may often be used in concert with computational models such as finite element models to enhance predictive capabilities.

14.4.1 BABT CLAY TEST METHODOLOGY

A study performed by Prather et al. (1977) is the basis for clay-based test methodologies used by many nations for the assessment of blunt trauma risk from BFD for both soft and hard body armours (e.g. NIJ0010.06). This work was derived from two principal foundations. First, assassinations and assassination attempts on politicians and other VIPs in the 1960s and 1970s led to the desire to provide unobtrusive ballistic protection to counterthreats. Second, advances in fibre technology provided a basis for protection against low-velocity handgun rounds using layers of woven fibres. Injury biomechanics investigations leading to the Prather study focused on protection from low-velocity handgun threats with soft body armour. The Prather study included an injury assessment methodology developed using simulant materials correlated with live-animal chest deformation responses for handgun rounds at typical muzzle velocities. A diagram of this process is shown in Figure 14.17.

To provide human injury risk, the Prather study used two essential analogies outlined in Figure 14.17. The first analogy asserts the similarity between deformation behind armour in paired tests using soft body armour and handgun bullets with live goats and ballistic gelatine surrogates. During this phase, no goats died from backface trauma. Penetration depths were derived using only the gelatine. The second analogy uses the first gelatine study to develop a hard cylindrical impactor used with live goats. Tests at similar impact velocities were used with gelatine and clay behind body armour to correlate gelatine and clay depths of penetration. The development of this process is discussed below.

14.4.1.1 Background

In response to armed attacks on public officials in the United States before 1973, the Law Enforcement Assistance Administration (LEAA) asked the US Army Land Warfare Laboratory (LWL)

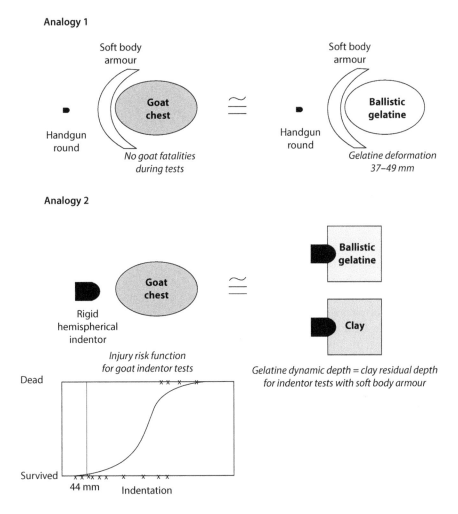

FIGURE 14.17 Overview of Prather clay ballistic test methodology (Prather et al. 1977). Analogy 1 correlates deformation behind soft body armour in goats with deformation in gelatine. Analogy 2 correlates a fatality risk assessment using goat tests with rigid impactors with gelatine *dynamic deformation* and clay *residual deformation.*

at Aberdeen Proving Ground, MD to develop lightweight, inconspicuous protective garments for public officials, eventually extended to law enforcement personnel (cf. Montanarelli et al. 1973), in collaboration with other agencies including US Army Natick Laboratories and the Aerospace Corporation. During that time, threats from civilian handguns were predominantly 0.38 calibre handgun rounds and below (Prather 2010, personal communication). Principal rounds chosen for this effort included the 0.38 calibre, lead round nose (LRN) bullet (158 grain) with muzzle velocity of 244 m/s and the 0.22 calibre, long rifle high-velocity (LRHV) bullet (40 grain) with muzzle velocity of 305 m/s. The design criteria for these ballistic protective vests included light in weight, inconspicuous and limited interference with body movement. Additional specifications included preventing bullet penetration, limiting the risk of fatality to less than 10% and allowing adult males to survive any ballistic attack. The last two requirements are not necessarily contradictory when viewed in the context of the assumption that victims would receive medical treatment for potential haemorrhage or other sequelae within an hour of the attack.

For such non-penetrating impacts, the kinetic energy and velocity must be dissipated by the armour through armour deformation, momentum transfer from bullet to the material, bullet deformation or fragmentation and, finally, the deformation of the body in contact with the armour. The momentum/energy transfer to the body can cause serious injury or death, even without penetrating the skin. This backface traumatic impact injury is called behind armour blunt trauma.

Numerous materials were assessed to determine the optimum weight to strength ratio, flexibility, cost, availability, ballistic qualities (ballistic limit and behind armour deformation) and tailorability. Materials investigated included high-tenacity nylon, nylon felts, high-tenacity rayon, graphite yarns, XP (an experimental plastic developed by Phillips Petroleum), Monsanto fibres and DuPont Kevlar 29 and 49. These materials were ranked, and a thickness of seven plies of 400/2 denier Kevlar 29 (K29) was selected for further investigation (Figure 14.18).

An animal model for BABT, the adult 40–50 kg Angora goat, was used to represent a 70-kg adult male. Goldfarb et al. (1975) studied the mechanical response of the lung, liver, kidney and spleen from both the goat and human cadavers under impact with a water jet. Based on qualitatively similar response of the goat and human collapsed lung and the determination that the goat liver and kidney were more robust for mechanical trauma than human organs, they concluded that the goat was a conservative BABT model. Paired injury tests in the goat model and deformation response tests in 20% gelatine were performed with *armour* samples against the 0.38 calibre, LRN bullet at 244 m/s. The lung, liver, heart, spine, gut and spleen were specifically targeted during these tests. Matched tests with the same round and body armour into a 20% gelatine backing material were used to develop a profile of the

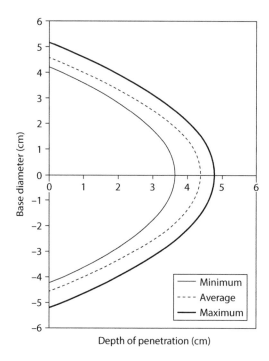

FIGURE 14.18   Blunt deformation profiles from a 0.38-calibre round at 213 m/s measured behind 7 ply Kevlar (k29) armour into ballistic gelatine using the same test parameters as in the goat model protected with the same armour. (Adapted from Prather, R.N., et al., *Backface Signatures of Soft Body Armors and the Associated Trauma Effects*, U.S. Army Armament Research and Development Command Technical Report ARCSL-TR-77055, Aberdeen Proving Ground, MD, 1977. Copyright clearance from National Academies.)

behind armour deformation (Figure 14.19). Since gelatine is transparent, high-speed film was used to determine the rate of deformation, deformation depth, volume and area. These were correlated with the trauma in the liver, lung, spleen and heart in the goat model. The gelatine profiles for this condition had an average dynamic deformation of 44 mm. This value was chosen for the BABT standard injury tolerance. *There were no deaths from backface trauma in the paired tests with the 0.22 calibre or 0.38 calibre rounds corresponding with this 44 mm deformation in gelatine* (e.g. Goldfarb et al. 1975). Hence, for soft body armours with handgun rounds similar to those used in the studies, this value is likely conservative and may be over-conservative for the desired risk levels.

At the same time, Clare et al. (1975a) developed analytical blunt trauma models formulated using tests in animal models not protected by body armour. They used tests for which the projectile impacts were similar to the assumed backface profile using larger impactors with slower velocity and larger impact mass (typically 50–200 g) than handgun rounds. One model uses projectile mass, M (grams), projectile impact velocity, V (m/s), projectile diameter, D (cm) and experimental animal weight, W (kg), with two discriminants, x_1 and x_2, defined as

$$x_1 = \ln[MV^2], x_2 = \ln[WD].$$

In the x_1, x_2 plane, three zones of lethality, low, medium and high, were established with increasing impact 'dose'.

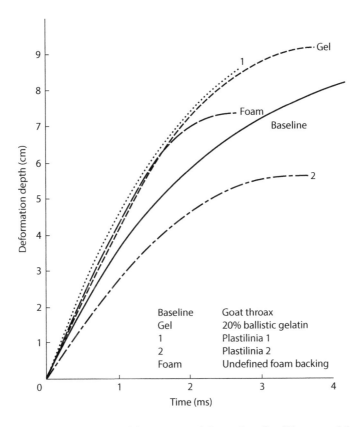

FIGURE 14.19 The hunt for a material with permanent deformation. Candidate material response for material compared with deformation of a goat thorax, blunt impactor at 55 m/s. (Derived from Prather, R.N., et al., *Backface Signatures of Soft Body Armors and the Associated Trauma Effects*, U.S. Army Armament Research and Development Command Technical Report ARCSL-TR-77055, Aberdeen Proving Ground, MD, 1977. Copyright clearance from National Academies.)

14.4.1.2 Clay Versus Gelatine for Biological Response

Before the development of the clay standards in the mid-1970s, 20% ballistic gelatine was used as the human surrogate/backface target for armour deformation testing (Russell Prather, personal communication); use of ballistic gelatine required a dynamic deformation measurement to assess BFD, often high-speed photography owing to gel elasticity. The ballistic gelatine approximately returned to its original state after the dynamic event was complete.

Since high-speed photography was expensive, a material was sought that had a permanent deformation following a backface event. Material selection criteria for a new backing material from LEAA include a response that was similar to gelatine, inexpensive, reusable, minimal post-event recovery. The intent was that the material be easy enough to use that individual law enforcement agencies could conduct their own testing at their facilities. The response was compared with hemispherical impact data on the goat thorax, intended to replicate behind armour response for soft armour rounds. These goat tests used an impact velocity of 55 m/s with a solid 200 g, 80 mm hemispherical impactor. Response to several materials was compared for similar impact conditions (Figure 14.19).

Although there were no materials that duplicated thoracic response, one modelling clay, Roma Plastilina No. 1, had similar deformation time histories to gelatine and was selected as the new candidate material. The material was inexpensive, repeatable and did not require high-speed photography. However, both the clay and gelatine response at this velocity for this impactor (55 m/s) is generally softer than for the goat thorax.

This deformation response was linked with fatality using blunt impactor data on live goats (e.g. Clare 1975b). The relationship was modelled using a logistic regression model between deformation and the probability of lethality (Figure 14.20). Using the model, 44 mm indentor impact deformation

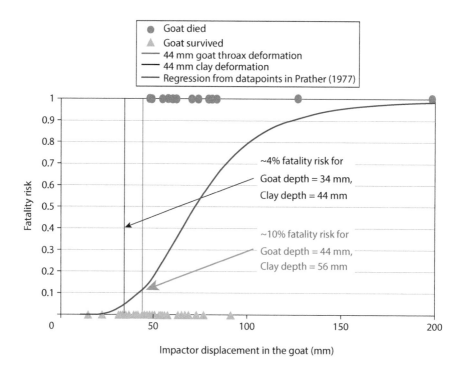

FIGURE 14.20 Logistic regression model of fatality against blunt impactor deformation into goat chests. (Based on Prather, R.N., et al., *Backface Signatures of Soft Body Armors and the Associated Trauma Effects*, U.S. Army Armament Research and Development Command Technical Report ARCSL-TR-77055, Aberdeen Proving Ground, MD, 1977.)

in the goats is associated with ~10% risk of death, similar to the initial program requirement of less than or equal to 10% mortality. This deformation level in the goats did not, however, produce any deaths in the goats. Hence, this depth was selected as a conservative injury reference value for the clay.

An important note is that this lethality risk assessment is indirect; it is not based on actual body armour testing. Use of non-compliant blunt impactor tests to model the deformation and injuries from BFD is likely conservative for this impact velocity range.

Outside this typical handgun/soft body armour velocity range, however, the injury risk function in terms of deformation response in the clay is uncertain. Prather's original study comments on the injury regression that:

> Attempts have been made using the original blunt impact data to correlate deformation depth with the probability of lethality. A depth of penetration greater than 500 mm is associated with a probability of lethality of approximately 15%. *However, the available data is limited and hence no solid conclusions can be drawn as yet regarding the effect of deformation depth* (emphasis added).

This emphasises the quite limited validation of the original injury correlation, even for soft body armour with handgun rounds. For a given impact, gelatine and clay depth of penetration were found to be generally greater than the deformation in the goat. Hence, Figure 14.19 can be used to determine the goat deformation compared with the clay. In the range of deformations 3–60 mm in the goats, the ratio of clay to goat deformation is approximately constant as:

$$D_{\text{clay}}/D_{\text{goat}} = 1.28.$$

In this deformation range, the ratio varies by less than 0.3%. This has implications for the current level of 44 mm in clay. *The 44 mm deformation in the clay is similar to 34 mm deformation in the goat (4% fatality risk in the model). Alternatively, 44 mm deformation in the goat (10% fatality risk in the model) is similar to 56 mm deformation in the clay as* shown in Figure 14.20.

14.4.1.3 Soft Body Armour – High-Energy Handgun Threats

In the late 1970s, as civilian handgun threats evolved to be more powerful rounds [e.g. 0.357 semi-wad cutter at 396 m/s (158 grain) and the 9 mm full metal jacket (FMJ) at 350 m/s (124 grain)], more protective soft body armours were studied. These investigations showed that a construction of 16 plies of K29 Kevlar was sufficient to prevent penetration and that the BFD profiles into gelatine were similar to those from the earlier soft armour studies with lower velocity rounds. Limited goat model studies showed similar injuries to those seen in the original 7 ply Kevlar tests with handgun rounds. The program was terminated before sufficient tests could be conducted to verify these preliminary conclusions. There are no formal reports on this work, but participants recall that these studies did not produce deaths with the more powerful round in the animal model behind the enhanced soft armour (Prather, 2010, personal communication).

14.4.1.4 Hard Body Armour – Rifle Threats and Clay Response

There were no systematic investigations of rifle threats with hard body armours in the Prather study. Ad hoc investigations included assessment of 50 cal. antipersonnel (AP) threats to helicopter pilots in animal models (Prather, 2010, personal communication). These limited studies, however, *produced no hard armour BABT risk methodology.*

For rifle round velocities, the correlation of structural BABT in human cadavers was compared with clay deformation behind a UHMWPE hard body armour system (Bass et al. 2006). The clay tests were based on NIJ standard 101.04 (reference) at a commercial test laboratory with a 7.62 × 54 mm M80 ball projectile. The test velocities ranged from ~670 to ~800 m/s. Clay BFD over this range of velocities (Figure 14.21) were very poorly correlated with peak deformation (Figure 14.22). In contrast, the human cadavers demonstrated a wide range of traumatic outcomes ranging from mild to very severe injuries over the same velocity range with the same round and body armour. The severity of these BABT injuries in the human cadavers generally scaled well with measured impact force and velocity. When the volume of the indentation or the cross-sectional area was used, the correlation improved, but the R^2 was still only 0.6 (Figure 14.23). *Since the clay depth correlates poorly with cadaveric injury rifle round threats with this body armour type, this emphasises concerns that the clay methodology may be inappropriate for hard armours with low areal density that may produce large velocity backface response.*

Furthermore, the NIJ 0101 (current version 06) evaluation measures static residual penetration depth into the clay. Bir (2000b) analysed dynamic clay deformation for non-lethal baton rounds finding that there was up to 20% greater dynamic deformation than residual quasi-static deformation. Also, this dynamic deformation may be rate sensitive and contact-area sensitive. *Both Bir et al. and Bass et al. raised concerns for using clay at the current depth for assessment of rifle rounds in hard body armour.* Depending on the characteristics of both the body armour and the clay response, it is uncertain whether a clay-based methodology for body armour based on the Prather methodology is overconservative for injury assessment.

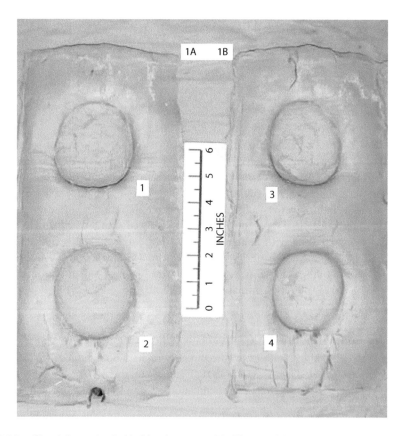

FIGURE 14.21   Clay deformation behind hard armour with rifle round threats. (From Bass, C.R., et al., *Int. J. Occup. Saf. Ergon.*, 12, 429–442, 2006. Copyright clearance from JOSE.)

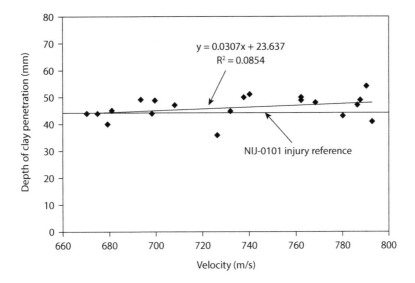

FIGURE 14.22 Variation of clay penetration depth with velocity for behind body armour deformation (7.62 mm NATO round, UHMPE body armour). (From Bass, C.R., et al., *Int. J. Occup. Saf. Ergon.*, 12, 429–442, 2006. Copyright clearance from JOSE.)

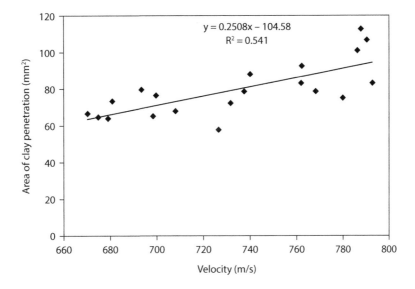

FIGURE 14.23 Variation of clay penetration area with velocity for behind body armour deformation (7.62 mm NATO round, UHMPE body armour). (From Bass, C.R., et al., *Int. J. Occup. Saf. Ergon.*, 12, 429–442, 2006. Copyright clearance from JOSE.)

14.4.1.5 Clay BABT Methodology Summary

The current clay methodologies for both soft and hard body armour (e.g. NIJ0101.06) are fundamentally based on the Prather study (1977). However, the Prather study was intended as a rapid expedient to field a test methodology for soft body armour against handgun rounds. Key limitations include:

1. The use of an impactor with a single profile at a single velocity to simulate ballistic deformation rather than ballistic behind armour deformation.

2. The lack of fatalities in the live goat tests at the levels used for the assessments.
3. The use of materials (clay and gelatine) that are softer than the goat thoraces.
4. Owing to changes in material response (National Academies Report 2014), the sculpting clay known as Roma Plastilina No. 1 has stiffened considerably since the early 1970s. This dramatic qualitative change requires that the clay must be preheated to the vicinity of 100°F to sufficiently soften the wax components so that the clay will have qualitatively similar response to the clay used by Prather et al. (1977). There is no guarantee that changes will not continue to occur in the constituents of this clay, although initial efforts have been made to standardise a variant of this clay for military use.

The Prather methodology provides no direct injury criterion for backface velocities typical of hard body armour, although it continues to be used by militaries to this date. Strengths and weaknesses of the current Prather methodology are listed in Table 14.6. *In view of these limitations, it is uncertain how conservative or liberal the current hard armour criteria are for human response.*

TABLE 14.6
Strengths and Weaknesses of the Prather Clay Methodology

Strengths	Weaknesses
Ease of use	Clay variability (handling, thixotropy, temperature effects, etc.)
Immediate results	Clay constituents have changed considerably since original study
Relatively low cost	Current methodology requires elevated clay temperatures
Large historical database of results	Method has limited validation for soft body armour
Apparent success in the field for soft body armour	Method has no validation for hard body armour
Apparent success in the field for hard body armour	Pass/fail criterion

14.4.2 INSTRUMENTED ALTERNATIVES TO CLAY FOR BABT TESTING

14.4.2.1 DERA BABT Simulator

The UK Defence Evaluation and Research Agency (DERA) has developed a test device that is intended to evaluate the injury effect of BABT, called the DERA BABT rig (Tam, W. 2000, personal communication). A half cylinder silicone rubber chest wall (GE Silicones RTV 428), providing rotation and vertical positioning, was developed by Cooper et al. (1996). This response element also allows varying thickness response elements. The physical model was derived from a finite element model that included high-rate blast and blunt response of the thorax. The test device uses a novel laser deformation sensor array system to measure the time history of displacement. The deformation sensor is kept outside the bullet trajectory and may provide velocity and acceleration response of the backface of the test device. The assumed BFD injury mechanism for this physical model is that injuries are a function of chest wall motion, including displacement, velocity, acceleration and deformation profile. This fixture has the potential for correlating porcine organ damage with the impact pressure wave and subsequent displacement.

The BABT simulator has been tuned to lateral eviscerated pig baton data and has been tested with 12.7 mm armour piercing round against the UK Enhanced Body Armour (EBA) and with 7.62 NATO ball against Improved Northern Ireland Body Armour (INIBA). Validation studies included both lateral and anterior shots. Peak accelerations were comparable between the BABT rig and lateral pig shots. These accelerations were approximately 20,000 g for the 7.62 mm test round and

INIBA armour. However, the response of the anterior porcine chest wall varied somewhat from the rig behaviour. As the rig was developed using lateral impacts, this is not surprising.

Details of eviscerated pig data 7.62 rounds at 3 kJ behind commercial ceramic body armour with six eviscerated and six intact pigs are reported by Cannon et al. (2000). The velocity of the physical model and the eviscerated model was seen to be similar (9.6 m/s for eviscerated and 15.2 m/s for physical model). A peak displacement of 12.6 mm was seen in the eviscerated pig, and a peak displacement of 7.8 mm was seen in the intact pigs. The local peak acceleration was approximately 13,000 g. The mean time to peak displacement was found to be approximately 2.7 ms. A viscous criterion of 0.29 m/s was seen for both. The DERA BABT device has been used in cooperative testing with Natick Soldier Center. The tests include body armour with a ceramic plate and a UHMWPE laminate. The test round was a 7.62 mm M80 ball round at 838 m/s nominal velocity.

There are two significant drawbacks to this system. First, the system has been tuned to a specific thoracic wall velocity. Outside this regime, it is unlikely to respond appropriately to rear face impact. Second, the simulator is shaped like a cylinder, and the laser system relies on this cylindrical shape to operate properly. As the human body is not cylindrical, it is difficult to use it to evaluate the actual body armour. To enhance the system, a different displacement measuring system could be developed using multiple instances of a diffuse laser time of flight system.

14.4.2.2 DREV Torso Injury Assessment Rig (Bourget et al. 2002)

Defense Research Establishment Valcartier (DREV) has developed a thoracic injury assessment rig that is similar to the DERA BABT simulator as shown in Figure 14.24. However, both the material of the simulator and the geometry have been altered and tuned to material response of Bir for non-lethal baton impacts (Bir2000b). It has a similar character and performance to the DERA BABT rig. Testing has been performed to optimise the mechanical properties to the material response of Bir.

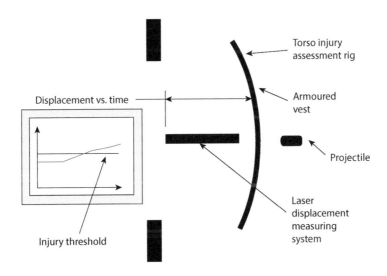

FIGURE 14.24 DREV BABT simulator displacement sensor system. (Based on Bourget 2002.)

14.4.2.3 Anter Company (Mirzeabassov et al. 2000)

A private company in St. Petersburg, Russia has patented a multilayer thorax simulator (Figure 14.25). This simulator is also covered under US Patent 5850033. The simulator is proposed for both penetrating and blunt injuries. It includes an outside skin layer, a layer of muscle-simulating material

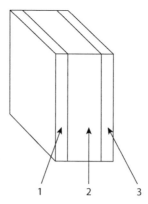

FIGURE 14.25 Overall schematic of the Anter thorax simulator: (1) skin simulant, (2) muscle simulant and (3) bone simulant. (From Mirzeabassov, T., et al., Further investigation of modelling for bullet-proof vests, *Personal Armor Safety Symposium*, Colchester, UK, 2000.)

and a layer of bone-simulating material. The muscle layer comprises several additional components. Additional detail in the model not represented in Figure 14.25 includes a component of muscle-simulating tissue, a layer of carbon paper or other indicating paper, a layer of cellular material such as nylon-6 net, two layers of paper and a separate muscle simulant from the other parts. The paper is used as a brittle strain-sensitive material to indicate the level of strain inside the simulant to approximate the temporary cavity during penetrating injury and to determine the extent of local deformation for blunt trauma.

The skin simulant is composed of vulcanised rubber with mechanical properties that approximate those of skin tissue. The muscle layer comprises unvulcanised rubber. The properties of the layers were selected to match low-rate physical properties of human tissues. Frontal area of the target is 40 cm × 40 cm square, with a muscle layer thickness of approximately 10–12 cm and the target weight is approximately 20 kg.

There is no electronic instrumentation for this simulator; penetration or injury is indicated by paper between layers. The construction is reusable; the paper indicator layers may be replaced. For additional information, these layers could be augmented with Fuji pressure-sensitive film. The basis for this injury model is experimentation using small dog subjects. Validation rounds include 7.62 mm with two impact energies, a 5.56 M16A1 and a 5.45 mm AK74. Validation included local thoracic deformation evaluated using flash radiography.

14.4.2.4 3-Rib Chest Structure (3RCS)

Bir (2000a) developed a physical model, as shown in Figure 14.26, based on a BioSID rib structure augmented with a high-rate (to 10 m/s) linear displacement transducer. The three ribs are mounted to an 18.1-kg spine box opposite the impact side, and viscoelastic damping material is incorporated into the ribs to provide bending resistance and energy dissipation. Nylon supports are mounted between the ribs to provide additional lateral support and to prevent large transverse rib motions. The impact surface is a urethane plate with a vertical dimension of 15.5 cm and a horizontal dimension of 23 cm. A nitrile shock damping padding covered the urethane bib for preliminary tests (Bir 2000a).

The model was validated against the UK baton round (instrumented baton – 140 g) at 20 m/s, 40 m/s and a reduced scale 30–40 g at 60 m/s impact velocity with corresponding initial impact energies of 28, 112 and 72 J, and initial impact momentum of 2.8, 5.6 and 2.4 kg m/s. The impact location was sternal directly on the thorax.

FIGURE 14.26   3-Rib chest structure (3RCS). (From Bir, C., *Comparison of Injury Evaluation in Non-Lethals, Non-Lethal Technology and Academic Research Symposium*, Portsmouth, New Hampshire, 2000.)

Thirteen cadaveric specimens were used, approximately 15th percentile male anthropometry. The data were normalised standard techniques often used as a source of validation data for relatively low-energy, high-momentum impacts. The scaling is based on geometric similarity (e.g. Eppinger et al. 1984), and chest depth was used as the scaling factor ℓ, where

$$\ell = 50\text{th \% male chest depth/subject chest depth.}$$

All dynamic quantities may be scaled using this factor. For instance, displacements are scaled as *normalised displacement* = ℓ × *measured displacement*. Bir developed response corridors (force vs. deflection) for each velocity impact; they are significantly different. Three thoracic ribs are used. This assumes localisation of impact with limited effect outside area. The model was originally developed for use with baseball impacts in evaluating commotio cordis injuries.

The model has sensor limitations at higher kinetic energy. The measured displacement and video displacement were not always the same, and this was exacerbated at higher kinetic energy. As the energy delivered to the thorax is relatively low and dispersed, this effect is expected to increase with increasing kinetic energy. In addition, the materials used were tuned for impacts with a specific energy and momentum. It is unlikely that they will be appropriate for use outside the specified range.

14.4.2.5   Hybrid III Crash Test ATD

The standard automobile frontal crash test anthropomorphic test device (ATD), the Hybrid III, has been used by several investigators for high-rate impact. In addition to the use discussed above by Bir et al., the Hybrid III ATD has been used by Bass et al. and Chichester et al. for mine blasts and high-rate impacts from blasts in structures (Bass 2001a, 2001b). The rib structure of the Hybrid III consists of six ribs constructed of steel overlaying a viscoelastic damping material. The ribs are connected in the front to a sternal bib, and a single displacement sensor is standard instrumentation. This single displacement sensor has significant limitations (Butcher 2001). Enhanced instrumentation is available with an array of string potentiometers. Use of the Hybrid III ATD has several advantages. The ATD is widely used and is manufactured in several different sizes representative of various anthropometries.

There are several substantial drawbacks to the use of the Hybrid III as a ballistic BFD test device (e.g. Bir 2000b). The ATD is validated for low-velocity frontal impacts typical of automobile crashes and not for high-velocity impacts typical of armour backface response. Furthermore, the local response of the Hybrid III is likely not biofidelic for thoracic impacts, even for these low-velocity impacts (cf. Kent et al. 2006) (Figure 14.27).

FIGURE 14.27 Hybrid III 50th percentile male ATD (Courtesy of Humanetics Innovative Solutions, Plymouth, MI).

14.4.2.6 ATM – US Army Medical Research and Materiel Command

This surrogate was developed by the US Army Medical Research and Materiel Command as an instrumented response element as a part of an injury assessment methodology for BFD. The shoulders of the torso are not instrumented but provide an anthropometrically correct platform for mounting body armour as worn by soldiers. The response of the Anthropometric Test Module (ATM) element is measured using a multiple accelerometer array implanted within a polymer with approximately cylindrical form. The initial peak impact response is mitigated using a rubber pad over the surface of the response element.

Essential elements of this methodology are:

1. Development of a detailed anatomical finite element model of the human
2. Identification of the global mechanical response and injury response in animal and cadaver tests for impact with hard projectiles
3. Development of a simple response element with mechanical response that is correlated with the animal and cadaver tests
4. Validation of the human finite element model with the animal and cadaver tests
5. Correlation of the response element deformation with finite element model calculations and hence animal and cadaver injury response.

The FE model is based on 30 moderate rate porcine and 12 low rate blunt impactor cadaver tests. Generally, test impact velocities did not reach those typical of high rate impact behind hard body armour. A limited model validation was performed using the animal and cadaver tests for the finite element model. The response included global response of the model and surrogates for a limited number of instrumentation locations.

Although the authors showed good correlation with automobile impact corridors (Kroell et al. 1971, 1974), the principal limitation is the lack of robust validation data for rates typical of impact behind hard body armour at typical rifle round velocities. Further limitations of the model involve the beam formulation of the ribs, which must be derived from validation data global stiffness.

Research questions raised by the authors include response to penetration, repeatability of measurements including the potential for material properties changes in the rubber with repeated shots and the basis of the model validation. These issues are the subject of ongoing research. Owing to the use of the complex finite element model for injury assessments, the ATM model and associated finite element model are more complex than the typical response elements considered above and likely represent a transition device between instrumented response elements and detailed anatomical surrogates.

14.4.3 Instrumented Detailed Anatomical Surrogates

The final class of instrumented surrogates is detailed anatomical surrogates. These include anatomical features that are similar to humans, generally including some representation of internal organs. When validated, these may be appropriate for the investigation of more complex behind armour phenomena. These devices are unlikely to directly form the basis for large-scale testing of body armour owing to their complexities in both construction and sensoring.

14.4.3.1 AUSMAN (Australian – Defence Science and Technology Group)

AUSMAN is a 21-kg mass reusable mechanical torso surrogate developed by the Defence Science and Technology Group (DST Group) of the Australian Department of Defence. The surrogate is shown in Figure 14.28 with body armour. As tested, AUSMAN includes a polymeric skeletal system with a simple cardiopulmonary system and liver. The skeletal system is anthropomorphic with a realistic spine. The covering of the thorax contains a polymer with a gel between internal viscera and rib structure. The lungs are designed with an open cell foam. In 2008, the design incorporated a mount for a Hybrid III head/neck intended to provide neck injury assessment from blunt impact.

Accelerometers are used to assess mechanical response of the torso in both the sternum and mid-thoracic spine. Pressure transducers are placed in the lungs. The implanted sensor structure was designed for easy access and replacement. In addition, the 2008 version of AUSMAN has limited validation for blast (cf. Bass, unpublished experimental results), but an earlier model was correlated against human cadaver sternal ballistic response (Bass et al. 2006).

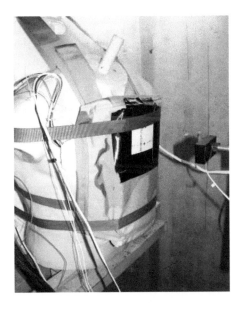

FIGURE 14.28   AUSMAN thorax with body armour in place, prior to testing (Bass unpublished).

14.4.3.2   National Defense Research Institute's Anthropometric Dummy (Sweden)

Jonsson et al. (1986) at the Swedish National Defense Research Institute developed an ATD thorax intended for use in ballistic blunt impact and direct missile impact to the thorax. This model is constructed inside an elliptical rubber tube (6 mm thick × 35 cm laterally × 25 mm in the anterior/posterior direction). A water-filled cavity with cylindrical foam rubber lungs was used to simulate internal viscera. The lung cylinders have a length of 20 cm axially and a diameter of 10 cm in the anteroposterior direction. A thin sheet of rubber is used to surround the lungs, preventing intrusion of water into the lungs. Structurally, the lungs are mounted rigidly at each end of the model and are maintained in the chest using strings and wire mesh. Each includes a centre-mounted pressure transducer (Figure 14.29).

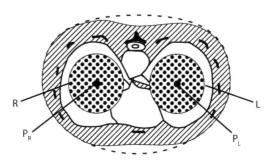

FIGURE 14.29   Swedish National Defense Research Institute Anthropomorphic Dummy's internal cross-section compared with human cross-section (dummy in dashed lines). (From Jonsson, A., et al., *J. Trauma.*, 28, S125–S131, 1986.)

Experiments using this model were performed with a shock tube, a pendulum device and ballistic impact behind body armour. Test rounds included 7.62 mm at 870 m/s in both soft point and FMJ, 9 mm submachine gun at 430 m/s and 37 mm anti-riot baton round at 76 m/s. The body armour for the rifle rounds was a 10 mm thick polyethylene trauma pack, 17 ply Kevlar 29 and 10 mm ceramic plate. For the 9 mm round, the body armour used a 2.3 mm steel plate in place of the 10 mm ceramic plate. For the baton round, no body armour was used. Impacts were performed at the mediastinum with energies that range from 0.2 to 5.6 kJ.

Intrathoracic pressure was measured for impact with the various rounds tested and compared with lateral rabbit impact experiments using the pendulum impactor. An injury scale for interlung pressure was developed using pendulum impacts with velocities of approximately 5 m/s that are stroke limited to 10% thoracic compression. They found that the blast pressure peaks were far larger in the blast experiments than in the impact experiments for similar levels of lung damage. Hence, they concluded that dummies validated for blast pressure will not be calibrated for blunt impacts using the same pressure measurements. From the scale, it was concluded that the impact of a 7.62 mm test round at 3.2 kJ behind a trauma pack, ceramic plate and Kevlar-29 body armour has a low risk of lung injury.

The researchers saw small (~10% of peak) oscillations in the pressure signal at approximately 3 kHz. They concluded that the oscillations were related to structural forcing. There are significant questions regarding impact injury with mediastinal forcing using injury indices developed in a different velocity regime for lateral impacts. Further development of this surrogate is unknown.

14.4.3.3   Instrumented Model Human Torso (Johns Hopkins Applied Physics Laboratory)

A further research device for ballistic impact has been developed by a team at the Johns Hopkins Applied Physics Laboratory. There have been at least two versions of this model, one frangible and a more recent reusable model (e.g. Bevin et al. 2001; Merkle et al. 2008).

14.4.3.3.1 Frangible JHU Model

The physical structure of the frangible model is shown in Figure 14.30. This proof-of-concept model includes rib, sternum and spinal structures, while the internal organs are represented as homogeneous viscera of urethane. The model has a fat layer and outer skin with a sensor pad between the skin and fat layer. A strong attempt was made to produce biofidelic materials. There has been no reported validation at high rates. The bone material was chosen to be frangible with a failure modulus similar to human values. The design is intended to allow rapid replacement of frangible components (Bevin, M. 2002, personal communication).

The instrumentation includes a sensor pad with an array of eight piezoelectric elements, which are sensitive in bending. An additional four sensor sheets were used: a resistive flexure grid twinned with the anterior pad, additional piezoelectric and resistive flexure grids are located 0.5 in. inside the rib structure. An additional piezoelectric pad was located internally at the location of the heart. Data from the anterior piezoelectric sensor array were sampled at 10 MHz and the remaining arrays were sampled at 25 kHz. This lower data rate does not appear to be an intrinsic limitation in data rate of the piezoelectric sensor arrays.

Preliminary testing was performed with soft body armour (Type II). The rounds tested include 9 mm to 330 m/s and 0.357 Magnum JSP to 420 m/s. Analysis of only the anterior piezoelectric element was reported. The researchers draw a correlation between measured peak voltage of the array and entering kinetic energy. There was no reported analysis of additional arrays.

The advantages of this model are that it is potentially extensible and potentially well instrumented; however, the calibration of the sensors may be difficult. Extensive experience with integration of bending elements across linear and planar structures (Bass et al. 1998) suggests that it is difficult to perform this integration without substantial error. In addition, the piezoelectric materials used for compressive force are generally extremely sensitive in bending. Furthermore, the model is frangible. It is difficult to produce a cost-effective model with large frangible components.

FIGURE 14.30 Johns Hopkins Instrumented Model Human Torso. (From Bevin, M., et al., *Instrumented Model of the Human Torso for Ballistic Impact Testing*, Presentation, Personal Communication, 2001. Copyright clearance from Bevin.)

14.4.3.3.2   Reusable JHU Model (Human Surrogate Torso Model)

The reusable JHU model HSTM (Human Surrogate Torso Model) is shown in Figure 14.31, and it has two versions with 5th and 50th percentile human anthropometry. It has a detailed anthropomorphic skeletal structure with overlying skin, internal organs and viscera. The internal organs include heart, lungs, stomach and intestinal mass. The model is constructed from ribs with fracture and bending properties of bone, while the organs are constructed of silicone polymers. High-rate material properties of these organs have not yet been compared with high-rate human material properties.

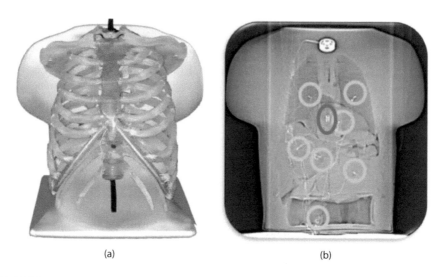

(a)   (b)

FIGURE 14.31   (a) Johns Hopkins Human Surrogate Torso Model (HSTM) diagram. (b) Radiograph of HSTM instrumentation (red = accelerometer, blue = pressure sensors) (From Merkle, A., Johns Hopkins applied physics laboratory, Adelphi, MD, personal communication, 2012).

Sensors used in the torso include sternal and spinal accelerometers, and pressure sensors at various locations as shown in Figure 14.31. Additional instrumentation may include a vertebral load cell, surface pressures and loads, strain gauges for bone simulants and the potential for displacement sensing.

The HSTM has been used in NIJ soft body armour tests to characterise mechanical response (Roberts et al. 2007; Merkle et al. 2008). Tests include 9 mm threats to 436 m/s incoming velocity. The responses have further been correlated with clay response. The device has no validation against any injury metric and is still under development.

14.4.4   Summary – Instrumented Alternatives to Clay

There are several existing test devices that are potentially suitable for use in the development of a test methodology for ballistic BFD. Each may have advantages for a given test condition, though none are widely used in BFD hard armour testing, and all are likely to need further testing and validation. For such testing, two aspects must be considered: the first is the biofidelic response of the surrogate and the second is the validation of the mechanical correlate from this model with an injury model. Each surrogate class has advantages. For example, many of the instrumented response elements and all of the anatomical surrogates have an anthropomorphically appropriate thoracic form, reducing the risk potentially misleading appliqué response in the clay or gelatine for body armour that is meant to be worn. A substantial drawback to instrumented surrogates is the potential for higher cost of surrogates and sensor replacement. However, careful design can likely minimise both handling and sensoring costs.

14.5 CONCLUSIONS

In the modern era of effective personal protection from many small arms threats, BFD becomes an important trade-off for injuries and mortality on both the battlefield and for general personal protection threats. Substantial research has been conducted on experimental testing of animals, less work on experimental studies of human cadavers, but the current work does not allow the development of a thoracic BFD injury criterion from existing studies. Additional animal and/or cadaveric experimentation is necessary to develop a BFD injury criterion. Furthermore, there is a strong need for a robust and widely used ballistic trauma injury classification scale. Although there are a number of existing injury scales, including a widely used scale for automobile injuries promulgated by AAAM, the AIS, none are well suited to ballistic trauma. To provide data for a new ballistic injury scale, there is a strong need to collect military epidemiological data on a large scale.

The preeminent methodology used for BABT testing, especially for hard body armour threats, is based on the Prather methodology. Unfortunately, the biomechanical and pathophysiological bases for this methodology are weak. An improved criterion would likely help optimise personal protection for the actual personal protection threats. Especially important is the development of a methodology with a repeatable material component. The clay used in the Prather methodology has changed dramatically and qualitatively since its development in the 1970s, requiring testing to be performed with the clay at temperatures approximately 15°C higher than originally tested to soften the response. The dynamic effect of this elevated temperature is still unclear.

Finally, test methodologies with more anatomical fidelity than the current single response elements such as clay or gelatine are likely required to investigate the effect of BFD on various organ systems. Although there has been substantial work done in the development of anatomical, physical and mathematical finite element models simulating the human thorax, heart, lungs, liver and kidneys, the fidelity of these models is limited at the present time. Thus, prediction of damage from transmitted pressures associated with blunt trauma to organs such as intestines, spinal cord, brain and the vascular system cannot be made as yet.

REFERENCES

AIS (Abbreviated Injury Scale). 2008. Association for the Advancement of Automotive Medicine. (AIS 2005 Update 2008), Chicago: IL.

Akimov, G.A., M.M. Odinak, S.A. Zhivolupov, B.S. Glushkov, T.I. Milovanova. 1993. (The mechanisms of the injuries to the nerve trunks in gunshot wounds of the extremities (experimental research)) (Article in Russian). *Voen Med Zh (Voennomeditsinskii zhurnal (Military Med J))*. 80(9):34–36.

Baker, S.P., B. O'Neill, W. Haddon, Jr., W.B. Long. 1974. The injury severity score: A method for describing patients with multiple injuries and evaluating emergency ca. *J Trauma*. 14:187–196.

Barwood, M.J., P.S Newton, M.J. Tipton. 2009. Ventilated vest and tolerance for intermittent exercise in hot, dry conditions with military clothing. *Aviat Space Environ Med*. 80(4):353–359.

Bass, C.R. 2001a. Development of a procedure for evaluating demining protective equipment. *J Mine Action*. 4:18–23.

Bass, C.R., B. Boggess, M. Davis, E. Sanderson, G. Di Marco, C. Chichester. 2001b. Effectiveness of personal protective ensembles for the demining of AP landmines. *UXO/Countermine Conference*, April, New Orleans, LA.

Bass, C.R., M. Bolduc, S. Waclawik. 2002. Development of a nonpenetrating, 9 mm, ballistic trauma test method. *Personal Armor Systems Symposium*, Colchester, UK.

Bass, C.R., J.R. Crandall, C. Wang, W. Pilkey, C. Chou. 1998. Open-loop chestbands for dynamic deformation measurements. *J Pass Cars*. 107:1380–1387.

Bass, C.R., K.A. Rafaels, R.S. Salzar. 2008. Pulmonary injury risk assessment for short-duration blasts. *J Trauma*. 65:604–615.

Bass, C.R., R. Salzar, S. Lucas, M. Davis, L. Donnellan, B. Folk, E. Sanderson, S. Waclawik. 2006. Injury risk in behind armor blunt thoracic trauma. *Int J Occup Saf Ergon*. 12:429–442.

Beekley, M.D., J. Alt, C.M. Buckley, M. Duffey, T.A. Crowder. 2007. Effects of heavy load carriage during constant-speed, simulated, road marching. *Mil Med*. 172(6):592–595.

Bensel, C.K., J.M. Lockhart. 1975. *The effects of body armor and load-carrying equipment on psychomotor performance (No. CEMEL-141).* Army Natick Research and Development Labs Ma Clothing Equipment and Materials Engineering Lab, Natick, MA.

Bevin, M., P. Biermann, R. Cain, B. Carkhuff, M. Kleinberger, J. Roberts. 2001. *Instrumented model of the human torso for ballistic impact testing, JHU Applied Physics Laboratory, Adelphi, MD.*

Bir, C. 2000a. *Comparison of injury evaluation in non-lethals, Non-lethal technology and academic research Symposium,* Portsmouth, New Hampshire.

Bir, C. 2000b. *The evaluation of blunt ballistic impacts of the thorax.* PhD Dissertation, Wayne State University, Detroit, MI.

Birrell, S.A., R.H. Hooper, R.A. Haslam. 2007. The effect of military load carriage on ground reaction forces. *Gait Posture.* 26(4):611–614.

Bourget, D., B. Anctil, D. Doman. 2002. Development of a surrogate thorax for BABT studies. *Personal Armor Systems Symposium,* Colchester, UK.

Bowen, I.G., E.R. Fletcher, D.R. Richmond, F.G. Hirsch, C.S. White. 1966. Biophysical mechanisms and scaling procedures applicable in assessing responses of the thorax energized by air-blast overpressures or by nonpenetrating missiles. *Ann N Y Acad Sci.* 152(1):122–146.

Butcher, J.T., G. Shaw, C.R. Bass, R.W. Kent, J.R. Crandall. 2001. Displacement measurements in the hybrid III chest. *J Pass Cars.* 110:26–31.

Cadarette, B.S., L. Blanchard, J.E. Staab, M.A. Kolka, M.N. Sawka. 2001. *Heat stress when wearing body armor (USARIEM-TR-T-01/9).* Army Research Institute of Environmental Medicine, Natick, MA.

Cannon, L. 2001. Behind armour blunt trauma: An emerging problem. *J Roy Army Med Corps.* 147:87–96.

Cannon, L., W. Tam, P. Dearden, P. Watkins. 2000. Thoracic wall response to nonpenetrating balistic injury: The validation of a physical model of behind armor blunt trauma. *12th Conference of the European Society of Biomechanics,* August, Dublin.

Carroll, A.W., C.A. Soderstrom. 1978. A new nonpenetrating ballistic injury. *Ann Surg.* 188:753–757.

Carstensen, E.L., D.S. Campbell, D. Hoffman, S.Z. Child, E.J. Ayme-Bellegarda. 1990. Killing of Drosophila larvae by the fields of an electrohydraulic lithotripter. *Ultrasound Med Biol.* 16:687–698.

Cavanaugh, J.M. 1993. The biomechanics of thoracic trauma. In *Accidental Injury, biomechanics and prevention,* edited by A.M. Nahum, J.W. Melvin, NY:Springer-Verlag. pp. 405–453.

Chamberlin, F.T. 1966. Gunshot wounds. In *Handbook for shooters and reloaders, Vol. II.* ed. P.O. Ackley, Plaza, Salt Lake City, UT. pp. 21–41.

Clare, V., J. Lewis, A. Mickiewicz, L.M. Sturdivan. 1975a. *Body Armor- Blunt Trauma Data.* Report EB-TR-75016, US Army, Aberdeen Proving Ground, MD.

Clare, V., J. Lewis, A. Mickiewicz, L.M. Sturdivan. 1975b. *Blunt Trauma Data Correlation,* EB-TR-75016, US Army, Aberdeen Proving Ground, Edgewood Arsenal, NTIS AD-A012 761.

Cooper, G.J., R.L. Maynard, B.P. Pearce, M.C. Stainer, D.E.M. Taylor. 1984. Cardiovascular distortion in nonpenetrating chest impacts. *J Trauma.* 24:188–200.

Cooper, G., B. Pearce, A. Sedman, I. Bush, C. Oakley. 1996. Experimental evaluation of a rig to simulate the response of the thorax to blast loading. *J Trauma.* 40:38S–41S.

Cooper, G.J., B.P. Pearce, M.C. Stainer, R.L. Maynard. 1982. The biomechanical response of the thorax to nonpenetrating impact with particular reference to cardiac injuries. *J Trauma.* 22:994–1008.

Crandall, J.R., C.R. Bass, S.M. Duma, S.M. Kuppa. 1998. *Evaluation of 5th percentile female Hybrid III thoracic biofidelity during out-of-position tests with a driver air bag (No. 980636).* SAE Technical Paper. Society of Automobile Engineers, Warrendale, PA.

DeMaio, M. 2001. Closed chest trauma: Evaluation of the protected chest. *Presentation to the NATO Task Group on Behind Armor Blunt Trauma,* Copenhagen.

Drobin, D., D. Gryth, J. Persson, D. Rocksen, U. Arborelius, L. Olssen, J. Bursell, B.T. Kjellstrom. 2007. Electroencephalograph, circulation and lung function after high-velocity behind armor blunt trauma. *J Trauma.* 6:405–413.

Eppinger, R., J. Marcus, R. Morgan. 1984. Development of dummy and injury index for NHTSA's thorax side impact protection research program. *Proceedings of the 28th Stapp Car Crash Conference,* SAE Paper Number 840885, Society of Automobile Engineers, Warrendale, PA.

Fackler, M., J. Malinowski. 1985. The wound profile: A visual method of quantifying gunshot wound components. *J Trauma.* 25:522–529.

Fung, Y.C., R.T. Yen, Z.L. Tao, S.Q. Liu. 1988. A hypothesis on the mechanism of trauma of lung tissue subjected to impact load. *Trans ASME*. 110:50–56.

Goldfarb, M., T. Clurej, M. Weinstein, L. Metker. 1975. *A method for soft body armor evaluation: Medical assessment*. EB-TR-74073. Aberdeen Proving Ground, MD, Edgewood Arsenal.

Gryth, D., D. Rocksen, J.K.E. Persson, U.P. Arborelius, D. Drobin, J. Bursell, L. Olsson, T.B. Kjellstrom. 2007. Severe lung contusion and death after high-velocity behind-armor blunt trauma: Relation to protection level. *Mil Med*. 172:1110–1116.

Haley, J., N. Alem, B.J. McEntire, J. Lewis. 1997. *Force transmission through chest armor for the defeat of .50 caliber rounds*. US Army Aeromedical Research Laboratory Report 96–14. Ft Rucker, AL.

Harman, E., K.H. Han, P. Frykman, M. Johnson, F. Russell, M. Rosenstein. 1992. The effects on gait timing, kinetics and muscle activity of various loads carried on the back. *Med Sci Sports Exerc*. 24:S129.

Harvey, E., J. McMillen, E. Butler. 1962. *Mechanism of wounding, in wound ballistics*. edited by J. Coates, pp. 143–235, Office of the Surgeon General, Department of the Army, Washington, DC.

Hawkins, J.L., L.M. Koonin, S.K. Palmer, C.P. Gibbs. 1997. Anesthesia-related deaths during obstetric delivery in the United States, 1979–1990. *Anesthesiology*. 86:277–284.

Hinsley, D.E., W. Tam, D. Evison. 2002. Behind Armour Blunt Trauma to the Thorax - Physical and Biological Models. *Personal Armour Systems Symposium*, November 18–22. The Hague, Netherlands.

Jonsson, A., E. Arvebo, B. Shantz. 1986. Intrathoracic pressure variations in an anthropomorphic dummy exposed to air blast, blunt impact, and missiles. *J Trauma*. 28:S125–S131.

Kent, R., C.R. Bass, W. Woods, R. Salzar, S. Lee, J. Melvin. 2006. The role of muscle tensing on the force-deflection response of the thorax and a reassessment of frontal impact thoracic biofidelity corridors. *J Auto Engr*. 220:853–868.

Knapik, J.J., K.L. Reynolds, E. Harman. 2004. Soldier load carriage: Historical, physiological, biomechanical, and medical aspects. *Mil Med*. 169(1):45–56.

Knudsen, P.J., H. Gøtze. 1997. *Behind armour blunt trauma*. Trials report 02/97, Danish Army Combat School, Oksbøl, Denmark.

Krajsa, J. 2009. *Causes of pericapillary brain hemorrhages caused by gunshot wounds*. PhD thesis in Czech. Institute of Forensic Medicine, Masaryk University, Brno, Czech Republic.

Kroell, C.K., D.C. Schneider, A.M. Nahum. 1971. Impact tolerance and response to the human thorax. *Proceedings of 15th Stapp Car Crash Conference*. SAE 710851.

Kroell, C.K., D.C. Schneider, A.M. Nahum. 1974. Impact tolerance and response to the human thorax. *Proceedings of 18th Stapp Car Crash Conference*. SAE 741187.

Lidén, E., R. Berlin, B. Janson, B. Schantz, T. Seeman. 1988. Some observations relating to behind body-armour blunt trauma effects caused by ballistic impact. *J Trauma*. 27:145–148.

Link, M.S., P.J. Wang, B.J. Maron, N.A. Estes. 1999. What is commotio cordis? *Cardiol Rev*. 7:265–269.

Liu, B., Z. Wang, H. Leng, Z. Yang, X. Li. 1996. Studies on the mechanisms of stress wave propagation in the chest subjected to impact and lung injuries. *J Trauma*. 40:53S–55S.

Mayorga, M.A., I. Anderson, J.L.M.J. van Bree, P. Gotts, J. Sarron, P. Knudsen. 2010. *Thoracic response to undefeated body armor*. Research and Technology Organisation, North Atlantic Treaty Organization (NATO), BP 25, F-92201 Neuilly-sur-Seine Cedex, France. Task Group HFM-024 TG-001.

Merkle, A.C., E. Ward, J. O'Connor, J. Roberts. 2008. Assessing behind armor blunt trauma (BABT) under NIJ standard-0101.04 conditions using human torso models. *J Trauma*. 64:1555–1561.

Miller, P.R., M.A. Croce, P.D. Kilgo, J. Scott, T.C. Fabian. 2002. Acute respiratory distress syndrome in blunt trauma: Identification of independent risk factors. *Am Surg*. 68:845–851.

Mirzeabassov, T., D. Belov, M. Tyurin, L. Klyaus. 2000. Further investigation of modelling for bullet-proof vests. *Personal Armor Safety Symposium*, Colchester, UK.

Mitchell, S.W., G.R. Morehouse, W.W. Keen, Jr. 1864. *Reflex paralysis*. Circular No. 6, Surgeon General's Office, U.S. Government, Washington, DC.

Montanarelli, N., C. Hawkins, M. Goldfarb, T. Ciurej. 1973. Protective garments for public officials. LWL-TR-30B73. Aberdeen Proving Ground, MD, Edgewood Arsenal.

National Academies Committee to Review the Testing of Body Armor Materials for Use by the US Army – Phase II. 2010. *Phase II Report on Review of the Testing of Body Armor Materials for Use by the U.S. Army*. National Academies Press, Washington, DC.

National Academies Report. 2014. https://www.nap.edu/catalog/13390/testing-of-body-armor-materials-phase-iii

Nechaev, E., A. Gritsanov, N. Fmoin, I. Minnullin. 1995. *Mine blast trauma*. Russian Ministry of Public Health and Medical Industry, St. Petersburg, FL.

Peterson, H.L. 1950. Body armor in the Civil War. *Ordnance*. 34:432–433.

Prat, N., F. Rongieras, E. Voiglio, P. Magnan, C. Destombe, E. Debord, F. Barbillon, T. Fusai, J.C. Sarron. 2010. Intrathoracic pressure impulse predicts pulmonary contusion volume in ballistic blunt thoracic trauma. *J Trauma*. 69:749–755.

Prather, R.N., C.L. Swann, C.E. Hawkins. 1977. *Backface signatures of soft body armors and the associated trauma effects*. U.S. Army Armament Research and Development Command Technical Report ARCSL-TR-77055. US Army, Aberdeen Proving Ground, MD.

Raeman, C.H., S.Z. Child, D. Dalecki, C. Cox, E.L. Carstensen. 1996. Exposure-time dependence of the threshold for ultrasonically induced murine lung hemorrhage. *Ultrasound Med Biol*. 22:139–141.

Roberts, J.C., E. Ward, A. Merkle, J. O'Connor. 2007. Assessing behind armor blunt trauma in accordance with the national institute of justice standard for personal body armor protection using finite element modeling. *J Trauma*. 62:1127–1133.

Sarron, J.C., J.P. Caillou, C. Destombe, T. Lonjon, J. Da Cunha, P. Vassout, J.C. Allain. 2001. Displacements of thoracic wall after 7.62 non penetrating impact. *Presentation to the NATO BABT Task Group*, Copenhagen.

Sarron, J.C., C. Destombe, H. Gøtze, M. Mayorga. 2000. Physiological results of NATO BABT experiments. *Personal Armour Systems Symposium (PASS 2000)*, Colchester, UK, 5–8 September.

Sendowski, I., P. Martin, J. Muzellec, J. Breteau. 1994. *Rear Side Effects of Protective Vests in Non-Perforating Model. Tolerance Norms Research*. Institute of Tropical Medicine of the Armed Forces Health Service, DGA, France.

Sondén, A., D. Rocksén, L. Riddez, J. Davidsson, J. Persson, D. Gryth, J. Bursell, U.P. Arborelius. 2009. Trauma attenuating backing improves protection against behind armor blunt trauma. *J Trauma*. 67:1191–1199.

Sperry, Y.K. 1993. Scleral and conjunctival hemorrhages arising from a gunshot wound of the chest. *J Forensic Sci*. 1:203–209.

Sturdivan, L.M. 1976. *Handbook of human vulnerability criteria, chapter 5, multiple impacts*. CSL-SP-81005, U.S. Army Chemical Systems Laboratory, Aberdeen Proving Ground, MD.

Suneson, A., H.A. Hansson, T. Seeman. 1987. Peripheral high-energy missile hits cause pressure changes and damage to the nervous system: Experimental studies on pigs. *J Trauma*. 27:782–789.

van Bree, J., P. Gotts. 2000. The 'Twin Peaks' of BABT. *Personal Armor Systems Symposium*, Colchester, UK.

Viano, D.C., I.V. Lau. 1985. Thoracic impact: A viscous tolerance criterion. *Tenth International Conference on Experimental Safety Vehicles*, pp. 104–114, Oxford, England.

Walsh, P.J., X. Hu, P. Cunniff, A.J. Lesser. 2006. Environmental effects on poly-p-phenylenebenzobisoxazole fibers. I. Mechanisms of degradation. *J Appl Polymer Sci*. 102(4):3517–3525.

Wilhelm, M., C. Bir. 2008. Injuries to law enforcement officers: The backface signature injury. *Forensic Sci Int*. 174:6–11.

15 The Lower Extremities

Cadaveric Testing, Anthropomorphic Test Device Legs and Injury Criteria

Cynthia Bir and Brian McKay

CONTENTS

15.1 INTRODUCTION AND ANATOMY OVERVIEW

The extremities are reported to be the most commonly injured body regions in modern warfare (Ramasamy et al. 2011c; Owens et al. 2007). In each armed conflict since World War II, more than 54% of all United States military combat wounds were extremity wounds (Owens et al. 2007). Specifically, lower extremity injuries accounted for 26% to 48% of all combat injuries sustained during this time period. While the majority of extremity wounds were soft tissue injuries, a large proportion, ranging from 23% to 39% in recent armed conflicts, resulted in fractures (Owens et al. 2007).

In an underbelly (UB) blast event, the lower extremity of a vehicle occupant is highly susceptible to injury because of its proximity to the primary impact point. The magnitude and rate of loading can cause both axial compressive and shear loading depending on the positioning of the lower extremity and load path. In the seated position, the pelvis and thighs rest on the seat, while the feet rest on the floorboard. During a UB blast impact, the foot is the first anatomical structure loaded in a classical UB landmine explosive impact scenario because of its immediate vicinity to the blast. Subsequently, loading is transferred from the foot through the ankle and onto the leg, which contains the tibia and fibula.

15.1.1 ANATOMY

The lower extremity can be divided into five regions: pelvic, thigh, leg, foot and ankle (Huelke 1986). The foot, ankle and leg regions are commonly referred to as the lower leg, while the knee, thigh and pelvis regions are referred to as the upper leg (Figure 15.1).

15.1.1.1 Pelvis

The pelvic bone transfers the body weight from the vertebral column to the pelvic girdle. From the pelvic girdle, weight is transferred to the femur when in a standing position. The pelvis comprises three fused bones: ilium, ischium and pubis (Moore et al. 2011). This fusion forms a socket cavity, or acetabulum, on the lateral aspect for articulation with the head of the femur (Huelke 1986). An extensive review of the pelvic anatomy has been presented in Chapter 12.

Three gluteal muscles, the gluteus maximus, medius and minimus, reside posteriorly on the pelvis and act to rotate the thigh region. The gluteal muscles originate from the ischium and ilium and attach to the proximal femur. The muscles are supplied by the internal iliac arteries and are innervated by the superior and inferior gluteal nerve (Moore et al. 2011).

15.1.1.2 Thigh

The thigh encompasses the area between the pelvic bone and the knee. The femur, the largest and heaviest bone of the body, resides in the thigh and transmits the body weight from the pelvis to the tibia. The cylindrical shaped femoral head, which lies at the proximal end of the femur, articulates with the acetabulum. The femoral head is connected to the femoral shaft by a narrow neck, which lies at an approximate angle of 125° with the shaft (Moore et al. 2011). The distal portion of the femur diverges into a medial and a lateral condyle. The femoral condyles articulate with the proximal tibia condyles to form the knee joint.

The musculature of the thigh can be divided into three parts: anterior, medial and posterior compartments (Huelke 1986). The anterior compartment consists of hip flexors and knee extensors. The quadriceps femoris, which include the rectus femoris, vastus lateralis, vastus medialis and vastus

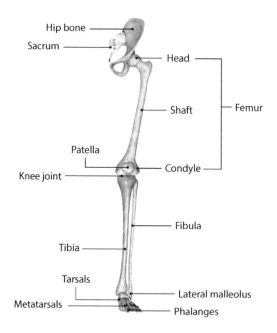

FIGURE 15.1 Skeletal anatomy of the lower extremity.

intermedius, are the most powerful extensors of the knee joint. The medial compartment features a heavy group of adductor muscles, including the adductor longis, adductor brevis, adductor magnus, gracilis and obturator externus. The hamstrings, located in the posterior compartment, are responsible for flexing the knee and extending the hip and thigh. The hamstrings consist of the semitendinosus, semimembranosus and bicep femoris (Moore et al. 2011).

15.1.1.3 Knee Joint

The patella, or knee cap, is a large sesamoid bone that is formed intratendinously after birth. The patella is positioned anterior to the knee joint and serves as a fulcrum, giving the quadriceps more power when extending the leg (Huelke 1986). The knee joint is a hinge type of synovial joint that allows flexion and extension of the leg. The knee joint consists of two joints forming three articulations: The femoro-patellar joint is the articulation between the patella and femur. The femoro-tibial joint links the lateral and medial articulations between the femoral and tibial condyles. The knee joint is held intact by ligaments connecting the tibia and femur as well as surrounding muscles and tendons (Moore et al. 2011).

15.1.1.4 Lower Leg

The leg is supported by two bones: tibia and fibula. The tibia, the second longest bone of the body, resides in the medial portion of the leg while the fibula resides in the lateral portion of the leg and posterolateral of the tibia. The tibia is longer, larger in diameter and stronger than the fibula. The tibia is the primary weight-bearing bone of the leg and accounts for 85% to 90% of weight transfer depending on the position of the foot and ankle (Moore et al. 2011). The tibia articulates with the medial and lateral femoral condyles superiorly, with talus inferiorly and laterally, and with the fibula at its proximal and distal ends (Moore et al. 2011). The distal portion of the tibia and fibula diverges into medial malleolus of the tibia and lateral malleolus of the fibula (Figure 15.2).

The tibia forms two joints with the fibula. The proximal tibiofibular joint is an arthodial joint that links a facet located on the posterolateral portion of the lateral tibia condyle with a facet located on fibular head (Huelke 1986). This proximal joint enables slight gliding movements during plantar flexion and dorsiflexion of the foot. The distal tibiofibular joint is a fibrous joint linking the medial distal fibula to the lateral malleolus of the tibia. The distal joint accommodates the talus during dorsiflexion of the foot, which is critical to ensuring ankle stability. Both joints are strengthened by anterior and posterior tibiofibular ligaments (Moore et al. 2011).

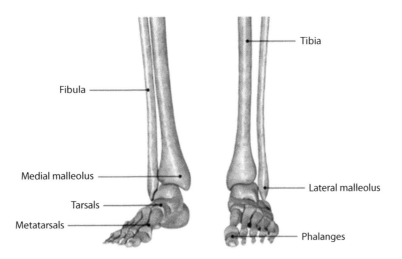

FIGURE 15.2   Foot and ankle skeletal anatomy.

Similar to the thigh, the musculature of the leg can be divided into three compartments: anterior compartment, which is responsible for extending the ankle and phalanges and includes the tibialis anterior, extensor hallicis longus and extensor digitorum longus; lateral or peroneal compartment, which is responsible for everting the ankle and includes the peroneus longus and peroneus brevis; and posterior compartment, which is responsible for flexing the ankle and phalanges and includes the gastrocnemius, soleus, tibialis posterior, flexor hallicis longus and flexor digitorum longus (Huelke 1986).

15.1.1.5 Foot

The foot, located distal to the leg, comprises three main parts: hindfoot, midfoot and forefoot (Figure 15.3). These parts work in harmony to support the body weight transferred from the leg and facilitate locomotion. The hindfoot includes talus and the calcaneus. The midfoot includes navicular, cuboid and three medial cuneiforms. The forefoot includes the five metatarsals and 14 phalanges (Moore et al. 2011).

The talocrural joint, or ankle joint, is a synovial joint that links the distal tibia and fibula to the proximal articular surfaces of the talus. The talus, the second largest bone in the foot, resides distal to the tibia and fibula and proximal to the calcaneus (or heel). The talus is the only bone in the foot that articulates with bones of the leg. The talus articulates with the tibia in two locations: (1) medial malleolus articulates with the medial aspect of the talus and (2) inferior surface of the distal tibia articulates with the superior surface, or trochlea, of the talus (Moore et al. 2011). The lateral malleolus of the fibula articulates with the lateral aspect of the talus. The articulation between the talus and tibia supports more weight than the talus and fibula articulation (Moore et al. 2011). The fibrous capsule of the ankle joint is strengthened laterally and medially by multiple ligaments (Moore et al. 2011).

The talus rests on the anterior two-thirds of the calcaneus (Moore et al. 2011). The calcaneus is the largest and strongest bone of the foot and transmits the majority of the body weight from the talus to the ground. The posterior portion of the calcaneus serves as the insertion point for the Achilles tendon. The head of the talus is supported by the talar shelf of the calcaneus. It also articulates with the navicular. The navicular bone resides on the medial side of the foot and has three strongly concave proximal articular surfaces for each of the three cuneiform bones. The medial,

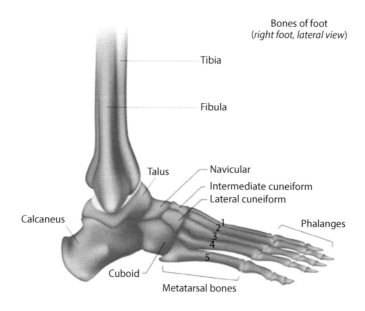

FIGURE 15.3   Detailed foot skeletal anatomy.

middle and lateral cuneiforms articulate with the first, second and third metatarsal bones, respectively, via a tarsometatarsal joint. Residing medial of the cuneiform is the cuboid bone. The lateral cuneiform and navicular bones articulate with the medial surface of the cuboid bone. The cuboid bone also articulates with the fourth and fifth metatarsal bones forming the tarsometatarsal joint and with the calcaneus proximally at the calcaneocuboid joint. The metatarsals connect the tarsus to the 14 phalanges. Each phalange is constructed of three bones except the first phalange which consists of two bones (Moore et al. 2011).

15.2 UNDERBELLY BLAST MECHANISMS AND LOWER EXTREMITY INJURIES

The mechanism of injury for all wounded combatants has been reported to be predominantly due to explosive events (79%) (Owens et al. 2008). Military vehicle occupants, or mounted soldiers, face the ever-present threat of UB explosive devices. Wang et al. (2001) noted that the average velocity and acceleration of a medium-sized armoured vehicle floorplate exceeded 12 m/s and 100 g respectively following a UB blast. As previously stated, the foot–ankle–tibia region is typically the initial impact point of an occupant in an UB blast event. Detonations occurring under a vehicle produce localised floorboard deformation and transmit high axial loads onto the foot/ankle/tibia complex of the occupant causing injuries to the lower leg. In addition to the deformation of the floor, there is global vehicle motion that occurs due to the kinetic energy from the soil ejecta and blast dynamics. This can create significant loading to the lower extremities as well. If the vehicle is breached, then soft tissue injuries are also likely.

Owens et al. (2007) reported that the most common fracture of the lower extremity is the tibia and fibula (48%). This was based on injury data from the Joint Theater Trauma Registry (JTTR) on all U.S. service members treated between October 2001 and January 2005 and included all injuries. Ramasamy et al. (2013) used the same database but queried it to find all UK service members treated between January 2006 and December 2008 who sustained lower leg injuries from explosive events. The most severely injured region in the body was reported to be the lower extremity, representing 89% of the total cases. Tibia and fibula fractures were reported to represent 71% of these lower extremity injuries (Ramasamy et al. 2013).

The primary mechanism of injury for the lower extremities was confirmed by Ramasamy et al. (2011a) to be due to the displacement of limb or objects near it, commonly known as tertiary blast injuries. Of the 48 lower extremity causalities reported, 96% were due to this type of mechanism of loading. The fracture characteristics included three-point bending, spiral fractures and those consistent with axial loading.

15.3 DEVELOPMENT OF INJURY CRITERIA AND BIOMECHANICAL RESPONSE CORRIDORS USING PMHS

In an effort to reduce the number of lower extremity injuries related to UB landmine explosive events, the North Atlantic Treaty Organization Human Factors and Medicine (NATO HFM) Task Group (090/TG-25) initiated a coordinated research effort to characterise impulses seen during these events (Horst et al. 2005). Initially, full-scale vehicle blast tests were simulated in a controlled test rig using scaled detonations (Dosquet et al. 2004). The Test Rig for Occupant Safety Systems (TROSS™) utilises scaled charges to produce reproducible blast loads that propel an elastic deformable floorplate towards the lower limb of a biomechanical surrogate (Dosquet et al. 2004). The interior of the TROSS™ is designed to represent a military vehicle crew cabin. An occupant, as simulated by a full-body anthropomorphic test device (ATD) biomechanical surrogate, is seated and belted with the lower limbs of the surrogate set on the moveable floorplate. A series of tests were conducted to characterise three relatively low-energy UB landmine blast events. Each event was characterised using floorplate acceleration, velocity and dynamic displacement measurements. The effect on the vehicle occupant was measured using the tibia axial force measured by Hybrid III and THOR-Lx biomechanical lower extremity surrogates.

The three parametric floorplate impact conditions were then carefully replicated and validated in a laboratory-scale setup using a Hybrid III (Barbir 2005). Specifically, force-time trajectories measured by the Hybrid III lower limb in the TROSS™ for each condition were replicated on the laboratory setup by tuning impact conditions. The laboratory-scale setup utilised a piston-driven linear impactor to propel a stainless steel footplate into the plantar aspect of the Hybrid III lower extremity biomechanical surrogate foot at a targeted velocity. In addition to supplying a desired loading rate, the impactor was designed to simulate floorplate intrusion. Three impact conditions were established to match the live fire data: Condition 1 (36.8 kg mass impactor at 3.7 m/s), Condition 2 (23.0 kg mass impactor at 4.8 m/s) and Condition 3 (36.8 kg mass impactor at 9.3 m/s). A foot displacement of 12 to 27 mm was identified during the Hybrid III laboratory-to-bench scale correlation and used for each condition (Barbir 2005).

After establishing UB landmine parametric loading rates in the laboratory-scale setup, Barbir (2005) ran a series of tests measuring the biomechanical responses of PMHS at two of the three UB blast loading rates. Ten instrumented PMHS lower limbs were axially impacted at the two higher loading conditions: Conditions 2 and 3. The lower limbs of all specimens were harvested at the femur approximately 20 cm from the knee. The specimen was potted into a device designed to interface with the Hybrid III surrogate. A six-axis tibia load cell was implanted into the mid-shaft of the tibia (Denton, Inc, Humanetics, Plymouth, MI). Five PMHS lower limbs were impacted at Condition 2 loading rates, producing an average tibia axial force of 3.2 kN. The tibia axial compression of the five specimens ranged from 2.3 to 4.4 kN. The tibia was loaded for a duration of approximately 10 to 15 ms. None of the PMHS sustained skeletal injury (Barbir 2005).

Similarly, five PMHS lower limbs were impacted at Condition 3 loading rates, producing an average tibia axial force of 4.5 kN. The tibia axial compression of the five specimens ranged from 3.3 to 6.4 kN. Again, the tibial loading lasted approximately 10 to 15 ms. One PMHS specimen, impacted at Condition 3 loading rates, sustained a plafond fracture. A second PMHS specimen sustained a fracture at the tibia pot and was deemed artificial (Barbir 2005).

The average PMHS lower limb tibia axial force of 4.5 kN was less than the 5.4 kN tolerance limits specified by Yoganandan et al.'s (1996) foot and ankle injury risk model for 45-year-old male subjects. The impact severity utilised by Barbir (2005) was not severe enough to cause a high probability of lower limb fracture.

Further research was conducted by McKay and Bir (2009) in an effort to develop both biomechanical response corridors and associated injury criteria at higher rates of loading. Data were analysed from full-scale blast-test data provided by the U.S. Army Tank-Automotive and Armaments Command (TACOM) and the U.S. Army Tank-Automotive Research, Development and Engineering Center (TARDEC). Examining the vehicle and mid-tibial accelerations validated the presence of axial impacts. Three distinct levels of impact severity, which correlated with floorplate velocities of 7.0, 10.0 and 12.0 m/s, were identified. These velocities were compared to the TROSS™ data associated with high-severity blast loading. Based on this combined data set, three incrementally severe experimental impact conditions were established. The impact conditions, termed WSU Conditions 1, 2 and 3 (abbreviated as WSU C1, WSU C2 WSU C3), targeted an impacting floorplate velocity of 7.0, 10.0 and 12.0 m/s, respectively. The mass of the impactor remained constant at 36.7 kg to produce kinetic energies of 900, 1837 and 2645 J for the three conditions.

A total of 18 PMHS lower extremities were included in this research and averaged 68 years of age. Given the 'older' population of the specimens, bone density dual-energy X-ray absorptiometry (DEXA) was conducted on each specimen to evaluate for osteoporosis. In addition, an implantable tibia triaxial load cell (Denton, Inc, Model 3786J) was inserted into the mid-shaft of each PMHS lower extremity using previously established techniques (Dean and Boyse 2003). Strain gauge rosettes 031RB (Vishay Micro-Measurements, Shelton, CT) were attached to the medial aspect of the calcaneus and the medial surface of the lower portion of the tibia. The specimens were harvested and the femur was potted into an interface plate such that the Hybrid III surrogate served as the body mass for testing (Figure 15.4). The specimens were positioned in front of the impactor in such a

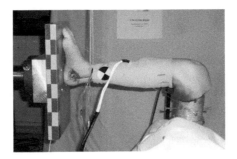

FIGURE 15.4 Pre-test positioning of post-mortem human specimen with linear impactor.

manner so as to produce 24 mm of loading after contact. PMHS were impacted with each of the three conditions. After testing, the specimens were evaluated for injuries using radiographic and forensic techniques. A board-certified orthopaedic surgeon classified all of the injuries according the Ankle and Foot Injury Scale for Severity (AFIS-S) injury scale definitions (Levine et al. 1995).

For the first impact condition (7 m/s), there were no skeletal injuries noted. However, the remaining two conditions (10 and 12 m/s) produced at least one skeletal injury in each of the specimens, with 50% (n = 6) having injuries that were classified to be incapacitating. All of the injured specimens had varying levels of calcaneus fractures (e.g. simple, undisplaced, closed, comminuted, displaced and open fractures). The talus bone was the second most injured skeletal segment with both undisplaced and displaced fractures. Tibia fractures were noted in four of the specimens. It was noted that the highest loading level (12 m/s) produced more comminuted, crush-type injuries than the 10 m/s loading condition.

Based on these injury data, risk curves were developed, and it was determined that an axial tibial force of 5,931 N corresponded to a 50% probability of an incapacitating injury (Figure 15.5). Both age and gender were evaluated to determine if they had a significant effect on the probability of injury, and both were found to be non-significant.

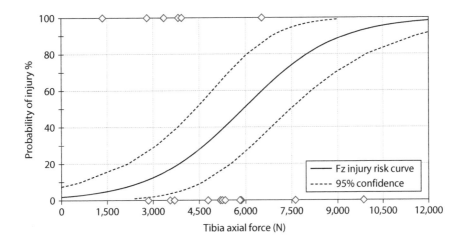

FIGURE 15.5 Injury risk curve based on tibia axial force.

15.3.1 BIOMECHANICAL RESPONSE CORRIDORS

In an effort to provide the ability to validate surrogates, biomechanical response corridors were developed using data from PMHS testing. The above (McKay and Bir, 2009) was used to develop three distinct corridors. The corridors were developed using the force-time trajectories with the trajectory of each impact aligned using the peak force. The mean average PMHS response was calculated with a plus and minus standard deviation providing the upper and lower bounds of the corridor.

Barbir (2005) produced two corridors with data generated from the two higher conditions (Condition 2 (4.8 m/s) and Condition 3 (9.3 m/s)). These conditions resulted in two force-time corridors. For Condition 2, the peak tibial forces within the corridor ranged between 2,882 and 4,329 N. The peak duration ranged from 8 to 15 ms (Figure 15.6). For Condition 3, the corridor established a peak tibia axial force in the range between 3,681 and 5,834 N (Figure 15.7). The duration of the event ranged between 8 and 16 ms.

Only the non-injurious data from McKay and Bir (2009) were used to develop a biomechanical response corridor. This corridor reflected the average peak tibia axial force response (5,377 N) plus and minus the standard deviation of the peak loads (408 N) (Figure 15.8). The corridors demonstrate an approximate 3 ms time to peak. All of these corridors, as well as others currently being developed, are essential in establishing the biofidelity of surrogates.

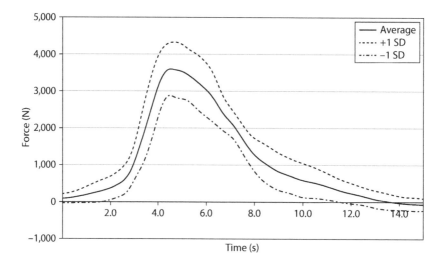

FIGURE 15.6   Force versus time corridor for Condition 2.

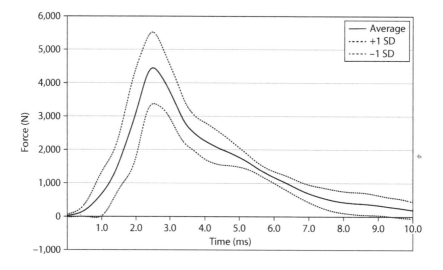

FIGURE 15.7   Force versus time corridor for Condition 3.

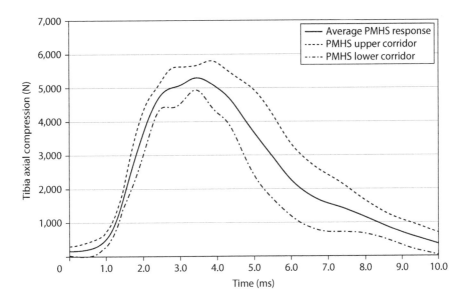

FIGURE 15.8   Force-time corridor based on PMHS responses.

15.4   SURROGATE DEVELOPMENT

The Hybrid III 50th percentile dummy is an ATD that is widely used in the automotive industry and, until recently, has been used for the evaluation of UB blast threats. The lower leg consists of a pin joint at the knee and a ball joint at the ankle. It also contains upper and a lower tibia load cells capable of measuring moments and forces. The shaft of the tibia in the Hybrid III is translated anteriorly at its proximal end and slightly posteriorly just above the ankle.

The Test Device for Human Occupant Restraint (THOR) was developed by NHTSA with the aim to offer increased biofidelity and measurement capability compared to the Hybrid III. Within the lower leg portion (THOR-Lx), some of the improvements over the Hybrid III include a compliant tibia element that modulates the response to the axial impact more realistically. Three independent axes of rotation for the ankle were also added based on flexion and extension properties in human ankle tests. A tensioned wire, simulating the Achilles tendon, provided more realistic axial forces.

In order for any biomechanical surrogate to be considered biofidelic, it must respond the same as the human anatomy responds. The establishment of the biomechanical response corridors in the UB blast loading condition allows for this evaluation to occur. By testing the surrogates with similar impact conditions, their response can be compared to established corridors. Surrogates with a response that falls within the corridor are considered to be biofidelic.

Utilising the same high-rate linear impactor and simulated UB landmine axial impact conditions, Barbir (2005) evaluated the biomechanical response of a Hybrid III lower limb and THOR-Lx. The biomechanical responses of the surrogates were compared to the response of instrumented PMHS lower limbs impacted at equivalent impact conditions. At relatively low severity simulated UB blast impact conditions (Condition 2), the tibia axial force of the THOR-Lx compared favourably with the PMHS lower limbs (Barbir 2005). In contrast, the Hybrid III lower limb measured peak tibia axial force nearly three times greater than the PMHS and THOR-Lx (Barbir 2005).

As simulated UB blast impact severity increased (Condition 3), the THOR-Lx measured a peak tibia axial force 62% larger than the PMHS. Hence, the THOR-Lx and Hybrid III lower limb biomechanical surrogates failed to produce a biofidelic response for UB landmine blast loading rates.

The simulated UB blast impacts of the Hybrid III and THOR-Lx conducted by Barbir (2005) revealed severe shortcomings in the design of each biomechanical surrogate. The biofidelity of each surrogate decreased substantially when impact severity was increased, resulting in an overestimation of peak tibia axial force. This trend suggested that the surrogates are too rigid in comparison to a human lower limb. The performance variability between the Hybrid III and THOR-Lx indicated that improvements in biofidelity could be achieved by modifying the surrogate geometry, components and materials.

In an effort to improve the biofidelity of the biomechanical surrogates, McKay (2010) expanded upon the work of Barbir (2005) to develop the Military Lower Extremity (Mil-Lx) surrogate. The Mil-Lx was designed with the partnership of Denton ATD, Inc and Humanetics. The core components of the THOR-Lx were utilised in the construction of the Mil-Lx. The tibia shaft was incorporated into the design of the Mil-Lx because of its human-like geometry.

In addition, the THOR-Lx tibia compliant element design attribute was also adopted into the Mil-Lx design. The compliant element was doubled in length from 5 to 10 cm in the Mil-Lx to enable additional room for compression. The longer tibia compliant element was selected to prevent the full compression or 'bottoming out' of the elastomer element at UB blast loading rates. A fully compressed element would often generate two tibia axial force peaks. The first peak was formed as a product of the initial impact. Subsequently, fully compressed compliant element would recoil producing a sudden tension force. The remaining loading from the footplate would then generate a second compressive force on the element.

The Mil-Lx foot and ankle closely resembles the structure of the Hybrid III and includes several improvements. The Mil-Lx incorporates a more durable polyurethane foot cover to enhance the recovery of the elastomer from compression impact to ensure repeatable performance for a longer duration. The cover also includes a slot to install a replaceable compliant footpad. Similar to the tibia compliant element, the footpad dampens or tunes the force transmission from the heel to the ankle joint and tibial shaft. A carbon fibre plate extending from the heel to the foot simulates the bones of the foot. An accelerometer mounting site is located at the midfoot of the Mil-Lx assembly. The Hybrid III ankle ball joint is utilised in the Mil-Lx to simulate the articulation of the foot and ankle. The ankle joint rotates around the x- and y-axes providing inversion/eversion and dorsiflexion/plantarflexion. The ankle joint was designed to reproduce the static and dynamic moment–angle response characteristics in flexion and inversion/eversion measured by Portier (1997) and Petit (1996) in PMHS lower limb studies (Olson et al. 2007).

Over 100 impact tests were conducted on the Mil-Lx to identify the optimal combination of compliant material stiffness and diameter to produce a biofidelic biomechanical response to simulated UB landmine impacts. After the final design was determined, validation impacts were conducted and compared to the previously established non-injury corridor by McKay and Bir (2009). The Mil-Lx peak upper tibia axial force averaged 5,377 N and ranged from 5,190 to 5,528 N. The biomechanical response of the optimised design demonstrated high repeatability, with a standard deviation of 181 N for peak tibial load. The biomechanical response of the Mil-Lx compared favourably with the PMHS non-injury corridor with respect to peak tibia axial force, loading rate, loading duration and intrusion (Figure 15.9).

The surrogates are considered to be biomechanical and can be reused for multiple testing with calibrations conducted when recommended. There is a second type of surrogate, which is frangible. These surrogates are meant to fracture in a manner similar as a human. Some of the initial surrogates have been used to evaluate injury from anti-personnel landmines. Simple models include the use of wooden rods to represent the bone and light concrete to represent soft tissue (Meppen Artificial Leg). Another basic frangible surrogate involves red deer tibia encased in gelatine with the dimensions of the human leg (Red Deer Lower Limb Model) (Hinsley et al. 2003). A more advanced model, the complex lower leg (CLL), was developed in Canada using synthetic materials having properties to represent bone and soft tissue of the human leg. The geometry was based on the human male although simplified to ensure consistency. The Frangible Surrogate Leg (FSL) contains even more details and was developed in Australia under the guidance of Defence Science and Technology Group (DST Group). The bone geometry was based on a cast made from 50th percentile

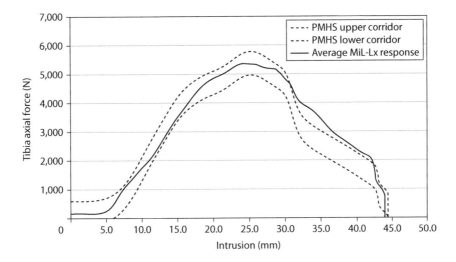

FIGURE 15.9 Average Mil-Lx response compared to PMHS corridors.

Australian male cadaver. Synthetic materials were used to create the bones with adhesive and simulated materials used to connect them. Gelatine was then poured around the skeletal structures to form the outer dimensions of the lower leg.

The enhanced Frangible Surrogate Lower Leg – Modified (eFSLLM) was a modification of the original FSL designed to improve the biofidelity of the original design. This design incorporated an improved soft tissue and bone simulant material and geometries. To provide the ability to measure load, a cylindrical section was added to the bony design in this final design iteration. Four strain gauges were bonded to this section in the anterior, posterior, medial and lateral surfaces to produce a low-cost tibia load cell.

Six eFSLLM specimens (Adelaide T&E Systems, Torrensville, South Australia) were acquired for impacts at the three previously determined velocities to compare their response to previous PMHS (Bir and Andrecovich 2011). Two eFSLLM surrogates were tested using the WSU C1 condition. The average velocity during the tests was 7.0 m/s and the average maximum tibial force experienced was 1,006 N. No fractures were noted during WSU C1; therefore, three eFSLLM surrogates were tested using the WSU C2 condition. The average velocity during the tests was 9.9 m/s, with an average maximum tibial force of 1,427 N. The last three eFSLLM surrogate tests were performed at an average velocity of 11.7 m/s, with an average maximum tibial force of 1,778 N. These forces were significantly lower than those seen in the PMHS testing.

Post-testing analysis of WSU C1 indicated that no fractures occurred during the impact event. Under testing condition WSU C2, there were distal tibial injuries in two of the three surrogates. In all three surrogates, there was evidence of a Chopart injury (dislocation of the mid-tarsal), without any evidence of calcaneal fractures and only one surrogate received a navicular fracture (Table 15.1). At least one metatarsal fracture was found in all three surrogates. Under testing condition WSU C3, one surrogate displayed a comminuted pilon fracture and subluxation of the tibiotalar joint. Another surrogate showed evidence of a posterior and lateral malleolus fracture. The last surrogate of the group did not show any evidence of ankle fracture; however, it did show evidence of metatarsal and navicular fractures.

Observations of the fractures in eFSLLM testing indicate they are not entirely consistent with the type of fractures experienced with the previous PMHS data. The sub-fracture condition, WSU C1, yielded similar results as there were no fractures induced either in the PMHS or in the eFSLLM surrogates. However, the injuries under conditions WSU C2 and C3 were not similar. The injuries observed in the eFSLLM involved the high-energy type of fractures involving the tibia and the fibula;

TABLE 15.1
Description of Injuries Seen in eFSLLM with UB Landmine Loading Events

Cadaver ID	Left/Right	WSU Condition	Fractures	Description of Injury
FSL232	Left	1	n/a	No skeletal injury observed
FSL227	Right	1	n/a	No skeletal injury observed
FSL228	Left	2	M_5	Fifth metatarsal fracture; non-displaced
FSL231	Right	2	N	Chopart joint injury
FSL232	Left	2	M_5	Base of 5th metatarsal fracture, Lisfranc injury
FSL230	Left	3	T	Subluxation or strain tibiotalar joint
FSL225	Right	3	M_1, M_5, Cn	First metatarsal fracture, base of 5th metatarsal fracture; non-displaced. Chopart.
FSL227	Right	3	M_5, Cn, F	Chopart and tibiotalar subluxation. Base of the 5th metatarsal; non-displaced

F = fibula, T = tibia, M = metatarsal, Cn = cuneiform, N = navicular.

however, the calcaneus was never fractured. Every PMHS tested with WSU C2 and C3 conditions resulted in a calcaneal fracture in addition to similar tibial and fibular fractures. Furthermore, these injuries do not correlate with those seen in the field where calcaneal injuries are prevalent (Ramasamy et al. 2011b).

15.5 SURROGATE USAGE FOR THE ASSESSMENT OF UNDERBELLY THREATS

Military vehicle occupant protection systems have been developed to reduce the injury potential caused by a UB blast event. These can include blast-mitigating padding material within the vehicle (Newell et al. 2013), designs of the vehicle underbelly or even the boots worn by the soldiers (Pandelani et al. 2015). The use of the validated Mil-Lx provides the opportunity to evaluate a variety of these devices in terms of their ability to reduce blast injury.

Footwear may be utilised to safeguard military occupants involved in explosive blast events. Barbir (2005) showed that a standard issue U.S. Army combat boot may decrease peak tibia axial force in a lower extremity biomechanical surrogate by as much as 35%. Whyte (2007) measured the static elastomer properties of numerous military combat boots from several NATO countries, revealing a broad range of boot padding properties including stiffness. The variance among elastomer properties suggests that each boot has the potential to provide a different level of protection to the user.

McKay (2010) conducted a series of impact tests to quantify the protective capability of military footwear to attenuate axial blast threats. Three military combat boots were evaluated in the study: US Army and Marine Infantry Combat Boot (ICB), US Army Desert Combat Boot (DCB) and AP Mine Protective Over Boot (MOB). The combination of the ICB and MOB was also tested (Figure 15.10). The validated Mil-Lx was dressed in each combat boot or combination of boots and impacted at the plantar aspect of the boot under WSU C1, C2 and C3 impact conditions. Each boot was impacted two times at each impact condition (McKay 2010).

The internal padding of each combat boot varied. The ICB padding was found to be the softest in benchtop testing followed by the DCB. The MOB was the stiffest boot. The MOB is designed for protection against AP landmines and contains a Kevlar sole to protect against projectiles.

The average peak tibia axial force for WSU C1 ranged from 3,859 to 4,865 N. The average peak tibia axial force for WSU C2 ranged from 6,128 to 7,387 N. The average peak tibia axial force for WSU C3 ranged from 7,325 to 8,320 N. The ICB was found to provide (alone or in combination with MOB) the lowest peak tibia axial force.

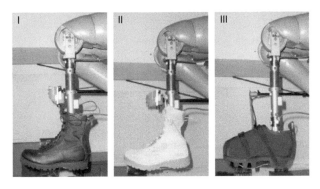

FIGURE 15.10 Various footwear (I-ICB, II-DCB and III-MOB) tested using Mil-Lx surrogate.

The Mil-Lx with ICB provided the greatest average reduction in peak tibia axial force at 28%, 15% and 8% at WSU C1, C2 and C3 impact severity, respectively. The Mil-Lx with MOB and Mil-Lx with DCB generated the least amount of peak tibia axial force reduction when exposed to WSU C2 and C3 loading severity (Table 15.2).

TABLE 15.2

Percent Change (Negative – Reduction, Positive – Increase) in Force with Booted System Over Non-Booted Condition

Boot System	WSU C1	WSU C2	WSU C3
Infantry Combat Boot (ICB)	−28%	−15%	−8%
Desert Combat Boot (DCB)	−20%	2%	0%
AP Mine Protective Over Boot (MOB)	−9%	−6%	3%
ICB with MOB combined	–	–	−10%

Research was also conducted at the Council for Scientific and Industrial Research (CSIR) in South Africa in an effort to evaluate two different types of combat boots: Meindl Desert Fox (Kirchanschoring, Germany) and Lowa Desert Fox (Jetzendorf, Germany) (Pandelani 2014). A Modified Lower Limb Impactor (MLLI) was used to create the loading condition. This experimental setup uses a spring-powered plate (32.8 kg) designed to strike the bottom of the foot of a surrogate in the seated position. The loading velocity was varied from 2.7 to 10.2 m/s. The Mil-Lx was attached to a standard Hybrid III ATD seated in a 90-90-90 posture.

Pandelani (2014) reported that the lower tibia load cell exhibited a decrease in forces during loading in the booted versus non-booted conditions. This effect of the boots increased with increasing loading rate. However, the upper tibia load cell did not exhibit a similar response, but both did demonstrate an increase in the time to peak compared to the non-booted surrogate.

15.6 FUTURE RESEARCH/WIAMAN

Further research is currently being conducted to improve upon the Mil-Lx design under the main Warrior Injury Assessment Mankin (WIAMan) effort. This effort is being led by the U.S. Army Research, Development and Engineering Command. The overall goal of this effort is to produce an ATD that has the ability to accurately measure the accelerative loads seen with underbelly, UB blast event. The prediction of injury risk and ability to assess blast-mitigating devices are the key outcomes of this effort.

ACKNOWLEDGEMENTS

This chapter would not have been possible without the support of the following individuals: Ana Barbir, Nate Dau, Samantha Staley, Todd Frush and Paul Dougherty. In addition, we would like to recognise the support from both the NATO HFM-90 and HFM-148 groups for their insights.

REFERENCES

Barbir, A. 2005. Validation of lower limb surrogates as injury assessment tools in floor impacts due to anti-vehicular landmine explosions. MS, Biomedical Engineering, Wayne State University.

Bir, C. A., and C. Andrecovich. 2011. A comparison of eFSLLM in underbelly blast events to post mortem human specimens. NATO HFM 198 Injury assessment methods for vehicle active and passive protection systems, Fall 2011 Meeting.

Dean, E. I., and H. Boyse. 2003. *Implantation design guidelines for instrumenting the cadaveric lower extremity to tranducer femur loads and tibial forces and moments.* Society of Automotive Engineers, Warrendale, PA.

Dosquet, F., O. Nies, and C. Lammer. 2004. Test methodology for protection of vehicles occupants against IED. *18th Symposium of Military Aspects of Shock and Blast,* Bad Reichenhall, Germany.

Hinsley, D. E., I. Softley, and S.R. Garrick. 2003. Assessment of the severity of land mind injury using a lower limb physical model. *J Bone Joint Surg Br* 85B(Supp II). pp. 177.

Horst, M. J., C. K. van der Simms, and R. Maasdam. 2005. Occupant lower leg injury assessment in landmine detonations under a vehicle. *IUTAM Symposium on Biomechanics of Impact: From Fundamental Insights to Applications,* Dubline, Ireland.

Huelke, D. F. 1986. *Anatomty of the lower extremity—An overview.* Society of Automotive Engineers (861921), Warrendale, PA.

Levine, R. S., A. Manoli, and P. Prasad. 1995. *Ankle and foot injury scales: AFIS-S & AFIS-I.* American Orthopaedic Foot and Ankle Trauma Committee, Rosemont, IL.

McKay, B. 2010. Development of lower extremity injury criteria and biomechanical surrogate to evaluate military vehicle occupant injury during an explosive blast event PhD, Biomedical Engineering Wayne State University (Paper 146).

McKay, B. J., and C. A. Bir. 2009. Lower extremity injury criteria for evaluating military vehicle occupant injury in underbelly blast events. *Stapp Car Crash J* 53:229–249.

Moore, K. L., A. M. R. Agur, and A. F. Dalley. 2011. *Essential clinical anatomy.* 4th ed. Baltimore, MD: Lippincott Williams & Wilkins.

Newell, N., S. Masouros, and A. M. Bull. 2013. *A comparison of Mil-Lx and Hybrid III response in seated and standing postures with blast mats in simulated under-vehicle explosions.* Gothenbert, Sweden: International Research Council on the Biomechanics of Injury.

Olson, C., S. Rouhana, and B. Spahn. 2007. *Comparison of the THOR and Hybrid III lower extremities in laboratory testing.* Society of Automotive Engineers, Warrendale, PA.

Owens, B. D., J. F. Kragh, Jr., J. Macaitis, S. J. Svoboda, and J. C. Wenke. 2007. Characterization of extremity wounds in Operation Iraqi Freedom and Operation Enduring Freedom. *J Orthop Trauma* 21(4):254–7. doi: 10.1097/BOT.0b013e31802f78fb.

Owens, B. D., J. F. Kragh, Jr., J. C. Wenke, J. Macaitis, C. E. Wade, and J. B. Holcomb. 2008. Combat wounds in operation Iraqi Freedom and operation Enduring Freedom. *J Trauma* 64(2):295–299. doi: 10.1097/TA. 0b013e318163b875.

Pandelani, T. 2014. An investigation of the forces within the tibiae at the typical blast loading rates—With different boots. MS, Department of Mechanical Engineering, University of Cape Town.

Pandelani, T., T. J. Sono, J. D. Reinecke, and G. N. Nurick. 2015. Impact loading response of the MiL-Lx leg fitted with combat boots. *International Journal of Impact Engineering.* 92(June 2016):26–31. doi:10. 1016/j.ijimpeng.2015.03.007.

Petit, P. 1996. Quasistatic characterization of the human foot-ankle joint in a simulated tensed state and updated accidentological data. *International IRCOBI Conference of Biomechanics of Impact,* Dublin, Ireland.

Portier, L. 1997. Dynamic biomechanical dorsiflexion responses and tolerances of the ankle joint complex. *41st Stapp Car Crash Conference,* Lake Buena Vista, FL, USA.

Ramasamy, A., A. M. Hill, S. Masouros, I. Gibb, A. M. Bull, and J. C. Clasper. 2011a. Blast-related fracture patterns: A forensic biomechanical approach. *J Roy Soc Interface* 8(58):689–698. doi: 10.1098/rsif. 2010.0476.

Ramasamy, A., A. M. Hill, R. Phillip, I. Gibb, A. M. Bull, and J. C. Clasper. 2011b. The modern "deck-slap" injury—Calcaneal blast fractures from vehicle explosions. *J Trauma* 71(6):1694–1698. doi: 10.1097/TA. 0b013e318227a999.

Ramasamy, A., A. M. Hill, S. Masouros, I. Gibb, R. Phillip, A. M. Bull, and J. C. Clasper. 2013. Outcomes of IED foot and ankle blast injuries. *J Bone Joint Surg Am* 95(5):e25. doi: 10.2106/JBJS.K.01666.

Ramasamy, A., S. D. Masouros, N. Newell, A. M. Hill, W. G. Proud, K. A. Brown, A. M. Bull, and J. C. Clasper. 2011c. In-vehicle extremity injuries from improvised explosive devices: Current and future foci. *Philos Trans Roy Soc Lond B Biol Sci* 366(1562):160–170. doi: 10.1098/rstb.2010.0219.

Wang, J., R. Bird, B. Swinton, and A. Krstic. 2001. Protection of lower limbs against floor impact in army vehicles experiencing landmine explosion. *J Battlefield Technol* 4(3):8–12.

Whyte, T. 2007. Investigation of factors affecting surrogate limb measurements in the testing of landmine protected vehicles. MS, University of Cape Town.

Yoganandan, N., F. Pintar, and M. Boynton. 1996. *Dynamic axial tolerance of the human foot-ankle complex.* Society of Automotive Engineers, Warrendale, PA.

16 The Lower Extremities
Computational Modelling Attempts to Predict Injury

Grigoris Grigoriadis, Spyros Masouros, Arul Ramasamy and Anthony M. J. Bull

CONTENTS

16.1 INTRODUCTION

When an explosive detonates beneath a vehicle, the blast wave from the explosion causes the release of a cone of super-heated gas and soil which impacts the undersurface of the travelling vehicle. This leads to rapid deflection of the vehicle floor, transmitting a very short duration (less than 10 ms), high amplitude load into anything that is in contact with it, as discussed in greater detail in Chapter 3. Most frequently, it is the leg and particularly the foot and ankle complex that is injured (Ramasamy et al. 2010, 2011).

Injuries to the foot and ankle complex are of particular interest as it has been demonstrated that patients with such injuries have significantly greater disability scores – Short form 36 (SF-36) < 50, Western Ontario and McMaster Universities Osteoarthritis Index (WOMAC) > 30 and the modified Boston Children's Hospital Grading System < 65 – than those without foot and ankle injuries (Tran and Thordarson 2002; Turchin et al. 1999). Especially for a young military population, which is likely to place significant functional demands on the foot and ankle complex, the impact of such disability on the quality of life will be greater than that of the average population. In a recent analysis of UK Service Personnel, Ramasamy et al. (2013) demonstrated that casualties who suffered lower limb injuries from under-vehicle explosives were frequently associated with multi-segmental foot and ankle injuries and resulted in an amputation rate of 30%. In addition, 75% of injured limbs were noted to have a poor clinical outcome three years following injury (Ramasamy et al. 2013).

Full-vehicle blast experiments on full-body cadavers or anthropometric testing devices (ATDs) offer the most realistic simulation of underbelly (UB) mine blasts. Bird (2001) placed UB landmines containing 7.3 kg of Plastic Explosive 4 (PE-4), equivalent to 10 kg of trinitrotoluene (TNT), under a series of military vehicles and mounted displacement transducers to the floor of the vehicles; unfortunately, no data on deformation or velocity of deformation were reported. Due to their nature, full-vehicle blast experiments are expensive, may result in instrumentation failure, show poor repeatability, have many uncontrolled variables and are always limited to small sample sizes due to cost limitations. In addition, the reality of defence UB mine research is that full-body cadaveric experiments involving blast are not ethically or politically supported in many countries. Alternative routes have therefore been sought in order to replicate the effects of blast on the vehicle occupant. In this chapter, computational approaches to model the effects of blast on skeletal injury are reviewed, research gaps presented, and ways forward proposed.

16.2   COMPUTATIONAL MODELLING

The development of accurate, validated numerical/computational models can allow a cost-effective alternative to expensive experimental set-ups, as well as allow the simulation of multiple scenarios by altering modelling input parameters, specifically, geometric, material, loading and boundary parameters.

Computational modelling of the lower limb at high strain rates has been developed historically for safety in the automotive industry, often aimed at simulating pedestrian or vehicle-occupant lower limb injury. A comprehensive review of models developed before 2006 can be found in Yang et al. (2006). All such models utilise explicit dynamic finite element codes, such as Pam-Crash (ESI Group, Paris, France), Radioss (Altair Engineering Inc., MI, USA) and LS-Dyna (Livermore Software Technology Corp. CA, USA).

Geometrical data are acquired from anthropometric studies or MR and CT imaging data of specific subjects (either living or post-mortem) and therefore vary among studies. The pathway from the imaging data to the end model may also vary, resulting in loss of geometric accuracy in every step, from image acquisition, image segmentation, to surface fitting and meshing. Material-property data for hard and soft tissues vary significantly among studies. Most studies utilise simplified material models, such as linearly elastic, linearly plastic and isotropic or transversely isotropic, with only very few modelling viscoelastic effects in soft tissues for which literature values are available. Importantly, the material coefficients assigned in most of the models have been derived from experimental data conducted at quasi-static strain rates. It has been shown that hard and soft tissue material behaviour is dependent upon strain rate (this is discussed in detail below) and that at least some measured outcomes of mechanical behaviour are strongly dependent upon the numerical value of material parameters (Beillas et al. 2007).

These points highlight the fundamental challenges in the development of accurate numerical models: the lack of data on material behaviour of musculoskeletal tissues under high loading rate conditions, the compromise between geometric biofidelity and model simplicity, and the difficulty of validating the models against appropriate experimental data.

16.3   FE MODELS SIMULATING ROAD TRAFFIC ACCIDENTS

FE models of the foot and ankle aimed at simulating impact were first developed in order to examine road traffic accidents. Injuries to the lower limb are common in road traffic accidents; tibial pilon, calcaneal and talar fractures are the most severe and difficult to treat injuries associated with frontal crashes (Owen et al. 2001). The most common modelling differences between high and low loading rate simulations are associated with: (a) the material properties, as biological tissues are sensitive to strain rate; (b) the meshing procedure whereby hexahedral elements are preferred over tetrahedral elements; and (c) the numerical scheme used where an explicit numerical scheme is preferred over an implicit one. In contrast to an implicit (Newton-Raphson-type) iterative, dynamic scheme, an explicit scheme does not require equilibrium between the internal structural forces and the externally applied loads to be satisfied at each time step. The time-step dependence of explicit schemes can lead to long run times compared to equivalent implicit approaches, but the lack of dependency on convergence often allows the representation of more complex material interactions (such as those associated with component contact). Furthermore, explicit schemes are used in scenarios involving highly transient phenomena and in dynamic problems where, at times, the strength of the loaded material is substantially lower than the stresses that develop during the simulation.

Impact loading of the foot in inversion and eversion was examined in a study where a 3D model consisting of membrane (ankle ligaments and Achilles tendon) and bar elements (foot ligaments) was developed (Beaugonin et al. 1996). Bones were modelled as rigid surfaces. The numerical results were compared to data from human cadaveric tests (Begeman et al. 1993) where feet, fixed at the distal tibia, were impacted in offset locations of the plantar heel pad to cause dynamic inversion and eversion. Numerical and experimental kinematic data were in better agreement than load and moment measurements; rigid bones and the lack of the plantar soft tissues did not permit realistic force transmission.

The same model was updated (Beaugonin et al. 1997) to include a deformable tibia, fibula, calcaneus and talus; hexahedral solid elements represented the trabecular structures of these bones covered by the cortex modelled as thin shell elements. The heel pad was modelled by soft contact springs 2 mm in length. A cadaveric experiment was simulated using this model whereby a ball impacted on the rigid surface where the foot was positioned. As the position of impact was at the level of the cuneiforms, foot dorsiflexion was caused. Compared to the previous model (Beaugonin et al. 1996), the peak force was less and was a better match to the experimental data. This highlights the importance of including the structure of the heel pad and modelling the bones as deformable structures.

Beillas et al. (2001) developed a 3D model (Lower Limb Model for Safety – LLMS) of the lower limb from MRI and CT scans of a subject of similar dimensions to the 50th percentile male. The model was adapted from a previous study (Beillas et al. 1999) where bones were modelled as rigid surfaces. The femur, tibia, patella, calcaneus and talus were meshed as hexahedral elements (trabeculae) covered by shell elements (cortex), while the rest of the bones of the foot were modelled as rigid bodies. Although the cartilage of the knee was modelled using solid elements, the cartilage of the ankle joint complex was not included in the model, and contact interfaces with specific attributes were set between the bones. The bones were represented by elastoplastic materials (Tables 16.1 and 16.2); a shear modulus associated with a relaxation function was used for the plantar pad (Table 16.3), while ankle ligaments were assigned viscoelastic properties (Kelvin-Voigt model) (Table 16.4). In a preliminary foot and ankle model (Beillas et al. 1999) on which this study was based, the fat pad was modelled as a linearly elastic material ($E = 40$ MPa). Assigning non-linearly elastic properties, instead of linearly elastic, to the heel fat pad was shown to be critical in improving the accuracy of the model by predicting more closely the reaction force of the impactor when simulating axial tibial impact loading. Other soft tissues including skin and flesh were modelled as linearly elastic materials. In total, the model consisted of 25,000 finite elements.

The same model was used to simulate static axial compression, axial tibial impact, patellar impact and knee high-rate shear and bending loading scenarios. The numerical results of these simulations

---

TABLE 16.1

Elastic Material (E – Young's Modulus and ν – Poisson's Ratio) and Physical (ρ – Density) Properties of Bone Implemented in Foot and Ankle FE Models Used to Simulate Automotive Impact Loading Scenarios

Study	Type and Location	E [GPa]	ν	ρ [tn/m^3]
Beillas et al. (2001)	Tibia/fibula cortex	12–17.5	0.3	1.8–2.1
Beaugonin et al. (1997)	Tibia/fibula cortex	18.4–18.9	0.29	1.81
Neale et al. (2012)	Tibia/fibula cortex	16	0.3	2
Beillas et al. (2001)	Tarsal bones cortex	5–6	0.3	1.8
Beaugonin et al. (1997)	Tarsal bones cortex	15	0.3	1.81
Neale et al. (2012)	Tarsal bones cortex	13.5	0.3	2
Shin et al. (2012)	Cortex	17.5	0.3	2
Beillas et al. (2001)	Trabeculae	0.075–0.45	0.3	0.13–0.185
Beaugonin et al. (1997)	Trabeculae	0.531	0.3	0.5
Neale et al. (2012)	Trabeculae	0.3	0.3	1.1
Shin et al. (2012)	Trabeculae	0.445	0.3	1.1

TABLE 16.2

Plastic Material Properties (σ_y – Yield Stress, E_t – Tangential Young's Modulus in Plastic Region, σ_f – Stress at Failure, ε_f – Strain at Failure) of Bone Implemented in Foot and Ankle FE Models Used to Simulate Automotive Impact Loading Scenarios

Study	Type	σ_y [MPa]	E_t [GPa]	σ_f [MPa]	ε_f [%]
Beillas et al. (2001)	Tibia/fibula cortex	80–120		110–135	3
Neale et al. (2012)	Tibia/fibula cortex	120	1		
Shin et al. (2012)	Tibia/fibula cortex	125		134	1.6
Beillas et al. (2001)	Tarsal bones cortex	80–100		100–125	2–3
Neale et al. (2012)	Tarsal bones cortex	90	2.25		
Shin et al. (2012)	Tarsal bones cortex	165	0	175	2.2
Beillas et al. (2001)	Trabeculae	10		15	3
Neale et al. (2012)	Trabeculae	10	0.056		
Shin et al. (2012)	Trabeculae	5.3		13.4	25

TABLE 16.3

Material Properties of Soft Tissues Implemented in Foot and Ankle FE Models Used to Simulate Automotive Impact Loading Scenarios

Study	Tissue	Type	Constants E, C_{ij}, G, K [MPa], ρ [tn/m^3], τ, t [ms], A_i, ν [~]
Beillas et al. (2001)	Flesh and skin	Linearly elastic	$E = 0.001$ ($\nu = 0.3$), $\rho = 1$
Neale et al. (2012) Shin et al. (2012)	Flesh	QLV	$C_{10qs} = 1.2\text{e-}4$, $C_{20qs} = 2.5\text{e-}4$, $K = 20$, $A_1 = 1.2$, $A_2 = 0.8$, $\tau_1 = 23$, $\tau_2 = 63$, $\rho = 0.8$
Shin et al. (2012)	Fat pad	Ogden	$\mu_1 = 0.0165$, $\alpha_1 = 6.82$
Beillas et al. (2001)	Fat pad	Non-linearly viscoelastic	$G = 0.6 + (6.3 - 0.6)e^{-\frac{t}{100}}$, $\rho = 1$

C_{ijqs} = material constants at equilibrium (quasi-static), A_i = relaxation constants, τ_i = relaxation time constants, μ_i, α_i = material constants, G = shear modulus, E = Young's modulus, K = bulk modulus, ρ = density, ν = Poisson's ratio.

were compared to measurements from experiments replicating the same load cases on cadaveric specimens (Banglmaier et al. 1999; Haut and Atkinson 1995; Kajzer et al. 1990, 1993); the specimens were impacted or compressed at different areas and in different directions by a tup fixed on a load cell. The force values of the tup predicted by the FE simulations were within the experimental corridors for most of the tests. Since each experiment loaded different structures of the lower limb, the model was validated region by region. For the case of an impact axial loading to the tibia, however, which is more relevant than the others to UB blast loading, the predicted tibial force oscillated over time with an amplitude equal to half of the peak force. It is not known if this response is physiological or a modelling artefact.

Another 3D model of the foot and ankle was isolated from a full-body FE model used by automotive companies [Total Human Model for Safety – THUMS (Iwamoto et al. 2002), Toyota Central R&D Labs., Inc., Nagakute, Japan] and modified by Morimoto and Yamazaki (2004) in an attempt to develop a tool which can predict foot and ankle fractures of occupants in frontal vehicle collisions. The plantar soft tissues and the plantar fascia were added, and non-linearly elastic properties were assigned through a force-displacement graph. Material properties of other tissues were not reported.

TABLE 16.4

Material Properties of Ankle and Foot Ligaments and the Plantar Fascia Implemented in Foot and Ankle FE Models Used to Simulate Automotive Impact Loading Scenarios

Study	Tissue	Constants k [N/mm], E [MPa], ρ [tn/m^3], μ [N·ms·mm^2]
Shin et al. (2012)	ATiF ligament	Non-linearly elastic, $k = 5400$*
Shin et al. (2012)	PTT ligament	Non-linearly elastic, $k = 1600$*
Shin et al. (2012)	CF ligament	Non-linearly elastic, $k = 1240$*
Shin et al. (2012)	ATaF ligament	Non-linearly elastic, $k = 210$*
Shin et al. (2012)	PTaF ligament	Non-linearly elastic, $k = 130$*
Shin et al. (2012)	PTiF ligament	Non-linearly elastic, $k = 70$*
Shin et al. (2012)	CTi ligament	Non-linearly elastic, $k = 50$*
Shin et al. (2012)	Midfoot ligaments	Scaled from ATaF properties based on each ligament's dimensions taken from (Corazza et al. 2003)
Shin et al. (2012)	TiNav ligament	$k = 39.1$
Shin et al. (2012)	DTarMt ligament	$k = 115$
Shin et al. (2012)	PTarMt ligament	$k = 90$
Shin et al. (2012)	IoTarMt ligament	$k = 189.7$
Shin et al. (2012)	DInMt ligament	$k = 125$
Shin et al. (2012)	Plantar fascia	$k = 203.3$
Beillas et al. (2001)	Ankle ligaments	Kelvin-Voigt $k = 50$, $\mu = 10$
Beillas et al. (2001)	Other ligaments	Kelvin-Voigt $k = 250$, $\mu = 10$
Beaugonin et al. (1997)	Ankle ligaments	$E = 88.2$, $v = 0.22$, $\rho = 1$

Ligaments: ATiF = anterior tibiofibular, PTT = posterior talotibial, CF = calcaneofibular, ATaF = anterior talofibular, PTaF = posterior talofibular, PTiF = posterior tibiofibular, CTi = calcaneotibial, TiNav = tibionavicular, DTarMt = dorsal tarsometatarsal, PTarMt = plantar tarsometatarsal, IoTarMt = interosseous tarsometatarsal, DInMt = dorsal intermetatarsal. E = Young's modulus, v = Poisson's ratio, ρ = density, k = stiffness, μ = viscosity coefficient.

* The stiffness values of the ligaments modelled with non-linearly elastic material properties correspond to the linear region of the force–displacement curve.

Compared to the full THUMS model, the posture was also corrected by aligning the bones accurately, as shown in Figure 16.1a. The modified model was validated against kinematic and kinetic data from drop tests on cadaveric lower limbs (Figure 16.1b). This set-up would be relevant to UB blast if the tibia was not totally constrained; such a loading environment refers to the case where the driver's foot is trapped underneath the dashboard of the vehicle following a frontal collision. The response of the modified model was stiffer than that of the original model (Figure 16.1c). Even though plantar soft tissues – that were expected to make the response more compliant – were added in the modified model, the effect on the overall response of altering the position of bones was probably greater. In addition, the modified model predicted high stress concentration on the medial side of the calcaneus as well as the diaphyseal tibia close to the distal end; this was probably due to the boundary conditions implemented, as the tibia was fixed at this level.

Similar to the study conducted by Morimoto and Yamazaki 2004, another foot and ankle model from a full-body FE model (H-dummy, ESI Group, Paris, France) was modified to simulate the axial tibial loading experienced by the driver in case of frontal impact (Murakami et al. 2004). All bones apart from the metatarsals and the phalanges were modelled as deformable bodies consisting of trabecular solid and cortical shell elements. The ligaments, plantar fat pad and Achilles tendon were also included, although the material properties assigned were not stated. In order to replicate the case where a driver is braking prior to an expected collision, the authors applied a tensile force through the Achilles tendon, both to the numerical model and the experiment, which was similar to the configuration shown in Figure 16.1b.

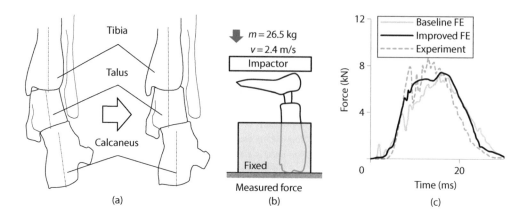

FIGURE 16.1   (a) The position of the bones was modified to ensure realistic proximal transmission of the force. The schematic was produced based on a figure from Morimoto and Yamazaki (2004). (b) The tibia was fixed while drop tests were performed on the foot. The same boundary conditions were applied to the computational simulation. The schematic was produced based on a figure from Morimoto and Yamazaki (2004). (c) The force measured at the proximal tibia was compared to the response of the initial (baseline) and the modified (improved) foot and ankle models. The modified model's response matched the experimental data better. The graph was plotted based on the results reported by Morimoto and Yamazaki (2004).

The tibia was fixed at its proximal end where it was potted and secured on a load cell while an Achilles tension pulley mechanism was added. The forces at the tibial load cell predicted by the simulation were similar to those measured experimentally, albeit with a small time delay. Similarly to the numerical response predicted by Beillas et al. (2001), multiple peaks of equal amplitude, however, were reported that were not attributed to multiple impacts, oscillations or fractures.

Using 3D digital data from a commercial library model, Neale et al. (2012) developed a 3D foot and ankle model. As shown in Figure 16.2a, trabecular and cortical parts of the tibia, fibula and tarsal bones were modelled with tetrahedral solid and shell elements, respectively, and elastoplastic material properties were assigned to them (Tables 16.1 and 16.2). The rest of the bones of the foot were modelled as rigid. Similar to the studies described earlier, cartilaginous tissues were not included, and sliding contacts with an initial gap with a friction coefficient of 0.01 were defined between the outer surfaces of articulating bones. Despite the addition of 26 ankle ligaments as non-linear spring and damper components between the nodes of shell elements, results from a preliminary simulation of the dorsiflexion motion showed a non-physiological rear gap between the talus and the tibia; this inaccurate behaviour was resolved by introducing two 'hypothetical' ligaments as shown in Figure 16.2b. A solid representation of the cartilage may have had a similar effect on the dorsiflexion motion. The Achilles tendon was represented by 1D cable elements, while flesh was represented by assigning 'tied' contact between skin and bones; stiffness and damping parameters of this interface were based on data from tests on cadaveric anterior thigh muscles and fat samples (Untaroiu et al. 2005) (Table 16.3). The foot was not covered by flesh and the plantar pad was not included in this model.

Different impact loadings were simulated to assess the validity of the response of the model. Pendulum impact tests to the ball of the foot were firstly applied based on the experiments performed by Wheeler et al. (2000) on cadaveric lower limbs. The tibia was potted and fixed at its proximal end, while the velocity at impact and the mass of the pendulum were attributed to a rigid geometry. Because the forefoot was not covered by flesh in the original model, a skin pad underneath the metatarsals was added; this ensured a less abrupt load transfer. Good correlation between simulation and experiment in acceleration of the pendulum was shown (Figure 16.3b). There was, however, no reporting of the measured and predicted forces at the proximal end of the tibia, which would indicate

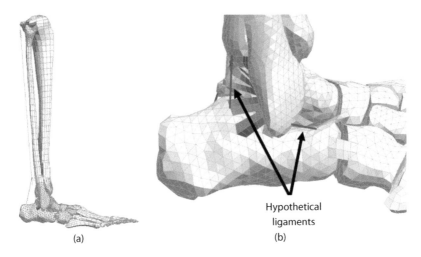

(a)   (b)

FIGURE 16.2 (a) An overview of the model with all bones (solid elements) and ligaments (shell) elements. (Reproduced from Neale, M., et al., *A Finite Element Modelling Investigation of Lower Leg Injuries. 07-0077*, TRL Limited, 2012. Available from: http://www-nrd.nhtsa.dot.gov/pdf/esv/esv20/07-0077-W.pdf. With permission.) (b) 'Hypothetical' ligaments were added to ensure better articulation of the tibiotalar joint. (Modified and reproduced from Neale, M., et al., *A Finite Element Modelling Investigation of Lower Leg Injuries. 07-0077*, TRL Limited, 2012. Available from: http://www-nrd.nhtsa.dot.gov/pdf/esv/esv20/07-0077-W.pdf. With permission.)

whether the force transmission through the limb was valid. Sled impact tests were also simulated based on the experiments performed by Hynd et al. (2003) (Figure 16.3c). Because the contact area was greater at this test, skin pads were added to both the areas underneath the metatarsals and the heel. Due to great discrepancies between experimental and numerical results, the authors executed a sensitivity analysis to assess the influence of the properties of the plantar ligaments and the heel pad on the resulting response. Figure 16.3d–e summarises the outcome of the sensitivity analyses; both tissues were found to be critical. The lack of solid representation of the plantar flesh of the foot was also deemed critical; the impact loading was transferred to the two regions of the padding instead of being spread over the whole plantar surface. Furthermore, the added pads were not constrained on the foot bones and remained fixed on the impact surface. This permitted unrealistic slipping and secondary contacts between the bones of the foot and the soft tissues of the plantar foot represented by the pads.

Aiming to simulate combined loading, Shin et al. (2012) developed a 3D model of the foot and ankle at different postures before applying impact loading through the forefoot. This study was funded by the Global Human Body Models Consortium (GHBMC); the consortium was created by automotive companies with the aim to consolidate and coordinate human body modelling efforts in order to improve the current technology in automotive safety. The deliverables of the consortium are a variety of biofidelic FE models of the human body corresponding to different percentiles and ages of both sexes to be used for simulations of road traffic accidents. The 3D FE model of the lower limb developed by Shin et al. (2012) was incorporated into the phase I GHBMC FE model of the 50th percentile male (http://www.ghbmc.com/GHBMCStatusPhase1.pdf, accessed on 30/3/16).

The geometries of the tissues were obtained from CT scans of a subject with anthropometry close to that of the 50th percentile male. Apart from the diaphyseal region of the tibia where cortex was represented by solid elements, trabecular and cortical structures were modelled as solid hexahedral and shell elements, respectively, with elastoplastic material properties (Tables 16.1 and 16.2). Mid- and forefoot bones were represented by rigid bodies. Both flesh and skin were included and modelled separately, while viscoelastic properties were assigned to the heel fat pad based on results

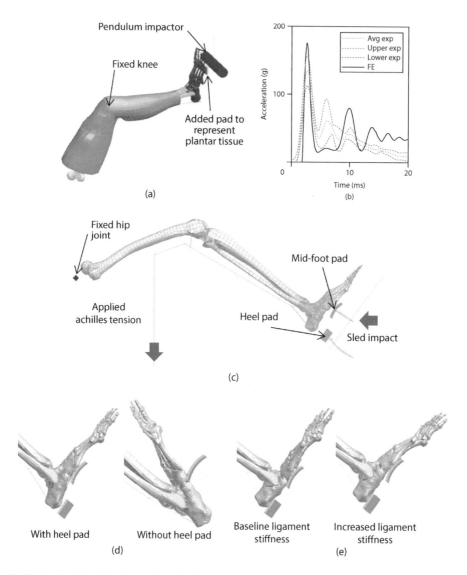

FIGURE 16.3 Details of the foot and ankle FE model developed by Neale et al. (2012). (a) Boundary conditions of the simulated pendulum impacts. (b) Accelerations of the pendulum were compared to validate the model. The predicted response was slightly stiffer than the experiment. It was not clarified whether secondary impacts, corresponding to the multiple peaks, were allowed. (c) Boundary conditions of the simulated sled tests. Inclusion/enhancement of (d) the heel pad and (e) the stiffness of the plantar ligaments improved the model response. (Modified and reproduced from Neale, M., et al., *A Finite Element Modelling Investigation of Lower Leg Injuries. 07-0077*, TRL Limited, 2012, Available from: http://www-nrd.nhtsa.dot.gov/pdf/esv/esv20/07-0077-W.pdf. With permission.)

from *in vivo* indentation tests (Table 16.3). Ligaments were modelled as 1D truss elements with material properties based on literature values (Table 16.4).

Forefoot impact was simulated by fixing the tibia and impacting with a cylindrical rigid body of known mass and initial velocity according to the experimental apparatus used by Wheeler et al. (2000) (Figure 16.4a). As in the study conducted by Neale et al. (2012), the limitation of this validation attempt is associated with the comparison of measured and predicted accelerations of the impacting structure (Figure 16.4b) instead of force and moments in the tibia. Other simulations included

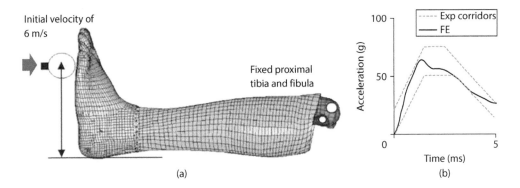

FIGURE 16.4 (a) Boundary conditions of the simulated forefoot impact. (Modified and reproduced from Shin, J., et al., *Ann. Biomed. Eng.*, 40(12), 2519–2531, 2012. With permission.) (b) The predicted acceleration of the impacting cylinder was within the experimental corridors. The graph was plotted based on the results reported by Shin et al. (2012).

quasi-static axial rotation, dynamic dorsiflexion and a combined load case where initial dorsiflexion was followed by axial compression and then external rotation. Despite the limitations regarding the lack of cartilaginous structures as well as the validation procedure, this is the most detailed and complete 3D model of the foot and ankle for automotive impact applications to date. The model, however, cannot be considered valid for impact through the hind foot.

In the case of simulating a driver involved in a road traffic accident, the impact loading is expected to be transferred to the lower limb through the metatarsal heads due to the foot being placed on the pedals while applying the brakes prior to the accident. Such a load, as it is offset to the long axis of the tibia, causes dynamic dorsiflexion of the ankle joint. In both studies described above (Neale et al. 2012; Shin et al. 2012), the ankle joint was allowed to reach dorsiflexion angles prior to failure greater than the maximum ankle dorsiflexion, which is reported to vary between 8° and 26° (Rome 1996). The ankle of the model developed by Shin et al. (2012) was dorsiflexed at 37° prior to failure of the posterior talotibial ligament. Neale et al. (2012) did not incorporate a failure mechanism for ligaments, and, as a result, the model predicted the unrealistic posture shown in Figure 16.3d without reporting fracture or failure. In addition, a non-physiological alignment of the ankle joint during the simulation affects the biofidelity of the model; the locations where high stress concentration is observed and therefore are prone to injury as well as the load transmission through the bones depend on the alignment of the ankle joint complex. Therefore, for load cases where rotation of the ankle joint is critical, the validation of the kinematic response of the FE model would eliminate the possibility of reaching unrealistic postures.

16.4 FE MODELS SIMULATING UNDERBODY BLAST

Compared to automotive road traffic accidents, the loading environment associated with the 'deck-slap' injury (Ramasamy et al. 2011) that occurs when the floor of a vehicle above an explosion deforms rapidly upwards and into the lower limbs of the occupants and other UB blast events is described by faster rates, shorter durations and different boundary conditions (McKay and Bir 2009). The first 3D FE model of the foot and ankle that was used to simulate axial impulsive loading, representative of blast loading scenarios, was developed by Tannous et al. (1996). Twenty-six bones were included from the plantar foot to the mid-tibia and fibula. Both cortical and trabecular structures were represented by hexahedral solid elements. The bones were set in a neutral posture based on anatomical studies from the literature, while the mid- and forefoot bones were fused. The plantar pad of the foot consisted of three layers: the ligamentous, the muscle and the fat layer (Table 16.7). As the geometric accuracy was limited, Bandak et al. (2001) modified the model by including cartilage and adding viscoelastic properties to the muscle and fat layers (Table 16.7) as well as the ligaments (Table 16.8). The model was used to simulate the pendulum impacts on cadaveric limbs performed by Yoganandan et al. (1996) for impact velocities of 2.23–7.6 m/s (Figure 16.5).

Iwamoto et al. (2005) isolated and modified the lower limb model of the THUMS full-body 3D FE model with the aim to predict tibial fractures more accurately. The tibial cortex was represented by solid elements, and a constitutive material formulation was implemented to account for anisotropy and strain-rate dependency. The custom constitutive equation was capable of predicting the response of the tibia under different loadings and under a variety of loading rates. It was recognised, however, that the implementation of the custom material laws rendered the model impractical to be used by others, and therefore, a simplified model was proposed whereby strain-rate dependence of the plastic behaviour of bone was implemented through the Cowper-Symonds law (Cowper et al. 1957) (Tables 16.5 and 16.6). The rest of the bones of the ankle joint complex were represented by solid hexahedrals (trabeculae) surrounded by shell elements (cortex) and were attributed with elastoplastic strain-rate-dependent (Cowper-Symonds law) material properties (Tables 16.5 and 16.6). Despite the detailed material representation of the bone structures, all soft tissues, including cartilage, were modelled as linearly elastic materials (Tables 16.7 and 16.8).

Two series of dynamic axial impacts on cadaveric specimens were simulated. In the experimental set-up used by Kitagawa et al. (1998), the Achilles tension was applied to the cadaveric heels, and the forefoot was impacted by a mass representing the braking pedal. The model was also used to simulate the pendulum impacts on cadaveric specimens performed by Yoganandan et al. (1995) in the same configuration as in Figure 16.5. The model with the simplified material properties of the tibia was predicting accurately the forces measured at the impactor (Figure 16.6a and c) and at the proximal tibia (Figure 16.6b and d). By altering the velocity, the angle and the allowed intrusion distance of the impactor as well as the Achilles tension, the authors conducted a parametric sensitivity analysis to identify the most vulnerable bone for each case. The most common injury predicted by the simulations was tibial pilon fractures. Although this is expected for frontal vehicle collisions where the leg is trapped under the dashboard (Funk et al. 2002), calcaneal fractures are the most common injury in laboratory-simulated UB blast (McKay and Bir 2009).

The simulation of a similar load case as presented in Figure 16.5 was the aim of another 3D foot and ankle model developed by Kraft et al. (2012). This model consisted of more than 500,000 tetrahedral elements; this is a much larger number of elements compared to hexahedral meshes used in the studies presented above (range of 15,000–60,000 elements). To facilitate the simulation, the talus was fused with the bones of the mid- and forefoot. This resulted in overestimation of the amount of talar fractures. To eliminate this artefact, the model was modified and all articulations were allowed (Kraft et al. 2013). Both cortex and trabeculae were represented by tetrahedral solid elements with linearly elastic material properties (Table 16.5). Despite this improvement in fidelity, the input to simulate

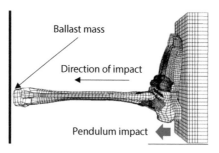

FIGURE 16.5 Boundary conditions of the simulated pendulum impact. Tibia is restricted to move only in the direction of the impact. (Modified and reproduced from Bandak, F.A., et al., *Int. J. Solids Struct.*, 38(10), 1681–1697, 2001. With permission.)

TABLE 16.5

Elastic Material (E – Young's Modulus and ν – Poisson's Ratio) and Physical (ρ – Density) Properties of Bone Implemented in Previous Foot and Ankle FE Models Used to Simulate Axial Impact Loading Scenarios Similar to UB Blast

Study	Type and Location	E [GPa]	ν	ρ [tn/m^3]
Tannous et al. (1996) Bandak et al. (2001)	Cortex	12.15	0.3	1.882
Iwamoto et al. (2005)	Tibia cortex	17	0.3	
Iwamoto et al. (2005)	Fibula cortex	18.5	0.3	
Iwamoto et al. (2005)	Tarsal bones cortex	15	0.3	
Kraft et al. (2012, 2013)	Cortex	19	0.3	1.81
Dong et al. (2013) Suresh et al. (2012) Gabler et al. (2014)	Tibia/fibula cortex	16	0.3	1.85
Dong et al. (2013)	Tarsal bones cortex	17.5	0.3	2
Tannous et al. (1996) Bandak et al. (2001)	Trabeculae	0.53	0.3	1
Iwamoto et al. (2005)	Trabeculae	0.1456	0.45	
Kraft et al. (2012, 2013)	Trabeculae	0.3	0.45	0.6
Dong et al. (2013)	Trabeculae	0.445	0.3	1.1
Gabler et al. (2014)	Trabeculae	1.07	0.3	1.1

TABLE 16.6

Plastic Material Properties (σ_y – Yield Stress, E_t – Tangential Young's Modulus in Plastic Region, σ_f – Stress at Failure, ε_f – Strain at Failure) of Bone Implemented in Previous Foot and Ankle FE Models Used to Simulate Axial Impact Loading Scenarios Similar to UB Blast

Study	Type	σ_y [MPa]	E_t [GPa]	σ_f [MPa]	ε_f [%]
Iwamoto et al. (2005)	Tibia cortex			150	
Iwamoto et al. (2005)	Fibula cortex			146	
Iwamoto et al. (2005)	Tarsal bones cortex			140	
Kraft et al. (2012, 2013)	Cortex			132	10
Dong et al. (2013)	Tibia/fibula cortex	125	1	134	
Dong et al. (2013)	Tarsal bones cortex	165	0.79	175	
Gabler et al. (2014)	Cortex	140		214	2.2
Kraft et al. (2012, 2013)	Trabeculae			1.5	10
Dong et al. (2013)	Trabeculae	53	0		12.2

axial impact was selected to be a bilinear velocity profile, namely a positive and a negative slope for the duration of the acceleration and the deceleration, respectively, instead of an initial velocity; this represents an impactor with constant accelerations. This is unrealistic as the acceleration of the floor of a vehicle involved in an explosion increases to reach a maximum when the floor plate is fully deformed and then decreases (Bailey et al. 2013). The results were not compared to experimental

TABLE 16.7

Material Properties of Soft Tissues Implemented in Foot and Ankle FE Models Used to Simulate Axial Impact Loading Scenarios Similar to UB Blast

Study	Tissue	Type	Constants E, C_{ij}, G, K, A [MPa], ρ [tn/m^3], τ, t [ms], A_i, B, ν [~]
Tannous et al. (1996)	Muscle	Linearly elastic	$E = 0.152$ ($\nu = 0.3$), $\rho = 1$
Tannous et al. (1996)	Fat	Linearly elastic	$E = 20$ ($\nu = 0.3$), $\rho = 1$
Bandak et al. (2001)	Muscle	Viscoelastic	$E = 0.152$ ($\nu = 0.49$), $K = 2.5$, $\rho = 1$
			$G = 0.05 + (0.04 - 0.05)e^{-\frac{t}{\tau}}$
Bandak et al. (2001)	Fat	Viscoelastic	$E = 20$ ($\nu = 0.49$), $K = 335$, $\rho = 1$
			$G = 6.7 + (5 - 6.7)e^{-\frac{t}{\tau}}$
Iwamoto et al. (2005)	Cartilage	Linearly elastic	$E = 100$–500 ($\nu = 0.3$)
Iwamoto et al. (2005)	Skin	Linearly elastic	$E = 22$ ($\nu = 0.4$)
Iwamoto et al. (2005)	Flesh	Linearly elastic	$E = 2$ ($\nu = 0.42$)
Kraft et al. (2012)	Muscle	QLV	$C_{10} = 0.94$, $C_{01} = 0.94$, $K = 167$, $A_1 = 0.465$, $A_2 = 0.2$,
			$A_3 = 0.057$, $A_4 = 0.066$, $A_5 = 0.089$, $\tau_1 = 600$, $\tau_2 = 6e3$,
			$\tau_3 = 30e3$, $\tau_4 = 60e3$, $\tau_1 = 300e3$, $\rho = 1$
Kraft et al. (2013)	Flesh and muscle	Neo-Hookean	$C_{10} = 3.35$, $K = 333$, $\rho = 1$
Dong et al. (2013)	Flesh	QLV	$C_{10qs} = 1.2e\text{-}4$, $C_{20qs} = 2.5e\text{-}4$, $K = 20$, $A_1 = 1.2$, $A_2 = 0.8$,
			$\tau_1 = 23$, $\tau_2 = 63$, $\rho = 1.3$
Dong et al. (2013)	Skin	QLV	$C_{10qs} = 1.9e\text{-}4$, $C_{20qs} = 1.8e\text{-}4$, $K = 20$, $A_1 = 1$, $A_2 = 0.9$,
			$\tau_1 = 10.4$, $\tau_2 = 84.1$, $\rho = 0.8$
Gabler et al. (2014)	Fat pad	QLV	$A = 0.00551$, $B = 11.9$, $A_1 = 0.741$, $A_2 = 0.125$,
			$A_3 = 0.059$, $A_4 = 0.02$, $A_5 = 0.015$, $\tau_1 = 1$, $\tau_2 = 10$,
			$\tau_3 = 100$, $\tau_4 = 1000$, $\tau_5 = 10000$

C_{ijqs} = material constants at equilibrium (quasi-static), A_i = relaxation constants, τ_i = relaxation time constants, μ_i, α_i, A, B = material constants, G = shear modulus, E = Young's modulus, K = bulk modulus, ρ = density, ν = Poisson's ratio.

data while the fracture was initiating from the trabecular, and not the cortical, structure of the talus. This is likely unrealistic as the cortex protects the inner trabecular structure and is expected to fail first.

Dong et al. (2013) modified the lower limb section (Figure 16.7a) of a previously developed FE model of the full body (Wayne State University Human model – WSUHM) (Belwadi et al. 2012) to use it for assessing the effect of posture and protective equipment on injury. Bones were modelled as elastoplastic materials (Tables 16.5 and 16.6) with strain-rate-dependent plasticity (Cowper-Symonds law). As cartilage was not present, contact interactions were assigned between surfaces of the articulating bones, while other soft tissues were modelled with a non-linearly viscoelastic formulation (Table 16.7). The material properties used for the flesh were attributed to the heel pad although they were based on data from tests on cadaveric anterior thigh muscles and fat samples (Untaroiu et al. 2005). The ligamentous properties were not stated. The experimental configuration used by McKay and Bir (2009) was simulated. A cadaveric leg with a load cell implanted in the mid-tibia was mounted on an ATD before being axially impacted by a rigid plate. The predicted mid-tibia force values, after being filtered, were in good agreement with the experimental measurements for the case of an impact with a velocity of 7.2 m/s (Figure 16.7b); however, the predicted pulse shape that presented multiple peaks was not discussed. Furthermore, at higher velocities, it was difficult to draw conclusions about the validity of the model due to the available experimental measurements; the tested cadavers sustained multiple fractures, and the experimental force–time curves did not show a consistent pattern with the peak forces varying by up to 8 kN. Dong et al. (2013) also attempted to simulate an unpublished sled impact test that was not described thoroughly.

TABLE 16.8
Material Properties of Ankle and Foot Ligaments and the Plantar Fascia Implemented in Previous Foot and Ankle FE Models Used to Simulate Axial Impact Loading Scenarios Similar to UB Blast

Study	Tissue	Constants E,K [MPa], ρ [tn/m^3]
Tannous et al. (1996)	Plantar ligaments	$E = 6$, $\nu = 0.49$, $\rho = 1$
Bandak et al. (2001)	Plantar ligaments, viscoelastic	$E = 6$, $K = 100$, $\nu = 0.3$, $\rho = 1$
		$G(t) = 2 + (1.5 - 2)e^{-\frac{t}{10}}$
Iwamoto et al. (2005)	CF and CTi ligaments	$E = 512$, $\nu = 0.49$
Iwamoto et al. (2005)	ATaF ligament	$E = 255.5$, $\nu = 0.49$
Iwamoto et al. (2005)	PTaF ligament	$E = 216.5$, $\nu = 0.49$
Iwamoto et al. (2005)	TiNav ligament	$E = 320.7$, $\nu = 0.49$
Iwamoto et al. (2005)	ATT ligament	$E = 184.5$, $\nu = 0.49$
Iwamoto et al. (2005)	PTT ligament	$E = 99.5$, $\nu = 0.49$
Kraft et al. (2013)	All ligaments	$C_{10} = 2.5$, $K = 250$, $\rho = 1$
	Neo-Hookean	

Ligaments: PTT = posterior talotibial, CF = calcaneofibular, ATaF = anterior talofibular, PTaF = posterior talofibular, CTi = calcaneotibial, TiNav = tibionavicular. E = Young's modulus, ν = Poisson's ratio, ρ = density, G = shear modulus, K = bulk modulus, C_{10} = material constant, t = time.

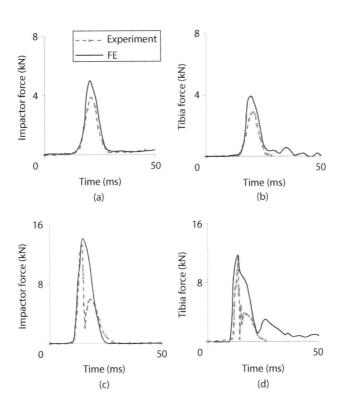

FIGURE 16.6 Experimental and numerical force–time curves measured at the impactor for velocities at impact of (a) 2.2 and (c) 7.6 m/s and at the proximal tibia for velocities at impact of (b) 2.2 and (d) 7.6 m/s. The graphs were plotted based on the results reported by Iwamoto et al. (2005).

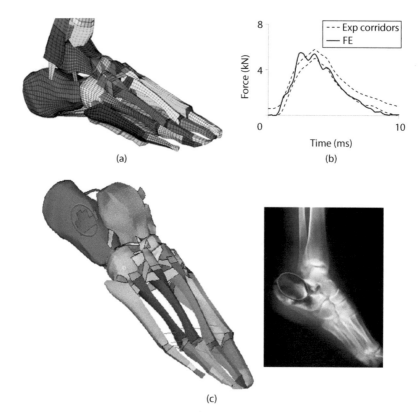

FIGURE 16.7 (a) Detailed view of the foot and ankle section of the WSUHM FE model after removing the flesh. (Reproduced Dong, L., et al., *J. Mechanical Behav. Biomed. Mater.*, 28, 111–124, 2013. With permission.) (b) The mid-tibia axial force predicted was close to the experimental corridors presented from McKay and Bir (2009). The graph was plotted based on the results reported in Dong et al. (2013). (c) The model predicted lateral calcaneal fractures when a sled experiment was simulated in agreement with the experimental outcome. (Reproduced from Dong, L., et al., *J. Mechanical Behav. Biomed. Mater.*, 28, 111–124, 2013. With permission.)

Computed tomography (CT) scans showing fractures from impacted cadavers were compared to predicted sites of failure by the model for the same conditions (Figure 16.7c). Despite the limited explanation of the experimental procedure and respective simulation, the prediction of calcaneal fractures is qualitatively in-line with laboratory-simulated underbody blast and battlefield experience.

Dong et al. (2013) also simulated axial loading with the leg set at different seating postures; while the foot was flat on the ground, the knee was flexed between 60° and 120°. In order to reduce the complexity, the input to the model was a bilinear velocity profile over time which does not represent realistically the insult, as explained above. As the angle increased, the damage less severe for the tibia and the site of fracture was shifted posteriorly (Figure 16.8). Although tibia fractures were less likely, it was not reported whether the calcaneus was more vulnerable at different postures; with the knee flexed at 60° and the foot flat, the subtalar joint is locked, which may affect the calcaneal fracture patterns expected in UB blast.

Finally, different blast mats were modelled and included in the simulation to assess their efficacy for the same loading scenario. The minimum thickness of a foam and a honeycomb blast mat able to protect the lower limb from fracturing was found to be 30 and 20 mm, respectively. Apart from reducing the peak tibia force by 41 and 11%, respectively, the foam and the honeycomb layers managed to delay the event and increase the time to peak force. Similar to the posture analysis, however,

FIGURE 16.8   The predicted tibia fracture when the knee was flexed at (a) 120° and (b) 60°. (Reproduced from Dong, L., et al., *J. Mechanical Behav. Biomed. Mater.*, 28, 111–124, 2013. With permission.)

the validity of this outcome is questionable due to the unrealistic boundary condition of a bilinear velocity profile that was applied.

The WSHU model used by Dong et al. (2013) was implemented by Suresh et al. (2012) in a full-body model. Two postures were simulated (Figure 16.9). The achieved acceleration pulse was reported to reach a peak value of 120 g within 5 ms. The peak values of the axial force predicted at the lower tibia were 4.5 and 3.9 kN when the knee was flexed at 120° and 70°, respectively, whereas at the upper tibia the prediction was reversed with 2.3 and 3.4 kN when the knee was flexed at 120° and 70°, respectively; this contradicts the results presented from the same model above (Dong et al. 2013). This discrepancy might be explained by the different initial conditions applied in each simulation. The predicted response of the model had multiple peaks, which is a common observation in many other studies; these were attributed to the lack of cartilaginous tissue.

The effect of posture on the outcome of a 'deck-slap' injury was examined in both studies presented above (Dong et al. 2013; Suresh et al. 2012). The models were set at different angles of knee flexion prior to simulating the impact load case. None of the studies, however, described the process of rotating and translating the bones and the structures of the foot to imitate a specific posture. Whether the subtalar, tibiofemoral or patellofemoral joint alignment was modified when the knee joint was flexed at 70° or 120° and the foot was set flat on the ground is not explained or compared to physiological measurements. Similarly, no information is given about the initial loading condition of ligaments and tendons apart from the Achilles tendon, yet these are likely to vary across lower extremity postures. Whereas in quasi-static load cases accurate joint alignment is critical to load

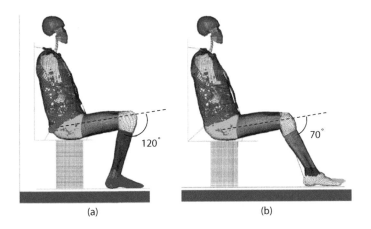

FIGURE 16.9   The WSUHM was set at different postures before simulating axial impact. In both postures, the foot was flat on the ground while the knee was flexed at (a) 120° and (b) 70°. (Modified and reproduced from Suresh, M., et al., *Blucher Mechanical Eng. Proc.*, 1(1), 1809–1818. With permission.)

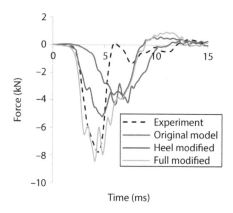

FIGURE 16.10 The graph was plotted based on the results reported in Gabler et al. (2014). The measured force at the proximal tibia of the modified model was in greater agreement to the experimental data than the response of the original model. After modifying the material properties of the heel, the phase of the response of the model was shifted to match the experimental data better. After refining the mesh, the response of the model achieved good correlation with the experiment in the loading phase.

transfer through the various tissues, it remains allusive whether this is also important in high-rate load cases through the foot.

The 3D model of the foot and ankle that was developed by Shin et al. (2012) for automotive impacts was used by Gabler et al. (2014) to simulate drop tests on cadaveric specimens (Henderson et al. 2013). The proximal tibia was fixed on a load cell while the plantar foot was impacted. The response of the model was not accurate, and so the authors created a finer mesh for the same geometry and modified the material properties of the heel pad (Table 16.7) based on high-rate compressive tests on cylindrical samples extracted from cadaveric heels. The predictions of the improved model for one simulation can be seen in Figure 16.10. Despite the oscillatory response of the model and the large difference compared to the experiment in the unloading phase, this study demonstrated that, as loading in road traffic accidents and UB blast are dissimilar, different modelling approaches need to be employed in order to achieve good correlations between experiment and simulation.

16.5 DISCUSSION/CONCLUSION

Computational modelling is an extraordinary engineering tool that allows the inexpensive simulation of multiple physical phenomena and look into intricacies of mechanical response of structures that cannot be measured physically. Its usefulness and accuracy, however, rely on the engineering judgement of the user with regard to input parameters and interpretation of results. It is important, therefore, to ascertain the utility, credibility, predictive ability and interpretation of each computational model. Specifically, with regard to computational models of the human body, both geometry of and material models assigned to the various tissues involved are always a balance between complexity and computational cost. In addition, the inhomogeneity, anisotropy and time-dependent behaviour of all human tissue means that almost any type of material model used will be inaccurate to some extent. Furthermore, the material properties assigned to the material model of choice will depend on the quality and quantity – including the variety of loading modes – of the experimental data used to derive them. These observations are manifested in the wide range of geometry, material models and assigned material properties (i.e. the values used for the same material model of the same tissue) in the FE models reviewed here. Therefore, the decision on complexity of FE models of the lower extremity should depend on, and interpreted in terms of, the research question associated with pathology, physiology, treatment, repair or protection.

The geometry of the foot and ankle of the models that were used to simulate impact loading scenarios represented the 50th percentile of male, either based on scans from subjects of similar dimensions (Beillas et al. 2001; Gabler et al. 2014; Shin et al. 2012) or by using scaling approaches (Bandak et al. 2001; Dong et al. 2013; Iwamoto et al. 2005; Suresh et al. 2012; Tannous et al. 1996). The loading environment experienced in UB blast is reported mostly in the battlefield. Therefore, an appropriate model would reflect the range in body dimensions of soldiers. The average anthropometry of service personnel in the UK (Simpson and Bolton 1970), the USA (White and Churchill 1971) and Australia (Coleman and Blanchonette 2000) has been reported to deviate from that of the 50th percentile male.

In the FE models of the foot and ankle presented here, the cortical and trabecular structures of the bones of the ankle joint complex have been modelled in different ways. Either both structures were modelled with 3D elements (Bandak et al. 2001; Dong et al. 2013; Kraft et al. 2013; Tannous et al. 1996) or only trabecular bone was represented by 3D elements while cortical bone was modelled with 2D shell elements (Beaugonin et al. 1997; Beillas et al. 2001; Murakami et al. 2004; Neale et al. 2012). As the intent of FE models of the foot and ankle developed to simulate UB blast is to predict injury that is primarily skeletal, the load transmission predicted by the model has to be accurate. Hence, the more detailed representation of the structures increases the ability of the model to predict vulnerable bone areas.

The bones of the mid- and forefoot were represented by rigid structures in some of the studies where impact loading cases were simulated (Beillas et al. 2001; Gabler et al. 2014; Neale et al. 2012). In the case of UB blast, loading is mainly transferred through the heel. Therefore, the deformation of the bones of the forefoot as well as the kinematic response of the respective articulations is not critical and so some authors fused the joints to a single structure (Neale et al. 2012).

The material properties of cortical and trabecular bone implemented in the majority of the models presented were elastoplastic (Beillas et al. 2001; Dong et al. 2013; Gabler et al. 2014; Iwamoto et al. 2005; Kraft et al. 2013; Neale et al. 2012; Shin et al. 2012). The Young's moduli assigned to the cortical tibia and cortical fibula did not vary significantly in the above studies with a mean value of 16.6 (SD 2.1) and 16.8 (SD 2.2) MPa, respectively. Similarly, the yield stress of the cortex of both long bones was set between 120 and 125 MPa. The Young's modulus as well as the yield stress applied for the cortex of the tarsal bones, however, varied from 5.5 to 19 MPa and 80 to 165 MPa, respectively. The tarsal bones were stiffer and less vulnerable to yield than the tibia and the fibula in some FE models of the foot and ankle (Dong et al. 2013; Shin et al. 2012), while the opposite was suggested by other studies (Beillas et al. 2001; Iwamoto et al. 2005; Neale et al. 2012). This discrepancy is due to the limited knowledge of the behaviour of irregular bones such as the talus and the calcaneus at high rates of loading; this requires further examination.

Articular cartilage of the foot and ankle joints was not physically represented in any of the studies reviewed in this chapter. Instead, contact interactions and kinematic constrains were set between contacting bones. This was due to the small thickness of the cartilaginous tissue in the joints of the foot and ankle, which is less than 1.5 and 2 mm at the talocrural (Millington et al. 2007) and subtalar joint (Akiyama et al. 2012), respectively. In order to represent such a thin structure, a fine finite element mesh is required, which would reduce significantly the maximum time step allowed in FE models that use explicit numerical schemes. The lack of cartilage reduced the computational cost but meant that bones could impact each other, thus inducing several peaks in the captured force while missing out the load distributing and load absorbing characteristics of articular cartilage. The effect of this limitation in model biofidelity on predicted response has not been investigated.

The importance of the material parameters selected to represent the heel fat pad for modelling high load-rate cases was highlighted by sensitivity studies (Beillas et al. 2001; Neale et al. 2012). Gabler et al. (2014) modified an existing model of the foot and ankle (Shin et al. 2012), by attributing stiffer properties to the heel fat pad tissue in an attempt to simulate accurately axial impact tests of the leg (Henderson et al. 2013). Apart from affecting the model's response, the wide variety of different material properties implemented in the various studies indicates the knowledge gap that exists regarding the behaviour of the heel fat pad at high loading rates, relevant to UB blast.

Many of the FE models presented here (Beillas et al. 2001; Gabler et al. 2014; Murakami et al. 2004; Shin et al. 2012; Suresh et al. 2012) reported force–time curves with numerous oscillations or filtered numerical data prior to comparing with experimental results. Suresh et al. (2012) attributed this to the lack of cartilaginous tissue causing artefactual impulse loads due to direct contact between the stiff bones. This, however, may be associated also with the explicit numerical solving scheme used; in contrary to the implicit scheme, when contact occurs in explicit FE modelling high-frequency modes are excited and propagate through the mesh causing oscillations. This problem can be magnified when using a small time step. Therefore, if the time step of an explicit FE simulation is not appropriate for the mesh of the model and the simulated load case, instead of not converging, as it would happen if an implicit scheme was selected, unrealistic oscillations could be induced.

Albeit many of the FE studies reviewed here claim to be validated, it is important for the reader and any potential user to appreciate the extent over which this is true. The range over which an FE model has been validated depends on the case and, although it should be, unfortunately it is not always communicated appropriately. It is quite uncommon in FE models of the lower extremity, but good practice overall, to design validation experiments that are well controlled aiming at testing specific predictive abilities of the FE model; the intention is not to carry out the complex experiment that the FE model is trying to simulate, but to build confidence in the predictions of parts of the FE model (e.g. the use of a specific material model) or at specific key locations for specific variables (e.g. strain).

As experiments with human tissue are expensive, variable (due to anatomical variability and tissue condition) and not necessarily very well controlled, sensitivity studies of the FE models offer a great means of understanding the envelope of validity of the FE model as well as the effect of key parameters on predicted behaviour. Sensitivity studies involve altering key input parameters of the FE model over a sensible/expected/physiological range in order to assess the dependency of key outputs (e.g. strain) on them. It is very common in human body modelling for a number of input parameters to be associated with uncertainty, for example, the material properties. It is important to assess the dependency of the prediction on especially uncertain input data; this will dictate the range over which the model is valid. Unfortunately, the vast majority of FE models of the lower extremity include very few, if any, sensitivity studies.

With biofidelity and usefulness of FE models of the lower extremity in UB blast in mind, the following modelling aspects/targets are suggested:

- Appropriate geometries of the lower limb should reflect the range of body dimensions of the persons most likely to be exposed in the simulated UB blast loading scenario.
- Accurate representation of the cartilaginous tissue as well as the cortex of bones is key to the model's predictive ability.
- The material properties of the heel fat pad affect highly the predictive ability of the model. Therefore, the material model and the material parameters used to represent the fat pad should be selected to represent accurately the non-linearly viscoelastic behaviour of the tissue under high loading rates.
- Sensitivity analysis should be carried out to identify the effect of parameters obtained from the literature such as material properties of bone, ligaments and cartilage on model predictions.
- The experimental data used to test the model for validity should be obtained from test conditions that can be defined precisely in the FE model. The extent over which the model has been validated should be communicated clearly.

Accurate, validated computational models of the lower extremity are a cost-effective, repeatable alternative to expensive experimental set-ups. As computational power is ever increasing and the explosive threat does not seem to be receding in military and civilian theatres alike, it is important that FE models of the lower extremity under a blast insult continue to be developed and enhanced. The intention and the drive should be for these models to become standard tools for designing protection, both personal and vehicular.

REFERENCES

Akiyama, K., T. Sakai, N. Sugimoto, H. Yoshikawa, and K. Sugamoto. 2012. Three-Dimensional Distribution of Articular Cartilage Thickness in the Elderly Talus and Calcaneus Analyzing the Subchondral Bone Plate Density. *Osteoarthritis and Cartilage* 20(4): 296–304. doi: 10.1016/j.joca.2011.12.014.

Bailey, A. M., J. C. John, K. Henderson, F. Brozoski, and R. Salzar. 2013. Comparison of Hybrid-III and PMHS Response to Simulated Underbody Blast Loading Conditions. In *Proceedings of IRCOBI Conference.* http://www.ircobi.org/downloads/irc13/pdf_files/25.pdf (accessed on December 20, 2015).

Bandak, F. A., R. E. Tannous, and T. Toridis. 2001. On the Development of an Osseo-Ligamentous Finite Element Model of the Human Ankle Joint. *International Journal of Solids and Structures* 38(10): 1681–1697.

Banglmaier, R. F., D. Dvoracek-Driksna, T. E. Oniang'o, and R. C. Haut. 1999. Axial Compressive Load Response of the 90° Flexed Human Tibiofemoral Joint. SAE Technical Paper 99SC08. Warrendale, PA: SAE Technical Paper. http://papers.sae.org/99SC08/ (accessed on December 20, 2015).

Beaugonin, M., E. Haug, and D. Cesari. 1996. A Numerical Model of the Human Ankle/Foot under Impact Loading in Inversion and Eversion. SAE Technical Paper 962428. Warrendale, PA: SAE Technical Paper. http://papers.sae.org/962428/ (accessed on December 20, 2015).

Beaugonin, M., E. Haug, and D. Cesari. 1997. Improvement of Numerical Ankle/Foot Model: Modeling of Deformable Bone. SAE Technical Paper 973331. Warrendale, PA: SAE Technical Paper. http://papers.sae.org/973331/ (accessed on December 20, 2015).

Begeman, P., P. Balakrishnan, R. Levine, and A. King. 1993. Dynamic Human Ankle Response to Inversion and Eversion. SAE Technical Paper 933115. Warrendale, PA: SAE Technical Paper. http://papers.sae.org/933115/ (accessed on December 20, 2015).

Beillas, P., P. C. Begeman, K. H. Yang, A. I. King, P. J. Arnoux, H. S. Kang, K. Kayvantash, C. Brunet, C. Cavallero, and P. Prasad. 2001. Lower Limb: Advanced FE Model and New Experimental Data. *Stapp Car Crash Journal* 45: 469–494.

Beillas, P., F. Lavaste, D. Nicolopoulos, K. Kayventash, K. H. Yang, and S. Robin. 1999. Foot and Ankle Finite Element Modeling Using Ct-Scan Data. SAE Technical Paper 99SC11. Warrendale, PA: SAE Technical Paper. http://papers.sae.org/99SC11/ (accessed on December 20, 2015).

Beillas, P., S. W. Lee, S. Tashman, and K. H. Yang. 2007. Sensitivity of the Tibio-Femoral Response to Finite Element Modeling Parameters. *Computer Methods in Biomechanics and Biomedical Engineering* 10: 209–221.

Belwadi, A., J. H. Siegel, A. Singh, J. A. Smith, K. H. Yang, and A. I. King. 2012. Finite Element Aortic Injury Reconstruction of near Side Lateral Impacts Using Real World Crash Data. *Journal of Biomechanical Engineering* 134(1): 011006. doi: 10.1115/1.4005684.

Bird, R. 2001. Protection of Vehicles against Landmines. *Journal of Battlefield Technology* 4(1): 14–17.

Coleman, J. L., and P. Blanchonette. 2000. Preliminary Comparison of Anthropometric Datasets for the Australian Defence Force. http://ergonomics.spyro.ddsn.net/downloads/EA_Journals/2011_Conference_Edition/Coleman_J.pdf (accessed on December 20, 2015).

Corazza, F., J. J. O'Connor, A. Leardini, and V. Parenti Castelli. 2003. Ligament Fibre Recruitment and Forces for the Anterior Drawer Test at the Human Ankle Joint. *Journal of Biomechanics* 36(3): 363–372.

Cowper, G. R., P. S. Symonds, United States. Office of Naval Research, and Brown University. Division of Applied Mathematics. 1957. *Strain-Hardening and Strain-Rate Effects in the Impact Loading of Cantilever Beams.* Providence, RI: Division of Applied Mathematics, Brown University.

Dong, L., F. Zhu, X. Jin, M. Suresh, B. Jiang, G. Sevagan, Y. Cai, G. Li, and K. H. Yang. 2013. Blast Effect on the Lower Extremities and Its Mitigation: A Computational Study. *Journal of the Mechanical Behavior of Biomedical Materials* 28: 111–124. doi:10.1016/j.jmbbm.2013.07.010.

Funk, J. R., J. R. Crandall, L. J. Tourret, C. B. MacMahon, C. R. Bass, J. T. Patrie, N. Khaewpong, and R. H. Eppinger. 2002. The Axial Injury Tolerance of the Human Foot/ankle Complex and the Effect of Achilles Tension. *Journal of Biomechanical Engineering* 124(6): 750–757.

Gabler, L. F., M. B. Panzer, and R. S. Salzar. 2014. High-Rate Mechanical Properties of Human Heel Pad for Simulation of a Blast Loading Condition. In. IRC-14-87 *IRCOBI Conference 2014*. http://www.ircobi. org/downloads/irc14/pdf_files/87.pdf (accessed on December 20, 2015).

Haut, R. C., and P. J. Atkinson. 1995. Insult to the Human Cadaver Patellofemoral Joint: Effects of Age on Fracture Tolerance and Occult Injury. SAE Technical Paper 952729. Warrendale, PA: SAE Technical Paper. http://papers.sae.org/952729/ (accessed on December 20, 2015).

Henderson, K., A. Bailey, J. Christopher, F. Brozoski, and R. Salzar. 2013. Biomechanical Response of the Lower Leg under High Rate Loading. In *Proceedings of IRCOBI Conference*. http://www.ircobi.org/ downloads/irc13/pdf_files/24.pdf (accessed on December 20, 2015).

Hynd, D., C. Willis, A. Roberts, R. Lowne, R. Hopcroft, P. Manning, and W. A. Wallace. 2003. The Development of an Injury Criteria for Axial Loading to the THOR-LX Based on PMHS Testing. http://trid.trb.org/ view.aspx?id=677978 (accessed on December 20, 2015).

Iwamoto, M., Y. Kisanuki, I. Watanabe, K. Furusu, K. Miki, and J. Hasegawa. 2002. Development of a finite element model of the total human knee model for safety (THUMS) and application to injury reconstruction. http://trid.trb.org/view.aspx?id=705796 (accessed on December 20, 2015).

Iwamoto, M., K. Miki, and E. Tanaka. 2005. Ankle Skeletal Injury Predictions Using Anisotropic Inelastic Constitutive Model of Cortical Bone Taking into Account Damage Evolution. *Stapp Car Crash Journal* 49: 133.

Kajzer, J., C. Cavallero, J. Bonnoit, A. Morjane, and S. Ghanouchi. 1993. Response of the Knee Joint in Lateral Impact: Effect of Bending Moment. In *Proceedings of the International Research Council on the Biomechanics of Injury Conference* 21:105–116. http://www.safetylit.org/citations/index.php?fuseaction=citations. viewdetails&citationIds[]=citjournalarticle_247779_38 (accessed on December 20, 2015).

Kajzer, J., C. Cavallero, S. Ghanouchi, J. Bonnoit, and A. Ghorbel. 1990. Response of the Knee Joint in Lateral Impact: Effect of Shearing Loads. http://trid.trb.org/view.aspx?id=1175127 (accessed on December 20, 2015).

Kitagawa, Y., I. Hideaki, A. I. King, and R. S. Levine. 1998. A Severe Ankle and Foot Injury in Frontal Crashes and Its Mechanism. SAE Technical Paper 983145. Warrendale, PA: SAE Technical Paper. http://papers. sae.org/983145/ (accessed on December 20, 2015).

Kraft, R. H., M. L. Lynch, W. Pino, A. Jean, R. Radovitzky, and S. Socrate. 2013. Computational Failure Modeling of Accelerative Injuries to the Lower Leg below the Knee. DTIC Document. http://oai.dtic.mil/oai/oai? verb=getRecord&metadataPrefix=html&identifier=ADA575839 (accessed on December 20, 2015).

Kraft, R. H., M. L. Lynch, and E. W. Vogel. 2012. Computational Failure Modeling of Lower Extremities. DTIC Document. http://oai.dtic.mil/oai/oai?verb=getRecord&metadataPrefix=html&identifier=ADA562360 (accessed on December 20, 2015).

McKay, B. J., and C. A. Bir. 2009. Lower Extremity Injury Criteria for Evaluating Military Vehicle Occupant Injury in Underbelly Blast Events. *Stapp Car Crash Journal* 53: 229–249.

Millington, S. A., M. Grabner, R. Wozelka, D. D. Anderson, S. R. Hurwitz, and J. R. Crandall. 2007. Quantification of Ankle Articular Cartilage Topography and Thickness Using a High Resolution Stereophotography System. *Osteoarthritis and Cartilage* 15(2): 205–211. doi: 10.1016/j.joca.2006.07.008.

Morimoto, J., and K. Yamazaki. 2004. *Improvement of Foot FE Model Based on the Movement of Bones during Heel Impact*. SAE Technical Paper 2004-01-0313. Warrendale, PA: SAE Technical Paper. http://papers. sae.org/2004-01-0313/ (accessed on December 20, 2015).

Murakami, D., Y. Kitagawa, S. Kobayashi, R. Kent, and J. Crandall. 2004. Development and Validation of a Finite Element Model of a Vehicle Occupant. SAE Technical Paper 2004-01-0325. Warrendale, PA: SAE Technical Paper. http://papers.sae.org/2004-01-0325/ (accessed on December 20, 2015).

Neale, M., R. Thomas, H. Baterman, and D. Hynd. 2012. *A Finite Element Modelling Investigation of Lower Leg Injuries*. 07-0077. TRL Limited. http://www-nrd.nhtsa.dot.gov/pdf/esv/esv20/07-0077-W.pdf (accessed on December 20, 2015).

Owen, C., Lowne, R., Mcmaster, J., 2001. Requirements for the Evaluation of the Risk of Injury to the Ankle in Car Impact Tests. *Proceedings of the 17th International Technical Conference on the Enhanced Safety of Vehicles*.

Ramasamy, A., A. M. Hill, S. Masouros, I. Gibb, A. M. J. Bull, and J. C. Clasper. 2010. Blast-Related Fracture Patterns: A Forensic Biomechanical Approach. *Journal of the Royal Society Interface* 8(58): 689–698.

Ramasamy, A., A. M. Hill, S. Masouros, I. Gibb, R. Phillip, A. M. J. Bull, and J. C. Clasper. 2013. Outcomes of IED Foot and Ankle Blast Injuries. *The Journal of Bone and Joint Surgery. American Volume* 95(5): e25. doi: 10.2106/JBJS.K.01666.

Ramasamy, A., A. M. Hill, R. Phillip, I. Gibb, A. M. J. Bull, and J. C. Clasper. 2011. The modern 'deck-slap' injury—Calcaneal blast fractures from vehicle explosions. *Journal of Trauma and Acute Care Surgery* 71(6): 1694–1698.

Rome, K. 1996. Ankle Joint Dorsiflexion Measurement Studies: A Review of the Literature. *Journal of the American Podiatric Medical Association* 86(5): 205–211. doi: 10.7547/87507315-86-5-205.

Shin, J., N. Yue, and C. D. Untaroiu. 2012. A Finite Element Model of the Foot and Ankle for Automotive Impact Applications. *Annals of Biomedical Engineering* 40(12): 2519–2531. doi: 10.1007/s10439-012-0607-3.

Simpson, R. E., and C. B. Bolton. 1970. An Anthropometric Survey of 200 RAF and RN Aircrew and the Application of the Data to Garment Size Rolls: Royal Aircraft Establishment Technical Report 67125, Jul. 1968. *Applied Ergonomics* 1(3): 179.

Suresh, M., F. Zhu, K. H. Yang, J. L. Serres, and R. E. Tannous. 2012. Finite Element Evaluation of Human Body Response to Vertical Impulse Loading. *Blucher Mechanical Engineering Proceedings* 1(1): 1809–1818.

Tannous, R, E., F. A. Bandak, T. G. Toridis, and R. H. Eppinger. 1996. A Three-Dimensional Finite Element Model of the Human Ankle: Development and Preliminary Application to Axial Impulsive Loading. SAE Technical Paper 962427. Warrendale, PA: SAE Technical Paper. http://papers.sae.org/962427/ (accessed on December 20, 2015).

Tran, T., and D. Thordarson. 2002. Functional Outcome of Multiply Injured Patients with Associated Foot Injury. *Foot & Ankle International* 23(4): 340–343.

Turchin, D. C., E. H. Schemitsch, M. D. McKee, and J. P. Waddell. 1999. Do Foot Injuries Significantly Affect the Functional Outcome of Multiply Injured Patients? *Journal of Orthopaedic Trauma* 13(1): 1–4.

Untaroiu, C., K. Darvish, J. Crandall, B. Deng, and J.-T. Wang. 2005. A Finite Element Model of the Lower Limb for Simulating Pedestrian Impacts. *Stapp Car Crash Journal* 49: 157–181.

Wheeler, L., P. Manning, C. Owen, A. Roberts, R. Lowne, and W. A. Wallace. 2000. Biofidelity of Dummy Legs for use in Legislative Car Crash Testing. http://trid.trb.org/view.aspx?id=671997 (accessed on December 20, 2015).

White, R. M., and E. Churchill. 1971. The Body Size of Soldiers: US Army Anthropometry-1966. DTIC Document. http://oai.dtic.mil/oai/oai?verb=getRecord&metadataPrefix=html&identifier=AD0743465 (accessed on December 20, 2015).

Yang, K. H., J. Hu, N. A. White, A. I. King, C. C. Chou, and P. Prasad. 2006. Development of Numerical Models for Injury Biomechanics Research: A Review of 50 Years of Publications in the Stapp Car Crash Conference. *Stapp Car Crash Journal* 50: 429–490.

Yoganandan, N., F. A. Pintar, M. Boynton, P. Begeman, P. Prasad, S. M. Kuppa, R. M. Morgan, and R. H. Eppinger. 1996. Dynamic Axial Tolerance of the Human Foot-Ankle Complex. SAE Technical Paper 962426. Warrendale, PA: SAE Technical Paper. http://papers.sae.org/962426/ (accessed on December 20, 2015).

Yoganandan, N., F. A. Pintar, M. Boynton, A. Sances, R. M. Morgan, R. Eppinger, and S. Kuppa. 1995. Biomechanics of Foot and Ankle Fractures. In *International Conference on Pelvic and Lower Extremity Injuries Proceedings*, 201–209.

Index

T - #0482 - 071024 - C432 - 254/178/19 - PB - 9780367657949 - Gloss Lamination